Grundlehren der mathematischen Wissenschaften

A Series of Comprehensive Studies in Mathematics

Volume 365

Grundlehren der mathematischen Wissenschaften (subtitled Comprehensive Studies in Mathematics), Springer's first series in higher mathematics, was founded by Richard Courant in 1920. It was conceived as a series of modern textbooks. A number of significant changes appear after World War II. Outwardly, the change was in language: whereas most of the first 100 volumes were published in German, the following volumes are almost all in English. A more important change concerns the contents of the books. The original objective of the Grundlehren had been to lead readers to the principal results and to recent research questions in a single relatively elementary and accessible book. Good examples are van der Waerden's 2-volume Introduction to Algebra or the two famous volumes of Courant and Hilbert on Methods of Mathematical Physics.

Today, it is seldom possible to start at the basics and, in one volume or even two, reach the frontiers of current research. Thus many later volumes are both more specialized and more advanced. Nevertheless, most volumes of the series are meant to be textbooks of a kind, with occasional reference works or pure research monographs. Each book should lead up to current research, without over-emphasizing the author's own interests. Proofs of the major statements should be enunciated, however the presentation should remain expository. Examples of books that fit this description are Maclane's Homology, Siegel & Moser on Celestial Mechanics, Gilbarg & Trudinger on Elliptic PDE of Second Order, Dafermos's Hyperbolic Conservation Laws in Continuum Physics ... Longevity is an important criterion: a GL volume should continue to have an impact over many years. Topics should be of current mathematical relevance, and not too narrow.

The tastes of the editors play a pivotal role in the selection of topics.

Authors are encouraged to follow their individual style, but keep the interests of the reader in mind when presenting their subject. The inclusion of exercises and historical background is encouraged.

The GL series does not strive for systematic coverage of all of mathematics. There are both overlaps between books and gaps. However, a systematic effort is made to cover important areas of current interest in a GL volume when they become ripe for GL-type treatment.

As far as the development of mathematics permits, the direction of GL remains true to the original spirit of Courant. Many of the oldest volumes are popular to this day and some have not been superseded. One should perhaps never advertise a contemporary book as a classic but many recent volumes and many forthcoming volumes will surely earn this attribute through their use by generations of mathematicians.

Cristian Giardinà • Frank Redig

Duality for Markov Processes

A Lie Algebraic Approach

Springer

Cristian Giardinà
Mathematics
University of Modena and Reggio Emilia
Modena, Italy

Frank Redig
Delft Institute of Applied Mathematics
Delft University of Technology
Delft, The Netherlands

ISSN 0072-7830 ISSN 2196-9701 (electronic)
Grundlehren der mathematischen Wissenschaften
ISBN 978-3-032-04098-5 ISBN 978-3-032-04099-2 (eBook)
https://doi.org/10.1007/978-3-032-04099-2

Mathematics Subject Classification: 82C22, 60K35, 82C23, 82C70, 33C45, 60J25, 17B15

This Springer imprint is published by the registered company Springer Nature Switzerland AG
The registered company address is: Gewerbestrasse 11, 6330 Cham, Switzerland

Acknowledgements Research is often a joint effort of several people. Duality theory of Markov processes, being a mathematical topic which is at the interface of diverse research areas, perfectly fits this scenario. In the writing of this monograph, it has been crucial the insights from statistical physics (we especially thank Jorge Kurchan and Tomohiro Sasamoto), interacting particle systems (Anna de Masi and Errico Presutti were a constant source of inspiration), probability theory (we are in debt to Frank den Hollander for his continued support), special functions (Erik Koelink is gratefully acknowledged for sharing with us his expertise on Lie algebras), mathematical population genetics (Ellen Baake and Martin Möhle let us appreciate the importance of duality in that area).

Special thanks go to Gioia Carinci. She worked hard on commenting and revising a lot of material in the book. Especially for Chaps. 9, 10, and Appendix C, her contribution was considerable and important. She has been a long-time companion on the long road of exploration of duality for interacting particle systems, which we walked together. We are grateful for the many fruitful, enlightening, and pleasant discussions we have had with her in the course of the years.

This book would not have been possible without the help and encouragement of many more people. We would like to thank several colleagues and friends with whom we had the pleasure to work on or discuss the topic of duality, including Siva Athreya, Mario Ayala, Marton Balazs, Jochen Blath, Cedric Bernardin, Lorenzo Bertini, Francesco Casini, Jean-René Chazottes, Pasquale Cirillo, Cristina Caroli Costantini, Ivan Corwin, Anna De Masi, Bernard Derrida, Pablo Ferrari, Patrik Ferrari, Simone Floreani, Chiara Franceschini, Rouven Frassek, Davide Gabrielli, Claudio Giberti, Bart van Ginkel, Patricia Goncalvez, Andreas Greven, Wolter Groenevelt, Stephan Grosskinsky, Frank den Hollander, Sabine Jansen, Milton Jara, Jeffrey Kuan, Kevin Kuoch, Jorge Kurchan, Noemi Kurt, Thomas G. Kurtz, Claudio Landim, Joel Lebowitz, Christian Maes, Shubamoy Nandan, Alex Opoku, Stefano Olla, Errico Presutti, Matteo Ruggiero, Wioletta Ruszel, Ellen Saada, Tomohiro Sasamoto, Federico Sau, Dario Spanò, Gunter Schütz, Herbert Spohn, Anja Sturm, Rongfeng Sun, Jan M. Swart, Dimitrios Tsagkarogiannis, Kiamars Vafayi, Florian Völlering, Stephan Wagner, Anton Wakolbinger, Hidde van Wiechen, Maite Wilke Berenguer, Anita Winter, and Nikos Zigouras. To all of them we express our deep gratitude and appreciation.

Frank Redig dedicates this monograph to his beloved wife, Laura Maria De Gaspari, for continuous support, understanding and consolation.

Frank Redig thanks Dries De Vink for solving many practical things in the house while we were working on the book.

Cristian Giardinà dedicates this monograph to his family—Elisa, Edo, Alex, and Ali—for their constant love and unwavering support.

The University of Delft and the University of Modena and Reggio Emilia are acknowledged for their financial support of several exchange visits that made possible the cooperation of the authors. Ecole Polytechnique hosted the authors in 2016 and 2020 to work on this monograph for a few months. MFO Oberwolfach hosted the authors in a research-in-pairs program in 2025.

Competing Interests The authors have no competing interests to declare that are relevant to the content of this manuscript.

Contents

Chapter 1
Introduction

This book is about a systematic Lie algebraic approach to duality for Markov processes. In this introduction, we explain in words what is duality, what are the fundamental concepts in the algebraic approach to duality, and how they can help in building a solid basis to find new dualities, and to apply duality in various contexts. We also provide an overview of the material in the book, including a short summary of the other chapters. Finally, we formulate the main choices made in the book, the intended audience, and possible reading paths.

Duality

Duality in the theory of Markov processes refers to a way by which one can connect two Markov processes $\mathbb{X} = \{X(t) : t \geq 0\}$ (the process under study) on a state space Ω and a process $\mathbb{Y} = \{Y(t) : t \geq 0\}$ on a state space $\widehat{\Omega}$ (the dual process) via a so-called duality function $D : \widehat{\Omega} \times \Omega \to \mathbb{R}$. The connection is expressed via expectations, i.e. for all $t \geq 0$ and for all $y \in \widetilde{\Omega}$, $x \in \Omega$

$$\widehat{\mathbb{E}}_y\big(D(Y(t), x)\big) = \mathbb{E}_x\big(D(y, X(t))\big), \tag{1.1}$$

which in words means the following: evolving the dual process and fixing the variable of the original process has the same effect as evolving the original process and fixing the variable of the dual process. This connection between the two processes becomes very useful if the dual process is simpler and the functions $D(y, \cdot)$ provide full information about the process under study. We write the connection symbolically as

$$\mathbb{Y} \xrightarrow{D} \mathbb{X} \tag{1.2}$$

and think of it as a relation between two processes, parametrized by the duality function D.

C. Giardinà, F. Redig, *Duality for Markov Processes*, Grundlehren der mathematischen Wissenschaften 365, https://doi.org/10.1007/978-3-032-04099-2_1

This notation immediately suggests natural questions such as: for a given D, which processes can be connected to each other via D, i.e., which $\mathbb{Y}$ and $\mathbb{X}$ satisfy $\mathbb{Y} \xrightarrow{D} \mathbb{X}$? Or, for two given processes which are in a duality relation, $\mathbb{Y} \xrightarrow{D} \mathbb{X}$, which are the possible D that connect these processes? Finally, how does the duality relation compare to other ways of connecting processes, such as intertwining, stochastic flows and coupling?

The Use of Duality

The main reason for looking for duality, i.e., a dual process and a duality function, is simplification. There are many ways in which duality induces simplification. Let us mention a few.

1. **From continuous to discrete.** We can connect a process $\mathbb{X}$ on a continuous state space such as $\mathbb{R}^d$ to a much simpler "pure jump process" $\mathbb{Y}$ on a countable state space such as $\mathbb{N}$. Sometimes the converse: from discrete to continuous can also be a simplification, for instance the discrete system is stochastic and the continuous process is a simple system of ODEs.
2. **From many to few.** In the context of interacting particle systems this is one of the most important simplifications induced by duality. Well-chosen time-dependent multivariate moments of order n in a system of many, possibly infinitely many particles can be reduced to studying the motion of n dual particles. An example of this is the fact that the expected number of particles at time $t > 0$ in a system of interacting particles with duality can be computed using a single random walker. This implies, as a further consequence, that one can infer in a very simple manner the scaling limit of the particle system, via the scaling limit of a single dual particle. E.g. if the single particle dynamics scales to Brownian motion, then the macroscopic scaling limit of the particle system is the heat equation.
3. **From non-equilibrium reservoirs to absorbing boundary sites.** Interacting particle systems are often used to model transport in the realm of non-equilibrium statistical physics. In this setting, one uses systems which at the boundaries allow input and output of particles with specific rates, which physically corresponds to putting the system in contact with boundary reservoirs. When these reservoirs have different parameters (such as temperature, chemical potential, etc.) then the system reaches in the course of time a so-called "non-equilibrium steady state". Steady state, because it is a long time limit. Non-equilibrium because currents will flow from one reservoir to another, and this means that the system is not invariant under time-reversal, i.e., is not in equilibrium. In the study of such systems, duality has been a very successful technique, because it allows to connect the non-equilibrium system under study to a much simpler system where the reservoirs are replaced by absorbing sites. As a consequence, moments of order n in the non-equilibrium steady state can be expressed in terms of absorption probabilities of n dual walkers. In some cases, for so-called "integrable systems" these absorption probabilities can be computed in closed form. Even without closed form formulas for the absorption probabilities of dual

walkers, still the duality with the absorbing system gives a lot of information on the structure of the non-equilibrium steady state.

4. **From micro to macro.** In statistical physics one wants to understand how macroscopic behavior arises from microscopic constituents. In interacting particle systems, the large scale behavior is often described by a partial differential equation for macroscopic quantities such as the particle density. The rigorous derivation of such macroscopic equations from microscopic dynamics became an area of research called "hydrodynamic limits" (see e.g. [71], [151], [220]). As mentioned in item 2., duality is useful in understanding the hydrodynamic limit, which can be inferred from the scaling limit of a single dual particle. Duality is also useful in understanding the fluctuations around the macroscopic equation, and in the proof of the emergence and propagation of local equilibrium via the scaling limit of a finite number of dual particles. If one can show duality with orthogonal duality functions, one can go further and study a hierarchy of fluctuation fields which are microscopic analogues of "Wick powers" of the fluctuation field (see e.g. [7]). The approach of hydrodynamic limits via duality was pioneered in [72], and further refined in [71].
5. **Structure of invariant measures of infinite interacting systems.** For infinite interacting particle systems, it is rarely the case that one can have a complete characterization of the set of invariant measures. Duality allows to study the infinite system via a finite number of dual particles, and invariant measures can be related to bounded harmonic functions of the dual process. If one can show that all bounded harmonic functions are constants, e.g. via a successful coupling, then one has control on the full structure of the set of invariant measures. This road was followed in [172] for the symmetric exclusion process, and as we will see can be followed for many other systems with duality.
6. **Showing existence of processes.** Duality can also be used to prove the existence of interacting particle systems in infinite volume, see e.g. the martingale problem approach of Holley and Stroock [135], or the construction of particle systems via graphical representations [83]. The idea is that the dual process which exists because it is a finite or countable state space Markov process can be used to define expectations of moments in the to-be-constructed process, which then can be defined via these moments, see e.g. [71] for an illustration of this approach, and [74] for a recent result showing existence via duality.

Besides the above listed applications of duality, which are the most relevant ones for the context of this monograph, there are many applications in stochastic models of population genetics [182] where one relates forward population models to backwards coalescents via duality, superprocesses [86] where one uses duality to solve the basic martingale problem showing the existence of the process, and stochastic partial differential equations [184] where asymptotic behavior can be derived via the study of a finite dimensional dual. From the point of view of applications, in this book we focus on the context of statistical mechanics, i.e., non-equilibrium steady states and hydrodynamic limits (in Chaps. 11, 12, and 13).

Duality in the Literature
We mention three important areas where duality has been extensively discussed since it plays a crucial role.

1. **Interacting particle systems.** Duality is one of the crucial techniques in the area of interacting particle systems. It was already present in the foundational paper of Spitzer [217], and pursued by Holley and Stroock [136]. Then it became one of the main ingredients used in Liggett's book [172]. In the context of hydrodynamic limits duality was of fundamental importance in the approach outlined in the books [71, 72], where coupling and duality are the main tools of the so-called correlation functions approach (also called v-function approach) to hydrodynamic limits. For recent reviews and developments of duality, including the pathwise approach, see [223, 224, 226] and more analytic overview paper [142].
2. **Mathematical population genetics.** As mentioned already, duality also plays a crucial role in models of mathematical population genetics (see [149, 150] for an introduction into this vast area), such as Wright-Fisher diffusion, Moran model, stepping stone model, and many extensions and generalizations of such models [66, 87]. In the literature related to mathematical population genetics, one usually studies the duality between the forward process, describing genetic traits (alleles) forward in time, and backwards coalescents. In this literature, this form of duality and various refinements of coalescents (such as Λ-coalescents, Beta-coalescents) and ancestral graphs have been developed to go beyond the classical Kingman's coalescent and describe the influence of mutation and selection.
3. **Quantum spin chains, exactly solvable models.** In the context of what is at present referred to as "integrable probability" [221], i.e., the analysis of stochastic models which are exactly solvable by matrix ansatz [75, 77], Bethe ansatz [33, 62, 67, 141], quantum inverse scattering and Yang-Baxter equation [64], duality usually accompanies such models and plays an important role in their solvability. The original work [213] exploits the connection between the symmetric exclusion process and the XXX quantum spin chain. This was indeed the first paper showing a connection between duality and "non-abelian symmetries". This approach also leads to dualities in the context of the asymmetric simple exclusion process [210], using quantum deformation of $\mathfrak{su}(2)$ Lie algebra. This approach was mostly based on recognizing the connection between the Markov generator and the Hamiltonian of a quantum spin system. Studying then the operators commuting with this Hamiltonian (the "non-abelian symmetries") leads naturally to the identification of duality and duality functions. There are many works in the physics literature relating Markov processes to quantum Hamiltonians, and then subsequently studying these Hamiltonians with physics methods (such as path integrals). In particular, the Doi-Peliti formalism of creation and annihilation operators associated to reaction-diffusion systems is well-developed, see e.g. [1, 193]. After the appearance of [213], there were a few mathematical papers further exploring this line of thought in the context of models associated to $\mathfrak{su}(2)$ Lie algebra in spin $1/2$ representation [225].

Summary of the Lie Algebraic Approach

For a pictorial description of the Lie algebraic approach to duality we refer to Fig. 1.1. Here we provide an informal summary of the main ideas behind the Lie algebraic approach to duality developed in this book:

1. **Generator duality.** The relation $\mathbb{Y} \xrightarrow{D} \mathbb{X}$ introduced in (1.1) can be replaced by a similar relation between the Markov generators of the two processes, here denoted $\widehat{L}$ and L. In the finite or countable state space setting the generators as well as the duality function simply become matrices, and the duality relation reads
$$\widehat{L}D = DL^T, \tag{1.3}$$
where T denotes transposition. Indeed, the Markov generator fully encodes the Markov process, and therefore in most cases, a duality relation for expectations is derived from a duality relation between the generators.
2. **Abstract duality.** The relation (1.3) is a linear relation between two matrices in the finite state space setting, and can also be rewritten as
$$(\widehat{L} \otimes I)D = (I \otimes L)D, \tag{1.4}$$
where I is the identity matrix and $\otimes$ denotes matrix tensor product. In this form, the duality relation (1.4) can be formulated in a very general setting, i.e., beyond matrices and even beyond duality "functions".
3. **Operator algebra duality.** It is now important to realize that besides relating $\widehat{L}$ to L, the duality function D will relate many more operators to other operators. More precisely, there is a family of operators $\{\widehat{A}_j\}_{j\in J}$, and a family of operators

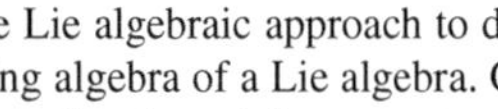

Fig. 1.1 Schematic picture for the Lie algebraic approach to duality: the abstract generator $\mathscr{L}$ is an object in the universal enveloping algebra of a Lie algebra. Considering two representations of the Lie algebra (denoted π_1 and π_2 in the picture) the generators $\widehat{L}$ and L of two Markov processes are obtained. The duality function D is related to the intertwiner between these two representations

$\{A_j\}_{j\in J}$, where J is an index set, such that

$$(\widehat{A}_j \otimes I)D = (I \otimes A_j)D \qquad \text{for all } j \in J. \tag{1.5}$$

The relation (1.5) then automatically extends to the algebras generated by the operators $\{\widehat{A}_j\}_{j\in J}$, resp. operators $\{A_j\}_{j\in J}$. More precisely, instead of only relating two Markov generators, the duality function D relates actually two algebras of operators.

4. **Intertwined representations.** In many cases of relevance in interacting particle systems, the operators $\{\widehat{A}_j\}_{j\in J}$ and $\{A_j\}_{j\in J}$ are in turn two intertwined representations of a Lie algebra, and in a sense to be described precisely in later chapters, the duality function is a kernel of the intertwiner between the two representations. In other words, the duality of Markov processes is nothing but a manifestation of, and a corollary of two intertwined representations of a given Lie algebra.
5. **Symmetries.** One immediately sees from (1.5) that for a given pair $\{\widehat{A}_j\}_{j\in J}$ and $\{A_j\}_{j\in J}$ such that $\widehat{A}_j \xrightarrow{D} A_j$, one can generate new duality functions $D' = \widehat{S}D$ from operators $\widehat{S}$ commuting with $\widehat{A}_j$ for all $j \in J$. Or new duality functions $D'' = DS$ from operators S commuting with A_j for all $j \in J$. In other words, using commuting operators ("symmetries") of the operator algebras, new duality functions can be generated from a given one.

These simple but very useful properties of the relation $\xrightarrow{D}$, and their applications, will be explained in much detail in Chap. 2. The identification of a Markov generator and its dual as two representations of an element of a Lie algebra (or more precisely of the universal enveloping algebra of a Lie algebra) immediately gives operators with which the generator (or the dual generator) commutes. With these commuting operators, from a given duality function new duality functions can be generated. In this way, for two given processes with duality we can also find several new duality functions, including e.g. orthogonal polynomials. To have a duality to "start from" is usually straightforward in the case of self-duality, where a simple D to start with is the matrix with elements the inverse of the reversible measure, as can be seen via detailed balance. In this sense, duality can be viewed as a non-diagonal generalization of detailed balance.

Realizing these facts goes far beyond a mathematically esthetic consideration. It means that in the search for dualities between Markov processes, or more generally in the search for processes which are related by duality, it is natural to start from Lie algebraic considerations and construct Markov generators from Lie algebra elements in a given representation. We will see that often the Markov generator is naturally connected to the Casimir element of a Lie algebra. The prototype examples will be the (partial) exclusion process which is related to the Lie algebra $\mathfrak{su}(2)$ corresponding to the compact Lie group $SU(2)$, and the inclusion process which is related to the Lie algebra $\mathfrak{su}(1,1)$ corresponding to the non-compact Lie group $SU(1,1)$. However, the property of being a Markov generator is dependent on the representation. Indeed, in the matrix setting it means non-negative off-diagonal elements and zero row sums, which is clearly dependent on the chosen basis.

Using the Lie Algebraic Approach

The key properties of the Lie algebraic approach discussed in the previous paragraph lead naturally to the following consequences.

1. **Families of processes and their duals.** Markov processes and their duals come in families, associated to Lie algebras and their representations. This leads often to a unification of various a priori unrelated dualities, but also to many new dualities, that can be obtained by varying a parameter labeling a representation (such as the spin value).
2. **Identification of symmetries of the generator.** Identifying the Markov generator as an element of the universal enveloping algebra of a Lie algebra naturally leads to the identification of symmetries, i.e., operators that commute with the generator. This is simply because Lie algebra generators come with commutation relations.
3. **Finding new duality functions via symmetries.** Symmetries applied to a duality function lead to new duality functions. In this way starting from a simple duality function, one can construct more useful (e.g. orthogonal) duality functions by acting with symmetries.
4. **Construction of new processes with symmetries.** The approach can also be used constructively, i.e., one can construct Markov processes from Lie algebra elements and then symmetries and associated dualities are built in the construction. E.g., in the setting of particle systems where particles hop over edges, one usually starts from a central element (such as the Casimir element) and then makes this into a generator acting on the two variables associated to an edge (such as the particle occupation numbers of the vertices associated to the edge) by using a co-product and a well-chosen representation (see e.g. [43, 44] for such constructions for asymmetric systems). In the context of discrete Markov chains, an algebraic construction from Hopf algebras has been considered in [80].

An overview of the dualities and the "families" of Markov processes associated with Lie algebras which we infer via this method can be found in Appendix C. The method also links processes which at first sight are unrelated, such as the symmetric inclusion process to the Kipnis-Marchiori-Presutti (KMP) process, and independent random walks to the Ginzburg Landau process with a quadratic potential. Once this link is provided via the Lie algebraic approach, i.e., by identifying the Lie algebraic form of the Markov generator, usually one obtains as a byproduct a whole one parameter family of processes related by duality, where the parameter labels the representation. E.g. the original KMP model [152] is based on a uniform redistribution of mass along edges of a graph. Once one realizes its underlying Lie algebraic form, one recognizes that it is related to the symmetric inclusion process, which carries a parameter $\alpha > 0$, and then one obtains automatically that mass redistribution models based on the Beta distribution with parameters (α, α) share all the duality properties of the original KMP model.

Choices Made in This Book

As already pointed out, duality of Markov processes is a vast subject and has a broad range of applications. We have chosen to limit the area which we cover in the book in the following way.

1. We treat only dualities related to classical Lie algebras, i.e., we do not cover the very exciting field of the processes associated to the q-deformed universal enveloping algebras (quantum groups). This would require another monograph. The dualities which come from classical Lie algebras are related to systems which satisfy detailed balance, and in the construction one can start from "single-edge" dynamics, which is then copied along the edges of a graph. I.e., these systems satisfy dualities in arbitrary graphs, whereas the dualities in their asymmetric q-deformed counterparts are necessarily limited to one-dimensional chains with nearest neighbor jumps.
2. We restrict to single-type particle systems, i.e., we do not consider higher rank Lie algebras that are associated to multi-type particle systems (e.g. the representations of the Lie algebra $\mathfrak{sl}(n)$ are associated to $n - 1$ types of particles [15, 53, 54, 156]).
3. We focus on simplicity, especially in the chapters on macroscopic fields, where we illustrate the use of duality in the setting of hydrodynamic limits. Here in principle, at the cost of more technicalities one can go much further (as already illustrated in the monograph [71]), but this would go beyond our aim of exposing the method, so we restrict to simple applications but provide enough ideas to convince the reader that there is also much beyond this.
4. We focus on interacting particle systems, in view of applications in non-equilibrium statistical physics. This means that we do not go into the very deep and well-developed field of dualities in mathematical population genetics. We do from time to time use the simplest examples from this area to illustrate duality in the setting of the Heisenberg algebra. One of the consequences is also that we will have dual processes conserving the number of particles. Many interesting systems, e.g. modelling reaction-diffusion systems can in principle be treated along the same line.
5. The theory of Lie groups, Lie algebras and their representations is a vast area. In this book, we have chosen to limit the use of Lie theory to what is strictly needed, i.e., we choose not to treat very interesting aspects of Lie representation theory such as Schur duality, root systems, Dynkin diagrams. This shows that to become familiar with and use the method presented in this book in the context of interacting particle systems, it is not at all necessary to become a specialist in Lie theory. In particular, we do not assume any prior knowledge of Lie theory of the reader. Appendix B is devoted to the basic background in Lie theory needed for the book.

Intended Audience

The book is mainly written for graduate students and researchers interested in the theory of Markov processes. The specific areas that we focus on are interacting particle systems and non-equilibrium statistical physics. People with an interest in those areas will directly find in the book several models and systems of immediate interest and application. However, the scope of the theory of duality that is developed in the book is larger and aimed to be used by mathematicians, physicists and biologists dealing with Markov processes in several other contexts such as integrable systems theory, representation theory, stochastic partial differential equations, branching processes, coalescence, population dynamics, etc.

The target audience is thus first of all researchers (from beginners to experienced ones) who want to learn duality for Markov processes and use it in their own area. On the other hand, Chap. 2 (which is an introduction to the concept of duality and the Lie algebraic approach) and a selection of later chapters (e.g. Chaps. 2 and 3 that describe several dualities for independent random walkers) can very well be used to teach a course at late undergraduate or graduate level.

Content Outline

We close this introduction by giving a panoramic view of the structure of the book and providing a succinct description of the content of each chapter. This allows the reader to construct its own reading path, we suggest two possibilities at the very end (one for researchers and one for graduate students). The reader may select the chapters he is interested in, it is not necessary to read all the chapters sequentially.

1. The material is organized into thirteen chapters and three appendices. Appendices on Markov processes theory (Appendix A) and Lie theory (Appendix B) aim at making the book as self-contained as possible. Some acquaintance with the theory of Markov processes, semigroups and generators in the spirit of [172] is desirable. On the contrary, acquaintance with Lie theory is not required. All the Lie algebraic concepts used are contained in the book and are gradually introduced when needed, whereas the necessary background is contained in Appendix B. The reader who wants to apply some of the new dualities can find a systematic overview in Appendix C.
2. Each chapter has an *Abstract* which describes the essential content of the chapter, and a closing section of *Additional notes* pointing to related literature and open research problems. Each chapter is as much self-contained as possible. The reader can e.g. choose to go straight to the chapters about particle system, i.e., he may skip at first reading the general material of Chap. 2.
3. As for the content, a road map is the following. We introduce and develop the algebraic approach to duality theory considering three basic examples from the theory of interacting particle systems: independent random walkers, the symmetric exclusion process and the symmetric inclusion process. We have chosen these three basic examples because they are associated to three classical Lie algebras: the Heisenberg algebra, the $\mathfrak{su}(2)$ algebra and the $\mathfrak{su}(1,1)$ algebra. We also provide applications in non-equilibrium statistical physics such as the

study of non-equilibrium steady states via duality and the study of hydrodynamic limits and fluctuation fields via duality.

4. More specifically, the content of each chapter is described in a concise way as follows.

   - Chapter 2 provides the basics of the algebraic approach to duality. It starts with two historical examples and then it introduces the definition of semigroup duality and of generator duality. The key principles of the algebraic approach (i.e., change of representation, symmetries, intertwining) are discussed by considering several contexts of increasing generality starting with finite state space Markov chains.
   - Chapters 3 and 4 develop the first main example process, that is the duality for independent random walkers. We spend quite some space and time on independent random walkers and related processes of independent particles. This material serves as the simplest context in which our approach can be explained. This includes both the Lie algebraic structure of the Markov generator, the various (self)-duality functions, the generating function method and intertwining. Some elements of ergodic theory, in particular the structure of invariant measures, are discussed as a first application of duality.
   - Chapters 5 and 6 introduce a second important example, which is then much studied in later chapters. They deal, respectively, with the duality properties of the symmetric inclusion process and with the duality properties of the Brownian energy process. These two processes share the same algebraic structure, as both are related to the Casimir element of the $\mathfrak{su}(1,1)$ Lie algebra. The first (inclusion process) arises when looking at this element in a discrete representation in terms of matrices, the second (Brownian energy process) arises in a continuous representation in terms of differential operators.
   - Chapter 7 treats the duality of the third example, namely the symmetric partial exclusion process. The process generator is now in direct relation with the Casimir of the $\mathfrak{su}(2)$ Lie algebra. As this process, especially the version with at most one particle per site, has been discussed in several textbooks, here we focus mostly on the aspects which are directly related to the algebraic description, such as the additive structure on ladders and the intertwining between different partial exclusion processes.
   - Chapter 8 contains the dualities of several other models. Via the Lie algebraic approach, we show that there is a large class of processes naturally connected to the three basic examples of previous chapters. This class includes interacting Markov diffusion processes such as the Ginzburg-Landau model with quadratic interaction potential, models of mass redistribution such as the Kipnis-Marchioro-Presutti (KMP) process or the Aldous averaging process, models from kinetic theory such as the Kac model and models from population genetics such as the Wright-Fisher diffusion and the Moran model.
   - Chapter 9 extends the previous duality results in the direction of orthogonal polynomials. Using the Lie algebraic structure and considering unitary equivalent representations, orthogonal polynomials arise as novel duality

functions, also useful for several applications. In fact, it is shown how the orthogonal dualities are produced via a Gram-Schmidt orthogonalization procedure acting on the triangular dualities.

- Chapter 10 highlights the link between the self-duality property of the particle processes and the property of consistency. This link emerges from a symmetry (the so-called annihilation operator) which easily follows from the algebraic perspective. A combinatorial interpretation and several applications of consistency are discussed.
- Chapter 11 illustrates the use and application of duality for non-equilibrium systems. This is first explained in the simple case of independent random walkers with reservoirs and then further developed for the other main processes by considering their boundary-driven set-up.
- Chapter 12 is focused on the use of duality to study macroscopic fields. We discuss how the dynamics of one dual particle is related to the hydrodynamic limit, the dynamics of two dual particles is related to the density fluctuation field, and the dynamics of n dual particles is related to the propagation of local equilibrium. Higher-order hydrodynamic fields are also introduced and applications to the Boltzmann Gibbs principle via orthogonal polynomial duality are also given.
- Chapter 13 studies the interplay between duality and integrability in a concrete family of exactly solvable processes with $\mathfrak{su}(1,1)$ symmetry. These are the harmonic processes related to integrable spin chains. The chapter shows how the combination of duality and integrability allows to establish properties which are rarely available, such as a full explicit description of the non-equilibrium steady state arising in the boundary-driven set-up. We also provide a more probabilistic approach to the exact solvability of the boundary driven harmonic processes, via a spatial Markov property.

5. In view of the description of the material given above, we suggest as a reading path for graduate students Chapters from 2 to 7, which are rather introductory and should be readable for everyone with a basic background in the theory of Markov processes. Researchers may proceed with the remaining Chapters from 8 to 13, which offer a perspective on the more recent advances of duality and its many applications to several problems in interacting particle systems and non-equilibrium statistical physics.

Chapter 2
Basics of the Algebraic Approach

In this chapter, after recalling two historical examples, we introduce the notion of duality between two Markov processes. We define both semigroup duality and generator duality, followed by a discussion on their reciprocal relation. We then consider the notion of duality between two algebras of operators in a more general context, i.e. beyond Markov processes. We proceed by increasing levels of generality: first we treat algebras of matrices, then algebras of bounded operator on a Hilbert space. Finally we give the abstract formulation of duality between two algebras which arises by considering two intertwined representations. The added value of this abstract formulation is that dualities of Markov processes can be understood from algebra representation theory. As an example, we show how this works for the duality between the Wright-Fisher diffusion and the Kingman's coalescent block counting process, which is explained using two representations of the Heisenberg Lie algebra.

2.1 Two Starting Examples

We start with two historical and important examples of duality between Markov processes, namely "Siegmund duality" between reflected and absorbed Brownian motion, and "moment duality" between the Wright-Fisher diffusion and the Kingman's coalescent block counting process.

2.1.1 Reflected and Absorbed Brownian Motion

This first example of duality goes back to Lévy [169], and in modern terminology is called "Siegmund duality" between reflected and absorbed Brownian motion. Let

C. Giardinà, F. Redig, *Duality for Markov Processes*, Grundlehren der mathematischen Wissenschaften 365, https://doi.org/10.1007/978-3-032-04099-2_2

$\{X(t) : t \geq 0\}$ denote Brownian motion on $[0, \infty)$ reflected at the origin, and $\{Y(t) : t \geq 0\}$ denote Brownian motion on $[0, \infty)$ absorbed at the origin. The transition densities of these processes are explicit and given by (see for instance [31])

$$p_t^R(x, y) = \frac{1}{\sqrt{2\pi t}} \left(e^{-\frac{(x-y)^2}{2t}} + e^{-\frac{(x+y)^2}{2t}} \right),$$
$$p_t^A(x, y) = \frac{1}{\sqrt{2\pi t}} \left(e^{-\frac{(x-y)^2}{2t}} - e^{-\frac{(x+y)^2}{2t}} \right), \tag{2.1}$$

for $x, y > 0, t > 0$. Let us denote by $\mathbb{P}_x$ the distribution of trajectories of $\{X(t) : t \geq 0\}$ starting from $X(0) = x \in [0, \infty)$ and $\widehat{\mathbb{P}}_y$ the distribution of trajectories of $\{Y(t) : t \geq 0\}$ starting from $Y(0) = y \in [0, \infty)$. Since the probability that the absorbed Brownian motion is at the origin at time $t > 0$ is given by

$$\widehat{\mathbb{P}}_y(Y(t) = 0) = \begin{cases} \text{Erfc}\left(\frac{y}{\sqrt{2t}}\right), & y > 0, \\ 1, & y = 0, \end{cases} \tag{2.2}$$

then one has the following relation between these two processes which follows from explicit computations

$$\int_0^x p_t^A(y, z)dz + \widehat{\mathbb{P}}_y(Y(t) = 0) = \int_y^\infty p_t^R(x, z)dz. \tag{2.3}$$

This can be rewritten as

$$\widehat{\mathbb{P}}_y(Y(t) \leq x) = \mathbb{P}_x(X(t) \geq y), \qquad x, y \geq 0, t \geq 0. \tag{2.4}$$

Denote $\mathbb{E}_x$ expectation in the process $\{X(t) : t \geq 0\}$ starting from $X(0) = x$, $\widehat{\mathbb{E}}_y$ expectation in the process $\{Y(t) : t \geq 0\}$ starting from $Y(0) = y$, and $D(y, x) = \mathbb{1}_{\{y \leq x\}}$ where $\mathbb{1}_A$ denotes the indicator of the set A. Then we can further rewrite (2.4) as

$$\widehat{\mathbb{E}}_y \left(D(Y(t), x)\right) = \mathbb{E}_x \left(D(y, X(t)\right) \qquad x, y \geq 0, t \geq 0. \tag{2.5}$$

Such a relation between two Markov processes is called "duality with duality function D". More precisely, we say that the process $\{Y(t) : t \geq 0\}$ is dual to the process $\{X(t) : t \geq 0\}$ with duality function

$$D(y, x) = \mathbb{1}_{\{y \leq x\}}. \tag{2.6}$$

Duality with the duality function $D(y, x) = \mathbb{1}_{\{y \leq x\}}$ is called Siegmund duality [214]. This duality function appears often in the context of monotone processes, i.e., processes for which the mapping

$$x \mapsto \mathbb{P}_x(X(t) \geq y)$$

is non-decreasing, for all y. E.g., Siegmund duality appears in birth and death processes, as well as in interacting particle systems and is always related to the preservation of some (partial) order [223].

2.1.2 *Wright-Fisher Diffusion and Kingman's Coalescent*

A second well-known historical example is "moment duality" between the Wright-Fisher diffusion and the Kingman's coalescent block counting process. Let $\{X(t) : t \geq 0\}$ denote the diffusion process on $[0, 1]$ which solves the stochastic differential equation

$$dX(t) = \sqrt{X(t)(1 - X(t))}dW(t), \tag{2.7}$$

where $\{W(t) : t \geq 0\}$ denotes standard Brownian motion. This process is one of the classical processes appearing in mathematical population genetics. In that area it models the fraction of individuals of type 1 in a population of two (allelic) types which is subject to random genetic drift, in the limit of large total population size (see [87] for more background). With probability 1 this process is eventually absorbed either at 0 or at 1, which corresponds to fixation of the allelic type.

To introduce the corresponding dual process, let $\{Y(t) : t \geq 0\}$ denote the Markov jump process on the natural numbers, which jumps from y to $y - 1$ at rate $\frac{1}{2}y(y-1)$. In this process the state 1 is absorbing, and from any initial state $y > 1$, the process only jumps down. This jump process is called the block counting process of the Kingman's coalescent. Let us denote by $\mathbb{E}_x$ expectation in the Wright-Fisher diffusion starting at x, and $\widehat{\mathbb{E}}_y$ denote expectation in the process $\{Y(t) : t \geq 0\}$ starting at $y \in \mathbb{N}$. Then we have the relation

$$\widehat{\mathbb{E}}_y(x^{Y(t)}) = \mathbb{E}_x(X(t)^y), \tag{2.8}$$

which can also be rewritten as

$$\widehat{\mathbb{E}}_y\left(D(Y(t), x)\right) = \mathbb{E}_x\left(D(y, X(t))\right), \tag{2.9}$$

where now

$$D(y, x) = x^y. \tag{2.10}$$

The relation (2.9) is called duality between the Kingman's coalescent block counting process and the Wright-Fisher diffusion, and the duality function (2.10) is called "moment duality function". We will come back to this example as a natural illustration of duality via a change of representation of an underlying algebra (in this case the Heisenberg algebra, see Sect. 2.5). The duality (2.9) provides full information about the original process, i.e., all moments of the Wright-Fisher diffusion can be obtained via the study of the much simpler discrete process counting the blocks of the Kingman's coalescent.

The duality relation (2.9) is actually a consequence of a duality relation between the generators of the two processes. The infinitesimal generator of the process $\{X(t) : t \geq 0\}$,

$$Lf(x) = \frac{1}{2}x(1-x)\frac{d^2 f}{dx^2}(x), \tag{2.11}$$

with domain the set of function on $C^2[0,1]$ functions such that $Lf(0^+) = Lf(1^-) = 0$. The set of polynomials in the variable x is a core for this generator. The infinitesimal generator of the jump process $\{Y(t) : t \geq 0\}$ reads, for $f : \mathbb{N} \to \mathbb{R}$

$$\widehat{L}f(y) = \frac{1}{2}y(y-1)(f(y-1) - f(y)). \tag{2.12}$$

The duality relation for expectations (2.9) is then implied (see Theorem 2.5 below) by the "generator duality"

$$(\widehat{L}D(\cdot, x))(y) = (LD(y, \cdot))(x). \tag{2.13}$$

Indeed, the action of $\widehat{L}$ on the y-variable is the same as the action of L on the x-variable:

$$\frac{1}{2}y(y-1)\left(x^{y-1} - x^y\right) = \frac{1}{2}x(1-x)\left(\frac{d^2(x^y)}{dx^2}\right)(x). \tag{2.14}$$

Let us give two simple applications of this duality, computing the fixation probabilities and the heterozygosity of the Wright-Fisher diffusion using the dual process. First, because in the Kingman's coalescent block counting process 1 is an absorbing state, we have:

$$\mathbb{E}_x(X(t)) = \mathbb{E}_x D(1, X(t)) = \widehat{\mathbb{E}}_1 D(Y(t), x) = D(1, x) = x, \tag{2.15}$$

which implies also that the fixation probabilities are given by

$$\lim_{t\to\infty} \mathbb{P}_x(X_t = 1) = x = 1 - \lim_{t\to\infty} \mathbb{P}_x(X_t = 0). \tag{2.16}$$

Similarly, for the second moment we have

$$\begin{aligned}\mathbb{E}_x(X(t)^2) &= \mathbb{E}_x D(2, X(t)) = \widehat{\mathbb{E}}_2 D(Y(t), x) \\ &= e^{-t} D(2, x) + (1 - e^{-t}) D(1, x) \\ &= e^{-t} x^2 + (1 - e^{-t}) x,\end{aligned} \tag{2.17}$$

where the third equality follows from the fact that in the dual process $Y(t)$, starting from initial state 2, the rate to move down to the state 1 is equal to 1, and 1 is an absorbing state. As a consequence, we obtain the so-called heterozygosity at time $t \geq 0$:

$$\mathbb{E}_x\big(X(t)(1 - X(t))\big) = x(1 - x)e^{-t}. \tag{2.18}$$

2.2 Semigroup Duality and Generator Duality

We now define the notion of duality and provide some basic properties. Let $\{X(t) : t \geq 0\}$ and $\{Y(t) : t \geq 0\}$ denote two Markov processes on state spaces Ω, resp. $\widehat{\Omega}$. We think of $\{X(t) : t \geq 0\}$ as "the process under study", and $\{Y(t) : t \geq 0\}$ as "the dual process". Usually, the existence of the dual process is useful because it is "a simpler process" yielding relevant (ideally full) information on the process $\{X(t) : t \geq 0\}$. Of course, the distinction between the process and the dual process is not relevant for the mathematical definition (see Remark 2.4).

We denote by $\mathbb{P}_x$ the path space measure of $\{X(t) : t \geq 0\}$, starting at $x \in \Omega$, with corresponding expectation $\mathbb{E}_x$, and by $\widehat{\mathbb{P}}_y$ the path space measure of $\{Y(t) : t \geq 0\}$, starting at $y \in \widehat{\Omega}$, with corresponding expectation $\widehat{\mathbb{E}}_y$.

Definition 2.1 (Semigroup Duality) Let $D : \widehat{\Omega} \times \Omega \to \mathbb{R}$ denote a measurable function. We say that D is a duality function for semigroup duality between the process $\{X(t) : t \geq 0\}$ and the dual process $\{Y(t) : t \geq 0\}$ if for all $x \in \Omega$, $y \in \widehat{\Omega}$ and $t \geq 0$, we have

$$\widehat{\mathbb{E}}_y(D(Y(t), x)) = \mathbb{E}_x(D(y, X(t))), \tag{2.19}$$

where it is implicitly assumed that the expectations are well-defined, i.e., the functions of which we take expectations in (2.19) are integrable. If the processes $\{X(t) : t \geq 0\}$ and $\{Y(t) : t \geq 0\}$ are the same in distribution, then we call (2.19) self-duality.

In case the duality functions are in the domain of the infinitesimal generator, we have the notion of "generator duality", which is defined as follows.

Definition 2.2 (Generator Duality) Let L_X denote the generator of the process $\{X(t) : t \geq 0\}$ and L_Y the generator of the process $\{Y(t) : t \geq 0\}$. Let $D : \widehat{\Omega} \times \Omega \to \mathbb{R}$ be a function, such that $D(y, \cdot)$ is in the domain of L_X and $D(\cdot, x)$ is in the domain of L_Y. We then say that D is a duality function for generator duality between the process $\{X(t) : t \geq 0\}$ and the dual process $\{Y(t) : t \geq 0\}$ if for all $x \in \Omega$, $y \in \widehat{\Omega}$, we have

$$(L_Y D(\cdot, x))\,(y) = (L_X D(y, \cdot))\,(x). \tag{2.20}$$

Remark 2.3 (Notation) The notion of semigroup duality in Definition 2.1 is equivalent to the notion of "process duality" given in the Introduction. As a concise notation for semigroup duality we will write

$$Y \xrightarrow{D} X. \tag{2.21}$$

The generator duality in Definition 2.2 will instead be denoted by

$$L_Y \xrightarrow{D} L_X. \tag{2.22}$$

Remark 2.4 (Role of the Processes) The role of the two processes in both duality definitions can be swapped, provided that also the variables of the duality function are swapped. More precisely, if $D : \widehat{\Omega} \times \Omega \to \mathbb{R}$ is a duality function such that

$$Y \xrightarrow{D} X \tag{2.23}$$

then

$$X \xrightarrow{D'} Y \tag{2.24}$$

where $D' : \Omega \times \widehat{\Omega} \to \mathbb{R}$ is defined by

$$D'(x, y) = D(y, x). \tag{2.25}$$

In case the state spaces of the process and its dual are finite, the generators are matrices, and also the duality function is a matrix, and the defining equality for generator duality can then be rewritten in matrix form as

$$L_Y D = D L_X^T, \tag{2.26}$$

where the superscript T denotes the transposed. Indeed, in that case we can rewrite (2.20) in terms of matrix elements of the generators L_Y, L_X as follows

$$\sum_{y' \in \widehat{\Omega}} L_Y(y, y') D(y', x) = \sum_{x' \in \Omega} L_X(x, x') D(y, x'), \tag{2.27}$$

which is exactly the element-wise version of (2.26). Starting from (2.26), one can iterate, and obtain

$$(L_Y)^n D = D(L_X^T)^n \tag{2.28}$$

for all $n \in \mathbb{N}$, and because the semigroups are simply the matrix exponentials $S_X(t) = \sum_{n=0}^{\infty} \frac{t^n}{n!} L_X^n$ (and similarly for $S_Y(t)$), we obtain

$$S_Y(t)D = DS_X(t)^T, \tag{2.29}$$

which is exactly (2.19).

In the following theorem we give implications between generator duality and semigroup duality in a more general context. The subtleties start to arise when the generators are unbounded operators. In order to explain the problem, and formulate a general theorem, we need some additional notation. We denote the semigroup of the process $\{X(t) : t \geq 0\}$ by

$$(S_X(t)f)(x) = \mathbb{E}_x[f(X(t))], \tag{2.30}$$

and similarly $(S_Y(t)f)(y) = \widehat{\mathbb{E}}_y[f(Y(t))]$. We denote by $\mathcal{B}_X$, respectively $\mathcal{B}_Y$, the Banach space on which $(S_X(t))_{t\geq 0}$, respectively $(S_Y(t))_{t\geq 0}$, is acting as a contraction semigroup. We recall that the domain of the generator L_X is the dense subset of $\mathcal{B}_X$ containing functions such that the limit

$$\lim_{t\to 0} \frac{\mathbb{E}_x f(X(t)) - f(x)}{t} = L_X f(x) \tag{2.31}$$

exists in the norm of $\mathcal{B}_X$. The domain of the generator L_Y is defined in an analogous way.

We can then introduce the contraction semigroups on the tensor product space $\mathcal{B}_Y \otimes \mathcal{B}_X$ equipped with a suitable norm

$$S_Y(t) \otimes I \qquad \text{and} \qquad I \otimes S_X(t), \tag{2.32}$$

where I denotes the identity operator. Here we remind the notation of tensor product of operators. If $A_Y : \mathcal{B}_Y \to \mathcal{B}_Y$ is a linear operator and $A_X : \mathcal{B}_X \to \mathcal{B}_X$ is another linear operator, then $A_Y \otimes A_X : \mathcal{B}_Y \otimes \mathcal{B}_X \to \mathcal{B}_Y \otimes \mathcal{B}_X$ is a linear operator defined via

$$(A_Y \otimes A_X)(g \otimes f) = A_Y g \otimes A_X f, \qquad g \in \mathcal{B}_Y, f \in \mathcal{B}_X,$$

and extended to linear combinations of tensors by linearity.

The choice of the norm on the tensor product space should be such that the two semigroups in (2.32) are contraction semigroups. In the case $\mathcal{B}_X$ and $\mathcal{B}_Y$ are Hilbert spaces, this norm is unique but in the Banach space case there are several choices

[204]. We further denote by $\mathcal{D}_Y$, resp. $\mathcal{D}_X$, the domains of the generators of the semigroups in (2.32), which are dense subsets of $\mathcal{B}_Y \otimes \mathcal{B}_X$. Notice that when f is in the domain of L_X, then $\psi(x, y) = f(x)$ is in the domain $\mathcal{D}_X$, but other elements not of tensor form will be in $\mathcal{D}_X$ as well. Also notice that the action of the generator of $I \otimes S_X(t)$, resp. $S_Y(t) \otimes I$, is always $I \otimes L_X$, resp. $L_Y \otimes I$.

Theorem 2.5 (Relation Between Semigroup Duality and Generator Duality) *Let L_X denote the generator of the process $\{X(t) : t \geq 0\}$ and L_Y the generator of the process $\{Y(t) : t \geq 0\}$. If $D : \widehat{\Omega} \times \Omega \to \mathbb{R}$ is such that for all $x \in \Omega$ the function $D(\cdot, x)$ is in the domain of L_Y and for all $y \in \widehat{\Omega}$ the function $D(y, \cdot)$ is in the domain of L_X, then semigroup duality*

$$(S_Y(t) \otimes I)D = (I \otimes S_X(t))D \tag{2.33}$$

implies generator duality

$$(L_Y \otimes I)D = (I \otimes L_X)D. \tag{2.34}$$

In the other direction, if

$$(S_Y(t) \otimes I)D \in \mathcal{D}_X \quad \textit{and} \quad (I \otimes S_X(t))D \in \mathcal{D}_Y \tag{2.35}$$

for all $t \geq 0$, then generator duality (2.34) *implies semigroup duality* (2.33).

Proof The first statement of the theorem is trivial: if (2.33) holds then we have for all $t > 0$

$$\frac{(S_Y(t) \otimes I) - (I \otimes I)}{t} D = \frac{(I \otimes S_X(t)) - (I \otimes I)}{t} D. \tag{2.36}$$

Taking the limit $t \to 0$ and using the assumption that for all $x \in \Omega$ the function $D(\cdot, x)$ is in the domain of L_Y and for all $y \in \widehat{\Omega}$ the function $D(y, \cdot)$ is in the domain of L_X, we arrive at (2.34).

To prove the second statement of the theorem, we start from the obvious identity (which is a consequence of Fubini's theorem)

$$(S_Y(t) \otimes I)(I \otimes S_X(s))D = (I \otimes S_X(s))(S_Y(t) \otimes I)D. \tag{2.37}$$

If the assumption (2.35) holds, we can now take the derivative at $s = 0$ of this identity, and obtain

$$(S_Y(t) \otimes I)(I \otimes L_X)D = (I \otimes L_X)(S_Y(t) \otimes I)D. \tag{2.38}$$

Notice that we can interchange $(S_Y(t) \otimes I)$ with the derivative d/ds because $(S_Y(t) \otimes I)$ is a contraction semigroup on the tensored space (and hence continuous).

Indeed, by assumption

$$\frac{(I \otimes S_X(s)) - I \otimes I}{s} D$$

converges, as $s \to 0$, in the norm of the tensored space, to

$$(I \otimes L_X)D$$

and as $(S_Y(t) \otimes I)$ is a bounded operator, the limit $s \to 0$ and $S_Y(t) \otimes I$ can be exchanged:

$$\begin{aligned}(S_Y(t) \otimes I) &\left(\lim_{s \to 0} \frac{(I \otimes S_X(s)) - I \otimes I}{s} D \right) \\ &= \lim_{s \to 0} (S_Y(t) \otimes I) \left(\frac{(I \otimes S_X(s)) - I \otimes I}{s} \right) D. \end{aligned} \tag{2.39}$$

On the other hand, by assumption (2.35), $(S_Y(t) \otimes I)D$ is in $\mathcal{D}_X$, hence

$$\left(\frac{(I \otimes S_X(s)) - I \otimes I}{s} \right) (S_Y(t) \otimes I)D$$

converges, as $s \to 0$, in the norm of the tensored space to $(I \otimes L_X)(S_Y(t) \otimes I)D$. From (2.38) we will now prove semigroup duality. Define $u(x, y, t) = (S_Y(t) \otimes I)D(y, x)$ and $v(x, y, t) = (I \otimes S_X(t))D(y, x)$. We have, by the domain assumptions and using generator duality

$$\begin{aligned}\frac{d}{dt} u(x, y, t) &= (S_Y(t) \otimes I)(L_Y \otimes I)D(y, x) \\ &= (S_Y(t) \otimes I)(I \otimes L_X)D(y, x) \\ &= (I \otimes L_X)(S_Y(t) \otimes I)D(y, x) \\ &= L_X u(x, y, t). \end{aligned} \tag{2.40}$$

Furthermore, by the definition of the generator

$$\frac{d}{dt} v(x, y, t) = L_X v(x, y, t). \tag{2.41}$$

Now, fixing y, we have that both $u(x, y, t)$ in (2.40) and $v(x, y, t)$ in (2.41) are solutions of the differential equation

$$\frac{d}{dt} \varphi(x, t) = L_X \varphi(x, t). \tag{2.42}$$

Given the initial condition, the solution of (2.42) is unique and given by $\varphi(x,t) = S_X(t)\varphi(x,0)$, provided $\varphi(x,0)$ is in the domain of L_X. Since we have $u(x,y,0) = v(x,y,0)$, we conclude $u(x,y,t) = v(x,y,t)$ for all $t \geq 0$. □

In the following proposition we collect a number of cases in which generator duality implies semigroup duality.

Proposition 2.6 *Generator duality implies semigroup duality in the following cases:*

1. *The dual process or the original process has a finite state space.*
2. *The generators of the original and dual processes are both bounded operators.*

Proof The second item is an immediate extension of the case where both processes have finite state space, i.e. the argument that (2.26) is equivalent to (2.29) via matrix exponentiation extends to bounded operators. For the first item, let us consider the case where the process Y is a finite state space Markov chain. Then $L_Y D(y,\cdot)$ is a finite sum, and therefore it can always be interchanged with $S_X(t)$ (by linearity), i.e., $S_X(t)L_Y D = L_Y S_X(t) D$. From here we can then further proceed as in the proof of Theorem 2.5. □

2.3 Dualities, Symmetries and Intertwinings for Finite Markov Chains

In this section, we assume that the Markov process $\mathbb{X} := \{X(t) : t \geq 0\}$ as well as $\mathbb{Y} := \{Y(t) : t \geq 0\}$ both have finite state spaces. As we saw before, in this context we automatically have the equivalence between generator duality and semigroup duality. We can then explain the basic principles relating dualities and symmetries, as well as dualities and intertwinings, without technicalities related to domains of unbounded operators. In words, these principles are formulated as follows.

1. *Self-duality and Symmetry.* Reversibility of a Markov process and the associated detailed balance relation is the simplest instance of self-duality: we call this "cheap self-duality". Non-trivial self-duality can be generated if the process has a symmetry, i.e. an operator commuting with the Markov generator. We obtain the new self-duality from the cheap one by acting on it with the symmetry. If the symmetry is in kernel operator form w.r.t. the reversible measure, then the corresponding kernel is a self-duality function. Conversely, to every self-duality function corresponds a symmetry by considering the corresponding kernel operator.
2. *Duality and Symmetry.* Stationarity of a Markov process is the simplest instance of duality: from a stationary measure one can construct a "cheap duality" between the process and its reversed process. Similarly to item 1, if one is given a duality relation between two processes and a symmetry of the dual process, then acting with this symmetry a new duality function can be produced.

3. *Duality and Intertwining.* This is a generalization of the symmetry principle stated in the previous items. Suppose that, instead of a symmetry of a Markov process, one has an intertwiner $\Lambda_{1,2}$ linking two processes $\mathbb{X}_1$ and $\mathbb{X}_2$. Assume moreover that the process $\mathbb{X}_2$ is dual to a Markov process $\mathbb{X}_3$ with duality function $D_{2,3}$. Then it is possible to produce a duality relation between the processes $\mathbb{X}_1$ and $\mathbb{X}_3$. The duality function is obtained as $D_{1,3} = \Lambda_{1,2} D_{2,3}$.

   We now explain these principles in full detail.

2.3.1 Self-duality and Symmetry

We start with self-duality. We remind the reader that in the whole of this section we assume that the Markov processes considered are continuous-time Markov chains with a finite state space, and as a consequence generators (and also duality functions) are finite matrices. For a generator matrix L indexed by the finite state space Ω, we denote its matrix elements by $L(x, x')$ where $x, x' \in \Omega$. We first define the notion of reversible measure.

Definition 2.7 (Reversible Measure) Let $\{X(t) : t \geq 0\}$ be a finite state space Markov process with generator L. A reversible measure (or detailed balance measure) is a map $M : \Omega \to (0, \infty)$ such that the detailed balance relation

$$M(x)L(x, x') = M(x')L(x', x) \tag{2.43}$$

holds for all $x, x' \in \Omega$.

As the reader might notice, we assume that $M(x) > 0$ for all $x \in \Omega$ which in the finite case amounts to irreducibility of the Markov chain. Moreover, remark that we do not assume that M is a probability measure. This is in view of later chapters, where we will consider countable infinite settings where we need a reversible σ-finite measure such as the counting measure.

As we will see below, the existence of a reversible measure implies a (trivial) self-duality relation. Next we define the notion of symmetry.

Definition 2.8 (Symmetry) Let $\{X(t) : t \geq 0\}$ be a finite state space Markov process with generator L. A matrix S indexed by the finite state space Ω is a symmetry of L if it commutes with L, i.e.,

$$[S, L] = SL - LS = 0. \tag{2.44}$$

More generally, for a matrix A indexed by the finite state space Ω (not necessarily a Markov generator), we define S to be a symmetry of A if it commutes with A. For a given matrix the set of symmetries is closed under matrix addition and multiplication.

In the following theorem we explain the relation between self-duality and symmetry.

Theorem 2.9 (Self-duality and Symmetry) *Let $\{X(t) : t \geq 0\}$ be a finite state space Markov process with generator L. Then we have the following:*

1. *Existence of a "cheap self-duality" made from the reversible measure. If M is a reversible measure then*

$$D(x, x') = \frac{1}{M(x)} \delta_{x,x'} \tag{2.45}$$

is a self-duality function. Here $\delta_{x,x'}$ denotes the Kronecker delta.

2. *Symmetries produce new self-duality functions from a given one. Assume D is a self-duality function. If S is a symmetry of L then SD and DS^T are self-duality functions as well.*

Proof For item 1:

$$(LD)(x, x') = \sum_{z\in\Omega} L(x, z)D(z, x') = L(x, x')\frac{1}{M(x')}, \tag{2.46}$$

and

$$(DL^T)(x, x') = \sum_{z\in\Omega} D(x, z)L^T(z, x') = \sum_{z\in\Omega} D(x, z)L(x', z) = \frac{1}{M(x)} L(x', x). \tag{2.47}$$

Therefore, by (2.43) we have $LD = DL^T$, i.e., D is a self-duality function (cf. (2.26)).

For item 2: if D satisfies $LD = DL^T$ and S commutes with L, then

$$L(SD) = SLD = SDL^T = (SD)L^T, \tag{2.48}$$

where in the first equality we used that S is a symmetry of L and in the second equality we used that D is a self-duality function. Similarly,

$$L(DS^T) = DL^T S^T = DS^T L^T = (DS^T)L^T. \tag{2.49}$$

Here we first used self-duality and then symmetry (the fact that S^T is a symmetry of L^T is an immediate consequence of S being a symmetry of L). □

Remark 2.10 In the setting of Theorem 2.9, SD can be viewed as the result of S acting on the left (y)-variable of D, whereas DS^T can be viewed as the result of S acting on the right (x)-variable of D.

The previous theorem shows that in the reversible setting, we can start from a cheap self-duality and produce non-trivial self-dualities by acting on it with symmetries of the generator. Conversely, as we will see in the next theorem, in the reversible setting every self-duality function is associated to two symmetries, i.e. those symmetries that have this self-duality function as the kernel in the L^2-space of the reversible measure. In other to introduce this, we need some more notation.

Let $M : \Omega \to (0, \infty)$ be a measure. We then define the inner product in $L^2(\Omega, M)$ as usual: for any two functions $f : \Omega \to \mathbb{R}$ and $g : \Omega \to \mathbb{R}$ we put

$$\langle f, g \rangle = \sum_{x \in \Omega} f(x) g(x) M(x). \tag{2.50}$$

Reversibility of a measure M, in the sense of Definition 2.7, is then equivalent with self-adjointness of L in the space $L^2(\Omega, M)$ (i.e., $L = L^*$, where L^* denotes the adjoint of L). Indeed, using (2.43), we have

$$\begin{aligned} \langle Lf, g \rangle &= \sum_{x, x' \in \Omega} L(x, x') f(x') g(x) M(x) \qquad (2.51)\\ &= \sum_{x, x' \in \Omega} L(x', x) g(x) f(x') M(x') = \langle f, Lg \rangle. \end{aligned}$$

We say that an operator S working on functions $f : \Omega \to \mathbb{R}$ is in kernel form with respect to a measure $M : \Omega \to (0, \infty)$, with *right kernel* $D : \Omega \times \Omega \to \mathbb{R}$, if

$$Sf(y) = \sum_{x \in \Omega} D(y, x) f(x) M(x). \tag{2.52}$$

We say that an operator S working on functions $f : \Omega \to \mathbb{R}$ is in kernel form with respect to a measure $M : \Omega \to (0, \infty)$, with *left kernel* $D : \Omega \times \Omega \to \mathbb{R}$, if

$$Sf(x) = \sum_{y \in \Omega} D(y, x) f(y) M(y). \tag{2.53}$$

We then have the following connection between symmetries and self-duality.

Theorem 2.11 (Equivalence Between Self-duality and Symmetry in Kernel Form: The Reversible Setting) *Let $\{X(t) : t \geq 0\}$ be a finite state space Markov process with generator L and reversible measure M, i.e.* (2.43) *holds.*

1. *Assume that S is an operator in kernel form with respect to M, with right kernel D in the sense of* (2.52). *Then S is a symmetry of L if and only if D is a self-duality function.*
2. *Assume that S is an operator in kernel form with respect to M, with left kernel D in the sense of* (2.53). *Then S is a symmetry of L if and only if D is a self-duality function.*

Proof We give only the proof of item 1, since the proof of item 2 is completely analogous. The symmetry property means that for all $f : \Omega \to \mathbb{R}$

$$LSf = SLf. \tag{2.54}$$

Writing this out it gives, for all $y \in \Omega$,

$$\begin{aligned} LSf(y) &= \sum_{y' \in \Omega} L(y, y')(Sf)(y') = \sum_{x, y' \in \Omega} L(y, y')D(y', x) f(x)M(x) \\ &= \sum_{x \in \Omega} (LD)(y, x) f(x)M(x). \end{aligned} \tag{2.55}$$

Furthermore, using the fact that $L = L^*$ with the inner product (2.50), we have

$$\begin{aligned} SLf(y) &= \sum_{x \in \Omega} D(y, x)Lf(x)M(x) = \langle D(y, \cdot), Lf \rangle \\ &= \langle LD(y, \cdot), f \rangle = \sum_{x, x' \in \Omega} L(x, x')D(y, x') f(x)M(x) \\ &= \sum_{x \in \Omega} (DL^T)(y, x) f(x)M(x). \end{aligned} \tag{2.56}$$

As a consequence, $LS = SL$ if and only if, for all $y \in \Omega$ and for all $f : \Omega \to \mathbb{R}$,

$$\sum_{x \in \Omega} LD(y, x) f(x)M(x) = \sum_{x \in \Omega} (DL^T)(y, x) f(x)M(x). \tag{2.57}$$

This is in turn equivalent with

$$LD = DL^T, \tag{2.58}$$

because we require (2.57) for all f and we assumed $M(x) > 0$ for all $x \in \Omega$. □

Theorem 2.11 identifies self-duality functions as kernels of symmetries in the L^2 space associated to a reversible measure. Notice that the proof can be repeated when sums are replaced by integrals, under the assumption that $L = L^*$ in the L^2 space associated to a reversible measure. As we will see later on, the equivalence between symmetries in kernel operator form and self-dualities is not limited to the finite state space case.

Example: The Symmetric Exclusion Process on Two Sites

As a well-known and also historical example, let us consider the symmetric exclusion process on two vertices. The state space of the process is {00, 01, 10, 11} where 0 codes for "the vertex has no particle" and 1 for "the vertex has a particle". A particle present on one vertex can hop at rate 1 to the other vertex if there is no particle. The generator matrix is then given by

$$L = \begin{pmatrix} 0 & 0 & 0 & 0 \\ 0 & -1 & 1 & 0 \\ 0 & 1 & -1 & 0 \\ 0 & 0 & 0 & 0 \end{pmatrix}. \tag{2.59}$$

Since this is a symmetric matrix, it follows that the uniform measure is a reversible measure. Therefore, since duality functions are defined up to multiplicative constants, a "cheap" self-duality function is given by the identity matrix

$$D((y_1, y_2), (x_1, x_2)) = \delta_{y_1,x_1}\delta_{y_2,x_2}. \tag{2.60}$$

Indeed this diagonal self-duality function is associated to a reversible measure M in the sense of (2.45). More precisely it is associated to the product Bernoulli measure with density $1/2$, for which all 4 configurations have the same probability $1/4$. The following "symmetry",

$$S((y_1, y_2), (x_1, x_2)) = \mathbb{1}_{\{y_1 \le x_1\}}\mathbb{1}_{\{y_2 \le x_2\}}, \tag{2.61}$$

is a matrix commuting with the generator, as can be verified by direct computation. Therefore, we have self-duality with self-duality function

$$D' = SD = S = \begin{pmatrix} 1 & 1 & 1 & 1 \\ 0 & 1 & 0 & 1 \\ 0 & 0 & 1 & 1 \\ 0 & 0 & 0 & 1 \end{pmatrix}. \tag{2.62}$$

Notice that with the point-wise ordering of configurations this is another example of Siegmund duality.

2.3.2 *Duality and Symmetry*

In the setting where there is no reversible measure, but still there is a stationary measure, the "cheap self-duality" is no longer available. Its analog is the "cheap duality", with diagonal elements given by the inverse of the stationary measure, which provides a duality between the forward and backward process. This cheap duality can then be turned into a non-trivial duality by acting with symmetries.

Definition 2.12 (Stationary Measure) Let $\{X(t) : t \geq 0\}$ be a finite state space Markov process with generator L. A measure $M : \Omega \to (0, \infty)$ is called stationary if, for all $x \in \Omega$, one has

$$\sum_{x' \in \Omega} M(x')L(x', x) = 0, \tag{2.63}$$

or equivalently, for all $x \in \Omega$, one has

$$\sum_{x' \in \Omega} \big(M(x')L(x', x) - M(x)L(x, x')\big) = 0. \tag{2.64}$$

From (2.64) we see that reversibility implies stationarity. When M is a stationary measure then we define the generator $\widetilde{L}$ of the time-reversed process via

$$\widetilde{L}(x, x') = \frac{M(x')L(x', x)}{M(x)}. \tag{2.65}$$

It is immediate to see that $\widetilde{L}$ is the generator of a Markov process on Ω, i.e., $\widetilde{L}(x, x') > 0$ for all $x, x' \in \Omega$ with $x \neq x'$, and $\sum_{x' \in \Omega} \widetilde{L}(x, x') = 0$ for all $x \in \Omega$. Furthermore, M is a stationary measure of the time-reversed process as well.

We can then formulate the analogue of Theorem 2.9.

Theorem 2.13 (Duality and Symmetry) *Let $\{X_1(t) : t \geq 0\}$ be a Markov process on a finite state space Ω_1, with generator L_1. Let $\{X_2(t) : t \geq 0\}$ be another Markov process on a finite state space Ω_2, with generator L_2.*

1. *Existence of a "cheap duality" made from the stationary measure. Fix $i \in \{1, 2\}$. If M_i is a stationary measure of the process $\{X_i(t) : t \geq 0\}$ then*

$$D(x, x') = \frac{1}{M_i(x)}\delta_{x,x'} \tag{2.66}$$

is a duality function between the time-reversed process and the process, i.e.:

$$\widetilde{L}_i D = D L_i^T. \tag{2.67}$$

2. *Symmetries produce new duality functions from a given one. If D_{12} is a duality function between L_1 and L_2, i.e.,*

$$L_1 D_{12} = D_{12} L_2^T, \tag{2.68}$$

then we have the following:

(i) If S_1 is a symmetry of L_1, then $S_1 D_{12}$ is a duality function between L_1 and L_2, i.e.

$$L_1(S_1 D_{12}) = (S_1 D_{12})L_2^T. \tag{2.69}$$

(ii) If S_2 is a symmetry of L_2 then $D_{12} S_2^T$ is duality function between L_1 and L_2 as well, i.e.

$$L_1(D_{12} S_2^T) = (D_{12} S_2^T)L_2^T. \tag{2.70}$$

Proof The proof is completely analogous to the proof of Theorem 2.9. □

Remark 2.14 (Example of Duality Between a Process and Its Time-Reversed) For an example of duality in the spirit of Theorem 2.13 we refer to Sect. 4.7, where the case of asymmetric random walkers is considered.

Theorem 2.15 (Equivalence Between Duality and Symmetry in Kernel Form: The Stationary Setting) *Let $\{X(t) : t \geq 0\}$ be a Markov process with finite state space Ω and stationary measure $M : \Omega \to (0, \infty)$, i.e.* (2.63) *holds. Let L denote the process generator and let $\widetilde{L}$ denote the generator of the reversed process.*

1. *Assume that S is an operator in kernel form with respect to M, with right kernel D in the sense of* (2.52)*. Then S is a symmetry of L if and only if D is a duality function between the process and its time reversed, i.e.*

$$LD = D\widetilde{L}^T. \tag{2.71}$$

2. *Assume that S is an operator in kernel form with respect to M, with left kernel D in the sense of* (2.53)*. Then S is a symmetry of $\widetilde{L}$ if and only if D is a duality function between the time-reversed process and the process, i.e.*

$$\widetilde{L}D = DL^T. \tag{2.72}$$

Proof The proof follows from the fact that $L^* = \widetilde{L}$ where $*$ denotes the adjoint in the space $L^2(\Omega, M)$, i.e., w.r.t. to the inner product

$$\langle f, g \rangle = \sum_{x \in \Omega} f(x)g(x)M(x). \tag{2.73}$$

After realizing this, the proof is analogous to the one of Theorem 2.11. □

2.3.3 Duality and Intertwining

In this section we introduce intertwining, which is a natural extension of the notion of symmetry. Let be given three continuous-time Markov processes with generators L_1, L_2, L_3 on finite state spaces $\Omega_1, \Omega_2, \Omega_3$.

Definition 2.16 (Intertwiner) An intertwiner between L_1 and L_2 is a matrix Λ_{12} with elements $\Lambda_{12}(x_1, x_2)$ indexed by $x_1 \in \Omega_1$ (row index) and $x_2 \in \Omega_2$ (column index), such that

$$L_1\Lambda_{12} = \Lambda_{12}L_2. \tag{2.74}$$

Remark 2.17 (Composing Intertwiners) Intertwining composes naturally in the following sense: if Λ_{12} is an intertwiner between L_1 and L_2 and Λ_{23} is an intertwiner between L_2 and L_3 then $\Lambda_{13} := \Lambda_{12}\Lambda_{23}$ is an intertwiner between L_1 and L_3.

Theorem 2.18 (Duality and Intertwining) *Let L_1, L_2, L_3 be generators of three Markov processes on finite state spaces $\Omega_1, \Omega_2, \Omega_3$. Assume that $D_{12} : \Omega_1 \times \Omega_2 \to \mathbb{R}$ is a duality function for duality between L_1 and L_2, i.e.,*

$$L_1 D_{12} = D_{12}L_2^T. \tag{2.75}$$

(i) If Λ_{31} is an intertwiner between L_3 and L_1, i.e.,

$$L_3\Lambda_{31} = \Lambda_{31}L_1, \tag{2.76}$$

then $D_{32} = \Lambda_{31}D_{12}$ is a duality function between L_3 and L_2, i.e.

$$L_3 D_{32} = D_{32}L_2^T. \tag{2.77}$$

(ii) If Λ_{32} is an intertwiner between L_3 and L_2, i.e.,

$$L_3\Lambda_{32} = \Lambda_{32}L_2, \tag{2.78}$$

then $D_{13} = D_{12}\Lambda_{32}^T$ is a duality function for duality between L_1 and L_3, i.e.

$$L_1 D_{13} = D_{13}L_3^T. \tag{2.79}$$

Proof We only prove item (i), as item (ii) is completely analogous. We have

$$L_3(\Lambda_{31}D_{12}) = \Lambda_{31}L_1D_{12} = (\Lambda_{31}D_{12})L_2^T. \tag{2.80}$$

Here in the first step we used that Λ_{31} is an intertwiner between L_3 and L_1, and in the second step that D_{12} is a duality function between L_1 and L_2. □

Looking back at the case of (self-)duality in the previous subsections, we see that symmetry is an instance of intertwining of a generator with itself and Theorem 2.18 includes as special cases the theorems of previous subsection. Indeed:

▷ if in Theorem 2.18 we choose $L_1 = L_2 = L_3 = L$, $D_{12} = D$ and $\Lambda_{31} = \Lambda_{32} = S$, we recover item 2 of Theorem 2.9;
▷ if in item 2(i) of Theorem 2.18 we choose $L_1 = L_3$ and $\Lambda_{31} = S_1$ we recover item 2(i) of Theorem 2.13;
▷ if in item 2(ii) of Theorem 2.18 we choose $L_2 = L_3$ and $\Lambda_{32} = S_2$ we recover item 2(ii) of Theorem 2.13.

The next theorem is the analog of Theorem 2.11 and of Theorem 2.15: an intertwiner in kernel operator form w.r.t. the reversible/stationary is equivalent to a duality.

Theorem 2.19 (Equivalence Between Duality and Intertwining in Kernel Form: The Reversible and the Stationary Setting) *Let $\{X_1(t) : t \geq 0\}$ be a Markov process on a finite state space Ω_1, with generator L_1. Let $\{X_2(t) : t \geq 0\}$ be another Markov process on a finite state space Ω_2, with generator L_2.*

1. *Assume that Λ_{12} working on functions $f : \Omega_2 \to \mathbb{R}$ is an operator in kernel form w.r.t. the measure $M_2 : \Omega_2 \to (0, \infty)$, with right kernel $D_{12} : \Omega_1 \times \Omega_2 \to \mathbb{R}$, i.e.*

$$\Lambda_{12} f(x_1) = \sum_{x_2 \in \Omega_2} D_{12}(x_1, x_2) f(x_2) M_2(x_2). \tag{2.81}$$

Then we have the following:

(1.a) If M_2 is a reversible measure for X_2, then Λ_{12} is an intertwiner between L_1 and L_2 (i.e. $L_1\Lambda_{12} = \Lambda_{12}L_2$) if and only if D_{12} is a duality function between L_1 and L_2 (i.e. $L_1 D_{12} = D_{12} L_2^T$).

(1.b) If M_2 is a stationary measure for X_2, then Λ_{12} is an intertwiner between L_1 and L_2 (i.e. $L_1\Lambda_{12} = \Lambda_{12}L_2$) if and only if D_{12} is a duality function between L_1 and $\widetilde{L}_2$, the generator of the time reversed process of X_2 (i.e. $L_1 D_{12} = D_{12}\widetilde{L}_2^T$).

2. *Assume that Λ_{21} working on functions $f : \Omega_1 \to \mathbb{R}$ is an operator in kernel form w.r.t. the measure $M_1 : \Omega_1 \to (0, \infty)$, with left kernel $D_{12} : \Omega_1 \times \Omega_2 \to \mathbb{R}$, i.e.*

$$\Lambda_{21} f(x_2) = \sum_{x_1 \in \Omega_1} D_{12}(x_1, x_2) f(x_1) M_1(x_1). \tag{2.82}$$

Then we have the following:

(2.a) If M_1 is a reversible measure for X_1, then Λ_{21} is an intertwiner between L_2 and L_1 (i.e. $L_2\Lambda_{21} = \Lambda_{21}L_1$) if and only if D_{12} is a duality function between L_1 and L_2 (i.e. $L_1 D_{12} = D_{12}L_2^T$).

(2.b) If M_1 is a stationary measure for X_1, then Λ_{21} is an intertwiner between L_2 and L_1 (i.e. $L_2\Lambda_{21} = \Lambda_{21}L_1$) if and only if D_{12} is a duality function between $\widetilde{L}_1$ (the generator of the time reversed process of X_1) and L_2 (i.e. $\widetilde{L}_1 D_{12} = D_{12}L_2^T$).

Proof For the proof of item (1.a), denote $\langle \cdot, \cdot \rangle_{M_2}$ the $L^2(\Omega_2, M_2)$ inner product. Start from

$$L_1\Lambda_{12}f(x_1) = \sum_{x_1' \in \Omega_1, x_2 \in \Omega_2} L_1(x_1, x_1')D_{12}(x_1', x_2)f(x_2)M_2(x_2). \tag{2.83}$$

Next, using reversibility of M_2, we write

$$\begin{aligned} \Lambda_{12}L_2 f(x_1) &= \langle D_{12}(x_1, \cdot), L_2 f\rangle_{M_2} = \langle L_2 D_{12}(x_1, \cdot), f\rangle_{M_2} \\ &= \sum_{x_2', x_2 \in \Omega_2} L_2(x_2, x_2')D_{12}(x_1, x_2')f(x_2)M_2(x_2). \end{aligned} \tag{2.84}$$

Because this holds for all f we conclude

$$\sum_{x_1' \in \Omega_1} L_1(x_1, x_1')D_{12}(x_1', x_2) = \sum_{x_2' \in \Omega_2} L_2(x_2, x_2')D_{12}(x_1, x_2'), \tag{2.85}$$

for all $x_1 \in \Omega_1$, $x_2 \in \Omega_2$, which is equivalent to duality between L_1 and L_2. To see item (1.b), use

$$\langle L_2 D_{12}(x_1, \cdot), f\rangle_{M_2} = \sum_{x_2', x_2 \in \Omega_2} \widetilde{L}_2(x_2, x_2')D_{12}(x_1, x_2')f(x_2)M_2(x_2), \tag{2.86}$$

since the generator $\widetilde{L}_2$ of the time reversal of X_2 is the adjoint of L_2 in $L^2(\Omega_2, M_2)$. The proof of the items (2.a) and (2.b) is analogous. □

2.4 Duality and Intertwining of Bounded Operators

In this section we extend the notion of duality beyond Markov generators by introducing the notion of duality between two algebras. As we will see later, it turns out that in most cases behind a duality of two Markov generators lies a duality relation between two algebras (or more precisely a duality between two representations of an algebra). Realizing this is very useful for two reasons. First, for two Markov generators related by duality it leads to a larger class of duality functions. Second, by considering other elements of the algebra it leads to many more dualities between Markov generators associated to a given algebra. Indeed if one has a duality relation between two algebras, then automatically all elements of one algebra are in a duality relation with corresponding elements of the other algebra.

The content of this section is organized as follows. We explain the duality of algebras by considering first the case of algebras of matrices. Then we generalize to algebras of bounded operators on a L^2-space and finally algebras of bounded operators on a general Hilbert space. Working with bounded operators is convenient as we can explain the concept of intertwined representations without technicalities related to domains of unbounded operators.

2.4.1 Duality and Intertwining of Matrices

The first step is to define duality for general square matrices, where we continue to think of matrices as indexed by finite sets, and identify matrices with linear maps as usual. I.e., a matrix with elements $A(x, x')$ with $x, x' \in \Omega$ is at the same time thought of as a linear map from functions $f : \Omega \to \mathbb{R}$ to functions $Af : \Omega \to \mathbb{R}$ via $Af(x) = \sum_{x' \in \Omega} A(x, x')f(x')$. In other words, we think of functions as column vectors.

In what follows we will consider matrices "with a hat" (such as $\hat{A}_1, \hat{A}_2, \ldots$) indexed by elements $y, y' \in \widehat{\Omega}$ and matrices "without hat" $A_1, A_2, \ldots$ indexed by elements $x, x' \in \Omega$. Here $\widehat{\Omega}$ and Ω are two fixed finite sets.

Definition 2.20 (Duality for Matrices) We say that a matrix $\widehat{A}$ is dual to a matrix A with duality function D (which is a matrix with elements $D(y, x)$ with $y \in \widehat{\Omega}$, $x \in \Omega$) if

$$\widehat{A}D = DA^T. \tag{2.87}$$

We denote the duality between two matrices by

$$\widehat{A} \xrightarrow{D} A \tag{2.88}$$

Remark 2.21 Equation (2.87) is the generalization of (2.26) to matrices which are not necessarily Markov generators. The notation $\widehat{A} \xrightarrow{D} A$ suggests to think of duality as a relation between two matrices $\widehat{A}, A$ parametrized by the duality function D. The element-wise version of (2.87) says that for all $y \in \widehat{\Omega}$, $x \in \Omega$

$$\sum_{y' \in \widehat{\Omega}} \widehat{A}(y, y')D(y', x) = \sum_{x' \in \Omega} A(x, x')D(y, x'). \tag{2.89}$$

This means that the action of $\widehat{A}$ on the y-variable is the same as the action of A on the x-variable

$$(\widehat{A}D(\cdot, x))(y) = (AD(y, \cdot))(x). \tag{2.90}$$

Definition 2.22 (Matrix Algebra) A set of matrices is called a matrix algebra if it is closed under linear combinations and multiplication of matrices. A matrix algebra $\mathcal{A}$ is generated by the matrices $A_1, \ldots, A_n$ if every element in the algebra is a linear combination of products of the form $A_{i_1} A_{i_2} \ldots A_{i_k}$ with $k \in \mathbb{N}$ and $i_1, \ldots, i_k \in \{1, \ldots, n\}$. We then call $A_1, \ldots, A_n$ the generators of the algebra $\mathcal{A}$.

The next theorem shows how to combine dualities with addition and multiplication of matrices. In particular it leads to a natural notion of duality between two *matrix algebras.*

Theorem 2.23 (Combining Dualities: The Matrix Case) *We have the following properties of the "* $\xrightarrow{D}$*-relation".*

1. *Linear combinations: if* $\widehat{A}_1 \xrightarrow{D} A_1$, $\widehat{A}_2 \xrightarrow{D} A_2$ *then for* $a, b \in \mathbb{R}$*:*

$$a\widehat{A}_1 + b\widehat{A}_2 \xrightarrow{D} aA_1 + bA_2. \tag{2.91}$$

2. *Products: if* $\widehat{A}_1 \xrightarrow{D} A_1$, $\widehat{A}_2 \xrightarrow{D} A_2$ *then*

$$\widehat{A}_1\widehat{A}_2 \xrightarrow{D} A_2 A_1. \tag{2.92}$$

3. *Duality of algebras from duality of algebra generators: If* $\widehat{A}_i \xrightarrow{D} A_i$ *for all* $i \in \{1, \ldots, n\}$ *then for every* $\widehat{A}$ *in the algebra* $\widehat{\mathcal{A}}$ *generated by* $\widehat{A}_i, i \in \{1, \ldots, n\}$ *there exists a corresponding element* A *in the algebra* $\mathcal{A}$ *generated by* $A_i, i \in \{1, \ldots, n\}$ *such that* $\widehat{A} \xrightarrow{D} A$.

Proof The first two items are straightforward from the definition of duality for matrices and the fact that $(A_1 A_2)^T = A_2^T A_1^T$. To see the last item consider an element in the algebra generated by $\widehat{A}_i, i \in \{1, \ldots, n\}$ of the form

$$\widehat{A}_{i_1} \widehat{A}_{i_2} \ldots \widehat{A}_{i_k} \tag{2.93}$$

for $i_1, \ldots, i_k \in \{1, \ldots, n\}$. Then, by iteratively applying the property in item 2, we see that this element is dual to the element

$$A_{i_k} A_{i_{k-1}} \ldots A_{i_1}. \tag{2.94}$$

Then we can apply the property of item 1 to extend to linear combinations of elements of the type (2.93). □

The last item of the proposition above indicates that a duality function translates elements of an matrix algebra to elements of a "conjugate algebra" obtained by removing hats from the $\widehat{A}_i$ and multiplying the elements in the reversed order. The source of this "reversed order of multiplication" is the fact that in the relation (2.87) there is a transposition involved.

Next, we also want to understand the relation between duality and intertwining for general matrices, i.e. beyond matrices that are generator of Markov processes.

Definition 2.24 (Intertwining of Matrices) A matrix $\widehat{A}$ is intertwined to a matrix A with intertwining matrix Λ if

$$\widehat{A}\Lambda = \Lambda A. \tag{2.95}$$

The following theorem is then a generalization of Theorem 2.19 which identifies the kernels of intertwiners as duality functions.

Theorem 2.25 (Equivalence Between Intertwining in Kernel Form and Duality: The Matrix Case) *Let* (2.95) *hold and assume that the intertwiner is of kernel operator form, i.e., that*

$$\Lambda f(y) = \sum_x D(y,x) f(x)\mu(x), \tag{2.96}$$

for some positive measure $\mu : \Omega \to (0\infty)$. *Then we have*

$$\widehat{A} \xrightarrow{D} A^*, \tag{2.97}$$

with A^* *the adjoint of* A *in* $L^2(\Omega, \mu)$.

Conversely, if the duality (2.97) *holds, then defining* Λ *via* (2.96), *the intertwining* (2.95) *holds.*

Proof Start from

$$\begin{aligned} \widehat{A}\Lambda f(y) &= \sum_{y'} \widehat{A}(y,y')\Lambda f(y') \\ &= \sum_{y',x} \widehat{A}(y,y')D(y',x) f(x)\mu(x). \end{aligned} \tag{2.98}$$

On the other hand

$$\begin{aligned} \Lambda A f(y) &= \sum_x D(y,x) A f(x)\mu(x) \\ &= \sum_{x',x} A^*(x,x')D(y,x') f(x)\mu(x), \end{aligned} \tag{2.99}$$

where in the last step we moved A to A^* working on the x-variable of D. As a consequence of (2.95) we obtain that for all $y \in \widehat{\Omega}$

$$\sum_{y',x} \widehat{A}(y, y')D(y', x)f(x)\mu(x) = \sum_{x',x} A^*(x, x')D(y, x')f(x)\mu(x). \tag{2.100}$$

Because this holds for all $f : \Omega \to \mathbb{R}$ and because μ is positive, then we obtain that for all $y \in \widehat{\Omega}$ and for all $x \in \Omega$ one has

$$\sum_{y'} \widehat{A}(y, y')D(y', x) = \sum_{x'} A^*(x, x')D(y, x'), \tag{2.101}$$

which is

$$\widehat{A} \xrightarrow{D} A^*. \tag{2.102}$$

The converse implication, i.e. from duality to intertwiner, follows from (2.98) and (2.99) combined with the fact that $(A^*)^* = A$. □

In view of the extension beyond finite matrices (e.g. differential operators), we face the problem that the notion of "transposition" is not available anymore. Instead, we will replace it by "moving in a tensor product from the left to the right."

In order to make this more clear, notice that in the matrix setting, the relation $\widehat{A} \xrightarrow{D} A$ can be rewritten element-wise as

$$\sum_{y',x'} \widehat{A}(y, y')\delta_{x,x'}D(y', x') = \sum_{y',x'} \delta_{y,y'}A(x, x')D(y', x'), \tag{2.103}$$

for all $y \in \widehat{\Omega}$ and $x \in \Omega$ where $\delta_{x,x'}$ is a Kronecker delta. This can be further written as

$$(\widehat{A} \otimes I)D = (I \otimes A)D, \tag{2.104}$$

which corresponds to moving $\hat{A}$ from the left to the right in the tensor product, and replacing it by A, which continues to make sense in a setting where there is no obvious notion of transposition. Equation (2.104) can be written equivalently as

$$\widehat{A}D(\cdot, x)(y) = AD(y, \cdot)(x), \tag{2.105}$$

which means that the action of $\widehat{A}$ on the y variable is the same as the action of A on the x variable.

2.4.2 Duality and Intertwining in the L^2-space Setting

Let us consider here the case of bounded operators on L^2-spaces, where there is still a natural notion of the adjoint of an operator, as well as a natural notion of operator in kernel form. This will in particular give us the analogue of Theorem 2.25 in this context.

We start from two measure spaces $(\widehat{\Omega}, \widehat{\mathcal{F}}, \nu)$ and $(\Omega, \mathcal{F}, \mu)$ with σ-finite measures ν, resp. μ. In the sequel, for simplicity of notation, we will omit the σ-algebras and write $L^2(\widehat{\Omega}, \nu) = L^2(\widehat{\Omega}, \widehat{\mathcal{F}}, \nu)$. Then $\widehat{\mathcal{H}} = L^2(\widehat{\Omega}, \nu)$, $\mathcal{H} = L^2(\Omega, \mu)$ are two Hilbert spaces, and their tensor product Hilbert space is $\widehat{\mathcal{H}} \otimes \mathcal{H} = L^2(\hat{\Omega} \times \Omega, \nu \otimes \mu)$.

We say that a bounded operator $\Lambda : \mathcal{H} \to \widehat{\mathcal{H}}$ is a kernel operator if there exists a function $D \in L^2(\widehat{\Omega} \times \Omega, \nu \otimes \mu)$ such that for all $f \in \mathcal{H}$

$$\Lambda f(y) = \int_{\Omega} D(y, x) f(x) d\mu(x), \tag{2.106}$$

where the equality as to be understood in $\widehat{\mathcal{H}}$, i.e., ν-almost surely.

Definition 2.26 (Intertwining of Bounded Operators in the L^2-space Case) Let $\widehat{A}$, resp. A, denote bounded operators on $\widehat{\mathcal{H}} = L^2(\widehat{\Omega}, \nu)$, resp. $\mathcal{H} = L^2(\Omega, \mu)$, and Λ a bounded operator from $\mathcal{H}$ to $\widehat{\mathcal{H}}$. We say that $\widehat{A}$ and A are intertwined with intertwiner Λ if

$$\widehat{A}\Lambda = \Lambda A. \tag{2.107}$$

Definition 2.27 (Duality of Bounded Operators in the L^2-space Case) Let $D \in \widehat{\mathcal{H}} \otimes \mathcal{H} = L^2(\hat{\Omega} \times \Omega, \nu \otimes \mu)$. We say that $\widehat{A}$ and A are in duality with duality function D if

$$(\widehat{A} \otimes I)D = (I \otimes A)\, D. \tag{2.108}$$

Remark 2.28 Notice that (2.108) is equivalent with

$$\widehat{A}D(\cdot, x)(y) = AD(y, \cdot)(x), \qquad \nu \otimes \mu - \text{almost surely},$$

which corresponds with the earlier definitions of semigroup duality (2.19) or generator duality (2.20), but now applied to general (bounded) operators A on $\mathcal{H}$ and general bounded operators $\widehat{A}$ on $\widehat{\mathcal{H}}$ instead of only to Markov semigroups or Markov generators.

As before we denote this by $\widehat{A} \xrightarrow{D} A$, and we think of it as a relation between operators, indexed by the duality function. Notice that in this setting, we have automatically that $\widehat{A} \otimes I$ and $I \otimes A$ are bounded operators on $\widehat{\mathcal{H}} \otimes \mathcal{H}$, and hence have full domain. Therefore, all the properties of the relation $\xrightarrow{D}$ derived in

Theorem 2.23 carry over to this setting of bounded operators on Hilbert spaces. We then have the following result connecting intertwining with duality.

Theorem 2.29 (Equivalence Between Intertwining in Kernel Form and Duality: The L^2-space Case) *In the L^2-space setting defined above, we have the equivalence between:*

1. *Λ is an intertwiner between $\widehat{A}$ and A, and is an operator with kernel D, i.e.,* (2.106) *and* (2.107) *hold.*
2. *$\widehat{A}$ and A^* are in duality with duality function D, where $*$ denotes adjoint in $\mathcal{H}$, i.e.,*

$$(\widehat{A} \otimes I)D = \left(I \otimes A^*\right) D. \tag{2.109}$$

Proof To prove item 1 $\Longrightarrow$ item 2, we write out $\widehat{A}\Lambda = \Lambda A$. Let $f \in \mathcal{H}$, using linearity and boundedness of $\widehat{A}$:

$$\widehat{A}\Lambda f(y) = \left(\widehat{A} \int_\Omega D(\cdot, x) f(x) d\mu(x)\right)(y) = \int_\Omega (\widehat{A} \otimes I) D(y, x) f(x) d\mu(x). \tag{2.110}$$

Next

$$\Lambda A f(y) = \int_\Omega D(y, x) A f(x) d\mu(x) = \int_\Omega (I \otimes A^*) D(y, x) f(x) d\mu(x). \tag{2.111}$$

Combination of (2.110) and (2.111) gives (2.109). For the implication item 2 $\Longrightarrow$ item 1 one starts from the assumed duality relation, i.e., for all $f \in \mathcal{H}$

$$\int_\Omega (\widehat{A} \otimes I) D(y, x) f(x) d\mu(x) = \int_\Omega (I \otimes A^*) D(y, x) f(x) d\mu(x), \tag{2.112}$$

and moves A^* to f in order to obtain $\widehat{A}\Lambda = \Lambda A$. □

Remark 2.30 The duality relation $(\widehat{A} \otimes I)D = (I \otimes A)D$ in this Hilbert-space setting is an equality of two elements of $L^2(\widehat{\Omega} \otimes \Omega, \nu \otimes \mu)$, and hence holds $(\nu \otimes \mu)$-almost surely. In many examples, the duality holds actually pointwise, i.e., for all $y \in \widehat{\Omega}$ and $x \in \Omega$ because the function D is continuous, and the measures μ, ν have full support.

A particular instance of Theorem 2.29 is when $\mathcal{H} = \widehat{\mathcal{H}}$, where we obtain that symmetries in kernel operator form lead to self-dualities. This is a more abstract version of Theorem 2.11.

Theorem 2.31 (Equivalence Between Symmetries in Kernel Form and Self-duality: The L^2-space Case) *Let A be a bounded self-adjoint operator on the Hilbert space $\mathcal{H} = L^2(\Omega, \mu)$, and assume that there exists a symmetry of A, i.e., a bounded operator $S : \mathcal{H} \to \mathcal{H}$, such that $[A, S] = 0$. Assume moreover that S is in* *kernel operator form, i.e.*

$$Sf(y) = \int_{\Omega} D(y, x) f(x) d\mu(x). \tag{2.113}$$

Then D is a self-duality function for A, i.e., $(A \otimes I)D = (I \otimes A)D$. Conversely, if D is a self-duality function, then S defined via (2.113) *is a symmetry of A.*

2.4.3 Duality and Intertwining: General Hilbert Space Case

We have seen how the equivalence between intertwining and duality extends to the setting of bounded operators on L^2-spaces. The first generalization which we present here is to abstract Hilbert space setting, i.e., not using the specific structure of the L^2-space. The main point is that we consider then an abstract form of duality where the duality "function" is replaced by an abstract element of the tensor product of two Hilbert spaces.

To prepare this, let us consider two Hilbert spaces $\widehat{\mathcal{H}}$ and $\mathcal{H}$ and bounded operators

$$\begin{aligned} \widehat{A} &: \widehat{\mathcal{H}} \to \widehat{\mathcal{H}} \\ A &: \mathcal{H} \to \mathcal{H} \\ \Lambda &: \mathcal{H} \to \widehat{\mathcal{H}} \end{aligned} \tag{2.114}$$

We denote by $\widehat{\mathcal{H}} \otimes \mathcal{H}$ the tensor product Hilbert space, uniquely defined by the scalar product

$$\langle \widehat{f} \otimes f, \widehat{g} \otimes g \rangle_{\widehat{\mathcal{H}} \otimes \mathcal{H}} = \langle \widehat{f}, \widehat{g} \rangle_{\widehat{\mathcal{H}}} \cdot \langle f, g \rangle_{\mathcal{H}} \tag{2.115}$$

In this setting we define duality and intertwining as follows.

Definition 2.32 (Duality of Bounded Operators in the Hilbert Space Case) Let $D \in \widehat{\mathcal{H}} \otimes \mathcal{H}$. We say that D is a duality element for duality between $\widehat{A}$ and A if

$$(\widehat{A} \otimes I)D = (I \otimes A)D. \tag{2.116}$$

Definition 2.33 (Intertwining of Bounded Operators in the Hilbert Space Case) We say that Λ is an intertwiner between $\widehat{A}$ and A if

$$\widehat{A}\Lambda = \Lambda A. \tag{2.117}$$

In order to establish the link between intertwining and duality, we need to construct an intertwiner from a duality element. If $D = \widehat{d} \otimes d$ is in tensor form then, inspired by the L^2-case, it is natural to define $\Lambda_D f$ for $f \in \mathcal{H}$ via

$$\Lambda_D f = (\langle d, f\rangle_{\mathcal{H}})\widehat{d}. \tag{2.118}$$

To extend this definition to general elements of $\widehat{\mathcal{H}} \otimes \mathcal{H}$, it is convenient to characterize this element $\Lambda_D f$ in an alternative way. Namely, as the unique element of $\widehat{\mathcal{H}}$ such that for all $\widehat{g} \in \widehat{\mathcal{H}}$ we have

$$\langle \widehat{g}, \Lambda_D f\rangle_{\widehat{\mathcal{H}}} = \langle \widehat{d}, \widehat{g}\rangle_{\widehat{\mathcal{H}}}\langle d, f\rangle_{\mathcal{H}} = \langle \widehat{g} \otimes f, D\rangle_{\widehat{\mathcal{H}}\otimes\mathcal{H}}. \tag{2.119}$$

The fact that this definition makes sense for all $D \in \widehat{\mathcal{H}} \otimes \mathcal{H}$ is proved in the following lemma.

Lemma 2.34 (Generalized Kernel Form) *For all $D \in \widehat{\mathcal{H}} \otimes \mathcal{H}$ and for all $f \in \mathcal{H}$ there exists a unique element $\Lambda_D f \in \widehat{\mathcal{H}}$ characterized by the fact that for all $\hat{g} \in \widehat{\mathcal{H}}$ we have*

$$\langle \widehat{g}, \Lambda_D f\rangle_{\widehat{\mathcal{H}}} = \langle \widehat{g} \otimes f, D\rangle_{\widehat{\mathcal{H}}\otimes\mathcal{H}}. \tag{2.120}$$

Moreover the assignment $f \to \Lambda_D f$ is a bounded operator from $\mathcal{H}$ to $\widehat{\mathcal{H}}$.

Proof Given $D \in \widehat{\mathcal{H}} \otimes \mathcal{H}$ and $f \in \mathcal{H}$ we define the linear map $\psi : \widehat{\mathcal{H}} \to \mathbb{R}$ via

$$\psi(\widehat{g}) = \langle \widehat{g} \otimes f, D\rangle_{\widehat{\mathcal{H}}\otimes\mathcal{H}}. \tag{2.121}$$

Using the Cauchy Schwarz inequality, together with the fact that $\|\widehat{g} \otimes f\|_{\widehat{\mathcal{H}}\otimes\mathcal{H}} = \|f\|_{\mathcal{H}}\|\widehat{g}\|_{\widehat{\mathcal{H}}}$, we obtain the estimate

$$|\psi(\widehat{g})| \leq \|D\|_{\widehat{\mathcal{H}}\otimes\mathcal{H}}\|f\|_{\mathcal{H}}\|\widehat{g}\|_{\widehat{\mathcal{H}}}, \tag{2.122}$$

and as a consequence ψ is a continuous linear map from $\widehat{\mathcal{H}}$ to $\mathbb{R}$. Therefore, using the Riesz representation theorem, there exists a unique element $\widehat{h} \in \widehat{\mathcal{H}}$ such that

$$\psi(\widehat{g}) = \langle \widehat{h}, \widehat{g}\rangle_{\widehat{\mathcal{H}}}. \tag{2.123}$$

Then we define $\Lambda_D f = \widehat{h}$.

From (2.122) it follows

$$|\langle \Lambda_D f, \widehat{g}\rangle_{\widehat{\mathcal{H}}}| \leq \|D\|_{\widehat{\mathcal{H}}\otimes\mathcal{H}}\|f\|_{\mathcal{H}}\|\widehat{g}\|_{\widehat{\mathcal{H}}}. \tag{2.124}$$

Therefore the dependence between D and Λ_D is linear and continuous, more precisely

$$\|\Lambda_D\|_{\mathcal{B}(\mathcal{H},\widehat{\mathcal{H}})} \leq \|D\|_{\widehat{\mathcal{H}}\otimes\mathcal{H}}, \tag{2.125}$$

where $\|\cdot\|_{\mathcal{B}(\mathcal{H},\widehat{\mathcal{H}})}$ denotes the operator norm. □

We then have the following equivalence between duality and intertwining.

Theorem 2.35 (Equivalence Between Duality and Intertwining in Generalized Kernel Form: The Hilbert Space Case) *Let the setting be as in* (2.114) *The following two properties are equivalent:*

1. *Duality between $\widehat{A}$ and A with duality element $D \in \widehat{\mathcal{H}} \otimes \mathcal{H}$, in the sense of* (2.116).
2. *Intertwining between $\widehat{A}$ and A^* with intertwiner Λ_D defined in Lemma 2.34.*

Proof For simplicity of notation, we will omit subindices in inner products, as it will be clear in every formula in which space the inner product is taken. Duality is characterized by

$$\langle(\widehat{A} \otimes I)D, \widehat{g} \otimes f\rangle = \langle(I \otimes A)D, \widehat{g} \otimes f\rangle, \tag{2.126}$$

for all $\widehat{g} \in \widehat{\mathcal{H}}$, $f \in \mathcal{H}$. This can be rewritten equivalently as

$$\langle D, (\widehat{A})^*\widehat{g} \otimes f\rangle = \langle D, \widehat{g} \otimes A^* f\rangle. \tag{2.127}$$

Intertwining of $\widehat{A}$ and A^* with intertwiner Λ_D defined in Lemma 2.34 is equivalent with the fact that for all $\widehat{g} \in \widehat{\mathcal{H}}$, $f \in \mathcal{H}$, we have

$$\langle\widehat{g}, \widehat{A}\Lambda_D f\rangle = \langle\widehat{g}, \Lambda_D A^* f\rangle, \tag{2.128}$$

which is equivalent with

$$\langle(\widehat{A})^*\widehat{g}, \Lambda_D f\rangle = \langle\widehat{g}, \Lambda_D A^* f\rangle, \tag{2.129}$$

which by the defining relation (2.120) reads

$$\langle(\widehat{A})^*\widehat{g} \otimes f, D\rangle = \langle\widehat{g} \otimes (A^* f), D\rangle. \tag{2.130}$$

This is the same as (2.127) and therefore the equivalence between duality and intertwiner is proven. □

Remark 2.36 We have extended the definition of duality with a duality function D (appearing in the context of L^2 spaces) to the definition of duality with a duality element D (appearing in the context of abstract Hilbert space). The advantage of this extension lies in the fact that we can then consider duality in the context of general representations of algebras, where the vector spaces on which the operators (representing the algebra elements) act are not necessarily function spaces. Of course, for finite dimensional representations this distinction between functions and elements is not relevant, because a finite column can always be viewed as a function.

Remark 2.37 If we come back to the L^2 case, i.e., $\widehat{\mathcal{H}} = L^2(\widehat{\Omega}, \nu)$, $\mathcal{H} = L^2(\Omega, \mu)$, then $\widehat{\mathcal{H}} \otimes \mathcal{H} = L^2(\widehat{\Omega} \times \Omega, \nu \otimes \mu)$, hence for an element $D \in \widehat{\mathcal{H}} \otimes \mathcal{H}$ the characterizing equation (2.120) for Λ_D can be written as

$$\int \widehat{g}(y)\Lambda_D f(y)d\nu(y) = \langle \widehat{g} \otimes f, D\rangle = \int \widehat{g}(y) \left(\int D(y,x) f(x) d\mu(x) \right) d\nu(y), \tag{2.131}$$

which implies that

$$\Lambda_D f(y) = \int D(y,x) f(x) d\mu(x), \tag{2.132}$$

which is the way in which we defined Λ_D before.

2.4.4 Intertwined Representations in Hilbert Spaces

Finally, we show in the Hilbert space context how dualities are related to intertwined representations of an algebra. This prepares for the later Sect. 2.6, where we will treat duality and intertwinings in the context of algebras. Let $\mathcal{A}$ denote an algebra, and $\widehat{\mathcal{H}}$, $\mathcal{H}$, and $\widehat{\mathcal{H}} \otimes \mathcal{H}$ denote Hilbert spaces as before. Also denote $\mathcal{B}(\widehat{\mathcal{H}}, \widehat{\mathcal{H}})$, resp. $\mathcal{B}(\mathcal{H}, \mathcal{H})$, the algebra of bounded linear operators on $\widehat{\mathcal{H}}$, resp. on $\mathcal{H}$. Let $\widehat{\rho}$ and ρ be two representations of $\mathcal{A}$, i.e. two algebra homomorphisms $\widehat{\rho} : \mathcal{A} \to \mathcal{B}(\widehat{\mathcal{H}}, \widehat{\mathcal{H}})$, resp. $\rho : \mathcal{A} \to \mathcal{B}(\mathcal{H}, \mathcal{H})$.

We say that two representations $\widehat{\rho}$, ρ are equivalent if they are related by an intertwiner, i.e. there exists a bounded linear operator $\Lambda : \mathcal{H} \to \widehat{\mathcal{H}}$ such that for all $a \in \mathcal{A}$

$$\widehat{\rho}(a)\Lambda = \Lambda\rho(a). \tag{2.133}$$

If this operator Λ is unitary, i.e., conserves inner products, then we call Λ a unitary intertwiner and we call the representations unitary equivalent. Unitary intertwiners play an important role in orthogonal polynomial dualities, as we will see in Chapter VIII.

The following theorem shows that from equivalent representations related by an intertwiner in kernel form one gets duality relations.

Proposition 2.38 (Equivalent Representations and Duality) *Let $\mathcal{A}$ be an algebra and $\widehat{\rho} : \mathcal{A} \to \mathcal{B}(\widehat{\mathcal{H}}, \widehat{\mathcal{H}})$ and $\rho : \mathcal{A} \to \mathcal{B}(\mathcal{H}, \mathcal{H})$ be two representations. Assume that $\widehat{\rho}$ and ρ are equivalent with intertwiner $\Lambda \in \mathcal{B}(\mathcal{H}, \widehat{\mathcal{H}})$. Assume moreover that Λ_D is in generalized kernel operator form with kernel $D \in \widehat{\mathcal{H}} \otimes \mathcal{H}$, i.e.,* (2.120) *holds. Then we have for all $a \in \mathcal{A}$, $\widehat{\rho}(a) \xrightarrow{D} (\rho(a))^*$, where $*$ denotes adjoint in $\mathcal{H}$.*

Proof This is an application of Theorem 2.35 with $\widehat{A} = \widehat{\rho}(a)$ and $A = \rho(a)$. □

2.5 Duality as a Change of Representation

In this section, we return to the example of duality between Wright-Fisher diffusion and the Kingman's coalescent block counting process, and show that it is an instance of duality obtained via a change of representation of an algebra. The process duality is then a consequence of the duality between the two representations. Thus the duality function does not only link the two Markov generators, but all elements of the algebra in one representation to corresponding elements of the conjugate algebra in another representation. This shows a key principle of the algebraic approach to duality described in this book: dualities at the level of the algebra (i.e. intertwining between two representations) can be used to obtain dualities of Markov processes.

In order to introduce this change of representation picture, we need some notation. In particular, recalling the notion of duality between two bounded operators on a L^2-space introduced in (2.108), we generalize it to the context of unbounded operators. For that we need a common set of functions on which the representation of the elements of the algebra are acting. E.g., if the algebra is generated by the multiplicative operator x and the differential operator $\frac{d}{dx}$, then we act on a common domain of smooth functions.

Definition 2.39 (Duality Between Two Unbounded Operators) We say that a function D of two variables y and x acts as duality function for duality between operators A and B if

$$(AD(\cdot, x))\,(y) = (BD(y, \cdot))\,(x), \tag{2.134}$$

where we implicitly assume that $D(\cdot, x)$ is in the domain of A and $D(y, \cdot)$ is in the domain of B.

In the next theorem we show how the relation $\stackrel{D}{\longrightarrow}$ is compatible with linear combinations and products of operators. This theorem is the generalized version of Theorem 2.23 which was restricted to square matrices.

Theorem 2.40 (Combination of Dual Operators) *Let $\widehat{\mathbb{O}}, \mathbb{O}$ be two algebras of operators working on a common domain of functions $f : \widehat{\Omega} \to \mathbb{R}$, resp. functions $f : \Omega \to \mathbb{R}$. Suppose that the algebra $\widehat{\mathbb{O}}$ is dual to the algebra $\mathbb{O}$ with duality function $D : \widehat{\Omega} \times \Omega \to \mathbb{R}$, denoted $\widehat{\mathbb{O}} \stackrel{D}{\longrightarrow} \mathbb{O}$, meaning that for all $\widehat{O} \in \widehat{\mathbb{O}}$ there exists $O \in \mathbb{O}$ such that*

$$(\widehat{O}D(\cdot, x))(y) = (OD(y, \cdot))(x), \qquad \text{for all } y \in \widehat{\Omega}, x \in \Omega\,. \tag{2.135}$$

Then for all $\widehat{O}_1, \widehat{O}_2 \in \widehat{\mathbb{O}}$ and $O_1, O_2 \in \mathbb{O}$ we have that if

$$\widehat{O}_i \stackrel{D}{\longrightarrow} O_i, \qquad \text{for} \quad i = 1, 2 \tag{2.136}$$

then

$$\widehat{O}_1\widehat{O}_2 \xrightarrow{D} O_2O_1,$$
$$\widehat{O}_1 + \widehat{O}_2 \xrightarrow{D} O_1 + O_2. \tag{2.137}$$

Proof We prove the first relation in (2.137), as the second one follows from a similar argument. In order not to overload the notation, we abbreviate and agree (in this proof) that operators with hats work on the first (left) variable of the function D, and operators without hats work on the second (right) variable of the function D. With this we can write

$$(\widehat{O}_1\widehat{O}_2)D = \widehat{O}_1 O_2 D = O_2\widehat{O}_1 D = O_2 O_1 D, \tag{2.138}$$

where we used that the operator O_2 working on the first variable of D and the operator $\widehat{O}_1$ working on the second variable of D commute. □

In words, Theorem 2.40 above says that a duality function translates elements of an algebra of operators $\widehat{O}_i$ to elements of a "dual" algebra of operators O_i *obtained by removing hats from the* $\widehat{O}_i$ *and multiplying the elements in the reversed order.*

2.5.1 Two Representations of the Heisenberg Algebra

To treat the Wright-Fisher/Kingman duality from the algebraic point of view, we will need the Heisenberg algebra. This is the Lie algebra generated by two elements $\mathbf{a}, \mathbf{a}^\dagger$ satisfying the canonical commutation relation $[\mathbf{a}, \mathbf{a}^\dagger] = I$ with I the identity. We will see that the duality of the Wright-Fisher generator and the generator of the Kingman's coalescent block-counting process is a consequence of a much more general duality between two representations of the Heisenberg algebra, with intertwiner the duality function $D(n, x) = x^n$.

We introduce the operators

$$A^\dagger f(x) = xf(x),$$
$$Af(x) = f'(x), \tag{2.139}$$

working on smooth functions of a real variable x. These operators satisfy the "canonical commutation relation" $[A, A^\dagger] = I$ where I denotes the identity:

$$[A, A^\dagger]f(x) = (xf)'(x) - xf'(x) = f(x) = (If)(x). \tag{2.140}$$

Next, we introduce the operators working on functions $f : \mathbb{N} \to \mathbb{R}$

$$a^\dagger f(n) = f(n+1),$$

$$af(n) = nf(n-1), \tag{2.141}$$

where by the last equation for $n = 0$ we mean $af(0) = 0$. These operators satisfy the negative of the "canonical commutation relations" $[a, a^\dagger] = -I$

$$[a, a^\dagger]f(n) = -f(n) = -(If)(n). \tag{2.142}$$

The operators $A, A^\dagger$ and $a, a^\dagger$ satisfy a duality relation via the monomials $D(n, x) = x^n$, i.e.,

$$\begin{aligned} \big(aD(\cdot, x)\big)(n) &= \big(AD(n, \cdot)\big)(x), \\ \big(a^\dagger D(\cdot, x)\big)(n) &= \big(A^\dagger D(n, \cdot)\big)(x). \end{aligned} \tag{2.143}$$

2.5.2 *The Duality Wright-Fisher/Kingman Revisited*

The operators $A, A^\dagger$ generates an algebra of differential operators acting on a common domain of functions. For the Wright-Fisher diffusion we may restrict to functions $f : [0, 1] \to \mathbb{R}$ which are polynomials. In terms of these operators the Wright-Fisher diffusion process generator reads

$$Lf(x) = \frac{1}{2}x(1-x)\frac{d^2}{dx^2}f(x) = \frac{1}{2}A^\dagger(I - A^\dagger)A^2 f(x), \tag{2.144}$$

whereas the generator of the Kingman's coalescent block counting process reads

$$\widehat{L}f(n) = \frac{1}{2}n(n-1)(f(n-1) - f(n)) = \frac{1}{2}\left(a^2(I - a^\dagger)a^\dagger f\right)(n). \tag{2.145}$$

We now see that we can pass from $A^\dagger(I - A^\dagger)A^2$ to $a^2(I - a^\dagger)a^\dagger$ by replacing A by a and writing the products of operators in the order from right to left.

The duality relation between the generators $\widehat{L}$ and L is indeed a consequence of the more general duality relation (2.143). This implies, using Theorem 2.40, that for every "word" of the form

$$W = W_1 W_2 \ldots W_n, \tag{2.146}$$

where $W_i \in \{A, A^\dagger\}$ there is a "dual word" given by

$$w = w_n w_{n-1} \ldots w_1, \tag{2.147}$$

i.e., the symbols of W written from right to left, and "translating" $w_i = a$, resp. $w_i = a^\dagger$, whenever $W_i = A$, resp. $W_i = A^\dagger$. E.g. $W = AA^\dagger AA^\dagger A^\dagger$ is "translated" to $a^\dagger a^\dagger a a^\dagger a$. The duality relation between a word W and its dual word w is then simply as in (2.143)

$$wD(\cdot, x)(n) = WD(n, \cdot)(x). \tag{2.148}$$

We indeed see that the word forming the Wright-Fisher diffusion generator $L = A^\dagger(I - A^\dagger)A^2$ has the dual word $\widehat{L} = a^2(I - a^\dagger)a^\dagger$ which is the generator of the Kingman's coalescent block counting process.

Summarizing, the duality Wright-Fisher/Kingman is a single instance of a much more general duality between a representation of the Heisenberg algebra and a representation of the conjugate Heisenberg algebra.

2.5.3 *Heisenberg Algebra: Further Examples of Intertwined Representations*

The Heisenberg algebra admits several intertwined representations which lead to dualities. Here we give a few examples. Later in the book, in the chapters on independent walkers and associated models in continuous variables, such representations and their intertwinings will play an important role.

We recall the "continuous" representation of the Heisenberg algebra (2.139), and the discrete representation of the conjugate Heisenberg algebra (2.141). We then have the following dualities.

1. Duality function linking the continuous and the discrete representation:

$$a \xrightarrow{D} A, \quad a^\dagger \xrightarrow{D} A^\dagger, \tag{2.149}$$

with

$$D(n, x) = x^n. \tag{2.150}$$

2. Duality function linking the discrete representation to its conjugate:

$$a \xrightarrow{D} a^\dagger, \quad a^\dagger \xrightarrow{D} a, \tag{2.151}$$

with

$$D(k, n) = n!\delta_{k,n}. \tag{2.152}$$

3. Duality function linking the continuous representation to its conjugate:

$$A \xrightarrow{D} A^\dagger, \quad A^\dagger \xrightarrow{D} A, \tag{2.153}$$

with

$$D(y, x) = e^{xy}. \tag{2.154}$$

More generally we have the following duality between a representation of the Heisenberg algebra and the discrete representation (2.141).

Theorem 2.41 *Let $A, A^\dagger$ denote any representation of the Heisenberg algebra on a space of functions $f : E \to \mathbb{R}$. Let $D(0, \cdot) : E \to R$ be a function such that*

$$AD(0, \cdot) = 0, \tag{2.155}$$

a so-called ground state. Next define the functions $D(n, \cdot)$ via

$$D(n, \cdot) = (A^\dagger)^n D(0, \cdot) \tag{2.156}$$

Then we have $a \xrightarrow{D} A$ and $a^\dagger \xrightarrow{D} A^\dagger$, where $a, a^\dagger$ are the discrete representation (2.141).

Proof By the commutation relations we have

$$A(A^\dagger)^n = (A^\dagger)^n A + n(A^\dagger)^{n-1}. \tag{2.157}$$

Applying this to $D(0, \cdot)$ yields, using $AD(0, \cdot) = 0$

$$AD(n, \cdot) = nD(n-1, \cdot). \tag{2.158}$$

Therefore, $A \xrightarrow{D} a$, and by definition (2.156) we automatically have $A^\dagger \xrightarrow{D} a^\dagger$.

□

2.5.4 Examples

1. The previously considered example $D(n, x) = x^n$ is a special case, when $A = \frac{d}{dx}$, $A^\dagger = x$ and so $D(0, x) = 1$, $D(n, x) = (A^\dagger)^n D(0, \cdot)(x) = x^n$.
2. As a more general example consider

$$A = c_1 x + c_2 \frac{d}{dx}, \quad A^\dagger = c_3 x + c_4 \frac{d}{dx}, \tag{2.159}$$

with $c_2c_3 - c_1c_4 = 1$ and $c_2 \neq 0$. Then we obtain $A \xrightarrow{D} a$ and $A^\dagger \xrightarrow{D} a^\dagger$ with

$$D(n,x) = \left(c_4 \frac{d}{dx} + c_3 x\right)^n e^{-\frac{c_1}{2c_2}x^2}. \tag{2.160}$$

This contains, when $c_1 = c_2 = \frac{1}{2}$, $c_3 = -c_4 = 1$, the case of the Hermite functions

$$D(n,x) = \left(x - \frac{d}{dx}\right)^n e^{-\frac{x^2}{2}}. \tag{2.161}$$

and, when $c_1 = 0$, $c_2 = c_3 = -c_4 = 1$, the case of Hermite polynomials

$$D(n,x) = \left(x - \frac{d}{dx}\right)^n 1. \tag{2.162}$$

Many more examples of these Heisenberg algebra dualities will follow in Chaps. 3, 4, and 8. See also the example of the duality between the squared Bessel process and a birth and death chain which is discussed in Sect. 2.7.

2.6 Duality and Intertwining of Algebras

In this section we introduce duality in the abstract sense, i.e., starting from an algebra and two representations. In the examples of later chapters, this algebra will always be the universal enveloping algebra of a Lie algebra. This framework will generalize both the matrix duality introduced before as well as the Hilbert space duality. We start with some basic definitions.

Definition 2.42 (Conjugate Algebra) Let $\mathcal{A}$ be an algebra i.e. a vector space over $\mathbb{R}$ (we will always consider $\mathbb{R}$ as the number field) and equipped with a multiplication operation. The conjugate algebra is the algebra with identical elements as the elements of $\mathcal{A}$ and with multiplication $*$ defined via

$$a * b = ba, \tag{2.163}$$

where ba denotes the multiplication of b and a in the algebra $\mathcal{A}$.

We will always denote by I the identity in $\mathcal{A}$, i.e., the element such that $aI = Ia = a$ for all $a \in \mathcal{A}$.

Definition 2.43 (Representation of an Algebra) A representation of $\mathcal{A}$ is given by an injective homomorphism

$$\rho : \mathcal{A} \to Gl(V), \tag{2.164}$$

where V is a vector space and $Gl(V)$ denotes the set of linear maps from V to V. More precisely ρ is an injection satisfying the additional properties:

1. ρ is linear: $\rho(\alpha a + \beta b) = \alpha\rho(a) + \beta\rho(b)$ for all $a, b \in \mathcal{A}$, $\alpha, \beta \in \mathbb{R}$.
2. ρ preserves multiplication: $\rho(ab) = \rho(a)\,\rho(b)$.

If we work in a representation, it means that we identify the elements $a \in \mathcal{A}$ with their images $\rho(a) \in Gl(V)$, or, equivalently think of the elements of $\mathcal{A}$ as acting on the vector space V.

We can now define the abstract notion of duality.

Definition 2.44 (Duality) Let $\mathcal{A}$ be an algebra and $\widehat{\rho} : \mathcal{A} \to Gl(\widehat{V})$, $\rho : \mathcal{A} \to Gl(V)$ be two representations. Let $a, \hat{a} \in \mathcal{A}$, then an element D of $\widehat{V} \otimes V$ is called a duality element between $\hat{a}$ and a in the representations $\widehat{\rho}$, ρ, if

$$(\widehat{\rho}(\hat{a}) \otimes I_V)D = (I_{\widehat{V}} \otimes \rho(a))D, \tag{2.165}$$

where $I_{\widehat{V}}$, resp. I_V denote the identity in $Gl(\widehat{V})$, resp. $Gl(V)$.

From now on we will omit the subindices V and $\widehat{V}$ from the identities, to alleviate notation. Furthermore we denote this property by $\hat{a} \xrightarrow{D} a$, thereby suppressing the dependence on the representation. We think of this property as a relation between certain elements of the algebra $\mathcal{A}$, induced by the two representations $\widehat{\rho}$, ρ, and parametrized by the duality function D.

There is a natural map $i : \widehat{V} \otimes V \to V \otimes \widehat{V}$ defined on tensors by $i(\widehat{v} \otimes v) = v \otimes \widehat{v}$. We then have the following properties of the relation "$\xrightarrow{D}$". This is the analogue of Theorem 2.23 in the current abstract setting.

Theorem 2.45

1. *Linearity: if $\hat{a} \xrightarrow{D} a$, $\hat{b} \xrightarrow{D} b$ then $\hat{a} + \hat{b} \xrightarrow{D} a + b$, and if $\lambda \in \mathbb{R}$, $\lambda\hat{a} \xrightarrow{D} \lambda a$.*
2. *Connecting the algebra to the conjugate algebra: $\hat{a} \xrightarrow{D} a$, $\hat{b} \xrightarrow{D} b$, then $\hat{a}\hat{b} \xrightarrow{D} a * b = ba$.*
3. *Commuting elements generate new duality functions: If $\hat{a} \xrightarrow{D} a$, and $[a, b] = ab - ba = 0$ then, if we define $D' = (I \otimes \rho(b))D$ we have that D' is also a duality function linking $\hat{a}$ and a. Also, if $[\hat{a}, \hat{b}] = 0$ then $D'' = (\widehat{\rho}(\hat{b}) \otimes I)D$ is a duality function linking a and $\hat{a}$.*
4. *If D is a duality function linking $\hat{a}$ and a then $D^* = i(D) \in V \otimes \widehat{V}$ is a duality function linking a and $\hat{a}$.*

Proof In order to facilitate the notation, we will identify the algebras with their representations, i.e., we write $\hat{a}$ instead of $\widehat{\rho}(\hat{a})$, and a for $\rho(a)$, i.e., we think of the elements of $\mathcal{A}$ as operators working on the vector spaces $\widehat{V}$, V. Then we can write e.g. $\hat{a} \otimes a$ working on a vector in $\widehat{V} \otimes V$ where we strictly speaking should write $\widehat{\rho}(\hat{a}) \otimes \rho(a)$. Item 1 follows from

$$(\hat{a} + \hat{b}) \otimes I = \hat{a} \otimes I + \hat{b} \otimes I, \tag{2.166}$$

and

$$(\lambda\hat{a}) \otimes I = \lambda(\hat{a} \otimes I). \tag{2.167}$$

For item 2, use that elements of the form $\hat{a} \otimes I$ always commute with elements of the form $I \otimes a$ because

$$(\hat{a} \otimes I)(I \otimes a) = \hat{a} \otimes a = (I \otimes a)(\hat{a} \otimes I). \tag{2.168}$$

Therefore

$$\begin{aligned}
(\hat{b}\hat{a} \otimes I)D &= (\hat{b} \otimes I)(\hat{a} \otimes I)D \\
&= (\hat{b} \otimes I)(I \otimes a)D \\
&= (I \otimes a)(\hat{b} \otimes I)D \\
&= (I \otimes a)(I \otimes b)D \\
&= (I \otimes ab)D.
\end{aligned} \tag{2.169}$$

For item 3

$$\begin{aligned}
(I \otimes a)(I \otimes b)D &= (I \otimes ab)D \\
&= (I \otimes ba)D \\
&= (I \otimes b)(I \otimes a)D \\
&= (I \otimes b)(\hat{a} \otimes I)D \\
&= (\hat{a} \otimes I)(I \otimes b)D \\
&= (\hat{a} \otimes I)D'.
\end{aligned} \tag{2.170}$$

The second statement of item 3 follows with similar proof. Finally, item 4 follows from

$$(a \otimes I)(i(D)) = i((I \otimes a)D) = i((\hat{a} \otimes I)D) = (I \otimes \hat{a})(i(D)). \tag{2.171}$$

□

An immediate consequence of this theorem is the following. We define for a given element $D \in \widehat{V} \times V$ the set

$$\mathcal{G}(D) = \{(\hat{a}, a) : \hat{a} \xrightarrow{D} a\}, \tag{2.172}$$

i.e. the set of all pairs whose components are in duality relation with duality element D. Then we have that $\mathcal{G}(D)$ is an algebra when equipped with multiplication

$$(\hat{a}, a) \cdot (\hat{b}, b) = (\hat{a}\hat{b}, a * b) = (\hat{a}\hat{b}, ba). \tag{2.173}$$

In other words $\mathcal{G}(D)$ behaves in the first component as the original algebra and in the second component as the conjugate algebra. As a consequence, the first and second projections of $\mathcal{G}(D)$ provide subalgebras of $\mathcal{A}$, resp. the conjugate algebra of $\mathcal{A}$.

Remark 2.46 The duality of algebras as described here deals with duality elements $D \in \widehat{V} \otimes V$. Very often in applications one has to go beyond finite linear combinations of tensors, i.e. one needs duality elements D in the closure of $\widehat{V} \otimes V$ with respect to a suitable norm. For instance, when the vector spaces $\widehat{V}, V$ are Banach spaces of functions, i.e., V, resp. $\widehat{V}$, is a set of functions $f : \Omega \to \mathbb{R}$, resp. $\widehat{f} : \widehat{\Omega} \to \mathbb{R}$, each equipped with its norm then the algebraic tensor product $\widehat{V} \otimes V$ is the set of (finite) linear combinations of the form $\sum_{i,j=1}^{n} c_{i,j} \widehat{f_i} f_j$. This space has to be equipped with a suitable norm which in general is not uniquely determined. Then one has to consider the closure of $\widehat{V} \otimes V$ in that norm, this is called the topological tensor product [204]. Duality functions are then allowed to be more general than elements of $\widehat{V} \otimes V$, i.e., can be limits of finite linear combinations of tensors under the chosen norm. This was already implicitly done in the Hilbert space case where the norm of the tensor product is uniquely determined by the scalar product on $\widehat{V} \otimes V$ via $\langle \widehat{f} \otimes f, \widehat{g} \otimes g \rangle = \langle \widehat{f}, \widehat{g} \rangle \cdot \langle f, g \rangle$ and bilinearity.

2.6.1 Intertwining and Duality: General Case

We consider vector spaces $V, \widehat{V}$ and their (algebraic) dual vector spaces $V^*, \widehat{V}^*$, i.e.

$$\begin{aligned} V^* &= \{v^* : V \to \mathbb{R},\ v^* \text{ linear}\}, \\ \widehat{V}^* &= \{\hat{v}^* : \widehat{V} \to \mathbb{R},\ \hat{v}^* \text{ linear}\}. \end{aligned} \tag{2.174}$$

Let us denote $\langle \cdot, \cdot \rangle$ both for the pairing between V^* and V and for the pairing between $\widehat{V}^*$ and $\widehat{V}$. I.e., $\langle v^*, v \rangle = v^*(v)$. We consider two linear maps

$$\begin{aligned} A &: V \to V, \\ \widehat{A} &: \widehat{V} \to \widehat{V}, \end{aligned} \tag{2.175}$$

then we automatically have two maps

$$\begin{aligned} A^* &: V^* \to V^*, \\ \widehat{A}^* &: \widehat{V}^* \to \widehat{V}^*, \end{aligned} \tag{2.176}$$

where $(A^*v^*)(v) = v^*(Av)$. We think of A and $\widehat{A}$ as two representations of two elements of an algebra $\mathcal{A}$, i.e., in the language of the previous paragraph, $A = \rho(a)$ and $\widehat{A} = \widehat{\rho}(\widehat{a})$, for some $a, \widehat{a} \in \mathcal{A}$. Assuming that A and $\widehat{A}$ are in duality with duality element $D \in \widehat{V} \otimes V$, i.e.

$$(\widehat{A} \otimes I)D = (I \otimes A)D. \tag{2.177}$$

we now want to find a corresponding intertwining as in Theorem 2.35.

First, we have the following (purely algebraic) analogue of Lemma 2.34.

Lemma 2.47 *For every D in $\widehat{V} \otimes V$ there exists a unique $\Lambda_D : V^* \to \widehat{V}^{**}$ such that*

$$\langle \Lambda_D v^*, \widehat{v}^* \rangle = \langle \widehat{v}^* \otimes v^*, D \rangle, \tag{2.178}$$

for all $\widehat{v}^* \in \widehat{V}^*$, $v^* \in V^*$.

Proof Fix $v^* \in V^*$. Consider the map $\psi : \widehat{V}^* \to \mathbb{R}$ defined via

$$\psi(\widehat{v}^*) = \langle \widehat{v}^* \otimes v^*, D \rangle. \tag{2.179}$$

This is a linear map between $\widehat{V}^*$ and $\mathbb{R}$ hence $\psi \in \widehat{V}^{**}$, then define

$$\psi = \Lambda_D \widehat{v}^*. \tag{2.180}$$

□

We then have the analogue of Theorem 2.35.

Theorem 2.48 (Duality and Intertwining: General Case) *Let the setting be as in* (2.175)*, then the following two properties are equivalent:*

1. *Duality between $\widehat{A}$ and A, in the sense* (2.177).
2. *Intertwining between $\widehat{A}$ and A^* with intertwiner Λ_D defined in Lemma 2.47, i.e.,*

$$\widehat{A}\Lambda_D = \Lambda_D A^*. \tag{2.181}$$

Proof The proof is an obvious modification of the Hilbert space case, i.e., the proof of Theorem 2.35, by replacing inner products by pairings between vector spaces and their dual. □

Remark 2.49 In this subsection we went along purely algebraic, i..e, we considered algebraic duals of vector spaces. The formulation of a result of the type of Theorem 2.48 above, as well as its proof, for topological duals would be much more complex as it would depend on the choice of norm on the considered vector space, and it would pose additional problems. For instance for a bounded operator A on a Banach space in general it is not true that $A \otimes I$ is bounded on the corresponding tensor product Banach space.

2.7 Additional Notes

In the context of birth and death processes, Siegmund duality was introduced in [214]. The use of duality in the context of interacting particle systems started in [217] which is also the very first paper that pioneered the area. In the context of spin flip dynamics (Glauber dynamics, or spin systems in the terminology of [172]), Holley and Stroock used duality as a form of Fourier transformation [136], i.e., by relating the dynamics of the Markov process on the state space $\Omega = \{-1, 1\}^{\mathbb{Z}^d}$ to the dynamics of the characters of the compact abelian group Ω. In [172] the concept of duality was systematized and applied in various interacting particle systems including the voter model, the contact process, general spin systems and the symmetric exclusion process. In chapter 8 of [172], the ergodic theory of the symmetric exclusion process is developed, based on duality. Recent review papers on duality are [142], [224]. Duality arising from graphical representations is due to Harris [133], see [83] for an approach to interacting particle systems driven by graphical representations. See also [223] for an approach to duality bases on monotonicity and graphical representation.

In the context of population dynamics, duality was studied in [182] and is a cornerstone technique in models of Wright-Fisher type, see e.g. [87] for a recent overview, see the book [66] for a complete account.

The connection between duality and symmetries was first discovered by Schütz in [213]. The "quantum formalism" notation introduced in that paper was further developed in [225] in the context of $SU(2)$ with spin $1/2$. In [116] the approach was systematized and examples from other groups including $SU(1, 1)$ were introduced. In the context of asymmetric processes, non-abelian symmetries were found in [210] using quantum Lie algebras, and in [44], [41] processes with self-duality properties were constructed from the coproduct applied to the Casimir element in the context of quantum Lie algebras. At present, many examples of such constructions e.g. in the context of higher rank Lie algebras have been produced, see e.g. [157].

The connection between duality and intertwining is discussed in [226]. Intertwining of Markov processes is a well-studied subject see e.g. [180], [179] for recent papers on intertwining and its applications. Reformulating duality in terms of intertwining is a key to define duality for processes in the continuum such as interacting Brownian motions, see [95]. The connection between self-duality and intertwining also leads to a new understanding of self-duality properties of the three basic processes studied in this book. Indeed, in Chap. 9 we will see that consistency, which is a special case of self-intertwining (i.e., symmetry) unifies the self-dualities for the independent walkers, and symmetric inclusion and exclusion processes. More precisely, these self-dualities correspond to the same intertwiner, but the reversible measures are different and therefore the duality functions are different.

The use of intertwining to study spectral properties and related relaxation properties of Markov generators is the subject of the recent works [178], [181], [192]. As a key example consider the Poisson intertwining between the squared Bessel process with index $\nu = 0$ and the corresponding discrete birth and death process (the case of a general index $\nu > -1$ is treated similarly). The diffusion process on $[0, \infty)$ has generator

$$Lf(x) = xf''(x) + f'(x). \tag{2.182}$$

The corresponding birth and death process on $\mathbb{N}_0 = \{0, 1, 2, \ldots\}$ has generator

$$\widehat{L}f(n) = (n+1)(f(n+1) - f(n)) + n(f(n-1) - f(n)). \tag{2.183}$$

Defining the Poisson intertwiner

$$\Lambda f(x) = \sum_{n=0}^{\infty} f(n) \frac{x^n e^{-x}}{n!}, \tag{2.184}$$

in [179] it is proved that

$$L\Lambda = \Lambda \widehat{L}. \tag{2.185}$$

Because the counting measure is reversible for the process with generator (2.183), the intertwining (2.185) implies that L and $\widehat{L}$ are dual with duality function

$$D(n, x) = \frac{x^n e^{-x}}{n!}. \tag{2.186}$$

This duality between L and $\widehat{L}$ corresponds to a duality on the level of the Heisenberg algebra with representations

$$A = \frac{d}{dx}, \qquad A^\dagger = x, \tag{2.187}$$

and

$$af(n) = (n+1)f(n+1), \qquad a^\dagger f(n) = f(n-1) - f(n), \tag{2.188}$$

with $f(-1) = 0$. More precisely, from a simple computation, we can see that

$$a \xrightarrow{D} A^\dagger, \qquad a^\dagger \xrightarrow{D} A, \tag{2.189}$$

with D given in (2.186), and as a consequence we obtain

$$\widehat{L} = (a^\dagger)^2 a + a^\dagger \xrightarrow{D} A^\dagger A^2 + A = L. \tag{2.190}$$

I.e., the birth and death chain with generator $\widehat{L}$ and the diffusion process with generator L are dual with duality function (2.186).

Chapter 3
Self-duality for Independent Random Walkers

In this chapter we illustrate how the Lie algebraic approach reproduces the basic self-duality relation for independent random walkers [71]. After providing the description of the process generator in terms of the generators of the *Heisenberg algebra*, we show how the triangular self-duality function (related to multivariate factorial moments) arises in two ways. The first method is by means of a *symmetry* acting on the diagonal self-duality function associated to reversibility. The symmetry is the total annihilation operator and is easily deduced from the algebraic description. The second method starts instead from basic dualities relating the generators of the Heisenberg algebra and then promotes them to self-duality of the Markov process by composing dualities. We close the chapter by showing how self-duality is used in the *ergodic theory* of the process on the infinite lattice $\mathbb{Z}^d$. In particular the infinite-volume process has products of Poisson distributions as reversible and ergodic measures. Under an appropriate moment growth condition, we show that these product Poisson distributions are the only ergodic distributions.

3.1 The Process on a Finite Set

We start by defining the independent random walkers process on a finite set V. Let $p : V \times V \to [0, \infty)$ be a non-negative symmetric and irreducible function. By this we mean that for all $x, y \in V$ it holds $p(x, y) = p(y, x) \geq 0$ and, furthermore, there exists $n \geq 2$ and a path $(x_1, x_2, \ldots, x_n)$ with $x_1 = x$ and $x_n = y$ such that $\prod_{i=1}^{n-1} p(x_i, x_{i+1}) > 0$. Alternatively we can think of a finite connected graph $G = (V, E)$ with vertex set V and edge set E. The edges $\{x, y\} \in E$ are un-oriented and weighted by edge weights $p(x, y)$.

We denote by $\eta : V \to \mathbb{N}$ the initial configuration of particles: at each site $x \in V$ we have initially $\eta_x \in \mathbb{N}$ particles. Each of these particles perform continuous-time random walk jumping at rate $p(x, y)$ from site x to site y, and different particles are

C. Giardinà, F. Redig, *Duality for Markov Processes*, Grundlehren der mathematischen Wissenschaften 365, https://doi.org/10.1007/978-3-032-04099-2_3

independent. More precisely, we consider a collection of such independent random walks $\{X^{x,i}(t)\}$, labeled by the initial position $x \in V$ and by $i = 1, \ldots, \eta_x$, and define the process $\{\eta_y(t),\ t \geq 0\}$ counting the number of walkers at location $y \in V$ at time $t > 0$ by

$$\eta_y(t) = \sum_{x \in V} \sum_{i=1}^{\eta_x} \mathbb{1}_{\{X^{x,i}(t)=y\}}. \tag{3.1}$$

In the above equation empty sums are interpreted as zero. We denote by $\Omega = \mathbb{N}^V$ the state space, that is made of configurations with a finite number of particles since we are on a finite set V. The process $\eta(t) = (\eta_x(t))_{x \in V}$ is a continuous-time Markov chain with countable state space and therefore is well-defined. Indeed the initial configuration is forced to have a finite number of particles and the dynamics conserves the number of particles. Therefore, at every time $t \geq 0$, the sum in (3.1) will stay finite.

We denote by $\mathbb{P}_\eta$ the path space measure of $\{\eta(t), t \geq 0\}$ starting from η, and by $\mathbb{E}_\eta$ the expectation w.r.t. $\mathbb{P}_\eta$. We denote by δ_x, the configuration with a single particle at $x \in V$ and no particles anywhere else, i.e.

$$(\delta_x)_y = \begin{cases} 1 \text{ if } y = x, \\ 0 \text{ if } y \neq x. \end{cases} \tag{3.2}$$

For a function $f : \Omega \to \mathbb{R}$ we define the semigroup

$$(S_t f)(\eta) = \mathbb{E}_\eta(f(\eta(t))) \,. \tag{3.3}$$

Then to the semigroup $\{S_t,\ t \geq 0\}$ is associated the generator of the process $\{\eta(t), t \geq 0\}$ via the usual formula

$$Lf(\eta) = \lim_{t \to 0} \frac{S_t f(\eta) - f(\eta)}{t} \,. \tag{3.4}$$

Using the explicit (3.1) one obtains the generator of the independent random walkers process in the form

$$\begin{aligned} L &= \frac{1}{2} \sum_{x,y \in V} p(x,y) L_{x,y} \\ &= \sum_{\{x,y\} \in E} p(\{x,y\}) L_{\{x,y\}}, \end{aligned} \tag{3.5}$$

where, with a slight abuse of notation $p(x, y) = p(y, x) = p(\{x, y\})$, and where $L_{x,y} = L_{y,x} = L_{\{x,y\}}$ is defined as

$$L_{x,y} f(\eta) = \eta_x (f(\eta^{x,y}) - f(\eta)) + \eta_y (f(\eta^{y,x}) - f(\eta)) . \tag{3.6}$$

Here $\eta^{x,y}$ is the configuration obtained from η by letting a particle move from x to y, i.e.

$$\eta^{x,y} = \begin{cases} \eta - \delta_x + \delta_y & \text{if } \eta_x \geq 1 \\ \eta & \text{otherwise.} \end{cases} \tag{3.7}$$

We call $L_{x,y}$ the *single-edge* generator. We will see that all the duality properties of the full generator L on the graph (3.5) follow from the duality properties of the single-edge generator.

As we have seen in the first chapter, since we are on a finite set V, the relation between the generator L and the semigroup S_t is matrix exponentiation, i.e.,

$$S_t = e^{tL} = \sum_{n=0}^{\infty} \frac{t^n L^n}{n!} . \tag{3.8}$$

In later sections we will also consider the process of independent random walkers on an infinite set V, such as the lattice $\mathbb{Z}^d$. In that setting, when the process is started from a configuration $\eta \in \Omega$ that contains infinitely many particles, at some time $t > 0$ the sum in (3.1) might be divergent (i.e., equal to $+\infty$). Thus on an infinite set V one needs to define a "good" set of initial configurations that guarantees that the process will not explode. This issue will be further discussed in Sect. 3.5. Furthermore the way to obtain the semigroup from the generator is also a delicate issue. In our current setting of independent random walkers, we can however circumvent this issue by taking increasing limits along sequences of finite configurations, see Sect. 3.5 for more details.

3.2 Symmetries of the Generator

In this section we provide the algebraic description of the generator (3.5) in terms of the Heisenberg algebra (more precisely the conjugate Heisenberg algebra). This will be the starting point of the algebraic approach to duality, since it will easily yield symmetries of the generator, that are crucial in the construction of the duality function.

Given a site $x \in V$, we define the *particle removal* operator a_x acting on function $f : \mathbb{N}^V \to \mathbb{R}$

$$a_x f(\eta) = \begin{cases} \eta_x f(\eta - \delta_x) & \text{if } \eta_x \geq 1, \\ 0 & \text{otherwise,} \end{cases} \tag{3.9}$$

and the *particle addition* operator $a_x^\dagger$ as

$$a_x^\dagger f(\eta) = f(\eta + \delta_x) . \tag{3.10}$$

We also define the commutator of these operators by

$$[a_x, a_x^\dagger] := a_x a_x^\dagger - a_x^\dagger a_x, \tag{3.11}$$

where, as usual, the product $a_x a_x^\dagger$ is understood as composition, i.e.

$$(a_x a_x^\dagger f)(\eta) := \big(a_x (a_x^\dagger f)\big)(\eta). \tag{3.12}$$

We denote by I the identity operator

$$I f(\eta) = f(\eta). \tag{3.13}$$

Proposition 3.1 (Commutation Relations and Abstract Form of the Generator) *The following holds:*

1. *The collection of operators* $(a_x)_{x\in V}$ *and* $(a_x^\dagger)_{x\in V}$ *defined in* (3.9) *and* (3.10) *satisfy the commutation relations*

$$\begin{aligned} [a_x, a_y^\dagger] &= -I\delta_{x,y}, \\ [a_x, a_y] &= 0, \\ [a_x^\dagger, a_y^\dagger] &= 0, \end{aligned} \tag{3.14}$$

   where $\delta_{x,y}$ *is the Kronecker delta.*
2. *The generator* (3.5) *of independent random walkers equals*

$$L = -\frac{1}{2} \sum_{x,y\in V} p(x,y)(a_y - a_x)(a_y^\dagger - a_x^\dagger). \tag{3.15}$$

   In other words, the single-edge generator (3.6) *equals*

$$L_{x,y} = -(a_y - a_x)(a_y^\dagger - a_x^\dagger) . \tag{3.16}$$

   We call this the abstract form of the generator, resp. the abstract form of the single-edge generator.

Proof For $x \neq y$, (3.14) is immediate. For $x = y$, the second and third commutators in (3.14) are also trivial, whereas the first commutator is verified by using the composition rule (3.12). One has:

$$(a_x^\dagger a_x f)(\eta) = (a_x f)(\eta + \delta_x) = (\eta_x + 1) f(\eta), \tag{3.17}$$

and

$$(a_x a_x^\dagger f)(\eta) = \eta_x (a_x^\dagger f)(\eta - \delta_x) = \eta_x f(\eta), \tag{3.18}$$

and, by taking the difference, the proof of (3.14) is complete.

To prove (3.15) we compute

$$(a_y a_x^\dagger f)(\eta) = \eta_y f(\eta + \delta_x - \delta_y) = \eta_y f(\eta^{y,x}), \tag{3.19}$$

Hence, combining (3.18) and (3.19) and using the symmetry of $p(x, y)$, one has

$$\begin{aligned} -\frac{1}{2} \sum_{x,y \in V} p(x, y)(a_y - a_x)(a_y^\dagger - a_x^\dagger) f(\eta) &= \frac{1}{2} \sum_{x,y \in V} \eta_x p(x, y)(f(\eta^{x,y}) - f(\eta)) \\ &+ \frac{1}{2} \sum_{x,y \in V} \eta_y p(x, y)(f(\eta^{y,x}) - f(\eta)) \\ &= Lf(\eta). \end{aligned} \tag{3.20}$$

□

Remark 3.2 (Tensor Product) Notice that $a_x^\dagger$, a_x are copies (labeled by $x \in V$ and acting on the variable η_x) of the operators $a^\dagger$, a defined on functions $f : \mathbb{N} \to \mathbb{R}$ via

$$a^\dagger f(n) = f(n+1), \qquad af(n) = nf(n-1), \tag{3.21}$$

that are represented by infinite dimensional matrices

$$a^\dagger = \begin{pmatrix} 0 & 1 & 0 & 0 & \dots \\ 0 & 0 & 1 & 0 & \ddots \\ 0 & 0 & 0 & 1 & \ddots \\ 0 & 0 & 0 & 0 & \ddots \\ \vdots & \ddots & \ddots & \ddots & \ddots \end{pmatrix}, \qquad a = \begin{pmatrix} 0 & 0 & 0 & 0 & \dots \\ 1 & 0 & 0 & 0 & \ddots \\ 0 & 2 & 0 & 0 & \ddots \\ 0 & 0 & 3 & 0 & \ddots \\ \vdots & \ddots & \ddots & \ddots & \ddots \end{pmatrix}. \tag{3.22}$$

In other words, the particle addition operator $a_x^\dagger$ has to be understood as the tensor product of identity operators I_y, labeled by $y \in V$, with $y \neq x$ and a copy of the operator $a^\dagger$, labeled by x. A similar remark applies to the particle removal operator a_x.

Remark 3.3 (Creation and Annihilation Operators) The particle addition and particle removal operators are related to so-called *creation* and *annihilation* operators of the physics literature. In physics, the creation and annihilation operators of the canonical commutation relations are usually defined by their action on the standard orthonormal basis $\{e_n\}_{n\geq 0}$ of the space $l_2(\mathbb{N})$ by

$$\begin{aligned} b^\dagger e_n &= e_{n+1}, \\ b e_n &= n e_{n-1}. \end{aligned} \tag{3.23}$$

Here e_n is the vector will all elements equal to zero, except the element at the nth position which is a 1. The action of the creation and annihilation operators on a general vector $f = \sum_{n\geq 0} f(n) e_n$ can be expressed as

$$b^\dagger f = \sum_{n\geq 0} (b^\dagger f)(n)\, e_n\,, \tag{3.24}$$

$$bf = \sum_{n\geq 0} (bf)(n)\, e_n.$$

On the other hand, using the linearity of operators and (3.23) we have

$$b^\dagger f = b^\dagger \sum_{k\geq 0} f(n) e_n = \sum_{n\geq 0} f(n) b^\dagger e_n = \sum_{n\geq 0} f(n) e_{n+1} = \sum_{n\geq 1} f(n-1) e_n\,, \tag{3.25}$$

$$bf = b \sum_{n\geq 0} f(k) e_n = \sum_{n\geq 0} f(n) b e_n = \sum_{n\geq 0} f(n) n e_{n-1} = \sum_{n\geq 0} (n+1) f(n+1) e_n.$$

Comparing (3.24) and (3.25) one finds

$$\begin{aligned} (b^\dagger f)(n) &= f(n-1), \\ (bf)(n) &= (n+1) f(n+1), \end{aligned} \tag{3.26}$$

with the convention $(b^\dagger f)(0) = 0$. These operators are represented by infinite dimensional matrices

$$b^\dagger = \begin{pmatrix} 0 & 0 & 0 & 0 & \cdots \\ 1 & 0 & 0 & 0 & \ddots \\ 0 & 1 & 0 & 0 & \ddots \\ 0 & 0 & 1 & 0 & \ddots \\ \vdots & \ddots & \ddots & \ddots & \ddots \end{pmatrix}, \qquad b = \begin{pmatrix} 0 & 1 & 0 & 0 & \cdots \\ 0 & 0 & 2 & 0 & \ddots \\ 0 & 0 & 0 & 3 & \ddots \\ 0 & 0 & 0 & 0 & \ddots \\ \vdots & \ddots & \ddots & \ddots & \ddots \end{pmatrix}. \tag{3.27}$$

The relation between the addition and removal operators defined in (3.21) and the creation and addition operators defined in (3.26) reads

$$a^\dagger = (b^\dagger)^*, \qquad a = b^*, \tag{3.28}$$

where $(b^\dagger)^*$ (resp. b^*) denotes the adjoint of $b^\dagger$ (resp. b) in the space $l_2(\mathbb{N})$. Indeed, comparing the matrices (3.22) and the matrices (3.27) one sees they are related by a transposition.

Remark 3.4 (Heisenberg Algebra and Its Conjugate) The creation and annihilation operators $b^\dagger, b$ in (3.26) form a representation of the Heisenberg algebra, i.e. they satisfy the canonical commutation relation

$$[b, b^\dagger] = I. \tag{3.29}$$

As already remarked in Proposition 3.1, the particle addition and particle removal operators form instead a representation of the conjugate Heisenberg algebra, i.e.

$$[a, a^\dagger] = -I.$$

This is consistent with the content of the previous remark, as "taking the transposed changes the sign of commutation relations".

An important consequence of having the generator L in abstract form is that we can identify operators that commute with L, which we call *symmetries*. More precisely we have the following result.

Proposition 3.5 (Symmetries of the Generator) *The generator L in* (3.5) *commutes with*

$$S^- := \sum_{x \in V} a_x, \qquad S^+ := \sum_{x \in V} a_x^\dagger. \tag{3.30}$$

Proof To prove that S^- commutes with L, use that for all $x, y, z \in V$ one has

$$[(a_y - a_x)(a_y^\dagger - a_x^\dagger), a_z] = (\delta_{z,y} - \delta_{z,x})(a_y - a_x), \tag{3.31}$$

which easily follows from the commutation rules (3.14). Therefore, using the expression (3.15) of the generator L one has

$$[L, S^-] = -\frac{1}{2} \sum_{z,x,y \in V} p(x, y)[(a_y - a_x)(a_y^\dagger - a_x^\dagger), a_z] = 0. \tag{3.32}$$

The proof for S^+ is similar. □

Remark 3.6 (More Symmetries and Co-product) Notice that the operators that commute with L naturally form a set which is closed under addition and multiplication. We show here how we can find more symmetries than S^+ and S^-. Denote by

$\mathfrak{g}$ the Lie algebra generated by $a^\dagger, a, I$. If we define

$$\Delta(a^\dagger) = \frac{a_1^\dagger + a_2^\dagger}{\sqrt{2}}, \qquad \Delta(a) = \frac{a_1 + a_2}{\sqrt{2}}, \tag{3.33}$$

then $\Delta(a^\dagger)$, $\Delta(a)$ satisfy the same commutation relations as $a^\dagger$ and a. Therefore, if we extend Δ as an algebra homomorphism to all elements of the universal enveloping algebra $U(\mathfrak{g})$ via

$$\Delta(g+h) = \Delta(g) + \Delta(h) \qquad \text{and} \qquad \Delta(gh) = \Delta(g)\Delta(h) \qquad \forall\, g, h \in U(\mathfrak{g}), \tag{3.34}$$

then we obtain a homomorphism $\Delta : U(\mathfrak{g}) \to U(\mathfrak{g}) \times U(\mathfrak{g})$ between $U(\mathfrak{g})$ and $U(\mathfrak{g}) \otimes U(\mathfrak{g})$ which is called a co-product. It is then the case that the single-edge generator

$$-(a_1 - a_2)(a_1^\dagger - a_2^\dagger) = L_{1,2} \tag{3.35}$$

commutes with all elements $\Delta(g)$ with $g \in U(\mathfrak{g})$. Note however that $L_{1,2}$ is not of the form $\Delta(g)$ for some $g \in U(\mathfrak{g})$. This is different from the cases which we will treat in the next chapters, such as the symmetric exclusion process or the symmetric inclusion process, where the generator itself is a co-product applied to a central element in $U(\mathfrak{g})$. Indeed, for the Heisenberg algebra, there is no non-trivial central element in $U(\mathfrak{g})$.

3.3 Self-duality

Having found symmetries of the generator, we know from the general principles illustrated in Chap. 2 that we are one step away from self-duality. The strategy is that one can produce useful self-duality relations by acting with symmetries on a trivial self-duality function, which is in turn provided by reversibility. We show here how the algebraic approach to duality recovers the basic self-duality relation of independent random walkers [71].

Notice that, because the configuration space $\Omega = \mathbb{N}^V$ is countable, the generator L working on functions $f : \Omega \to \mathbb{R}$ corresponds to an infinite matrix with matrix elements

$$L(\eta, \xi) = \delta_{\xi, \eta^{x,y}} \cdot \eta_x p(x, y) - \delta_{\xi, \eta} \cdot \sum_{x,y \in V} \eta_x p(x, y). \tag{3.36}$$

Thus we are in a set-up similar to the one of Sect. 2.3, where we were dealing with finite state space Markov chains. In particular we can still think of functions f :

$\Omega \to \mathbb{R}$ as column vectors and the action of generator is given by the multiplication of the matrix L with the vector f, i.e., $Lf(\eta) = \sum_{\eta' \in \Omega} L(\eta, \eta') f(\eta')$. The main difference compared to Sect. 2.3 is that now both the vectors and the matrices are infinite. In order to make this chapter self-contained we briefly recall the results of Sect. 2.3 by reformulating them for countable state space Markov chains.

We recall (cf. Definition 2.7) that a measure M on a countable set Ω is said to be reversible for the Markov process with generator L if we have for all $\xi, \eta \in \Omega$ that

$$M(\xi)L(\xi, \eta) = M(\eta)L(\eta, \xi). \tag{3.37}$$

The following proposition shows that a family of (un-normalized) Poisson product measures are reversible measures for independent particles. In what follows we use the convention $0! = 1$.

Proposition 3.7 (Reversible Measure of Independent Walkers) *For all $\rho > 0$ we have that the measure*

$$M(\xi) = \prod_{x \in V} \frac{\rho^{\xi_x}}{\xi_x!} \tag{3.38}$$

is reversible for independent walkers on a finite set V.

Proof This follows from the fact that for all $n, m \in \mathbb{N}, n \geq 1$

$$\frac{\rho^n}{n!} \frac{\rho^m}{m!} n = \frac{\rho^{n-1}}{(n-1)!} \frac{\rho^{m+1}}{(m+1)!} (m+1), \tag{3.39}$$

which implies (3.37). □

Remark 3.8 (Invariant Measures) As a consequence of the previous proposition, the homogeneous Poisson product measures are invariant, but of course not ergodic, because the total number of particles is preserved. In fact, starting the process from a configuration $\eta \in \mathbb{N}^V$ with $\sum_{x \in V} \eta_x = N$, the law of the process $\{\eta(t) : t \geq 0\}$ converges as $t \to \infty$ to the multinomial distribution

$$\nu^{(N)}(\eta) = \frac{1}{Z_{N,V}} \frac{N!}{\prod_{x \in V} \eta_x!}, \tag{3.40}$$

where $Z_{N,V} = |V|^N$ is the normalization constant.

We recall (cf. Definition 2.2) that a generator L is self-dual with self-duality function $D : \Omega \times \Omega \to \mathbb{R}$ if, for all $\xi, \eta \in \Omega$, we have

$$LD(\cdot, \eta)(\xi) = LD(\xi, \cdot)(\eta). \tag{3.41}$$

For Markov processes with a countable state space, such as independent random walkers on a finite graph, the duality function D can be viewed as an infinite matrix indexed by elements of Ω. Then (3.41) reads in matrix notation

$$LD = DL^T, \tag{3.42}$$

where L^T denotes the transposed of L.

The following result, which generalizes and combines Theorem 2.9 and Theorem 2.11, shows how to find self-duality functions from reversible measures and symmetries.

Theorem 3.9 (Self-duality via Symmetries) *Let the generator L be defined on functions $f : \Omega \to \mathbb{R}$, where Ω is a countable state space. We have:*

1. *Cheap self-duality functions. If M is a positive reversible measure, i.e. $M(\xi) > 0$ for all $\xi \in \Omega$, then*

$$D_{cheap}(\xi, \eta) = \frac{1}{M(\xi)}\, \delta_{\xi,\eta} \tag{3.43}$$

is a self-duality function.

2. *From symmetries to new self-duality functions. If D is a self-duality function and S is an operator working on functions $f : \Omega \to \mathbb{R}$ which commutes with the generator L, then also SD and DS^T are self-duality functions.*
3. *And back. Conversely, if D is a self-duality function, then there exists an operator S commuting with L such that $D = SD_{cheap}$.*

Proof The proof of item 1. and 2. is analogous to the one of Theorem 2.9. Item 3. is obtained as in the proof of Theorem 2.11. □

We now use Theorem 3.9 to construct the basic self-duality function for independent random walkers on a finite set V.

Theorem 3.10 (Self-duality of Independent Random Walkers) *Define for $\eta, \xi \in \Omega$*

$$D(\xi, \eta) = \prod_{x \in V} d(\xi_x, \eta_x), \tag{3.44}$$

with

$$d(k, n) = \begin{cases} \frac{n!}{(n-k)!} & \text{if } k \le n, \\ 0 & \text{otherwise.} \end{cases} \tag{3.45}$$

Then we have that the independent random walkers generator L in (3.5) *is self-dual with self-duality function D given in* (3.44)*. As a consequence, for all* $\xi, \eta \in \Omega$ *we have*

$$\mathbb{E}_\xi D(\xi(t), \eta) = \mathbb{E}_\eta D(\xi, \eta(t)). \tag{3.46}$$

Proof We apply Theorem 3.9. We start from the cheap self-duality function that is obtained from the reversible measure (3.38) with $\rho = 1$, i.e.

$$D_{\text{cheap}}(\xi, \eta) = \prod_{x \in V} d_{\text{cheap}}(\xi_x, \eta_x), \tag{3.47}$$

with

$$d_{\text{cheap}}(k, n) = k!\delta_{k,n}. \tag{3.48}$$

Since we want to obtain a self-duality function in factorized form, the symmetry we will choose will also be in factorized form. Therefore, we choose

$$S = e^{S^+} = \prod_{x \in V} e^{a_x^\dagger}, \tag{3.49}$$

where S^+ is defined in (3.30). Then from items 1 and 2 of Theorem 3.9 we have that

$$D := SD_{\text{cheap}}$$

is again a self-duality. To show that such D is of the form given in (3.44) and (3.45), it remains to show that

$$(e^{a^\dagger} d_{\text{cheap}}(\cdot, n))(k) = d(k, n). \tag{3.50}$$

For a function $f : \mathbb{N} \to \mathbb{R}$, using $a^\dagger f(k) = f(k+1)$, we find

$$(e^{a^\dagger} f)(k) = \sum_{\ell \in \mathbb{N}} \frac{f(k+\ell)}{\ell!}. \tag{3.51}$$

Combining together (3.48) and (3.51) we have

$$(e^{a^\dagger} d_{\text{cheap}}(\cdot, n))(k) = \sum_{\ell \in \mathbb{N}} \frac{d_{\text{cheap}}(k+\ell, n)}{\ell!} \tag{3.52}$$

$$= \sum_{\ell \in \mathbb{N}} \frac{(k+\ell)!\delta_{k+\ell,n}}{\ell!} \tag{3.53}$$

$$= \frac{n!}{(n-k)!} \mathbb{1}_{\{n \geq k\}}, \tag{3.54}$$

from which (3.50) follows. We thus have

$$LD(\cdot, \eta)(\xi) = LD(\xi, \cdot)(\eta). \tag{3.55}$$

Since we are working on a countable state space Ω and the particle number is conserved we can exponentiate this relation and find

$$S_t D(\cdot, \eta)(\xi) = S_t D(\xi, \cdot)(\eta), \tag{3.56}$$

for all $\xi, \eta \in \Omega$ and for all $t \geq 0$. This amounts to (3.46). □

The following simple proposition, establishes the relation between the self-duality function and the homogeneous Poisson product measure ν_λ with parameter $\lambda > 0$.

Proposition 3.11 (Expectation of the Self-duality Function in the Poisson Product Measure) *Let D be the self-duality function defined in Theorem 3.10. For all $\xi \in \Omega$ and for all $\lambda > 0$, we have*

$$\int D(\xi, \eta)\nu_\lambda(d\eta) = \lambda^{|\xi|} = \left(\int D(\delta_x, \eta)\nu_\lambda(d\eta)\right)^{|\xi|} \qquad \textit{for all } x \in V, \tag{3.57}$$

where we use the notation $|\xi| = \sum_{x\in V} \xi_x$.

Proof This immediately follows from the formula for the factorial moments of the Poisson distribution: for all $k \in \mathbb{N}$

$$\sum_{n=k}^{\infty} \frac{n!}{(n-k)!} \frac{\lambda^n}{n!} e^{-\lambda} = \lambda^k. \tag{3.58}$$

□

3.4 Self-duality as a Change of Representation

It is instructive to see that both the trivial (diagonal) self-duality associated to reversibility, as well as the triangular self-duality obtained from the symmetry $e^{S^+} = \prod_{x\in V} e^{a_x^\dagger}$, are consequences of more fundamental dualities relating the generators of the Heisenberg algebra. The dualities at the algebra level can in turn can be seen as a change of representation.

Proposition 3.12 (Cheap Self-duality as a Change of Representation) *Consider the representation of the conjugate Heisenberg algebra*

$$af(n) = nf(n-1),$$

$$a^\dagger f(n) = f(n+1)\,. \tag{3.59}$$

Then, as already remarked in (2.151) *and* (2.152) *one has the duality relations*

$$a \xrightarrow{d_{cheap}} a^\dagger, \quad a^\dagger \xrightarrow{d_{cheap}} a \tag{3.60}$$

with

$$d_{cheap}(k,n) = n!\delta_{k,n}\,. \tag{3.61}$$

As a consequence the process of independent random walkers with generator L in (3.5) *is self-dual with self-duality function*

$$D_{cheap}(\xi,\eta) = \prod_{x\in V} d_{cheap}(\xi_x,\eta_x)\,. \tag{3.62}$$

Proof The derivation of the process self-duality with self-duality function (3.62) from the dualities of the Heisenberg algebra generators follows from the composition rule for dualities described in Theorem 2.23. Indeed the "word"

$$(a_x - a_y)(a_x^\dagger - a_y^\dagger) \tag{3.63}$$

is not changed by replacing each element with its dual and inverting the order. □

Proposition 3.13 (Triangular Self-duality as a Change of Representation) *Consider the operators* $\mathrm{a}, \mathrm{a}^\dagger$ *defined on functions* $f:\mathbb{N}\to\mathbb{R}$ *by*

$$\begin{aligned}\mathrm{a} f(k) &= f(k) + kf(k-1),\\ \mathrm{a}^\dagger f(k) &= f(k+1).\end{aligned} \tag{3.64}$$

They form a representation of the conjugate Heisenberg algebra:

$$[\mathrm{a},\mathrm{a}^\dagger] = -I\,. \tag{3.65}$$

Furthermore the operators $\mathrm{a}, \mathrm{a}^\dagger$ *satisfy a duality relation with the operators* $a, a^\dagger$ *given in* (3.59)*:*

$$\mathrm{a} \xrightarrow{d} a^\dagger, \quad \mathrm{a}^\dagger \xrightarrow{d} a, \tag{3.66}$$

with

$$d(k,n) = \frac{n!}{(n-k)!}\mathbb{1}_{\{n\ge k\}}\,. \tag{3.67}$$

As a consequence independent random walkers with generator L in (3.5) are self-dual, with self-duality function

$$D(\xi, \eta) = \prod_{x \in V} d(\xi_x, \eta_x). \tag{3.68}$$

Proof We first verify that the operators a, $a^\dagger$ form a representation of the conjugate Heisenberg algebra. We have

$$\mathrm{a}\mathrm{a}^\dagger f(k) = \mathrm{a}^\dagger f(k) + k\mathrm{a}^\dagger f(k-1) = f(k+1) + kf(k), \tag{3.69}$$

and

$$\mathrm{a}^\dagger \mathrm{a} f(k) = \mathrm{a} f(k+1) = f(k+1) + (k+1) f(k). \tag{3.70}$$

Taking the difference it gives (3.65).

Next we verify that the operators a, $\mathrm{a}^\dagger$ satisfy a duality relation with the operators $a^\dagger, a$. We have

$$\begin{aligned}(\mathrm{a}d(\cdot, n))(k) &= d(k, n) + kd(k-1, n) \\ &= \frac{n!}{(n-k)!} + k\frac{n!}{(n-k+1)!} \\ &= \frac{(n+1)!}{(n+1-k)!},\end{aligned} \tag{3.71}$$

and

$$\begin{aligned}(a^\dagger d(k, \cdot))(n) &= d(k, n+1) \\ &= \frac{(n+1)!}{(n+1-k)!}.\end{aligned} \tag{3.72}$$

This shows that a $\xrightarrow{d} a^\dagger$. Similarly

$$\begin{aligned}(\mathrm{a}^\dagger d(\cdot, n))(k) &= d(k+1, n) \\ &= \frac{n!}{(n-k-1)!},\end{aligned} \tag{3.73}$$

and

$$\begin{aligned}(ad(k, \cdot))(n) &= nd(k, n-1) \\ &= \frac{n(n-1)!}{(n-1-k)!}.\end{aligned} \tag{3.74}$$

Thus $\mathrm{a}^\dagger \xrightarrow{d} a$ and the proof of (3.66) and (3.67) is concluded.

The derivation of the process self-duality with self-duality function (3.68) from the dualities of the algebra generators follows from the composition rule for dualities: starting from the sequence

$$(a_x - a_y)(a_x^\dagger - a_y^\dagger)\,, \tag{3.75}$$

replacing each element with its dual and inverting the order, one gets

$$(\mathrm{a}_x - \mathrm{a}_y)(\mathrm{a}_x^\dagger - \mathrm{a}_y^\dagger)\,, \tag{3.76}$$

which is equal $(a_x - a_y)(a_x^\dagger - a_y^\dagger)$ because

$$\mathrm{a} = a + I, \qquad\qquad \mathrm{a}^\dagger = a^\dagger. \tag{3.77}$$

□

3.5 Infinite Configurations

The self-duality derived so far for independent random walkers on a finite set extends to independent random walkers on a countable infinite set V. In particular we will be mostly interested in the case $V = \mathbb{Z}^d$. The extension of self-duality is strictly related to the proof of existence of the process with infinitely many particles. In this section we first review the standard process construction by increasing volume limits and then we address the self-duality with infinitely many particles.

Let $p : V \times V \to [0, \infty)$ be an irreducible transition function, and let η_x particles be placed initially at each site $x \in V$. The particles evolve independently, and we denote by $\eta_x(t)$ the number of particles at time $t > 0$ defined as in (3.1). In order for the individual walker to be well-defined we require

$$\sup_{x\in V} \sum_{y\in V} p(x, y) < \infty. \tag{3.78}$$

This guarantees that, with probability 1, the single particle will not escape to infinity in a finite time. As a consequence, under this condition the process of independent random walkers on the countable set V initialized with finitely many particles is also well defined.

In order to avoid explosions in finite time when the process of independent particles on V is initialized with infinitely many particles (explosion meaning that at a certain time there are infinitely many particles at a given site), one has to select the initial configuration from a suitable subset of the configuration space. We will call this the "set of allowed configurations" and we will denote it by $\Omega_{\text{alw}} \subseteq \Omega = \mathbb{N}^V$. We first introduce a truncation argument that will be useful in the process construction.

Process Started from a Truncated Configurations Let $(V_n)_{n\in\mathbb{N}}$ be an approximating increasing sequence of finite volumes. i.e. $V_n \nearrow V$ as $n \to \infty$. For $\eta \in \Omega$ and $n \in \mathbb{N}$ we denote by η^{V_n} the truncated configuration which coincides with η on V_n and is zero outside V_n. We denote by $\{\eta^{V_n}(t),\ t \geq 0\}$ the process of independent random walkers moving on V starting from η^{V_n}. For every $n \in \mathbb{N}$ this is a well defined process since the number of walkers is finite.

Increasing Limit We say that the process $\{\eta(t),\ t \geq 0\}$ with infinitely many particles is the increasing limit of the sequence of processes started from truncated configurations if there exists a coupling such that almost surely in this coupling we have $\eta^{V_n}(t) \leq \eta^{V_{n'}}(t)$ for all $t \geq 0$ and for $n \leq n'$ (where $\eta \leq \eta'$ means $\eta_x \leq \eta'_x$ for all $x \in V$). We denote this increasing limit by $\eta^{V_n}(t) \nearrow \eta(t)$ as $n \to \infty$.

Notice that at this point the limit $\{\eta(t) : t \geq 0\}$ exists but might still be infinite at certain vertices $x \in V$. In order to prevent this we have to choose initial configurations from a suitable set.

Definition 3.14 (Allowed Configurations) Let $\Omega = \mathbb{N}^V$, we define a set of allowed configurations Ω_{alw} to be any subset of Ω satisfying the following properties:

(i) the process $\{\eta(t),\ t \geq 0\}$ of independent random walkers starting from $\eta \in \Omega_{\text{alw}}$ is well-defined and is the increasing limit of $\eta^{V_n}(t)$ for a sequence $V_n \nearrow V$;
(ii) if $\eta \in \Omega_{\text{alw}}$, then $\eta(t) \in \Omega_{\text{alw}}$ for all $t \geq 0$.

As a consequence of Definition 3.14 the existence of the process with infinitely many particles amounts to showing that Ω_{alw} is a sufficiently rich set. Several choices for the set of allowed configurations have been proposed in the literature, see for instance [3].

Remark 3.15 (Example: $p(x, y)$ Translation Invariant) Following [71], we here give more details for the case $V = \mathbb{Z}^d$ and $p(x, y) = \pi(y - x)$ that is translation invariant and has finite second moment $\sum_{x\in\mathbb{Z}^d} \|x\|^2\pi(x) < \infty$. In this case we may choose Ω_{alw} as the subset of configurations growing at most polynomially

$$\Omega_{\text{alw}} = \cup_{c>0,n\in\mathbb{N}}\{\eta \in \Omega : \ \eta_x \leq c(\|x\|^n + 1)\ \forall x \in \mathbb{Z}^d\}. \tag{3.79}$$

Remark 3.16 (Process Generator) We do not address here the description of the process generator and its domain. Intuitively, the "generator" of the process of independent random walkers with infinitely many particles should be the analogue of the generator that we defined before in finite volume, i.e.,

$$Lf(\eta) = \sum_{x,y\in V} p(x, y)\eta_x(f(\eta^{x,y}) - f(\eta)), \tag{3.80}$$

but now acting on local functions f and evaluated in allowed configurations η. The relation between this informal generator and the semigroup $\{S(t) : t \geq 0\}$ of the process $\{\eta(t) : t \geq 0\}$ is pointwise and restricted to the set of allowed configurations, i.e.

$$\lim_{t \to 0} \frac{S(t)f(\eta) - f(\eta)}{t} = Lf(\eta), \tag{3.81}$$

for all f local bounded functions (i.e., depending only on a finite number of $\eta_x, x \in V$) and $\eta \in \Omega_{\text{alw}}$. Moreover, we have the analogue of the usual relation $\frac{d}{dt}S(t) = LS(t)$, namely:

$$S(t)f(\eta) - f(\eta) = \int_0^t LS(s)f(\eta)ds = \int_0^t S(s)Lf(\eta)ds. \tag{3.82}$$

See [3] for a detailed account on these issues.

Duality via Truncated Configurations Let $\Omega_{\text{finite}} := \{\xi \in \Omega : \sum_{x \in V} \xi_x < \infty\}$ denote the set of finite configurations. Then, via truncated configurations, one obtains the self-duality for the system with infinitely many particles. This is shown in the following

Proposition 3.17 (Infinitely-Many Particles Self-duality) *For every $\xi \in \Omega_{finite}$, $\eta \in \Omega_{alw}$ we have the self-duality relation*

$$\mathbb{E}_\eta D(\xi, \eta(t)) = \mathbb{E}_\xi D(\xi(t), \eta). \tag{3.83}$$

where D is the function defined in (3.44) *and* (3.45).

Proof We will exploit the monotonicity property of the duality function with respect to the second variable, i.e. $D(\xi, \eta) \le D(\xi, \eta')$ for $\eta \le \eta'$. Let $\mathbb{E}_\eta^c$ denote the expectation with respect to the entire sequence of approximating processes $\{\eta^{V_n}(t),\ t \ge 0, n \in \mathbb{N}\}$ started from the truncations η^{V_n} of $\eta \in \Omega_{\text{alw}}$ and coupled in such a way to have monotonic convergence $\eta^{V_n}(t) \nearrow \eta(t)$ (i.e. $\eta^{V_n}(t) \le \eta^{V_{n'}}(t)$ for $n \le n', t \ge 0$ almost-surely in this coupling). From monotonicity of $D(\xi, \cdot)$ and monotonicity of $\eta^{V_n}(t)$, we have that

$$D(\xi, \eta^{V_n}(t)) \le D(\xi, \eta^{V_{n'}}(t)), \qquad \text{for} \quad n \le n' \quad \text{a.s..} \tag{3.84}$$

Then, by the fact the $\eta^{V_n}(t) \nearrow \eta(t)$ we have

$$D(\xi, \eta(t)) := \lim_{n \to \infty} D(\xi, \eta^{V_n}(t)) \quad \text{a.s..} \tag{3.85}$$

Hence we have

$$\begin{aligned}
\lim_{n \to \infty} \mathbb{E}_{\eta^{V_n}} \left[D(\xi, \eta^{V_n}(t)) \right] &= \lim_{n \to \infty} \mathbb{E}_\eta^c \left[D(\xi, \eta^{V_n}(t)) \right] \\
&= \mathbb{E}_\eta^c \left[\lim_{n \to \infty} D(\xi, \eta^{V_n}(t)) \right]
\end{aligned}$$

$$= \mathbb{E}^c_\eta \left[D(\xi, \eta(t)) \right]$$
$$= \mathbb{E}_\eta \left[D(\xi, \eta(t)) \right], \tag{3.86}$$

where the second equality is justified by the monotone convergence theorem. Similarly we define $D(\xi(t), \eta)$ as the (monotonic) limit of $\lim_{n\to\infty} D(\xi(t), \eta^{V_n})$ and thus we can write

$$\lim_{n\to\infty} \mathbb{E}_\xi \left[D(\xi(t), \eta^{V_n}) \right] = \mathbb{E}_\xi \left[\lim_{n\to\infty} D(\xi(t), \eta^{V_n}) \right] = \mathbb{E}_\xi \left[D(\xi(t), \eta) \right]. \tag{3.87}$$

Now the duality in infinite volume (3.83) follows from the duality identity for the approximating processes

$$\mathbb{E}_{\eta^{V_n}} D(\xi, \eta^{V_n}(t)) = \mathbb{E}_\xi D(\xi(t), \eta^{V_n}) \qquad \forall \xi \in \Omega_{\text{finite}},\ \eta \in \Omega_{\text{alw}}, \tag{3.88}$$

by taking the limit $n \to \infty$. □

Remark 3.18 (Finiteness) Proposition 3.17 does not exclude the possibility that both expectations in (3.83) are infinite. The finiteness of expectations can be verified in the setting of translation invariant transition rates of [71] with Ω_{alw} as in (3.79). In such setting, additionally to the statement (3.83), we have that the r.h.s. satisfies

$$\mathbb{E}_\xi D(\xi(t), \eta) = \sum_{\xi'} D(\xi', \eta) p_t(\xi, \xi') < \infty, \tag{3.89}$$

and hence also the l.h.s. $\mathbb{E}_\eta D(\xi, \eta(t)) < \infty$.

Remark 3.19 (Process Construction via Duality) In view of Proposition 3.17, one can construct the process with infinitely many particles via duality. More precisely, we can *define* the expectations $\mathbb{E}_\eta D(\xi, \eta(t))$ for an infinite configuration $\eta \in \Omega$ via

$$\mathbb{E}_\eta D(\xi, \eta(t)) := \mathbb{E}_\xi D(\xi(t), \eta) = \sum_{\xi' \in \Omega_{\text{finite}}} p_t(\xi, \xi') D(\xi', \eta), \tag{3.90}$$

where $p_t(\xi, \xi')$ denotes the transition probabilities in the finite process. Of course the definition (3.90) only makes sense when

$$\sum_{\xi' \in \Omega_{\text{finite}}} p_t(\xi, \xi') D(\xi', \eta) < \infty. \tag{3.91}$$

This condition, that only depends on the initial configuration η and the countable state space Markov process $\{\xi(t), t \geq 0\}$, suggests to define a new set of allowed configurations that are the ones for which (3.91) holds for all $t \geq 0$ and $\xi \in \Omega_{\text{finite}}$. Hence, in this way we can *construct the infinite volume process* from the process with only finitely many particles by means of the self-duality relation. The expectations $\mathbb{E}_\eta D(\xi, \eta(t))$, $\xi \in \Omega_{\text{finite}}$ fix indeed all the moments of the process $\{\eta(t),\ t \ge 0\}$, and this in turn defines uniquely the process provided that these moments do not grow too fast (e.g. they satisfy the Carleman's condition). This is particularly useful when the existence of the process $\{\eta(t),\ t \ge 0\}$ cannot be obtained from general methods such as the construction of the semigroup via the Hille-Yosida theorem or monotonicity process.

We can use now the self-duality relation (3.83) to analyze the properties of the infinite-volume process $\{\eta(t),\ t \ge 0\}$ with infinitely many particles via the process $\{\xi(t),\ t \ge 0\}$ with finitely many particles. In particular, in the next section, we establish the following properties:

1. the infinite-volume process $\{\eta(t),\ t \ge 0\}$ has homogeneous products of Poisson distributions as reversible and ergodic measures;
2. these product Poisson distributions are the only ergodic distributions within a class of so-called tempered measures (see below for a precise definition). Intuitively the condition of temperedness means that all moments are finite and do not grow too fast as a function of the number of dual particles. The growth condition is needed in order to ensure that the moments uniquely determine the measure.

The restriction to finite moments distributions arises from the fact that we have to consider starting measures for which expectations of the self-duality functions are well-defined. We will extend this result in Sect. 4.6 by using duality of independent walkers with a deterministic system. There it will be shown that Poisson product measures are the only ergodic distributions.

3.6 The Set of Ergodic Tempered Measures in $\mathbb{Z}^d$

Our aim in this section is to show—using self-duality—that for the infinite system of independent random walkers on the lattice $\mathbb{Z}^d$ with a translation invariant $p(x, y)$ having a finite second moment, the homogeneous Poisson product measure ν_λ with parameter $\lambda > 0$ and marginals

$$\nu_\lambda(\eta_x = k) = \frac{\lambda^k}{k!} e^{-\lambda}, \qquad x \in \mathbb{Z}^d, k \in \mathbb{N}, \tag{3.92}$$

are the only measures that are invariant and ergodic, at least within a proper set of tempered measures.

We start with the following analogue of the Proposition 3.11 in the infinite volume setting, whose proof is identical to the finite volume case.

Proposition 3.20 (Expectation of the Self-duality Function in the Poisson Product Measure) *Let D be the self-duality function defined in Theorem 3.10. For all $\xi \in \Omega_{finite}$ and for all $\lambda > 0$, we have*

$$\int D(\xi, \eta)\nu_\lambda(d\eta) = \lambda^{|\xi|} = \left(\int D(\delta_0, \eta)\nu_\lambda(d\eta)\right)^{|\xi|}, \tag{3.93}$$

where we use the notation $|\xi| = \sum_{x\in\mathbb{Z}^d} \xi_x$.

As a consequence we obtain the following:

Theorem 3.21 (Invariance and Ergodicity of Product Poisson Measures) *The homogeneous Poisson product measures ν_λ with parameter $\lambda > 0$ are invariant and ergodic for the independent random walkers process on $\mathbb{Z}^d$.*

Proof To prove invariance, we start from the self-duality integrated against Poisson product measures

$$\int \mathbb{E}_\eta D(\xi, \eta(t))\nu_\lambda(d\eta) = \int \mathbb{E}_\xi D(\xi(t), \eta)\nu_\lambda(d\eta) \qquad \text{for} \quad \xi \in \Omega_{\text{finite}}.$$

Using Fubini theorem in the r.h.s. and applying Proposition 3.20 we have

$$\int \mathbb{E}_\eta D(\xi, \eta(t))\nu_\lambda(d\eta) = \mathbb{E}_\xi \lambda^{|\xi(t)|} = \lambda^{|\xi|}, \tag{3.94}$$

where the last equality follows from conservation of the number of particles in the process $\{\xi(t), t \geq 0\}$. Another application of the Proposition 3.20 yields $\lambda^{|\xi|} = \int D(\xi, \eta)\nu_\lambda(d\eta)$. which, together with (3.94) gives

$$\int \mathbb{E}_\eta D(\xi, \eta(t))\nu_\lambda(d\eta) = \int D(\xi, \eta)\nu_\lambda(d\eta). \tag{3.95}$$

Because the functions $D(\xi, \cdot)$ in Theorem 3.10 are measure determining, we conclude from (3.95) the invariance of the Poisson product measure ν_λ.

In what follows we prove mixing which implies ergodicity. To prove mixing, by density of the vector space spanned by linear combinations of the functions $D(\xi, \cdot)$ in $L^2(\nu_\lambda)$, it suffices to show that for all ξ, ξ' finite configurations, we have

$$\lim_{t\to\infty} \int \mathbb{E}_\eta D(\xi, \eta(t))D(\xi', \eta)\nu_\lambda(d\eta) = \int D(\xi, \eta)\, \nu_\lambda(d\eta) \int D(\xi', \eta)\, \nu_\lambda(d\eta). \tag{3.96}$$

By Proposition 3.20 this is equivalent to proving

$$\lim_{t\to\infty} \int \mathbb{E}_\eta D(\xi, \eta(t))D(\xi', \eta)\nu_\lambda(d\eta) = \lambda^{|\xi|+|\xi'|}, \tag{3.97}$$

for all $\xi, \xi' \in \Omega_{\text{finite}}$. Now, using self-duality and Fubini, we have

$$\int \mathbb{E}_\eta D(\xi, \eta(t)) D(\xi', \eta) \nu_\lambda(d\eta) = \mathbb{E}_\xi \int D(\xi(t), \eta) D(\xi', \eta) \nu_\lambda(d\eta). \tag{3.98}$$

Denote $\xi(t) \perp \xi'$ the event that the supports of $\xi(t)$ and ξ' are disjoint. If this event occurs then we have

$$D(\xi(t), \eta) D(\xi', \eta) = D(\xi(t) + \xi', \eta), \tag{3.99}$$

and $|\xi(t) + \xi'| = |\xi(t)| + |\xi'| = |\xi| + |\xi'|$. Hence if $\xi(t) \perp \xi'$ then Proposition 3.20 implies

$$\mathbb{1}_{\{\xi(t) \perp \xi'\}} \int D(\xi(t), \eta) D(\xi', \eta) \nu_\lambda(d\eta) = \mathbb{1}_{\{\xi(t) \perp \xi'\}} \lambda^{|\xi| + |\xi'|}, \tag{3.100}$$

where the statement holds $\mathbb{P}_\xi$-almost surely. Thus, the proof of (3.97) would follow if we could prove that the probability that the event $\xi(t) \perp \xi'$ does not occur goes to zero in the limit of large times. By translation invariance, this will be a consequence of the fact that the continuous-time random walkers eventually spread out all over the lattice. Denoting by $p_t(x, y)$ the transition probability of one continuous-time random walk, we have the estimate

$$\mathbb{P}_\xi(\xi(t) \not\perp \xi') \leq \sum_{x,y} \xi_x \xi'_y p_t(x, y), \tag{3.101}$$

and the r.h.s. of this inequality tends to zero because $p_t(x, y)$ tends to zero for all x, y and the sum over x, y is a finite sum because ξ and ξ' are finite configurations. Hence we have $\mathbb{P}_\xi(\xi(t) \not\perp \xi') \to 0$ as $t \to \infty$. Therefore

$$\begin{aligned} &\left| \int \mathbb{E}_\eta D(\xi, \eta(t)) D(\xi', \eta) \nu_\lambda(d\eta) - \lambda^{|\xi| + |\xi'|} \right| \\ &\leq \mathbb{E}_\xi \left(\mathbb{1}_{\{\xi(t) \not\perp \xi'\}} \int D(\xi(t), \eta) D(\xi', \eta) \nu_\lambda(d\eta) \right) \\ &+ \mathbb{P}_\xi(\xi(t) \not\perp \xi') \, \lambda^{|\xi| + |\xi'|}. \end{aligned} \tag{3.102}$$

Now we use the uniform estimate

$$\sup_{\substack{|\xi| = n \\ |\xi'| = m}} \int D^2(\xi, \eta) D^2(\xi', \eta) \nu_\lambda(d\eta) \leq K_{n,m}, \tag{3.103}$$

for some constant $K_{n,m}$. This estimate is obtained by noting that $D^2(\xi, \eta) D^2(\xi', \eta)$ is a polynomial of degree at most $2n + 2m$, and $d(i, k) \leq k^i$ for all $i \leq k$. Then, using Cauchy-Schwarz we find

$$\mathbb{E}_{\xi}\left(\mathbb{1}_{\{\xi(t)\not\perp\xi'\}}\int D(\xi(t),\eta)D(\xi',\eta)\nu_{\lambda}(d\eta)\right)\leq\sqrt{K_{|\xi|,|\xi'|}}\sqrt{\mathbb{P}_{\xi}(\xi(t)\not\perp\xi')}. \tag{3.104}$$

Hence, we conclude

$$\lim_{t\to\infty}\left|\int\mathbb{E}_{\eta}D(\xi,\eta(t))D(\xi',\eta)\nu_{\lambda}(d\eta)-\lambda^{|\xi|+|\xi'|}\right|=0, \tag{3.105}$$

which proves (3.97). □

To further characterize the invariant measures of the process $\{\eta(t):t\geq 0\}$, we define a class of so-called tempered measures.

Definition 3.22 (Tempered Measures) We call a probability measure ν on Ω tempered if all the moments $\int D(\xi,\eta)\nu(d\eta)$ exist and moreover we have the Carleman growth condition ensuring the fact that ν is determined by these moments. More precisely we require that for all n

$$c_n := \sup_{|\xi|=n}\int D(\xi,\eta)\nu(d\eta)<\infty, \tag{3.106}$$

and we have the growth condition

$$\sum_{n\geq 1}(c_n)^{-1/n}=\infty. \tag{3.107}$$

We denote by $\mathscr{P}$ the class of all tempered probability measures on Ω.

Notice that by self-duality and conservation of the number of particles, this condition is preserved in time, i.e., if the process is started from a measure ν with constants c_n then, for all $t>0$, the evolved measure ν_t has the same constants. The condition (3.107) ensures that if $\nu,\nu'\in\mathscr{P}$ have the same expectations for all functions $D(\xi,\cdot)$, then they are equal. This means that on $\mathscr{P}$ the D-transform defined via

$$\hat{\nu}(\xi)=\int D(\xi,\eta)\nu(d\eta),\qquad \xi\in\Omega_{\text{finite}} \tag{3.108}$$

is well-defined and determines the measure, i.e., $\nu,\nu'\in\mathscr{P}$ and $\hat{\nu}=\widehat{\nu'}$ implies that $\nu=\nu'$.

The following theorem shows that invariance of a measure in $\mathscr{P}$ is equivalent with the measure having harmonic D-transform.

Theorem 3.23 (Invariant Measures and D-transform) *Let $\nu \in \mathcal{P}$. The following two statements are equivalent*

(i) ν is an invariant measure.
(ii) $\hat{\nu}$ is harmonic, i.e., for all $\xi \in \Omega_{finite}$ we have

$$\mathbb{E}_\xi \hat{\nu}(\xi(t)) = \hat{\nu}(\xi). \tag{3.109}$$

Proof If ν is invariant then for all $t \geq 0$

$$\hat{\nu}(\xi) = \int \mathbb{E}_\eta D(\xi, \eta(t))\nu(d\eta) = \mathbb{E}_\xi \hat{\nu}(\xi(t)), \tag{3.110}$$

where in the first equality we used invariance and in the second equality we used self-duality and Fubini. Conversely if $\hat{\nu}$ is harmonic then, by self-duality

$$\hat{\nu}(\xi) = \mathbb{E}_\xi \hat{\nu}(\xi(t)) = \int \mathbb{E}_\eta D(\xi, \eta(t))\nu(d\eta), \tag{3.111}$$

which implies that ν and ν_t, that are both in $\mathcal{P}$, have the same D-transform, and are therefore equal. □

Definition 3.24 (Successful Coupling) We say that there exists a successful coupling if for all ξ, ξ' with $|\xi| = |\xi'|$, there exists a coupling of the path space measures (i.e., a measure $P_{\xi,\xi'}$ on trajectories $\{(\xi^{(1)}(t), \xi^{(2)}(t)) : t \geq 0\}$ with marginals $\mathbb{P}_\xi$, and $\mathbb{P}_{\xi'}$) such that the coupling time

$$\tau := \inf\{T > 0 : \xi^{(1)}(t) = \xi^{(2)}(t) \quad \forall t > T\} \tag{3.112}$$

is $P_{\xi,\xi'}$ almost surely finite.

The following lemma shows that the existence of a successful coupling implies that the D-transform $\hat{\nu}$ is a function of the number $|\xi|$ of dual particles.

Lemma 3.25 (Successful Coupling and Invariant Measure) *If there exists a successful coupling, and ν is a tempered invariant measure, then $\hat{\nu}(\xi) = \hat{\nu}(\xi')$ if $|\xi| = |\xi'|$.*

Proof Assume that ν is invariant and let ξ, ξ' be such that $|\xi| = |\xi'|$. Then

$$\begin{aligned}
\hat{\nu}(\xi) &= \mathbb{E}_\xi \hat{\nu}(\xi(t)) \\
&= \mathbb{E}_{\xi,\xi'} \hat{\nu}(\xi^{(1)}(t)) \\
&= \mathbb{E}_{\xi,\xi'} \left(\hat{\nu}(\xi^{(2)}(t))\mathbb{1}_{\{\xi^{(1)}(t)=\xi^{(2)}(t)\}}\right) + \mathbb{E}_{\xi,\xi'} \left(\hat{\nu}(\xi^{(1)}(t))\mathbb{1}_{\{\xi^{(1)}(t)\neq\xi^{(2)}(t)\}}\right).
\end{aligned} \tag{3.113}$$

By (3.106) we may bound

$$\mathbb{E}_{\xi,\xi'}\left(\hat{\nu}(\xi^{(i)}(t))\mathbb{1}_{\{\xi^{(1)}(t)\neq\xi^{(2)}(t)\}}\right) \leq c_{|\xi|}\mathbb{P}_{\xi,\xi'}\left(\xi^{(1)}(t)\neq\xi^{(2)}(t)\right) \qquad i=1,2. \tag{3.114}$$

Therefore, since the coupling is successful, by dominated convergence, (3.113) implies

$$\begin{aligned}\hat{\nu}(\xi) &= \mathbb{E}_{\xi,\xi'}\left(\hat{\nu}(\xi^{(2)}(t))\right)+o(1)\\ &= \mathbb{E}_{\xi'}\hat{\nu}(\xi(t))+o(1)\\ &= \hat{\nu}(\xi')+o(1).\end{aligned} \tag{3.115}$$

The result follows by taking the limit $t\to\infty$. □

The successful coupling in the case of independent random walks moving on $\mathbb{Z}^d$ is well-known and we describe it here for the case of nearest neighbor random walk. The general case is similar. If $\xi = \sum_{i=1}^n \delta_{x_i}$ and $\xi' = \sum_{i=1}^n \delta_{x_i'}$, we couple the dn dimensional random walks starting at $(x_1,\ldots,x_n)$ and $(x_1',\ldots,x_n')$ by the so-called Ornstein coupling. This coupling, denoted by $(X_1(t),\ldots,X_k(t);Y_1(t),\ldots,Y_k(t))$, is described as follows. Initially $(X_1(0),\ldots,X_k(0))=(i_1,\ldots,i_k)$, $(Y_1(0),\ldots,Y_k(0))=(i_1',\ldots,i_k')$ with $k=nd$. Then the processes run independently until the coordinates $X_1(t)$ and $Y_1(t)$ are equal for the first time. This will happen with probability one because the difference $|X_1(t)-Y_1(t)|$ is a one-dimensional continuous-time nearest neighbor random walk, which is recurrent. From that moment on, we let $X_1(t)$ and $Y_1(t)$ perform identical jumps, and wait now for the first moment until $X_2(t)$ and $Y_2(t)$ are equal, etc., until eventually $X_i(t)=Y_i(t)$ for all $i=1,\ldots,nd$.

The following theorem shows that the only ergodic invariant measures, within the set of tempered measures, are the Poisson product measures.

Theorem 3.26 (Uniqueness of Tempered Ergodic Invariant Measure) *If ν is tempered, invariant and ergodic then $\nu=\nu_\lambda$ for some $\lambda>0$.*

Proof If ν is tempered and invariant, by the existence of a successful coupling we conclude that $\hat{\nu}(\xi)=f(|\xi|)$ for some function $f:\mathbb{N}\to\mathbb{R}$. We will show that $f(n+m)=f(n)f(m)$. This implies $f(n)=\lambda^n$ and hence by uniqueness of the D-transform we deduce via (3.93) that ν is the Poisson product measure ν_λ. Let $\xi\in\Omega_{\text{finite}}$ with $|\xi|=n$ and $\xi'\in\Omega_{\text{finite}}$ with $|\xi'|=m$. Denote by $S_t\varphi(\eta)=\mathbb{E}_\eta(\varphi(\eta(t)))$ the semigroup and $\mathcal{S}_T\varphi=\frac{1}{T}\int_0^T S_t\varphi dt$ the ergodic average. Then by the ergodic theorem

$$\mathcal{S}_T D(\xi,\cdot)\to f(n) \qquad \text{as} \quad T\to\infty, \qquad \nu-\text{a.s.} \tag{3.116}$$

Hence by dominated convergence, using (3.106),

$$\int D(\xi',\eta)\mathcal{S}_T D(\xi,\cdot)(\eta)\nu(d\eta) \to f(m)f(n) \qquad \text{as} \quad T \to \infty. \tag{3.117}$$

On the other hand, using duality we have

$$\int D(\xi',\eta)\mathcal{S}_T D(\xi,\cdot)(\eta)\nu(d\eta) = \frac{1}{T}\int_0^T \mathbb{E}_\xi \int D(\xi',\eta)D(\xi(s),\eta)\nu(d\eta)ds. \tag{3.118}$$

On the event $\xi(s) \perp \xi'$ we have $D(\xi',\eta)D(\xi(s),\eta) = D(\xi'+\xi(s),\eta)$ and hence because then also $|\xi'+\xi(s)| = n+m$ we have $\int D(\xi',\eta)D(\xi(s),\eta)\nu(d\eta) = f(n+m)$. Therefore, using that $\mathbb{P}_\xi(\xi(s) \perp \xi') \to 1$ as $s \to \infty$, and dominated convergence we conclude

$$\frac{1}{T}\int_0^T \mathbb{E}_\xi \int D(\xi',\eta)D(\xi(s),\eta)\nu(d\eta)ds \to f(n+m), \tag{3.119}$$

as $T \to \infty$. Combining (3.117), (3.118), (3.119) then yields

$$f(n+m) = f(n)f(m), \tag{3.120}$$

and hence $f(n) = \lambda^n$ with $\lambda = f(1)$. □

Remark 3.27 (Beyond Tempered Measures) Because we use self-duality we restrict necessarily to measures in $\mathcal{P}$. At this point, whether there exist invariant measures not in $\mathcal{P}$ is still an open problem. It is related to the problem of the possible limit points (in time) of the time-evolved distribution when started from a distribution for which e.g. only a finite number of moments exist, such as a distribution with power law tails (e.g. a discrete Pareto distribution). In that case, the limiting distribution cannot be characterized via self-duality. We shall solve this problem in Sect. 4.6 where we show that Poisson product measure are the only ergodic invariant measure in $\mathbb{Z}^d$. In other words we can remove the assumption of temperedness by switching from the use of self-duality to the use of duality with uniformly bounded duality functions.

3.6.1 Invariant Measures Ergodic Under Translations

We have shown that for all tempered measures ν which are invariant and ergodic there is the relation

$$\int D(\xi,\eta)\nu(d\eta) = \lambda^{|\xi|} = \left(\int D(\delta_0,\eta)\nu(d\eta)\right)^{|\xi|}, \tag{3.121}$$

and as consequence ν is a homogeneous product Poisson measure: $\nu = \nu_\lambda$. Now we show that the same holds for the invariant tempered measures that are *ergodic under translations*. Hence every such measure is a homogeneous product Poisson measure, and thus also automatically ergodic under the time evolution of the independent walkers.

Theorem 3.28 (Stationary Measure Ergodic Under Translations) *Let ν be an invariant measure which is ergodic under translations and tempered. Then we have* (3.121)*. Hence ν is the Poisson product measure ν_λ with $\lambda = \int D(\delta_0, \eta)\nu(d\eta)$ and, as a consequence, ν is also ergodic for the process $\{\eta(t) : t \geq 0\}$.*

Proof Since ν is invariant, we have that its D-transform $\hat{\nu}$ is harmonic. Because for independent random walkers we have a successful coupling and ν is tempered, we conclude $\hat{\nu}(\xi) = f(|\xi|)$ for some function $f : \mathbb{N} \to \mathbb{R}$. But this implies that for all $x_1, \ldots, x_n$ and $y_1, \ldots, y_n \in \mathbb{Z}^d$ we have

$$\int D\left(\sum_{i=1}^{n} \delta_{x_i}, \eta\right) \nu(d\eta) = \int D\left(\sum_{i=1}^{n} \delta_{y_i}, \eta\right) \nu(d\eta). \tag{3.122}$$

Put $\Lambda_N = [-N, N]^d \cap \mathbb{Z}^d$, and $\lambda = \int D(\delta_0, \eta)\nu(d\eta)$. By the Birkhoff ergodic theorem we have, ν a.s.,

$$\frac{1}{|\Lambda_N|} \sum_{y \in \Lambda_N} D(\delta_y, \eta) \to \lambda \qquad \text{as } N \to \infty. \tag{3.123}$$

As a consequence of (3.122), for all $N \in \mathbb{N}$, we have

$$\int D\left(\sum_{i=1}^{n} \delta_{x_i}, \eta\right) \nu(d\eta) \tag{3.124}$$

$$= \frac{1}{\left|\{y_1, \ldots, y_n \in \Lambda_N : y_1 \neq \ldots \neq y_n\}\right|} \cdot \sum_{\substack{y_1, \ldots, y_n \in \Lambda_N \\ y_1 \neq \ldots \neq y_n}} \int \prod_{i=1}^{n} D(\delta_{y_i}, \eta)\nu(d\eta).$$

Since for all $n \in \mathbb{N}$ it holds

$$\lim_{N \to \infty} \frac{\left|\{y_1, \ldots, y_n \in \Lambda_N\}\right|}{\left|\{y_1, \ldots, y_n \in \Lambda_N : y_1 \neq \ldots \neq y_n\}\right|} = 1, \tag{3.125}$$

then we obtain

$$\begin{aligned}
&\int D\left(\sum_{i=1}^{n} \delta_{x_i}, \eta\right) \nu(d\eta) \\
&= \lim_{N\to\infty} \frac{1}{\left|\{y_1, \ldots, y_n \in \Lambda_N : y_1 \neq \ldots \neq y_n\}\right|} \cdot \sum_{\substack{y_1,\ldots,y_n \in \Lambda_N \\ y_1 \neq \ldots \neq y_n}} \int \prod_{i=1}^{n} D(\delta_{y_i}, \eta) \nu(d\eta) \\
&= \lim_{N\to\infty} \left(\frac{1}{|\Lambda_N|}\right)^n \sum_{y_1,\ldots,y_n \in \Lambda_N} \int \prod_{i=1}^{n} D(\delta_{y_i}, \eta) \nu(d\eta) \\
&= \lambda^n, \qquad (3.126)
\end{aligned}$$

where, in the last identity we used the ergodic theorem (3.123), dominated convergence and the fact that ν is tempered. □

3.7 Additional Notes

The reformulation of reaction-diffusion systems in terms of creation and annihilation operators stems originally from Doi and Peliti see [193], [81], [1]. In the physics literature, this formalism is known under names such as Doi-Peliti field theory, and path integral approach to reaction diffusion systems. The symmetries of the random walk generator and the corresponding dualities have been introduced in [116]. In [71] self-duality for independent random walkers is proved, using direct computations with the generator. In [71] a proof of the existence of the infinite systems of independent random walkers starting from appropriate initial configurations (growing at most polynomially at infinity) is given. In [71] duality is also used to prove macroscopic properties of reaction-diffusion systems including hydrodynamic limits and fluctuations around the hydrodynamic limit.

The use of duality combined with coupling to prove properties of the set of invariant measures, via a characterization of bounded harmonic functions of the dual process, was used in [172], Chapter 8, for the symmetric exclusion process. In the setting of independent random walkers, [3] uses monotonicity and coupling to characterize invariant measures of monotone zero range processes, which include independent random walkers. In the context of more general particle systems with duality, recently in [200] a general characterization of ergodic tempered measures is given, using the methodology of this chapter.

Chapter 4
Duality for Independent Random Walkers

In this chapter, we continue the exploration of dualities for independent random walkers. By taking the "many particle" limit, a *deterministic process* (a system of ODEs) arises as a dual process, with a duality function which is also obtained by a proper rescaling of the self-duality function of Chap. 3. This deterministic process is in turn self-dual with a very simple duality function. From the algebraic perspective, we show that behind all these dualities there is always the same abstract object, which is written in terms of the *Heisenberg algebra* generators. The dualities then arise by considering several representations of the algebra. From an analytic perspective we introduce *generating functions* to show the equivalence of all these dualities. When interpreted as Poisson averaging, the generating function also serves as *an intertwining operator*. Furthermore, the use of generating functions gives a full classification of all product self-dualities for independent random walkers, which can essentially be of two types: either the triangular single-site self-dualities of Chap. 3 or self-dualities involving Charlier polynomials. We close the chapter with two applications. First, by using the duality with the deterministic system, we complete the *ergodic theory* of independent random walkers on the infinite lattice $\mathbb{Z}^d$, removing the restrictive assumption of an appropriate moment growth condition that was necessary in Chap. 3. Second, for asymmetric random walkers, we state duality with the reversed process and use this to compute the joint moment generating function of currents along edges.

4.1 Many Particle Limit and New Dualities

Consider the independent random walk process $\{\eta(t), t \geq 0\}$ on a finite vertex set V driven by an irreducible and symmetric transition function $p(x, y)$ where $x, y \in V$. We define a sequence of initial configurations $\eta^{(N)}$, $N \in \mathbb{N}$ with a total number of

C. Giardinà, F. Redig, *Duality for Markov Processes*, Grundlehren der mathematischen Wissenschaften 365, https://doi.org/10.1007/978-3-032-04099-2_4

particles of order N, i.e. for all $x \in V$ and $N \in \mathbb{N}$,

$$\eta_x^{(N)} = \lfloor N\zeta_x \rfloor, \tag{4.1}$$

for some configuration $\zeta : V \to [0, +\infty)^V$. This implies that $\eta_x^{(N)}/N \to \zeta_x$ as $N \to \infty$. We then have the following result.

Theorem 4.1 (Scaling limit and duality of independent random walkers) *As $N \to \infty$, the process $\{\eta^{(N)}(t)/N : t \geq 0\}$ weakly converges (in the Skorohod topology) to a deterministic process $\{\zeta(t) : t \geq 0\}$ on $[0, \infty)^V$ which is the solution of the following system of linear ODE's:*

$$\frac{d\zeta_x(t)}{dt} = \sum_{y \in V} p(x, y)(\zeta_y(t) - \zeta_x(t)). \tag{4.2}$$

We call the process (4.2) *the "deterministic energy process" (DEP). Then we have the duality between the DEP $\{\zeta(t) : t \geq 0\}$ and the independent random walk process $\{\xi(t) : t \geq 0\}$ with duality function*

$$\mathcal{D}(\xi, \zeta) = \prod_{x \in V} \zeta_x^{\xi_x}. \tag{4.3}$$

Proof The proof of the scaling limit is a classical application of the Trotter Kurtz theorem [89], i.e., we show that the generator $L^{(N)}$ of the process $\zeta^{(N)}(t) := \eta^{(N)}(t)/N$ converges as $N \to \infty$ to the generator of the deterministic system (4.2) which equals

$$\mathcal{L} f(\zeta) = \sum_{x,y \in V} p(x, y)(\zeta_y - \zeta_x)\frac{\partial f(\zeta)}{\partial \zeta_x}, \tag{4.4}$$

or, using the symmetry $p(x, y) = p(y, x)$,

$$\mathcal{L} f(\zeta) = -\frac{1}{2}\sum_{x,y \in V} p(x, y)(\zeta_x - \zeta_y)\left(\frac{\partial f(\zeta)}{\partial \zeta_x} - \frac{\partial f(\zeta)}{\partial \zeta_y}\right). \tag{4.5}$$

For N fixed, the generator of $\{\zeta^{(N)}(t), t \geq 0\}$ reads

$$L^{(N)} f(\zeta) = \frac{N}{2}\sum_{x,y \in V} p(x, y)\Big(\zeta_x\Big(f\Big(\zeta - \tfrac{1}{N}\delta_x + \tfrac{1}{N}\delta_y\Big) - f(\zeta)\Big) \tag{4.6}$$
$$+\zeta_y\Big(f\Big(\zeta - \tfrac{1}{N}\delta_y + \tfrac{1}{N}\delta_x\Big) - f(\zeta)\Big)\Big).$$

Assuming now that $f : [0, \infty)^V \to \mathbb{R}$ is smooth, by Taylor expansion, we find

$$\lim_{N\to\infty} L^{(N)} f = \mathcal{L} f, \tag{4.7}$$

where the convergence is uniform on compact sets. Because such smooth f are a core of the generator $\mathcal{L}$, we conclude that $\{\zeta^{(N)}(t) : t \geq 0\} \to \{\zeta(t) : t \geq 0\}$ as $N \to \infty$, where the convergence is weak convergence in the Skorohod topology.

The duality relation between the deterministic system $\{\zeta(t) : t \geq 0\}$ and the independent random walk process $\{\eta(t) : t \geq 0\}$ could be proved by a direct computation by plugging in the duality function (4.3) into the generator (4.5). In this way we would find the generator duality

$$L\mathcal{D}(\cdot, \zeta)(\xi) = \mathcal{L}\mathcal{D}(\xi, \cdot)(\zeta), \tag{4.8}$$

which, by exponentiating, would give the semigroup duality. Alternatively, we can start from the self-duality relation of independent random walkers and then use the scaling limit. Let $\{\xi(t), t \geq 0\}$ (starting from $\xi \in \mathbb{N}^V$ at time zero) a copy of the process $\{\eta(t) : t \geq 0\}$ on V. Then

$$\mathbb{E}_\eta D(\xi, \eta(t)) = \mathbb{E}_\xi D(\xi(t), \eta), \tag{4.9}$$

with D given by (see (3.44)-(3.45))

$$D(\xi, \eta) = \prod_{x\in V} \frac{\eta_x!}{(\eta_x - \xi_x)!}. \tag{4.10}$$

Let $n := |\xi| = \sum_{x\in V} \xi_x$, then $D(\xi, \cdot)$ is a polynomial of degree n, hence we put $\eta^{(N)} = \lfloor N\zeta \rfloor$ and $\eta^{(N)}(t) = \lfloor N\zeta(t) \rfloor$, divide by N^n, take the limit $N \to \infty$ and find, using the convergence $\{\eta^{(N)}(t) : t \geq 0\} \to \{\zeta(t) : t \geq 0\}$, that

$$\frac{1}{N^n} D(\xi, \eta^{(N)}(t)) \to \prod_{x\in V} \zeta_x(t)^{\xi_x} \qquad \text{as } N \to \infty, \tag{4.11}$$

and

$$\frac{1}{N^n} D(\xi(t), \eta^{(N)}) \to \prod_{x\in V} \zeta_x^{\xi_x(t)} \qquad \text{as } N \to \infty. \tag{4.12}$$

As a consequence, we can take the limit as $N \to \infty$ in the self-duality relation (4.9) and obtain

$$\mathbb{E}_\zeta \mathcal{D}(\xi, \zeta(t)) = \mathbb{E}_\xi \mathcal{D}(\xi(t), \zeta),$$

with $\mathcal{D}$ given by (4.3). □

Remark 4.2 (Many particle limit) Theorem 4.1 illustrates a method that will be used several times in later chapters. Starting from a self-duality relation, and scaling one of the two processes, one obtains a duality relation between the limiting process and the original process. In this case the limiting process is deterministic, and so we have a duality relation between a continuous deterministic system and a discrete stochastic system.

Remark 4.3 (Mass conservation) From the symmetry of $p(x, y)$ it immediately follows that the deterministic dynamics (4.2) conserves the total "mass" $|\zeta(t)| := \sum_{x\in V} \zeta_x(t)$. This conservation law corresponds to the conservation of particle number in the independent random walkers process.

Remark 4.4 (The duality on two sites) Consider the system of equations (4.2) for simplicity in the context of two vertices, $V = \{1, 2\}$ and $p(1, 2) = 1$. Then it is clear that from an initial condition (ζ_1, ζ_2) this system converges exponentially fast to its stable fixed point $\zeta^* = (\frac{\zeta_1+\zeta_2}{2}, \frac{\zeta_1+\zeta_2}{2})$. Let us see how this is consistent with our duality. The dual system is a system of independent random walkers on two vertices: starting from the initial state (ξ_1, ξ_2), as $t \to \infty$, the process $(\xi_1(t), \xi_2(t))$ will converge in distribution to $(X, \xi_1 + \xi_2 - X)$ where X is a Binomial random variable $\mathrm{BIN}(n, p)$ with parameters $n = \xi_1 + \xi_2$ and $p = 1/2$. Therefore,

$$\begin{aligned}\lim_{t\to\infty} \mathbb{E}_\zeta \mathcal{D}(\xi, \zeta(t)) &= \lim_{t\to\infty} \mathbb{E}_\xi \mathcal{D}(\xi(t), \zeta) \\ &= \sum_{k=0}^{\xi_1+\xi_2} \binom{\xi_1+\xi_2}{k} \frac{1}{2^{\xi_1+\xi_2}} \zeta_1^k \zeta_2^{\xi_1+\xi_2-k} \\ &= \left(\frac{\zeta_1+\zeta_2}{2}\right)^{\xi_1+\xi_2} = \mathcal{D}(\xi, \zeta^*),\end{aligned} \tag{4.13}$$

which indeed shows that $\zeta(t) \to \zeta^*$ as $t \to \infty$.

4.2 Duality as a Change of Representation

We have seen in Sect. 3.4 that the self-duality of independent random walkers arises a change of representation between the Heisenberg algebra and the conjugate Heisenberg algebra. More precisely, recalling the representation of the conjugate Heisenberg algebra $[a, a^\dagger] = -I$ given by

$$af(n) = nf(n-1), \qquad a^\dagger f(n) = f(n+1), \tag{4.14}$$

then the self-duality of

$$L = -\frac{1}{2} \sum_{x,y \in V} p(x, y)(a_y - a_x)(a_y^\dagger - a_x^\dagger), \tag{4.15}$$

follows from the basic dualities $\mathrm{a} \xrightarrow{d} a^\dagger$ and $\mathrm{a}^\dagger \xrightarrow{d} a$ with $d(k, n) = \frac{n!}{(n-k)!} \mathbb{1}_{\{n \geq k\}}$ and

$$\mathrm{a} f(k) = f(k) + k f(k-1), \qquad \mathrm{a}^\dagger f(k) = f(k+1). \tag{4.16}$$

An important observation at this point is that also the generator (4.5) of the deterministic process $\{\zeta(t) : t \geq 0\}$ can be written in abstract form. Namely,

$$\begin{aligned} \mathcal{L} &= -\frac{1}{2} \sum_{x,y \in V} p(x, y)(\zeta_y - \zeta_x) \left(\frac{\partial}{\partial \zeta_y} - \frac{\partial}{\partial \zeta_x} \right) \\ &= -\frac{1}{2} \sum_{x,y \in V} p(x, y)(A_y^\dagger - A_x^\dagger)(A_y - A_x), \end{aligned} \tag{4.17}$$

with now

$$\begin{aligned} A_x^\dagger f(\zeta) &= \zeta_x f(\zeta), \\ A_x f(\zeta) &= \frac{\partial}{\partial \zeta_x} f(\zeta). \end{aligned} \tag{4.18}$$

Notice that these operators satisfy the commutation relations of the Heisenberg algebra:

$$[A_x, A_y^\dagger] = I \delta_{x,y}. \tag{4.19}$$

In fact, we show in this section that the duality between the generators L and $\mathcal{L}$ can be derived from a more elementary duality between the operators of the algebra generated by $a_x, a_x^\dagger, x \in V$ and the algebra generated by $A_x, A_x^\dagger, x \in V$.

We now apply Theorem 2.40 to our operators $a, a^\dagger$ and $A, A^\dagger$. We show that the duality between L and $\mathcal{L}$ is an instance of duality of two operator algebras.

Theorem 4.5 (Duality for the Heisenberg algebra) *For all $x \in V$ we have*

$$a_x \xrightarrow{D} A_x, \tag{4.20}$$

and

$$a_x^\dagger \xrightarrow{D} A_x^\dagger, \tag{4.21}$$

with duality function D *given by* (4.3)*. As a consequence for* L *in* (4.15) *and* $\mathcal{L}$ *in* (4.17) *we have* $L \xrightarrow{D} \mathcal{L}$*, with duality function* D *given by* (4.3).

Proof Put $a, a^\dagger$ the operators $af(n) = nf(n-1), a^\dagger f(n) = f(n+1)$ for functions $f : \mathbb{N} \to \mathbb{R}$ and $Af(z) = f'(z), A^\dagger f(z) = zf(z)$ for smooth functions $f : [0, \infty) \to \mathbb{R}$. Then, for $d(n, z) = z^n$, we have that $a \xrightarrow{D} A$ and $a^\dagger \xrightarrow{D} A^\dagger$. Thus, the result easily follows from Theorem 2.40. □

4.3 Generating Functions

4.3.1 Connecting Dualities and Discovering New Ones

The use of generating functions turns out to be very useful in the study of dualities. In particular it allows to deduce *equivalence* between different dualities. As a consequence, new dualities can be obtained from known dualities. In this section we discuss the application of generating functions to independent random walkers. This will immediately entail that the duality between independent walkers and the deterministic evolution is equivalent to the self-duality property of independent walker. Furthermore, via generating function we shall unveil the following additional dualities:

1. self-duality of independent random walkers with self-duality function given by a product of Charlier polynomials;
2. self-duality of the deterministic process. The self-duality is in a simple product form and is described below.

We start by explaining the main idea. Consider a self-duality function $D : \mathbb{N}^V \times \mathbb{N}^V \to \mathbb{R}$ for independent random walkers on a set V and assume it is in a product form with a "single-site self-duality function" d, i.e.

$$D(\xi, \eta) = \prod_{x \in V} d(\xi_x, \eta_x) . \tag{4.22}$$

We define $G : \mathbb{N}^V \times \mathbb{R}^V \to \mathbb{R}$ the generating function of D as

$$G(\xi, \zeta) = \prod_{x \in V} g(\xi_x, \zeta_x), \qquad \text{with} \qquad g(k, z) := \sum_{n=0}^{\infty} d(k, n) \frac{z^n}{n!} . \tag{4.23}$$

In the following theorem we show that a self-duality relation for independent random walkers is equivalent to a duality relation between independent random walkers and the deterministic system introduced in Sect. 4.1.

Theorem 4.6 (Duality and generating function, part 1) *Let* $D(\xi, \eta)$ *and* $G(\xi, \zeta)$ *be the two functions as in* (4.22) *and* (4.23). *Let* L *be the independent random walk generator in* (3.5) *and* $\mathcal{L}$ *be the generator of the deterministic process in* (4.5). *Then*

$$(LD(\xi, \cdot))(\eta) = (LD(\cdot, \eta))(\xi) \qquad \textit{for all} \quad \xi, \eta \in \mathbb{N}^V \tag{4.24}$$

is equivalent to

$$(\mathcal{L}G(\xi, \cdot))(\zeta) = (LG(\cdot, \zeta))(\xi) \qquad \textit{for all} \quad \xi \in \mathbb{N}^V,\ \zeta \in [0, \infty)^V. \tag{4.25}$$

Proof Due to the symmetry of $p : V \times V \to \mathbb{R}$, the generator L can be rewritten as the sum

$$L = \frac{1}{2} \sum_{x,y \in V} p(x, y) L_{x,y}, \tag{4.26}$$

where $L_{x,y}$ is the single-edge generator working on η_x, η_y (and not changing the other η_z's for $z \neq x, y$). Clearly, because of (4.26) and the product nature of the duality functions involved in the theorem, imposing the duality relation on a set V is equivalent to imposing it for each couple of sites, so, it is enough to prove the statement of the theorem for the single-edge generators. For functions $f : \mathbb{N}^2 \to \mathbb{R}$, let $L_{1,2}$ be defined by

$$\begin{aligned} L_{1,2} f(n_1, n_2) &= n_1(f(n_1 - 1, n_2 + 1) - f(n_1, n_2)) \\ &\quad + n_2(f(n_1 + 1, n_2 - 1) - f(n_1, n_2)), \end{aligned} \tag{4.27}$$

and for smooth functions $f : [0, \infty)^2 \to \mathbb{R}$ let $\mathcal{L}_{1,2}$ be defined by

$$\mathcal{L}_{1,2} f(z_1, z_2) = (z_2 - z_1) \left(\frac{\partial}{\partial z_1} - \frac{\partial}{\partial z_2} \right) f(z_1, z_2). \tag{4.28}$$

We prove the "if" part of the theorem. We thus assume that $d(k_1, n_1) d(k_2, n_2)$ is a self-duality function for $L_{1,2}$, i.e. for all natural numbers k_1, k_2 and n_1, n_2,

$$\big(L_{1,2}\, d(k_1, \cdot) d(k_2, \cdot)\big)(n_1, n_2) = \big(L_{1,2}\, d(\cdot, n_1) d(\cdot, n_2)\big)(k_1, k_2), \tag{4.29}$$

and would like to prove that for all $z_1, z_2 \in \mathbb{R}$ and for all $k_1, k_2 \in \mathbb{N}$,

$$\big(\mathcal{L}_{1,2}\, g(\cdot, n_1) g(\cdot, n_2)\big)(z_1, z_2) = \big(L_{1,2}\, g(\cdot, n_1) g(\cdot, n_2)\big)(k_1, k_2). \tag{4.30}$$

Using the definition of the generating function g given in (4.23) and using the assumed self-duality (4.29), the right hand side of (4.30) reads

$$
\begin{aligned}
&\big(L_{1,2}\, g(\cdot, z_1) g(\cdot, z_2)\big)(k_1, k_2) \\
&= \sum_{n_1=0}^{\infty} \sum_{n_2=0}^{\infty} \big(L_{1,2}\, d(\cdot, n_1) d(\cdot, n_2)\big)(k_1, k_2) \frac{z_1^{n_1}}{n_1!} \frac{z_2^{n_2}}{n_2!} \\
&= \sum_{n_1=0}^{\infty} \sum_{n_2=0}^{\infty} \big(L_{1,2}\, d(k_1, \cdot) d(k_2, \cdot)\big)(n_1, n_2) \frac{z_1^{n_1}}{n_1!} \frac{z_2^{n_2}}{n_2!} \\
&= \sum_{n_1=0}^{\infty} \sum_{n_2=0}^{\infty} n_1 (d(k_1, n_1 - 1) d(k_2, n_2 + 1) - d(k_1, n_1) d(k_2, n_2)) \frac{z_1^{n_1}}{n_1!} \frac{z_2^{n_2}}{n_2!} \\
&\quad + \sum_{n_1=0}^{\infty} \sum_{n_2=0}^{\infty} n_2 (d(k_1, n_1 + 1) d(k_2, n_2 - 1) - d(k_1, n_1) d(k_2, n_2)) \frac{z_1^{n_1}}{n_1!} \frac{z_2^{n_2}}{n_2!} .
\end{aligned}
\tag{4.31}
$$

For the left hand side of (4.30) we get

$$
\begin{aligned}
&\big((\mathcal{L}_{1,2}\, g(k_1, \cdot) g(k_2, \cdot)\big)(z_1, z_2) \\
&= \sum_{n_1=0}^{\infty} \sum_{n_2=0}^{\infty} \Big(z_1 \frac{\partial}{\partial z_2} - z_1 \frac{\partial}{\partial z_1}\Big) d(k_1, n_1) d(k_2, n_2) \frac{z_1^{n_1}}{n_1!} \frac{z_2^{n_2}}{n_2!} \\
&\quad + \sum_{n_1=0}^{\infty} \sum_{n_2=0}^{\infty} \Big(z_2 \frac{\partial}{\partial z_1} - z_2 \frac{\partial}{\partial z_2}\Big) d(k_1, n_1) d(k_2, n_2) \frac{z_1^{n_1}}{n_1!} \frac{z_2^{n_2}}{n_2!} \\
&= \sum_{n_1=0}^{\infty} \sum_{n_2=0}^{\infty} n_1 (d(k_1, n_1 - 1) d(k_2, n_2 + 1) - d(k_1, n_1) d(k_2, n_2)) \frac{z_1^{n_1}}{n_1!} \frac{z_2^{n_2}}{n_2!} \\
&\quad + \sum_{n_1=0}^{\infty} \sum_{n_2=0}^{\infty} n_2 (d(k_1, n_1 + 1) d(k_2, n_2 - 1) - d(k_1, n_1) d(k_2, n_2)) \frac{z_1^{n_1}}{n_1!} \frac{z_2^{n_2}}{n_2!} .
\end{aligned}
\tag{4.32}
$$

This proves that (4.29) implies (4.30). The implication in the other direction follows from a similar reasoning. □

As a first application of Theorem 4.6 we show that the self-duality function of independent random walkers found in Chap. 3 can be obtained from the duality function of Theorem 4.1. Indeed, the single site self-duality function that we found in (3.45) was

$$
d(k, n) = \frac{n!}{(n-k)!}\, \mathbb{1}_{\{k \le n\}} . \tag{4.33}
$$

The corresponding generating function is

$$g(k,z) = \sum_{n=0}^{\infty} d(k,n)\frac{z^n}{n!} = e^z z^k. \tag{4.34}$$

Using Theorem 4.6, proving that (4.33) is a single-site self-duality for $L_{1,2}$ amounts to prove that

$$\big(L_{1,2}\, g(\cdot, z_1)g(\cdot, z_2)\big)(k_1,k_2) = \big(\mathscr{L}_{1,2}\, g(k_1,\cdot)g(k_2,\cdot)\big)(z_1,z_2). \tag{4.35}$$

This follows easily by using the explicit expression (4.34) of $g(k,z)$. We have indeed

$$\begin{aligned}
&\big(L_{1,2}\, g(\cdot, z_1)g(\cdot, z_2)\big)(k_1,k_2)\\
&= \Big(k_1(z_1^{k_1-1}z_2^{k_2+1} - z_1^{k_1}z_2^{k_2}) + k_2(z_1^{k_1+1}z_2^{k_2-1} - z_1^{k_1}z_2^{k_2})\Big)\, e^{z_1+z_2}\\
&= \big(\mathscr{L}_{1,2})(g(k_1,\cdot)g(k_2,\cdot)\big)(z_1,z_2),
\end{aligned} \tag{4.36}$$

where in the last equality it has been used that the total mass is conserved for the process $(z_1(t), z_2(t))$ and thus $(\mathscr{L}_{1,2}f)(z_1,z_2) = 0$ for any function f that is a function of the sum of the coordinates, i.e. $f(z_1,z_2) = F(z_1+z_2)$ for some smooth function $F : \mathbb{R}_+ \to \mathbb{R}$.

We thus see that, by the use of generating functions, we have turned a computation to verify a self-duality relation into a computation to verify a duality relation.

As a second application of Theorem 4.6 we show that the product of Charlier polynomials is also a self-duality function for independent random walks (in Chap. 9 we will discuss in general terms the relation between duality, orthogonal polynomials and representation theory of Lie algebras).

Definition 4.7 (Charlier polynomial) The k-th order Charlier polynomial is defined by

$$C(k,n) = \frac{\partial^n}{\partial z^n}\Big((1-z)^k e^z\Big)\Big|_{z=0}. \tag{4.37}$$

It is well known (see for instance [153]) that these polynomials are orthogonal with respect to the Poisson distribution with parameter 1. We then have the following:

Corollary 4.8 (Charlier polynomials self-duality) *The independent random walk generator on a set V is self-dual with self-duality function*

$$D(\xi,\eta) = \prod_{x\in V} C(\xi_x,\eta_x), \tag{4.38}$$

where $C(k,n)$ are the Charlier polynomials given in (4.37).

Proof By (4.37) and the Taylor formula we have

$$g(k,z) = \sum_{n=0}^{\infty} C(k,n)\frac{z^n}{n!} = (1-z)^k e^z. \tag{4.39}$$

The corollary is then an immediate application of Theorem 4.6. □

We now proceed, again by using generating function, to find a self-duality of the system of ODE's which arises from the scaling limit of the independent random walkers. This will be the first example of a *self-duality of a process with continuous variables*. The idea is simple: discrete self-dualities can be "lifted" to continuous-discrete dualities by applying once a generating function, and to continuous-continuous self-dualities by applying the generating function twice.

Let $d(k,n)$ be a single-site self-duality polynomial for the generator of independent random walkers. Then from Theorem (4.6) we have

$$\big(L_{1,2}g(\cdot,z_1)g(\cdot,z_2)\big)(k_1,k_2) = \big(\mathcal{L}_{1,2}g(k_1,\cdot)g(k_2,\cdot)\big)(z_1,z_2). \tag{4.40}$$

If we now define

$$h(v,z) := \sum_{k=0}^{\infty} \frac{g(k,z)v^k}{k!} = \sum_{k=0}^{\infty}\sum_{n=0}^{\infty} d(k,n)\frac{z^n}{n!}\frac{v^k}{k!}, \tag{4.41}$$

then we have, as a consequence of (4.40), that

$$\big(\mathcal{L}_{1,2}h(\cdot,z_1)h(\cdot,z_2)\big)(v_1,v_2) = \big(\mathcal{L}_{1,2}h(v_1,\cdot)h(v_2,\cdot)\big)(z_1,z_2). \tag{4.42}$$

In other words, $h(v_1,z_1)h(v_2,z_2)$ is a self-duality function for the differential operator $\mathcal{L}_{1,2}$ in (4.28). In our concrete case, if we take the $d(k,n)$ as in (4.33), then we have that

$$h(v_1,z_1)h(v_2,z_2) = e^{v_1z_1+v_2z_2}e^{z_1+z_2}, \tag{4.43}$$

is a self-duality function of the two-site deterministic process with generator (4.28). Notice that in (4.43) we can also drop the factor $e^{z_1+z_2}$ (since this is a conserved quantity of the dynamics). From (4.42) we can once more extract a more general statement. If $h(v,z)$ satisfies (4.42) and is analytic in v, z with series expansion

$$h(v,z) = \sum_{k=0}^{\infty}\sum_{n=0}^{\infty} d(k,n)\frac{v^k z^n}{k!n!}, \tag{4.44}$$

then $d(k_1,n_1)d(k_2,n_2)$ is a self-duality function for the independent random walk generator on two sites. We can thus summarize our findings in the following theorem, relating self-duality of independent walkers to duality between independent walkers and the deterministic system with generator (4.5), and finally to self-duality of this deterministic system.

Theorem 4.9 (Duality and generating functions, part 2) *The following three statements are equivalent:*

1. $D(\xi, \eta) = \prod_{x\in V} d(\xi_x, \eta_x)$ *is a self-duality function for independent random walkers with generator* L *in* (3.5).
2. $G(\xi, \zeta) = \prod_{x\in V} g(\xi_x, \zeta_x)$ *with* $g(k, z) = \sum_{n=0}^{\infty} d(k, n)\frac{z^n}{n!}$ *is a duality function between independent random walkers and the deterministic process with generator* $\mathcal{L}$ *in* (4.5).
3. $H(\upsilon, \zeta) = \prod_{x\in V} h(\upsilon_x, \zeta_x)$ *with* $h(\upsilon, z) = \sum_{k=0}^{\infty}\sum_{n=0}^{\infty} d(k, n)\frac{\upsilon^k z^n}{n!k!}$ *is a self-duality function for the deterministic process with generator* $\mathcal{L}$ *in* (4.5).

4.3.2 From Reversible Measures to Self-duality Functions

In Proposition 3.11 we have proved that if $D(\xi, \eta)$ is a self-duality function in "simple factorized form", i.e., of the form $\prod_{x\in V} d(\xi_x, \eta_x)$ then we have, for the product Poisson measure ν_λ with parameter λ,

$$\int D(\xi, \eta)\nu_\lambda(d\eta) = \left(\int D(\delta_x, \eta)\nu_\lambda(d\eta)\right)^{|\xi|}, \qquad \text{for all } x \in V\,. \tag{4.45}$$

In this section we show that this relation between the duality functions and the reversible product measures *determines the possible single-site duality self-functions* $d(k, n)$ in terms of their one-particle value $d(1, n)$. This relies once more on the use of generating functions. In particular the result will follow for the fact that Poisson expectation acts like a generating function, and therefore fixes the coefficients.

Let us first illustrate how this works by showing that choosing $d(1, n) = n$ we may recover the single-site duality functions defined in (3.45), i.e. $d(k, n) = \frac{n!}{(n-k)!}\mathbb{1}_{k\leq n}$. Indeed, since $d(1, n) = n$, (4.45) yields

$$\sum_{n=0}^{\infty} d(k, n)\nu_\lambda(n) = \left(\sum_{n=0}^{\infty} n\nu_\lambda(n)\right)^k = \lambda^k. \tag{4.46}$$

Equivalently,

$$\sum_{n=0}^{\infty} d(k, n)\frac{\lambda^n}{n!} = \lambda^k e^\lambda, \tag{4.47}$$

and then (4.47) *determines* the function $d(k, n)$ for general k. Indeed Taylor's theorem applied to (4.47) gives

$$d(k, n) = \frac{d^n}{d\lambda^n}(\lambda^k e^\lambda)\Big|_{\lambda=0} = \frac{n!}{(n-k)!}\mathbb{1}_{\{k \le n\}}. \tag{4.48}$$

This gives a new procedure to obtain the single-site self-duality function from the reversible measure and the "first single-site self-duality function" $d(1, n)$. More precisely, if $d(1, n)$ is given, and we call

$$\theta(\lambda) := \sum_{n=0}^{\infty} d(1, n)\nu_\lambda(n), \tag{4.49}$$

then we have

$$\sum_{n=0}^{\infty} d(k, n)\nu_\lambda(n) = \theta(\lambda)^k, \tag{4.50}$$

and as a consequence we obtain

$$d(k, n) = \frac{d^n}{d\lambda^n}(\theta(\lambda)^k e^\lambda)\Big|_{\lambda=0}. \tag{4.51}$$

We summarize these findings in the following theorem.

Theorem 4.10 (From reversible measures to self-duality functions) *Let $D(\xi, \eta) = \prod_{x \in V} d(\xi_x, \eta_x)$ be a self-duality function for the system of independent walkers on a set V. Then*

$$d(k, n) = \frac{d^n}{d\lambda^n}(\theta(\lambda)^k e^\lambda)\Big|_{\lambda=0}, \tag{4.52}$$

where

$$\theta(\lambda) := \sum_{n=0}^{\infty} d(1, n)\nu_\lambda(n). \tag{4.53}$$

As a next step we show that independent symmetric random walkers is the only symmetric self-dual process (with factorized self-duality function) of zero range type and the only possible choice for the single-site duality functions is $d(1, n) = a + bn$ for some $a, b \in \mathbb{R}$. This, via Theorem 4.10 will then fix all the possible self-duality functions for the system of independent random walkers.

We recall that a zero range process for a particle system on two sites has a generator of the form

$$
\begin{aligned}
L_{1,2} f(n_1, n_2) &= c(n_1)(f(n_1 - 1, n_2 + 1) - f(n_1, n_2)) \\
&\quad + c(n_2)(f(n_1 + 1, n_2 - 1) - f(n_1, n_2)),
\end{aligned}
\tag{4.54}
$$

where $c : \mathbb{N} \to [0, \infty)$ is such that $c(0) = 0$. We then have the following

Theorem 4.11 (Self-duality and zero range) *Assume that the process with generator* (4.54) *is self-dual with self-duality functions of the form*

$$
D_{1,2}(k_1, k_2; n_1, n_2) = d(k_1, n_1) d(k_2, n_2), \tag{4.55}
$$

where $d(k, n)$ is not constant as a function of n for $k \geq 1$. Then $c(n) = c(1)n$ and hence the corresponding process is a system of independent random walkers. Moreover, $d(1, n) = a + bn$ for some $a, b \in \mathbb{R}$ and $b \neq 0$.

Proof Without loss of generality we assume $d(0, n) = 1$. Using the self-duality relation for $k_1 = 1, k_2 = 0$, and using $c(0) = 0$ we obtain the identity

$$
\begin{aligned}
&c(n_1)(d(1, n_1 - 1) - d(1, n_1)) + c(n_2)(d(1, n_1 + 1) - d(1, n_1)) \\
&\quad = c(1)(d(1, n_2) - d(1, n_1)).
\end{aligned}
\tag{4.56}
$$

Putting $n_1 = n_2 = n$ this yields,

$$
d(1, n + 1) + d(1, n - 1) - 2d(1, n) = 0, \tag{4.57}
$$

from which we derive $d(1, n) = a+bn$. Because $d(1, n)$ is not constant as a function of n, we have $b \neq 0$. Inserting $d(1, n) = a + bn$ in (4.56) yields

$$
c(n_1)(-b) + c(n_2)(b) = c(1)(b(n_2 - n_1)),
$$

from which we derive $c(n) = c(1)n$. □

4.3.3 Classification of Product Self-duality Functions

As a consequence of Theorem 4.10 and Theorem 4.11, we obtain the following

Theorem 4.12 (Classification of product self-dualities for independent random walkers) *The single-site self-dualities for the system of independent random walkers are given by*

$$
d_{a,b}(k, n) = \frac{d^n}{d\lambda^n}((a + b\lambda)^k e^{\lambda})\Big|_{\lambda=0}, \tag{4.58}
$$

for some $a, b \in \mathbb{R}$ and $b \neq 0$.

Proof Via the combination of Theorems 4.10, 4.11, we conclude that $d_{a,b}$ in (4.58) are the only possible single-site self-duality functions. So the only thing to be proved is that they are indeed single-site self-duality functions. For this we use the generating function method, more precisely Theorem 4.9. The two-variable generating function is given by

$$\sum_{k=0}^{\infty}\sum_{n=0}^{\infty} d_{a,b}(k,n)\frac{v^k}{k!}\frac{z^n}{n!} = e^{(a+bz)v}e^z, \tag{4.59}$$

and it is immediate to see that this function is a single-site self-duality function for the continuous deterministic dynamics. As a consequence of Theorem 4.9, we conclude that $d_{a,b}$ is then a single-site self-duality function for independent random walkers. □

From Theorem 4.12 we can find all self-duality functions $d_{a,b}$. An immediate computation gives

$$d_{a,b}(k,n) = \sum_{r=0}^{n}\binom{n}{r}k(k-1)\dots(k-r+1)b^r a^{k-r}. \tag{4.60}$$

In particular for $a = 0$ we recover the triangular duality function $d(n,k)$ defined in (3.45):

$$d_{0,b}(k,n) = \frac{n!}{(n-k)!}b^k \cdot \mathbb{1}_{\{n\geq k\}}, \tag{4.61}$$

while for $a \neq 0$

$$d_{a,b}(k,n) = a^k \sum_{r=0}^{\min\{k,n\}}\binom{n}{r}\binom{k}{r}\left(\frac{b}{a}\right)^k r! = a^k\, {}_2F_0\left(\begin{matrix}-k,-n\\ -\end{matrix};\frac{b}{a}\right). \tag{4.62}$$

In particular, for the choice $a \cdot b < 0$ we recover the Charlier polynomials duality functions:

$$d_{a,b}(k,n) = a^k C_{-\frac{a}{b}}(k,n), \tag{4.63}$$

$\{C_\rho(k,n),\ k \in \mathbb{N}\}$ being the Poisson-Charlier polynomials which are orthogonal w.r.t. the Poisson distribution with parameter $\rho > 0$, i.e. they satisfy

$$\int C_\rho(k,n)C_\rho(k',n)\nu_\rho(dn) = 0, \qquad \text{for} \quad k \neq k'. \tag{4.64}$$

4.4 Intertwining

The generating function approach discussed in the previous section can be turned into a statement about *intertwining* of the semigroup of independent random walkers with the semigroup of the deterministic dual dynamics. Moreover the generating functions that we used in the previous section can also be interpreted as *Poisson averaging*.

Let the set V and the symmetric irreducible transition function $p : V \times V \to \mathbb{R}$ be given, we recall the notation

$$Lf(\eta) = \sum_{x,y\in V} p(x,y)\eta_x \left(f(\eta^{x,y}) - f(\eta)\right), \tag{4.65}$$

$$\mathscr{L}f(\zeta) = -\sum_{x,y\in V} p(x,y)(\zeta_x - \zeta_y)\left(\frac{\partial}{\partial\zeta_x} - \frac{\partial}{\partial\zeta_y}\right), \tag{4.66}$$

for the generators of the independent random walkers, and the associated deterministic dynamics. We will denote the corresponding semigroups by $S(t)$, resp. $\mathcal{S}(t)$. For the deterministic dynamics, we denote by $Z^\zeta(t)$ the flow associated to the generator (4.66) started from $\zeta \in [0,\infty)^V$. We define, for $\zeta \in [0,\infty)^V$,

$$(\Lambda f)(\zeta) = \sum_{\eta\in\mathbb{N}^V} f(\eta)\prod_{x\in V}\frac{\zeta_x^{\eta_x}}{\eta_x!}e^{-\zeta_x}. \tag{4.67}$$

We think of Λ as an operator turning functions on the state space $\mathbb{N}^V$ of the discrete process into functions on the state space $[0,\infty)^V$ of the continuous process. Moreover, Λ has the probabilistic interpretation of averaging over an inhomogeneous product Poisson distribution, i.e.,

$$\Lambda f(\zeta) = \int f(\eta)\nu_\zeta(d\eta), \tag{4.68}$$

where ν_ζ is the product Poisson measure on $\mathbb{N}^V$ with parameters ζ_x for $x \in V$.

The following lemma shows that the Poisson averaging is an intertwiner between two representations of the Heisenberg algebra. We recall the notation

$$af(n) = nf(n-1), \qquad a^\dagger f(n) = f(n+1), \tag{4.69}$$

and

$$Af(z) = \tfrac{df}{dz}(z), \qquad A^\dagger f(z) = zf(z). \tag{4.70}$$

Lemma 4.13 (Poisson intertwining between two representations of the Heisenberg algebra) *Define (with small abuse of notation)*

$$(\Lambda f)(z) = \sum_{n=0}^{\infty} f(n)\frac{z^n}{n!} e^{-z}, \tag{4.71}$$

for $f : \mathbb{N} \to \mathbb{R}$ such that the series is convergent for all $z \geq 0$ and defines an analytic function of z, then

$$\Lambda a f = A^\dagger \Lambda f, \qquad \textit{and} \qquad \Lambda a^\dagger f = (A + I)\Lambda f. \tag{4.72}$$

Proof We have

$$\begin{aligned}(\Lambda a f)(z) &= \sum_{n=0}^{\infty} n f(n-1)\frac{z^n}{n!} e^{-z} = z\sum_{n=1}^{\infty} f(n-1)\frac{z^{n-1}}{(n-1)!} e^{-z} \\ &= z\Lambda f(z) = A^\dagger \Lambda f(z). \end{aligned} \tag{4.73}$$

On the other hand

$$\begin{aligned}(\Lambda a^\dagger f)(z) &= \sum_{n=0}^{\infty} f(n+1)\frac{z^n}{n!} e^{-z} = \sum_{n=1}^{\infty} f(n)\frac{z^{n-1}}{(n-1)!} e^{-z} \\ &= \frac{d}{dz}\left(\sum_{n=0}^{\infty} f(n)\frac{z^n}{n!}\right) e^{-z} = \frac{d}{dz}\left(e^z \Lambda f(z)\right) e^{-z} = (A+I)\Lambda f(z). \end{aligned} \tag{4.74}$$

□

Theorem 4.14 (Intertwiner between independent walkers and deterministic system) *Consider the independent random walk process with semigroup $S(t)$ and the deterministic system with semigroup $\mathcal{S}(t)$. Then we have the following:*

a) Λ defined in (4.67) is an intertwiner between the semigroups $\mathcal{S}(t)$ and $S(t)$, i.e. for all $t > 0$ and for all $f : \mathbb{N}^V \to \mathbb{R}$ bounded

$$\mathcal{S}(t)(\Lambda f) = \Lambda(S(t)f). \tag{4.75}$$

b) As a consequence, we recover Doob's theorem about propagation of inhomogeneous Poisson measures: for all $\zeta \in [0,\infty)^V$ and $f : \mathbb{N}^V \to \mathbb{R}$ bounded

$$\int S(t)f(\eta)\nu_\zeta(d\eta) = \int f(\eta)\nu_{Z^\zeta(t)}(d\eta), \tag{4.76}$$

with

$$Z_x^\zeta(t) = \mathbb{E}_x^{\mathrm{RW}}\left(\zeta_{X(t)}\right), \tag{4.77}$$

$\mathbb{E}_x^{\mathrm{RW}}$ *denoting expectation of the single random walk* $X(t)$ *jumping at rate* $p(x, y)$ *between* x *and* y *and starting from* $x \in V$.

Proof Item a) follows from the corresponding identity on the level of generators:

$$\mathscr{L}(\Lambda f) = \Lambda(Lf).$$

To prove this it is enough to prove the intertwining for the single-edge generators defined in (4.27) and (4.28). In their abstract form they read

$$L_{1,2} = -(a_1 - a_2)(a_1^\dagger - a_2^\dagger), \tag{4.78}$$

$$\mathscr{L}_{1,2} = -(A_1^\dagger - A_2^\dagger)(A_1 - A_2). \tag{4.79}$$

Then, calling

$$(\Lambda_{1,2} f)(\zeta_1, \zeta_2) = \sum_{\eta \in \mathbb{N}^2} f(\eta_1, \eta_2) \frac{\zeta_1^{\eta_1}}{\eta_1!} \frac{\zeta_2^{\eta_2}}{\eta_2!} e^{-(\zeta_1 + \zeta_2)}, \tag{4.80}$$

from Lemma 4.13 it follows that

$$\Lambda_{1,2} a_i f = A_i^\dagger \Lambda_{1,2} f \qquad \text{and} \qquad \Lambda_{1,2} a_i^\dagger f = (A_i + I)\Lambda_{1,2} f \qquad i = 1, 2. \tag{4.81}$$

As a consequence we have

$$\Lambda_{1,2}(a_1 - a_2)(a_1^\dagger - a_2^\dagger) = (A_1^\dagger - A_2^\dagger)(A_1 - A_2)\Lambda_{1,2}, \tag{4.82}$$

which implies

$$\Lambda_{1,2} L_{1,2} = \mathscr{L}_{1,2} \Lambda_{1,2}. \tag{4.83}$$

Item b) follows from item a). Indeed

$$\begin{aligned}
\int S(t) f(\eta) \nu_\zeta(d\eta) &= \Lambda(S(t) f)(\zeta) \\
&= \mathcal{S}(t)(\Lambda f)(\zeta) \\
&= \Lambda f(Z^\zeta(t)) \\
&= \int f(\eta)\, \nu_{Z^\zeta(t)}(d\eta).
\end{aligned} \tag{4.84}$$

It only remains to prove that $(Z^{\zeta}(t))_x = \mathbb{E}_x^{\mathrm{RW}}\left(\zeta_{X(t)}\right)$, but this follows immediately from the duality relation between the deterministic system and the independent random walkers system initialized with a single particle at site x (see Theorem 4.1). □

Remark 4.15 We remark that the intertwiner defined in (4.67) is in the form

$$(\Lambda f)(\zeta) = e^{-\sum_x \zeta_x} \sum_{\eta \in \mathbb{N}^V} D(\eta, \zeta) M(\eta) f(\eta), \tag{4.85}$$

where

$$M(\eta) = \prod_{x \in V} \frac{1}{\eta_x!}, \tag{4.86}$$

and

$$D(\eta, \zeta) = \prod_{x \in V} \zeta_x^{\eta_x}. \tag{4.87}$$

Notice that by the fact that $\sum_x \zeta_x$ is a conserved quantity we have that Λ being intertwiner between L and $\mathcal{L}$ is equivalent with $\widetilde{\Lambda}$ being an intertwiner between L and $\mathcal{L}$, where

$$(\widetilde{\Lambda} f)(\zeta) = \sum_{\eta \in \mathbb{N}^V} D(\eta, \zeta) M(\eta) f(\eta). \tag{4.88}$$

The fact that $\widetilde{\Lambda}$ is an intertwiner follows from Theorem 2.29 and the fact that $L^* = L$ in the Hilbert space $L^2(\mathbb{N}^V, M)$.

Remark 4.16 (Probabilistic interpretation) The intertwining relation (4.75) can be understood probabilistically as follows: starting independent walkers from ν_ζ and evolving at time t has the same distribution as $\nu_{Z^{\zeta}(t)}$, i.e., choosing η directly from the Poisson distribution with parameters $Z^{\zeta}(t)$. Notice that the fact that the dual dynamics in continuous variables is deterministic is special for the case of independent random walks, and this is the reason why inhomogeneous Poisson distributions are exactly reproduced in the course of the evolution of independent walkers. Later on, we will encounter similar intertwining results where the dual dynamics is *stochastic*, and therefore, the local stationary measures are not exactly reproduced but are replaced by convex combinations of local stationary measures.

Remark 4.17 (Inverse intertwiner) If f is an analytic function of ζ, then we can define

$$[\Lambda^{-1} f](\eta) = \frac{\partial^{\eta}}{\partial \zeta^{\eta}} f(\zeta)\Big|_{\zeta=0}, \qquad \text{with} \qquad \frac{\partial^{\eta}}{\partial \zeta^{\eta}} = \left(\frac{\partial^{\eta_x}}{\partial \zeta_x^{\eta_x}}\right)_{x \in V}, \tag{4.89}$$

i.e.

$$f(\zeta) = \sum_{\eta} [\Lambda^{-1} f](\eta) \frac{\zeta^{\eta}}{\eta!}. \tag{4.90}$$

This "operator" Λ^{-1} acts (formally) as the inverse of the intertwiner Λ and, as a consequence, it intertwines between $\mathcal{L}$ and L, i.e.

$$\Lambda^{-1}\mathcal{L} = L\Lambda^{-1}, \tag{4.91}$$

and then

$$\Lambda^{-1}\mathcal{S}(t) = S(t)\Lambda^{-1}. \tag{4.92}$$

In other words, if we Taylor expand the function which maps ζ to $[\mathcal{S}(t)f](\zeta) = f(Z^{\zeta}(t))$, then we have

$$[\mathcal{S}(t)f](\zeta) = \sum_{\eta} R_t(\eta) \frac{\zeta^{\eta}}{\eta!}, \tag{4.93}$$

where

$$R_t(\eta) = [S(t)R_0](\eta) = \mathbb{E}_{\eta} R_0(\eta(t)), \qquad \text{and} \qquad R_0(\eta) = [\Lambda^{-1} f](\eta), \tag{4.94}$$

meaning that the multivariate Taylor coefficients of $\mathcal{S}(t)f$ evolve as independent random walk expectations of the Taylor coefficients of f.

4.5 More General Independent Processes

In this section we give a more direct proof of the dualities and intertwinings that we have encountered before. The advantage is that no generators are used, and hence the results apply also to general processes (e.g. beyond Markov processes), provided a strong form of time-inversion symmetry is satisfied.

We denote by V the finite set on which the processes (of a single particle) will evolve. We call $\{X_t, t \geq 0\}$ the process of a single particle and denote by $\{X_t^x, t \geq 0\}$ the process conditioned on $X_0 = x \in V$, with

$$p_t(x, y) := \mathbb{P}(X_t = y | X_0 = x) . \tag{4.95}$$

Notice that we do not require the process to be Markov, hence $p_t(x, y)$ will not necessarily satisfy the Chapman Kolmogorov equation. We only require the symmetry property:

$$p_t(x, y) = p_t(y, x). \tag{4.96}$$

In particular, if the collection of processes $\{X_t^x, t \geq 0\}, x \in V$ is generated via a stochastic flow, then it is sufficient that this flow is time-reversible.

For an initial configuration $\eta \in \Omega = \mathbb{N}^V$ we denote $\{X_t^{x,i}, t \geq 0\}$ with $i = 1, \ldots, \eta_x$, the collection of η_x independent copies of the process X_t, starting from site $x \in V$. So by letting x vary in V we obtain $|\eta| = \sum_{x \in V} \eta_x$ independent copies of the same process starting according to the initial configuration η. The random configuration $\eta(t)$ at time $t > 0$ is then defined as in (3.1) via

$$(\eta(t))_x = \sum_{y \in V} \sum_{i=1}^{\eta_y} \mathbb{1}_{\{X_t^{y,i} = x\}}. \tag{4.97}$$

We denote by

$$S(t) f(\eta) = \mathbb{E}(f(\eta(t)) | \eta(0) = \eta) \tag{4.98}$$

the time-evolution operator for expectations (which is now not necessarily a semigroup). We define the analogue of the deterministic process $\{Z_t^z, t \geq 0\}$ taking values in $\mathbb{R}^V$, via

$$(Z^z(t))_x := \sum_{y \in V} p_t(x, y) z_y, \tag{4.99}$$

where the upper index z refers to the initial condition $z \in \mathbb{R}^V$. For $\eta \in \Omega$, $z \in \mathbb{R}^V$, we denote by

$$z^\eta := \prod_{x \in V} z_x^{\eta_x}, \qquad \eta! := \prod_x \eta_x!. \tag{4.100}$$

We call a labeled configuration of M particles a M-tuple $\mathbf{x} := (x_1, \ldots, x_M) \in V^M$. For $\mathbf{x}$ a labeled configuration we denote by $\Xi(\mathbf{x}) \in \Omega$ the corresponding configuration, i.e, for $y \in V$,

$$(\Xi(\mathbf{x}))_y := \sum_{i=1}^{M} \mathbb{1}_{\{x_i = y\}}. \tag{4.101}$$

The following theorem shows that, under the symmetry assumption (4.96), all the dualities and intertwinings derived for the system of independent random walkers still hold.

Theorem 4.18 (Duality for Independent Particles)

(a) The processes $\{Z(t) : t \geq 0\}$ and $\{\eta(t) : t \geq 0\}$ are dual with duality function $D(z, \eta) = \prod_{x \in V} z_x^{\eta_x}$, i.e., for all $\eta \in \Omega$, $z \in \mathbb{R}^V$,

$$\mathbb{E}_\eta D(z, \eta(t)) = D(Z^z(t), \eta). \tag{4.102}$$

(b) The processes $\{Z(t) : t \geq 0\}$ and $\{\eta(t) : t \geq 0\}$ are intertwined with intertwining operator

$$\Lambda f(z) = \sum_{\eta \in \Omega} \frac{z^\eta}{\eta!} f(\eta), \tag{4.103}$$

i.e., for all $t > 0$ and for all $f : \Omega \to \mathbb{R}$ bounded

$$\Lambda S(t) f = \mathcal{S}(t) \Lambda(f). \tag{4.104}$$

(c) The deterministic process $\{Z(t) : t \geq 0\}$ is self-dual with self-duality function $D(v, z) = \prod_{x \in V} d(v_x, z_x)$ where the single-site self-duality function is $d(v_x, z_x) = e^{z_x v_x + z_x}$.

(d) The process $\{\eta(t),\ t \geq 0\}$ is self-dual with single-site self-duality function $d(k, n)$ if and only if the corresponding generating function

$$g(v, z) = \sum_{k,n} d(k, n) \frac{z^k v^n}{k! n!} \tag{4.105}$$

is a single-site self-duality function for the deterministic process $\{Z(t),\ t \geq 0\}$. This is true, in particular, for $d(k, n) = \frac{n!}{(n-k)!} \mathbb{1}_{k \leq n}$.

Proof Item (a). Let $\eta \in \Omega$ be a configuration with n particles, and let $\mathbf{y} \in V^n$ be fixed such that $\Xi(\mathbf{y}) = \eta$. Then, for every permutation invariant $f : V^n \to \mathbb{R}$, and corresponding function on configurations $g : \Omega \to \mathbb{R}$ defined via $g(\Xi(\mathbf{y})) = f(\mathbf{y})$, we have

$$\mathbb{E}_\mathbf{y} f(\mathbf{y}_t) = \sum_{\mathbf{x} \in V^n} f(\mathbf{x}) \prod_{i=1}^n p_t(y_i, x_i) = \mathbb{E}_\eta g(\eta(t)) \,. \tag{4.106}$$

Therefore we may write

$$
\begin{aligned}
(Z_t^z)^\eta &= \prod_{y\in V}\left(\sum_{x\in V} p_t(y,x)z_x\right)^{\eta_y} \\
&= \prod_{y\in V}\prod_{i=1}^{\eta_y}\left(\sum_{x_i} p_t(y,x_i)z_{x_i}\right) \\
&= \prod_{i=1}^{n}\left(\sum_{x_i} p_t(y_i,x_i)z_{x_i}\right) \\
&= \mathbb{E}_\eta z^{\eta(t)},
\end{aligned}
\tag{4.107}
$$

where we used (4.106) for $f(x) = \prod z_{x_i}$ which is clearly permutation invariant and corresponds to z^η on configurations.
Item (b). Let us denote by $\mathcal{S}(t)f(z) = f(Z^z(t))$ the evolution of a function f under the deterministic dynamics. Let, for $\eta \in \Omega$ with $|\eta| = n$, $\mathbf{x}[\eta]$ denote a fixed element of V^n such that $\Xi(\mathbf{x}[\eta]) = \eta$. Notice that the set of $\mathbf{x} \in V^n$ such that $\Xi(\mathbf{x}) = \eta$ has cardinality $\frac{n!}{\prod_x \eta_x!}$. As a consequence, we compute:

$$
\begin{aligned}
[\mathcal{S}(t)(\Lambda f)](z) &= \sum_\eta \frac{z_t^\eta}{\eta!} f(\eta) \\
&= \sum_{n=0}^{\infty}\sum_{\eta:|\eta|=n} \frac{\prod_{i=1}^n (Z^z(t))_{\mathbf{x}[\eta]_i}}{\prod_x \eta_x!} f(\Xi(\mathbf{x}[\eta])) \\
&= \sum_{n=0}^{\infty}\frac{1}{n!}\sum_{\mathbf{x}\in V^n} f(\Xi(\mathbf{x}))\prod_{i=1}^n (Z^z(t))_{x_i} \\
&= \sum_{n=0}^{\infty}\frac{1}{n!}\sum_{\mathbf{x},\mathbf{y}\in V^n} f(\Xi(\mathbf{x}))\prod_{i=1}^n p_t(x_i,y_i)\, z_{y_i}.
\end{aligned}
\tag{4.108}
$$

If we now use the assumed symmetry in (4.96) we may continue and write

$$
\begin{aligned}
[\mathcal{S}(t)(\Lambda f)](z) &= \sum_{n=0}^{\infty}\frac{1}{n!}\sum_{\mathbf{x},\mathbf{y}\in V^n} f(\Xi(\mathbf{x}))\prod_{i=1}^n p_t(y_i,x_i)\, z_{y_i} \\
&= \sum_{n=0}^{\infty}\frac{1}{n!}\sum_{\mathbf{y}\in V^n}\prod_i z_{y_i}\ \mathbb{E}_{\Xi(\mathbf{y})}[f(\eta(t))]
\end{aligned}
$$

$$
\begin{aligned}
&= \sum_{\eta} \frac{z^{\eta}}{\eta!} \, \mathbb{E}[f(\eta(t))|\eta(0) = \eta] \\
&= [\Lambda(S(t)f)](z). \qquad (4.109)
\end{aligned}
$$

Item (c). Notice that, by conservation of total mass, the factor e^z can be added and removed without affecting the self-duality. Therefore, the only thing to be proved is the fact that, for all $v, z \in \mathbb{R}^V$ and $t > 0$,

$$
\sum_x v_x Z^z(t)_x = \sum_x z_x Z^v(t)_x, \qquad (4.110)
$$

which follows via (4.99):

$$
\sum_{x \in V} v_x Z^z(t)_x = \sum_{x,y \in V} v_x p_t(x, y) z_y = \sum_{x,y \in V} z_y p_t(y, x) v_x = \sum_{y \in V} z_y Z^v(t)_y. \qquad (4.111)
$$

Item *d*). It follows immediately from items *a*), *b*), *c*). □

Remark 4.19 (Point Process) Notice that Eq. 4.108 is an instance of the so-called "random displacement theorem" from point-process theory [161] which states the following. If we start from a Poisson point process, and randomly and independently displace the points, then we get a new Poisson point process. Translated to our context this means that if we start a configuration process from a product Poisson measure with parameter z_x at site $x \in V$, then, at time t we have again a product of Poisson measures with parameter $z_x(t) = \sum_y z_y p_t(y, x)$ at site $x \in V$. If $p_t(y, x)$ is symmetric $z_x(t)$ can be rewritten as $\mathbb{E}_x z_{X(t)}$.

4.6 Ergodic Measures in $\mathbb{Z}^d$

In Sect. 3.6, when we studied invariant and ergodic measures for independent random walkers on $\mathbb{Z}^d$, we had to impose a condition on the moments (i.e. we had to require the invariant-ergodic measures to be tempered) in order to identify them with homogeneous Poisson product measures. The main reason for that restriction is that we do not have a set of bounded self-duality functions, and we need integrability of all the self-duality functions. In order to circumvent this issue, we will use duality with the dual deterministic dynamics, where we can use *bounded duality functions*. More precisely we will show the following theorem.

Theorem 4.20 (Ergodic measures for independent random walkers) *For independent random walkers on the lattice $\mathbb{Z}^d$, let the rates $p(x, y)$ satisfy the following conditions:*

(a) translation invariance: $p(x, y) = \pi(y - x)$;
(b) finite second moment: $\sum_{x \in \mathbb{Z}^d} \|x\|^2 \pi(x) < \infty$.

Then the set of ergodic invariant measures of the system of independent random walkers coincides with the set of homogeneous Poisson product measures.

The rest of the section is devoted to the proof of this theorem, which is in the same spirit as the proof given in Sect. 3.6, but due to the fact that the dual dynamics is in the continuum and deterministic, some non-trivial new elements have to enter.

We start with some notations. We denote by $\mathcal{X}$ the set of finite configurations in $\mathbb{R}^{\mathbb{Z}^d}$, bounded between 0 and 1 i.e., configurations $u : \mathbb{Z}^d \to [0, 1]$ such that the support of u, i.e., the set of $i \in \mathbb{Z}^d$ such that $u_i \neq 0$, is finite. This set will be the analogue of the finite particle configurations in Sect. 3.6. For $u \in \mathcal{X}$ we define $M(u) = \sum_x u_x$ to be its mass (which is the same as its l_1-norm). We denote by $\mathcal{X}_1$ the set of configurations $u : \mathbb{Z}^d \to [0, 1]$ with $M(u)$ finite. Define, for $\xi \in \Omega$, $u \in \mathcal{X}_1$

$$D(\xi, u) = \prod_x (1 - u_x)^{\xi_x}. \tag{4.112}$$

Then, D acts as a duality function between the deterministic dynamics with generator

$$\mathcal{L} = -\sum_{x,y} p(x, y)(u_x - u_y)\left(\frac{\partial}{\partial u_x} - \frac{\partial}{\partial u_y}\right), \tag{4.113}$$

and the system of independent random walkers, i.e., we have, for $u \in \mathcal{X}_1$

$$\mathbb{E}_\xi D(\xi(t), u) = D(\xi, U^u(t)), \tag{4.114}$$

where U^u_t denotes the flow generated by the deterministic dynamics with generator $\mathcal{L}$ starting from $u \in \mathcal{X}$. Notice that we have, as we saw before

$$(U^u(t))_x = \sum_y p_t(x, y)u_y, \tag{4.115}$$

where $p_t(x, y)$ is the continuous-time random walk transition probability associated to the rates $p(x, y)$. This implies, in particular, that for a configuration $u \in \mathcal{X}_1$, also $U^u(t) \in \mathcal{X}_1$, and $M(U^u(t)) = M(u)$ for all $t > 0$. Notice that the duality function (4.112) satisfies the convex factorization property

$$D(\xi, \gamma u + (1 - \gamma)v) = D(\xi, \gamma u)D(\xi, (1 - \gamma)v), \tag{4.116}$$

for $u, v \in \mathcal{X}_1$, $\gamma \in [0, 1]$ and such that the supports of u and v are disjoint.

As before, for a probability measure μ on Ω we then define its D-transform

$$\widehat{\mu}(u) = \int D(\xi, u)\mu(d\xi). \tag{4.117}$$

This is well-defined for every μ because the functions $D(\cdot, u)$ are bounded for all $u \in \mathcal{X}$. Notice that, if $\mu = \nu_\rho$ is a homogeneous Poisson product measure with density ρ, then for $u \in \mathcal{X}_1$,

$$\widehat{\nu}_\rho(u) = e^{-\rho M(u)}. \tag{4.118}$$

This property characterizes the Poisson measures. More precisely, if μ is such that its D-transform ψ is a function of the mass of u only, and additionally satisfies the convex factorization property

$$\psi(\gamma u + (1-\gamma)v) = \psi(\gamma u)\psi((1-\gamma)v), \tag{4.119}$$

for all u, v with disjoint supports, then μ is a homogeneous Poisson product measure. We will use this characterization to show that all ergodic measures for the independent random walk system are homogeneous Poisson measures. By the assumptions on transition rates we obtain the following. Define the Fourier transform of the transition probabilities

$$P(k,t) := \sum_x e^{ikx} p_t(0,x). \tag{4.120}$$

Then we have the formula

$$P(k,t) = e^{-\Gamma(k)t}, \tag{4.121}$$

for $k \in (-\pi, \pi)^d$ where $\Gamma(k) \geq 0$, $\Gamma(k) \approx Ck^2$ for $k \to 0$ and $\Gamma(k) = 0$ if and only if $k = 0$. Indeed, putting $\sum_y \pi(y) = C$ we have that the discrete walk which makes steps from x to $x+y$ with probability $\pi(y)/C$ and the continuous-time random walk are connected via $X_t = X^{\text{discr}}_{N_t}$ where N_t is a rate C Poisson process. As a consequence $\Gamma(k) = \mathbb{E}(e^{ikX_1^{\text{discr}}} - 1)C$.

This, as we will see in Lemma 4.22 below, implies that the deterministic dynamics has the total mass as unique conserved quantity.

Lemma 4.21 *Let μ be an invariant measure, then its D-transform $\widehat{\mu}$ is invariant under the flow of the deterministic dynamics generated by $\mathcal{L}$.*

Proof This follows via duality (4.114) and invariance of μ. □

Lemma 4.22 *Let $F : \mathcal{X}_1 \to \mathbb{R}$ be a function which is invariant under the flow generated by $\mathcal{L}$. Then F is a function of the total mass, i.e., $F(u) = G(M(u))$ for some $G : \mathbb{R} \to \mathbb{R}$.*

Proof We can consider F to be a function of the Fourier transform of $\hat{u}(k) := \sum_x e^{ikx} u_x$ as well, because the Fourier transform determines uniquely u. On the Fourier transform, the dynamics acts as a multiplication with $e^{-t\Gamma(k)}$, cf. (4.120), (4.121), i.e.

$$[\widehat{U}^u(t)](k) = \sum_x e^{ikx} U^u(t)_x = P(k,t) \sum_y u_y e^{iky} = P(k,t)\hat{u}(k). \tag{4.122}$$

In that case, the invariance under the dynamics of F means that

$$F(\{\hat{u}(k), k\}) = F\left(\{\hat{u}(k) e^{-\Gamma(k)t}, k\}\right). \tag{4.123}$$

By taking the limit $t \to \infty$, we get, using $\Gamma(k) = 0$ only if $k = 0$, and $\Gamma(k) > 0$ for $k \neq 0$

$$F(\{\hat{u}(k), k\}) = F(\{\mathbb{1}_{\{k=0\}} \hat{u}(0), k\}). \tag{4.124}$$

Finally, notice that for $u \in \mathcal{X}_1$, $\hat{u}(0) = M(u)$. This concludes the proof. □

As a consequence of Lemmas 4.21 and 4.22, we have that the invariance of μ implies that its D-transform is a function of the total mass only. In order to prove that for μ invariant and ergodic, μ is a homogeneous Poisson measure, it suffices to show that for u, v with finite support, and $\gamma \in [0, 1]$

$$\hat{\mu}(\gamma u + (1-\gamma)v) = \hat{\mu}(\gamma u)\hat{\mu}((1-\gamma)v). \tag{4.125}$$

So let us fix $u, v \in \mathcal{X}$ with disjoint supports. By ergodicity of μ, duality, the Birkhoff ergodic theory and dominated convergence we have

$$\begin{aligned} &\lim_{T\to\infty} \frac{1}{T} \int_0^T \int D(\xi, U^{\gamma u}(t)) D(\xi, (1-\gamma)v)\, \mu(d\xi)\, dt \\ &= \lim_{T\to\infty} \int D(\xi, (1-\gamma)v) \cdot \frac{1}{T} \int_0^T \mathbb{E}_\xi[D(\xi(t), \gamma u)]\, dt\, \mu(d\xi) \qquad (4.126)\\ &= \hat{\mu}(\gamma u)\hat{\mu}((1-\gamma)v). \qquad (4.127) \end{aligned}$$

So we have to show that the LHS equals $\hat{\mu}(\gamma u + (1-\gamma)v)$. Define the configuration

$$V(t,u,v)_x = \begin{cases} 0 & \text{if } x \in \text{supp}(v), \\ U^{\gamma u}(t)_x & \text{otherwise.} \end{cases} \tag{4.128}$$

Then, since $p_t(x,y) \leq p_t(0,0) \to 0$ as $t \to \infty$ for all $x, y \in \mathbb{Z}^d$, we have

$$D(\xi, U^{\gamma u}(t)) - D(\xi, V(t,u,v)) \to 0, \tag{4.129}$$

as $t \to \infty$, uniformly in ξ. Furthermore,

$$D(\xi, V(t,u,v))D(\xi,(1-\gamma)v) = D(\xi, V(t,u,v)+(1-\gamma)v), \tag{4.130}$$

by disjointness of the supports of $V(t,u,v)$ and $(1-\gamma)v$. Therefore, using (4.129) twice we obtain, using that $\hat{\mu}$ depends only on the total mass, which is preserved in the course of the flow generated by $\mathcal{L}$:

$$\begin{aligned}
&\lim_{T\to\infty} \frac{1}{T}\int_0^T \int D(\xi, U^{\gamma u}(t))D(\xi,(1-\gamma)v)\,\nu(d\xi)\,dt\\
&= \lim_{T\to\infty} \frac{1}{T}\int_0^T \hat{\mu}(V(t,u,v)+(1-\gamma)v)\\
&= \lim_{T\to\infty} \frac{1}{T}\int_0^T \hat{\mu}(U^{\gamma u}(t)+(1-\gamma)v)\\
&= \lim_{T\to\infty} \frac{1}{T}\int_0^T \hat{\mu}(\gamma u+(1-\gamma)v)\\
&= \hat{\mu}(\gamma u+(1-\gamma)v),
\end{aligned} \tag{4.131}$$

which shows the desired property of $\hat{\mu}$, and shows that $\hat{\mu}$ is a homogeneous Poisson measure, and therefore, concludes the proof of Theorem 4.20.

4.7 Non-symmetric Case

In this section we consider independent random walkers with irreducible transition function $p(x,y) \geq 0$ for $x, y \in V$ and $p(x,y) \neq p(y,x)$ for at least one edge $\{x,y\}$. Furthermore we assume that p is doubly stochastic, i.e.,

$$\sum_y p(x,y) = \sum_x p(x,y) = 1. \tag{4.132}$$

What follows will also work when $\sum_y p(x,y) = \sum_x p(x,y)$, i.e., the sum does not necessarily have to be equal to one. We put $q(x,y) = p(y,x)$. We denote by $\mathbb{E}^p_\eta$, resp. $\mathbb{E}^q_\eta$ the expectation in the independent random walks process jumping according to p, resp. q. If necessary we will also provide the associated generators with a upper index p, resp. q, i.e. $L^{(p)}$ and $L^{(q)}$. Notice that the two processes are one the time-reversed of the other. The generator of the system of independent random walkers jumping at rate one according to the transition probability function $p(x,y)$ is still given by (3.5), but, as we cannot symmetrize it anymore, in terms of creation and annihilation operators in (3.10), the generator reads

$$L = \sum_{x,y \in V} p(x, y)(a_x a_y^\dagger - a_x a_x^\dagger). \tag{4.133}$$

We then have the following generalization of the self-duality result. It states that the self-duality function of the symmetric independent random walkers process become, in the non-symmetric case, a duality function between the process and its time-reversed.

Theorem 4.23 (Duality for non-symmetric independent random walkers) *For the polynomials defined in* (3.44), (3.45) *we have the following duality relation. For all* $\xi \in \Omega_{finite}$ *and allowed* $\eta \in \Omega_{alw} \subseteq \mathbb{N}^V$,

$$\mathbb{E}_\eta^p D(\xi, \eta(t)) = \mathbb{E}_\xi^q D(\xi(t), \eta). \tag{4.134}$$

Proof The proof, which is in the spirit of Theorem 2.13, follows the same steps as the proof of self-duality in Theorem 3.10.

Step 1: Cheap Duality Function
First we restrict to the case $\xi, \eta \in \Omega_{\text{finite}}$. We have then a "cheap duality" function between the generators $L^{(p)}$ and $L^{(q)}$. Define as before $M(\xi) = \prod_{x \in V} \rho^{\xi_x}/\xi_x!$, and $D_{\text{cheap}}(\xi, \eta) = \frac{\delta_{\xi,\eta}}{M(\xi)}$ then we have

$$\big(L^{(p)} D_{\text{cheap}}(\cdot, \eta)\big)(\xi) = \big(L^{(q)} D_{\text{cheap}}(\xi, \cdot)\big)(\eta). \tag{4.135}$$

Indeed, (4.135) follows from the simple "generalized" detailed balance relation

$$M(\xi) L^{(p)}(\xi, \eta) = M(\eta) L^{(q)}(\eta, \xi), \tag{4.136}$$

Note that the case $\eta = \xi$ requires $\sum_y p(x, y) = \sum_x p(x, y)$.

Step 2: Symmetries of the Generators $L^{(p)}, L^{(q)}$
The generators $L^{(p)}, L^{(q)}$ commute with S^+, S^-, the operators defined in (3.30). This follows by direct computation from (4.133) and the commutation relations in (3.14).

Step 3: Duality Function from Cheap Duality Function and Symmetry
It follows from (4.135) and the fact that both $L^{(p)}, L^{(q)}$ commute with S^+, S^- that for $D = e^{S^+} D_{\text{cheap}}$ we have

$$\big(L^{(p)} D(\cdot, \eta)\big)(\xi) = \big(L^{(q)} D(\xi, \cdot)\big)(\eta), \tag{4.137}$$

which implies (4.134) for finite ξ, η.

The generalization to the case $\xi \in \Omega_{\text{finite}}$ and $\eta \in \Omega_{\text{alw}}$ is then obtained, as in Sect. 3.5, by taking finite approximations of η. □

As a consequence we obtain—exactly as in the symmetric case—invariance and ergodicity of homogeneous Poisson product measures for the translation invariant independent random walkers on $\mathbb{Z}^d$. In the asymmetric case however, also non-translation invariant ergodic product measures can exist and therefore there is no analogue of the complete characterization of the ergodic measures via duality as we did in the symmetric case.

4.8 Joint Moment Generating Function of Currents

This section is yet another illustration of duality in concrete computations, where we use twisted creation and annihilation operators and their duality properties to compute a joint moment generating function of currents.

We consider asymmetric nearest-neighbor independent random walkers moving on the one dimensional lattice $\mathbb{Z}$, jumping at rate $p > 0$ to the right and at rate $q > 0$ to the left. The generator of this process is given by

$$[L^{p,q} f](\eta) = \sum_x \left\{ p\eta_x[f(\eta^{x,x+1}) - f(\eta)] + q\eta_{x+1}[f(\eta^{x+1,x}) - f(\eta)] \right\}, \tag{4.138}$$

or equivalently

$$L^{p,q} = \sum_x \left\{ p(a_x a^\dagger_{x+1} - a_x a^\dagger_x) + q(a_{x+1} a^\dagger_x - a_{x+1} a^\dagger_{x+1}) \right\}, \tag{4.139}$$

with a_x and $a^\dagger_x$ as in (3.10). We are interested in the behavior of the currents over nearest neighbor edges especially in their large deviations. For this it is essential to gain information about the joint moment generating function of multiple currents. Because we are in a setting of independent walkers, we will show that this joint moment generating function can be expressed in terms of local times of a single random walk. In this computation, we will encounter natural twisted creation and annihilation operators. In particular we will use a duality result for a deformed generator (Theorem 4.26 below) that naturally leads to the joint generating function of currents over different edges.

More specifically, in order to study joint moment generating functions of currents it is convenient to introduce the following operator.

Definition 4.24 (The Deformed Generator) Let $\lambda : \mathbb{Z} \to \mathbb{R}$ be a bounded function, and define its discrete gradient (on edges of nearest neighbouring sites) as

$$u_{x,x+1} = (\nabla\lambda)_x = \lambda_{x+1} - \lambda_x. \tag{4.140}$$

Then we define the operator $L^{\lambda,p,q}$ by

$$[L^{\lambda,p,q} f](\eta) = \sum_x \Big\{ p\eta_x [e^{u_{x,x+1}} f(\eta^{x,x+1}) - f(\eta)] \\ + q\eta_{x+1}[e^{-u_{x,x+1}} f(\eta^{x+1,x}) - f(\eta)] \Big\}, \tag{4.141}$$

i.e.

$$L^{\lambda,p,q} = \sum_x \Big\{ p(e^{u_{x,x+1}} a_x a^\dagger_{x+1} - a_x a^\dagger_x) + q(e^{-u_{x,x+1}} a_{x+1} a^\dagger_x - a_{x+1} a^\dagger_{x+1}) \Big\}. \tag{4.142}$$

For $\lambda = 0$ this operator reduces to the generator (4.139). In order to explain the role of the operator defined in (4.142), we introduce the current. For the process with generator $L^{p,q}$ we define the current $J_{x,x+1}(t)$ over the edge $(x, x+1)$ in the time interval $[0, t]$ as the number of jumps of a particle from x to $x+1$ minus the number of jumps of a particle from $x+1$ to x that occurred in the time interval $[0, t]$. We then have the following proposition which explains the role of $L^{\lambda,p,q}$.

Proposition 4.25 (The Deformed Semigroup) *For all λ bounded functions we have the following relation between the operator $L^{\lambda,p,q}$ and $L^{p,q}$. For $f : \Omega \to \mathbb{R}$ a local function*

$$e^{tL_{\lambda,p,q}} f(\eta) = \mathbb{E}^{p,q}_\eta \left[e^{-\sum_x u_{x,x+1} J_{x,x+1}(t)} f(\eta(t)) \right], \tag{4.143}$$

where $e^{tL_{\lambda,p,q}}$ denotes the semigroup generated by $L^{\lambda,p,q}$ and $\mathbb{E}^{p,q}_\eta$ denotes the expectation w.r.t. the Markov process generated by $L^{p,q}$.

Proof First notice that $\mathcal{J}_t := \sum_x u_{x,x+1} J_{x,x+1}(t)$ is an additive functional of the process $\{\eta(t),\ t \geq 0\}$, i.e., denoting by ω a path of the process $\{\eta(t),\ t \geq 0\}$ and by θ_t its time-shift $\theta_t(\omega)(s) = \omega(s+t)$ we have

$$\mathcal{J}_{t+s}(\omega) = \mathcal{J}_t(\omega) + \mathcal{J}_s(\theta_t \omega). \tag{4.144}$$

As a consequence, by the Markov property, the r.h.s. of (4.143) defines a semigroup. In order to prove (4.143) it thus suffices to prove the equality of the corresponding generators, i.e., to prove that for all f local we have

$$L^{\lambda,p,q} f = \frac{d}{dt} \left\{ \mathbb{E}^{p,q}_\eta \left[e^{-\sum_x u_{x,x+1} J_{x,x+1}(t)} f(\eta(t)) \right] \right\} \Big|_{t=0}. \tag{4.145}$$

Use the following identity

$$\sum_x u_{x,x+1} J_{x,x+1}(t) = \sum_x \lambda_x(\eta_x(t) - \eta_x(0)), \tag{4.146}$$

which follows from the discrete continuity equation,

$$\eta_x(t) = \eta_x(0) + J_{x,x+1}(t) - J_{x-1,x}(t), \tag{4.147}$$

we obtain that the r.h.s. of (4.145) is, for $\tilde{f}(\eta) = e^{-\sum_x \lambda_x \eta_x} f(\eta)$, equal to

$$e^{\sum_x \lambda_x \eta_x} \cdot [L^{p,q} \tilde{f}](\eta) = [L^{\lambda,p,q} f](\eta). \tag{4.148}$$

□

In order to understand better the semigroup generated by $L^{\lambda,p,q}$ we have to look for "cheap dualities" and symmetries. This is the content of the following theorem.

Theorem 4.26 (Duality for the Deformed Generator) *Define*

$$D_{cheap}(\xi, \eta) = \prod_x \xi_x! \delta_{\xi_x,\eta_x}. \tag{4.149}$$

Then we have the following results:

1. *$L^{\lambda,p,q}$ and $L^{-\lambda,q,p}$ are dual with duality function D_{cheap}.*
2. *$L^{\lambda,p,q}$ commutes with*

$$S^{\lambda,+} = \sum_x e^{\lambda_x} a_x^\dagger. \tag{4.150}$$

3. *$L^{\lambda,p,q}$ and $L^{-\lambda,q,p}$ are dual with duality function*

$$D(\xi, \eta) = \prod_x e^{\lambda_x(\eta_x - \xi_x)} \frac{\eta_x!}{(\eta_x - \xi_x)!}. \tag{4.151}$$

4. *For all η and ξ finite we have*

$$\mathbb{E}_\eta^{p,q}\left[e^{-\sum_x u_{x,x+1} J_{x,x+1}(t)} D(\xi, \eta(t))\right] = \mathbb{E}_\xi^{q,p}\left[e^{\sum_x u_{x,x+1} J_{x,x+1}(t)} D(\xi(t), \eta)\right]. \tag{4.152}$$

Proof

1. This follows from the fact that, if we put $\Lambda(\xi) = \prod_x \frac{\rho^{\xi_x}}{\xi_x!}$, then

$$\Lambda(\xi) L^{\lambda,p,q}(\xi, \eta) = L^{-\lambda,q,p}(\eta, \xi)\Lambda(\eta). \tag{4.153}$$

Putting $\rho = 1$ we find from (4.153) the cheap duality.

2. Define the modified creation and annihilation operators

$$\tilde{a}_x = e^{-\lambda_x} a_x,$$
$$\tilde{a}_x^\dagger = e^{\lambda_x} a_x^\dagger, \tag{4.154}$$

then we have (by bi-linearity of commutators) that $\tilde{a}_x, \tilde{a}_x^\dagger, x \in \mathbb{Z}$ satisfy the same commutation relations as $a_x, a_x^\dagger, x \in \mathbb{Z}$. Moreover, the generator $L^{\lambda,p,q}$ reads

$$L^{\lambda,p,q} = \sum_x p(\tilde{a}_x \tilde{a}_{x+1}^\dagger - \tilde{a}_x \tilde{a}_x^\dagger) + q(\tilde{a}_{x+1} \tilde{a}_x^\dagger - \tilde{a}_{x+1} \tilde{a}_{x+1}^\dagger). \tag{4.155}$$

Because this has the same form as (4.139) but with $a, a^\dagger$ replaced by the corresponding $\tilde{a}, \tilde{a}^\dagger$, we obtain that $L^{\lambda,p,q}$ commutes with $\sum_x \tilde{a}_x^\dagger$ which is (4.150).

3. From item 1 we have that

$$\big(L^{\lambda,p,q} D_{\text{cheap}}(\cdot, \eta)\big)(\xi) = \big(L^{-\lambda,q,p} D_{\text{cheap}}(\xi, \cdot)\big)(\eta). \tag{4.156}$$

Because $e^{\mathcal{S}^+}$ commutes with $L^{\lambda,p,q}$ we obtain that $D(\xi, \eta) = \big(e^{\mathcal{S}^+} D_{\text{cheap}}(\cdot, \eta)\big)(\xi)$ is also a duality function, i.e.

$$\big(L^{\lambda,p,q} D(\cdot, \eta)\big)(\xi) = \big(L^{-\lambda,q,p} D(\xi, \cdot)\big)(\eta). \tag{4.157}$$

Computing explicitly this function, we obtain the claimed (4.151).

4. This follows from the previous item via Proposition 4.25.

□

From (4.146) we have that the duality relation (4.152) reads

$$\mathbb{E}_\eta^{p,q}\left[e^{-\sum_x \lambda_x (\eta_x(t) - \eta_x(0))} \mathcal{D}(\xi, \eta(t))\right] = \mathbb{E}_\xi^{q,p}\left[e^{\sum_x \lambda_x (\eta_x(t) - \eta_x(0))} \mathcal{D}(\xi(t), \eta)\right]. \tag{4.158}$$

Unhappily, this does not contain more information than the relation holding for the case $\lambda = 0$. In order to obtain information about exponential moments of currents, we will exploit the fact that the operator $L^{\lambda,p,q}$ is, up to a multiplication operator, equal to the generator of a system of independent walkers. Indeed we have

$$\begin{aligned} L^{\lambda,p,q} &= \sum_x p e^{u_{x,x+1}} (a_x a_{x+1}^\dagger - a_x a_x^\dagger) + \sum_x q e^{-u_{x,x+1}} (a_{x+1} a_x^\dagger - a_{x+1} a_{x+1}^\dagger) \\ &+ \sum_x p(e^{u_{x,x+1}} - 1) a_x a_x^\dagger + q(e^{-u_{x,x+1}} - 1) a_{x+1} a_{x+1}^\dagger \\ &= \widetilde{L^{\lambda,p,q}} + \Psi, \end{aligned} \tag{4.159}$$

where $\widetilde{L^{\lambda,p,q}}$ is the generator of independent random walkers jumping at rate $pe^{u_{x,x+1}}$ from x to $x+1$, and at rate $qe^{-u_{x,x+1}}$ from $x+1$ to x, and where Ψ is the multiplication operator, $\Psi f(\eta) = \psi(\eta) \cdot f(\eta)$ associated to the function

$$\psi(\eta) = \sum_x p(e^{u_{x,x+1}} - 1)\eta_x + q(e^{-u_{x,x+1}} - 1)\eta_{x+1} = \sum_x \varphi(x)\eta_x, \tag{4.160}$$

with

$$\varphi(x) := p(e^{u_{x,x+1}} - 1) + q(e^{-u_{x-1,x}} - 1). \tag{4.161}$$

In order to have well-defined quantities here, we require either that $u = (u_{x,x+1})_x$ has finite support, or that we are on the torus $T_N = \mathbb{Z}/N\mathbb{Z}$. Let us denote by $\widetilde{\mathbb{E}}_\eta^{p,q,u}$ the expectation in the process with generator $\widetilde{L^{\lambda,p,q}}$. Then, using (4.143), (4.159) and the Feynman-Kac formula, we obtain

$$\begin{aligned}
&\mathbb{E}_\eta^{p,q}\left[e^{-\sum_x u_{x,x+1} J_{x,x+1}(t)} f(\eta(t))\right] \\
&= e^{tL^{\lambda,p,q}} f(\eta) \\
&= \widetilde{\mathbb{E}}_\eta^{p,q,u}\left[e^{\int_0^t \psi(\eta(s))ds} f(\eta(t))\right] \\
&= \widetilde{\mathbb{E}}_\eta^{p,q,u}\left[\exp\left\{\sum_x \varphi(x) \int_0^t \eta_x(s)ds\right\} f(\eta(t))\right].
\end{aligned} \tag{4.162}$$

If we are interested in the stationary current fluctuations, then we further want to average η over the Poisson measure ν_ρ. For this we use the following lemma which is an easy computation using the moment generating function of the Poisson distribution of which we therefore do not spell out the proof.

Lemma 4.27 (Generating Function Independent Poisson Random Variables) *Let η be distributed according to a product of Poisson measures with $\mathbb{E}(\eta_x) = \rho_x$. Let $G_{y,j}$, $y \in \mathbb{Z}$, $j = 1, \ldots, \eta_y$ be independent random variables with distribution only depending on y. Define $F_y = \log \mathbb{E}[e^{G_{y,1}}]$, then we have*

$$\mathbb{E}\left[\exp\left\{\sum_y \sum_{j=1}^{\eta_y} G_{y,j}\right\}\right] = \exp\left\{\sum_y \rho_y(e^{F_y} - 1)\right\}. \tag{4.163}$$

Let ν_ρ denote the product Poisson distribution with expectation ρ_x at site x. Then we continue from (4.162). We compute

$$
\begin{aligned}
&\int \mathbb{E}_\eta^{p,q}\left[e^{-\sum_x u_{x,x+1}J_{x,x+1}(t)}\right]\cdot \nu_\rho(d\eta)\\
&= \int \widetilde{\mathbb{E}}_\eta^{p,q,u}\left[\exp\left\{\sum_x \varphi(x)\int_0^t \eta_x(s)ds\right\}\right]\cdot \nu_\rho(d\eta)\\
&= \widetilde{\mathbb{E}}_{\nu_\rho}^{p,q,u}\left[\exp\left\{\sum_y \sum_{j=1}^{\eta_y}\sum_x \varphi(x)\int_0^t \mathbb{1}_{\{X^{y,j}(s)=x\}}ds\right\}\right],
\end{aligned}
\tag{4.164}
$$

where in the last step we used (3.1), and abbreviated the expectation over both walks and η by "$\widetilde{\mathbb{E}}_{\nu_\rho}^{p,q,u}$". We are now in the situation of Lemma 4.27, with

$$
G_{y,1} = \sum_x \varphi(x)\int_0^t \mathbb{1}_{\{X^{y,1}(s)=x\}}ds. \tag{4.165}
$$

As a consequence, we obtain

$$
\int \mathbb{E}_\eta^{p,q}\left[e^{-\sum_x u_{x,x+1}J_{x,x+1}(t)}\right]\cdot \nu_\rho(d\eta) = e^{\sum_y \rho_y (e^{F_y}-1)}, \tag{4.166}
$$

with

$$
F_y = \log \mathbb{E}_y^{RW,p,q,u}\left[\exp\left\{\sum_x \varphi(x)\int_0^t \mathbb{1}_{\{X^{y,1}(s)=x\}}ds\right\}\right]. \tag{4.167}
$$

We then finally obtain the following expression for the joint log moment generating function of currents:

$$
\begin{aligned}
&\log \int \mathbb{E}_\eta^{p,q}\left[e^{-\sum_x u_{x,x+1}J_{x,x+1}(t)}\right]\cdot \nu_\rho(d\eta)\\
&= \sum_y \rho(y)\cdot\left(\mathbb{E}_y^{RW,p,q,u}\left[e^{\sum_x \varphi(x) l_t(x)}\right]-1\right),
\end{aligned}
\tag{4.168}
$$

where now $\mathbb{E}_y^{RW,p,q,u}$ denotes expectation w.r.t a single random walker starting at y and jumping at rate $pe^{u_{x,x+1}}$ (resp. $qe^{-u_{x,x+1}}$) to the right (resp. left), and $l_t(x)$ denotes its local time at x, i.e., $l_t(x) = \int_0^t \mathbb{1}_{\{X(s)=x\}}ds$. This formula shows that the computation of the joint log moment generating function of currents reduces, in this case, to a single random walk computation.

4.9 Additional Notes

The duality between the deterministic systems of ODE's and independent walkers was proved in [116], the intertwining was discussed in [198]. These properties can also be seen as a consequence of Doob's theorem [71] which states that a product of Poisson measures is preserved under the evolution of independent particles. This in turn can be seen as a consequence of the random displacement theorem in point process theory see [161]. The use of duality in the characterization of invariant measures is again in the spirit of [172], Chapter 8. The use of creation and annihilation operators to compute current large deviation functions is in the spirit of [165].

Chapter 5
Duality for the Symmetric Inclusion Process

In this chapter we apply the Lie algebraic approach to the symmetric inclusion processes. This is a process describing particles that move on a lattice with an attractive interaction. We show that the generator of this process can be written in an abstract form in terms of the generators of the Lie algebra $\mathfrak{su}(1, 1)$. This allows to easily find symmetries of the generator. Applying these symmetries to the cheap self-duality function related to the reversible product measures, we find a non-trivial triangular self-duality function. We then show how the same result can be obtained via the change-of-representation method. As an application of this self-duality result we show how the self-duality relation provides information about the n-point correlations. In particular we show that, starting from local-equilibrium measures, the time-evolved measure under the inclusion dynamics has positive correlations.

5.1 Introduction

Just as systems of independent random walkers are related to the Heisenberg algebra, we shall introduce now a class of models whose underlying algebraic structure is based on the Lie algebra $\mathfrak{su}(1, 1)$. This is a rich family whose two basic models are the *symmetric inclusion process* (an interacting particle system) and the *Brownian energy process* (an interacting diffusions system). These are be the subjects of Chaps. 5 and 6, respectively. Their thermalization limits will produce several redistribution models with discrete and with continuous variables, including models well-known in the literature, as well as other novel models, some of which will be treated in Chap. 8. Finally, we will also discuss families of integrable processes with $\mathfrak{su}(1, 1)$ symmetry in Chap. 13.

C. Giardinà, F. Redig, *Duality for Markov Processes*, Grundlehren der mathematischen Wissenschaften 365, https://doi.org/10.1007/978-3-032-04099-2_5

The symmetric inclusion process with parameter $\alpha > 0$, denoted SIP(α), is a process where particles perform symmetric random walks at rate α and, on top of that, each pair of particles at neighboring positions can have an "inclusion event" after which they join at the same place. These "inclusion jumps" induce an attractive interaction between the particles that encourages particles to move to the same place. This is somehow the opposite of what happens in the standard exclusion process (with at most one particle per site) or in the partial exclusion process (allowing more than one particle per site) where particles are forbidden, or discouraged, to be at the same place. We say therefore that the symmetric inclusion process is a "bosonic" analogue of the symmetric exclusion process which is "fermionic". The use of the word "bosonic" can be further justified by the fact that the SIP(α) generator can indeed be written in terms of bona-fide bosonic creation and annihilation operators (as the symmetric exclusion process generator can be written using fermions).

Taking the scaling limit of the SIP(α) leads to a diffusion process known as the Brownian energy process with parameter α, denoted BEP(α). This process describes the continuous diffusion over a set of vertices of a positive quantity. We can think of this quantity as energy and then the model can be viewed as a model of heat conduction [115]. Remarkably, if we specialize to kinetic energies, then also the process describing the evolution of momenta turns out to evolve as a diffusion process, the so-called Brownian momentum process BMP(α) model.

A famous process for which duality was a key tool of analysis is the KMP process (after Kipnis, Marchioro and Presutti [152]) where, at random instances, energy is uniformly redistributed over the vertices of each edge of a graph. We shall show in Chap. 8 that the KMP process arises as the thermalization limit of the Brownian energy process with parameter $\alpha = 1$. Similarly, the thermalization of the Brownian momentum process with parameter $\alpha = 1/2$ leads to another famous process widely used in kinetic theory, the Kac model [143]. The redistribution model which emerges as the thermalization limit of the symmetric inclusion process is related to wealth redistribution models, which are known in the literature under the name "immediate exchange models". We refer again to Chap. 8 for a discussion of these and several other discrete redistribution models.

The SIP(α) family has other possible interpretations. For instance, in the context of *population genetics*, instead of lattice sites and number of particles located at different sites, one thinks of a finite graph of "types" and number of individuals of different types. These individuals can mutate (independent random walk part of the hopping rate) and can have a random mating event (inclusion part of the hopping rate, interpreted here as coalescence), while the total population size is constant. Models of this type are then known under the name of *Moran models*, and their diffusion limits as *Wright-Fisher diffusions*. There exists an extensive literature treating this class of models and their duality properties (see e.g. [66] for a general approach to duality in the context of spatial Fleming-Viot models, and [182] for an early paper on the subject of duality in exchangeable models of population dynamics). See the end of Chap. 8 for a discussion of the algebraic perspective on some of these dualities.

5.2 The Finite Symmetric Inclusion Process

Let V denote a finite set and $p : V \times V \to (0, \infty)$ a non-negative symmetric and irreducible function (as defined at the beginning of Sect. 3.1). For $\alpha > 0$ we define the SIP(α) as the process $\{\eta(t) : t \geq 0\}$ on the configuration space $\Omega = \mathbb{N}^V$ with generator

$$Lf(\eta) = \sum_{x,y \in V} p(x, y)\eta_x(\alpha + \eta_y)\left(f(\eta^{x,y}) - f(\eta)\right). \tag{5.1}$$

We can split the generator in two parts

$$L = \alpha L^{\text{irw}} + L^{\text{incl}}.$$

Here

$$\alpha L^{\text{irw}} = \sum_{x,y \in V} \alpha p(x, y)\eta_x \left(f(\eta^{x,y}) - f(\eta)\right) \tag{5.2}$$

describes independent random walkers moving at rate $\alpha p(x, y)$ between x and y, and

$$L^{\text{incl}} = \sum_{x,y \in V} p(x, y)\eta_x\eta_y \left(f(\eta^{x,y}) - f(\eta)\right) \tag{5.3}$$

describes the attractive interaction part which works as follows: each particle at site x sends an invitation to each particle at site y at rate $p(x, y)$ (different invitations are sent independently); then, after receiving the invitation, one of the particles at y joins the site x.

5.3 Symmetries of the Generator

In this section we are going to write the generator of the symmetric inclusion process in an abstract form, in order to infer commuting operators. The ultimate goal is to construct a non-trivial self-duality, using symmetries and a cheap self-duality (see Sect. 5.4), in the spirit of Theorem 2.9.

Two-Site System

As we did in Chap. 3 for independent random walkers, we first consider the two-site system with lattice vertices that we call 1 and 2. We denote by $(\eta_1, \eta_2) \in \mathbb{N}^2$ the corresponding particle numbers and, without loss of generality, we put $p(1, 2) =$

$p(2, 1) = 1$. Then we have

$$L_{1,2}f(\eta_1, \eta_2) = \eta_1(\alpha + \eta_2)(f(\eta_1 - 1, \eta_2 + 1) - f(\eta)) \\ + \eta_2(\alpha + \eta_1)(f(\eta_1 + 1, \eta_2 - 1) - f(\eta)). \tag{5.4}$$

We introduce the operators K^+, K^- and K^0 working on functions $f : \mathbb{N} \to \mathbb{R}$ as follows:

$$K^+ f(n) = (\alpha + n) f(n+1), \\ K^- f(n) = n f(n-1), \\ K^0 f(n) = \left(\tfrac{\alpha}{2} + n\right) f(n). \tag{5.5}$$

The following lemma establishes the commutation relations of the operators K^+, K^- and K^0 and show that they form a representation of the Lie algebra $\mathfrak{su}(1, 1)^*$, i.e. the conjugate of the Lie algebra $\mathfrak{su}(1, 1)$.

Lemma 5.1 (Commutation Relations) *The operators K^+, K^-, K^0 defined in* (5.5) *satisfy*

$$[K^\pm, K^0] = \pm K^\pm, \\ [K^+, K^-] = 2K^0. \tag{5.6}$$

Proof The proof is a simple explicit computation:

$$K^+K^- f(n) - K^-K^+ f(n) \\ = (\alpha + n)(n + 1) f(n) - n(\alpha + n - 1) f(n) \\ = (\alpha + 2n) f(n) = 2K^0 f(n). \tag{5.7}$$

Moreover

$$K^+K^0 f(n) - K^0K^+ f(n) \\ = (\alpha + n)(\tfrac{\alpha}{2} + n + 1) f(n + 1) - (\tfrac{\alpha}{2} + n + 1)(\alpha + n) f(n + 1) \\ = (\alpha + n) f(n) = K^+ f(n), \tag{5.8}$$

and

$$K^-K^0 f(n) - K^0K^- f(n) \\ = n(\tfrac{\alpha}{2} + n - 1) f(n - 1) - (\tfrac{\alpha}{2} + n) n f(n - 1) \\ = -n f(n - 1) = -K^- f(n). \tag{5.9}$$

□

Remark 5.2 The operators K^+, K^- and K^0 provide a discrete representation of the conjugate Lie algebra $\mathfrak{su}(1,1)^*$ defined by the commutation rules (5.6). The commutation relations of the Lie algebra $\mathfrak{su}(1,1)$ are instead obtained from (5.6) by inverting the signs on the right hand sides of (5.6).

We extend now the definition of the operators (5.5) to the two-site system. For $u \in \{-, 0, +\}$, we denote by $K_1^u = K^u \otimes I$, resp. $K_2^u = I \otimes K^u$, the copy of the operator K^u working on site 1, resp. on site 2. Notice that both K_1^u and K_2^u are operators acting on functions $f : \mathbb{N}^2 \to \mathbb{R}$. The following lemma shows the connection between these operators and the symmetric inclusion process generator $L_{1,2}$.

Proposition 5.3 (Abstract Form of the Generator) *The single-edge generator* (5.4) *of the SIP*(α) *is equal to:*

$$L_{1,2} = K_1^+ K_2^- + K_1^- K_2^+ - 2K_1^0 K_2^0 + \frac{\alpha^2}{2}. \tag{5.10}$$

Proof The proof is a simple explicit computation following from (5.5). Let $f : N^2 \to \mathbb{R}$, then we have

$$\begin{aligned}
&\left(K_1^+ K_2^- + K_1^- K_2^+ - 2K_1^0 K_2^0 + \frac{\alpha^2}{2}\right) f(\eta_1, \eta_2) \\
&= (\alpha + \eta_1)\eta_2 f(\eta_1 + 1, \eta_2 - 1) + \eta_1(\alpha + \eta_2) f(\eta_1 - 1, \eta_2 + 1) \\
&\quad -2\left(\frac{\alpha}{2} + \eta_1\right)\left(\frac{\alpha}{2} + \eta_2\right) f(\eta_1, \eta_2) + \frac{\alpha^2}{2} f(\eta_1, \eta_2) \\
&= L_{1,2} f(\eta_1, \eta_2).
\end{aligned} \tag{5.11}$$

This concludes the proof. □

Finally, we use this abstract form of the generator to infer its symmetries.

Lemma 5.4 (Symmetries of the Generator) *The generator L in* (5.10) *commutes with*

$$K_1^+ + K_2^+, \qquad K_1^- + K_2^-, \qquad K_1^0 + K_2^0, \tag{5.12}$$

with $K_1^u = K^u \otimes I$ and $K_2^u = I \otimes K^u$, for $u \in \{-, 0, +\}$.

Proof We prove the statement for $K_1^+ + K_2^+$, the computations for other cases being similar. We have

$$\begin{aligned}
[L_{1,2}, K_1^+ + K_2^+] &= [K_1^+ K_2^- + K_1^- K_2^+ - 2K_1^0 K_2^0, K_1^+ + K_2^+] \\
&= [K_1^-, K_1^+] K_2^+ - 2[K_1^0, K_1^+] K_2^0 + K_1^+ [K_2^-, K_2^+]
\end{aligned}$$

$$
\begin{aligned}
&-2K_1^0[K_2^0, K_2^+] \\
&= -2K_1^0 K_2^+ + 2K_1^+ K_2^0 - 2K_1^+ K_2^0 + 2K_1^0 K_2^+ = 0, \quad (5.13)
\end{aligned}
$$

where we used the fact that $[K_1^u, K_2^v] = 0$ for all $u, v \in \{-, 0, +\}$. □

General Case

We extend now the results obtained in the previous paragraph, to a general lattice V. The generator of the symmetric inclusion process with parameter α defined in (5.1) can be written in the form

$$
L = \frac{1}{2} \sum_{x,y \in V} p(x, y) L_{x,y} = \sum_{\{x,y\} \in E} p(\{x, y\}) L_{\{x,y\}}, \tag{5.14}
$$

where E is the edge set, $p(\{x, y\}) = p(x, y) = p(y, x)$ and $L_{\{x,y\}} = L_{x,y}$ is the single-edge generator, that is given by

$$
\begin{aligned}
L_{x,y} f(\eta) &= \eta_x(\alpha + \eta_y)(f(\eta - \delta_x + \delta_y) - f(\eta)) \\
&\quad + \eta_y(\alpha + \eta_x)(f(\eta + \delta_x - \delta_y) - f(\eta)). \quad (5.15)
\end{aligned}
$$

Here δ_x is as usual the configuration with only one particle at site $x \in V$.

We then define copies of the operators K^u, $u \in \{-, 0, +\}$ labeled by the sites of the lattice. More explicitly, for all $x \in V$ we introduce the operators K_x^+, K_x^- and K_x^0 working on functions $f : \mathbb{N}^V \to \mathbb{R}$ as follows:

$$
\begin{aligned}
K_x^+ f(\eta) &= (\alpha + \eta_x) f(\eta + \delta_x), \\
K_x^- f(\eta) &= \eta_x f(\eta - \delta_x), \\
K_x^0 f(\eta) &= \left(\tfrac{\alpha}{2} + \eta_x\right) f(\eta_x). \quad (5.16)
\end{aligned}
$$

In other words, the operator K_x^u, with $u \in \{-, 0, +\}$, has to be understood as the tensor product of identity operators I_y (labeled by $y \in V$ with $y \neq x$) and a copy of the operator K^u (labeled by x). Then, as a consequence of Lemma 5.1 we have that the following commutation relations hold

$$
\begin{aligned}
[K_x^{\pm}, K_y^0] &= \pm K_x^{\pm} \delta_{x,y}, \\
[K_x^+, K_y^-] &= 2K_x^0 \delta_{x,y}. \quad (5.17)
\end{aligned}
$$

The action of the generator L can be decomposed in the action of the single-edge generators $L_{x,y}$ working only on the two sites x and y and given, in its abstract form by

$$L_{x,y} = K_x^+ K_y^- + K_x^- K_y^+ - 2K_x^0 K_y^0 + \frac{\alpha^2}{2}. \tag{5.18}$$

As a consequence, we can repeat the argument used in the proof of Lemma 5.4 to find symmetries of $L_{x,y}$, and then generalize the result to the graph.

Proposition 5.5 (Symmetries of the Generator) *The generator L in* (5.14) *commutes with*

$$S^+ = \sum_{x \in V} K_x^+, \qquad S^- = \sum_{x \in V} K_x^-, \qquad S^0 = \sum_{x \in V} K_x^0, \tag{5.19}$$

where K_x^u, with $u \in \{-, 0, +\}$, is given in (5.16).

Proof We prove the statement for S^+, the computations for other cases being similar. Since both $L_{x,y}$ and $K_x^+ + K_y^+$ act only on the coordinates x and y, using Lemma 5.4 we deduce that

$$[L_{x,y}, K_x^+ + K_y^+] = 0 \qquad \text{for all} \qquad x, y \in V. \tag{5.20}$$

We compute

$$[L, S^+] = \frac{1}{2} \sum_{x,y,z \in V} p(x,y)[L_{x,y}, K_z^+] = \frac{1}{2} \sum_{x,y \in V} p(x,y)[L_{x,y}, K_x^+ + K_y^+] = 0, \tag{5.21}$$

where the last identity follows from the fact that $[L_{x,y}, K_z^+] = 0$ for all $z \notin \{x, y\}$. Then, using (5.20), it follows that $[L, S^+] = 0$. □

Remark 5.6 The last of the symmetries in (5.19) admits a direct interpretation for the process as it expresses the fact that the symmetric inclusion process conserves the number of particles. The other two symmetries are instead called "hidden symmetries" as they are found from the algebraic description.

5.4 Self-duality

In this section we follow the strategy of Theorem 3.9 to find a non-trivial self-duality function for the symmetric inclusion process. In the previous section we found non trivial symmetries $S^{\pm}$ of the generator that are of the form of particle addition, resp. particle removal operators. These are additive symmetries. Since we want duality functions in product form, we want symmetries which factorize. This is obtained by taking the exponential $e^{S^{\pm}}$, which leads to triangular symmetries of the generator. The only missing ingredient, in order to implement the procedure developed in Theorem 3.9 is a cheap self-duality function. We will see that this can be easily obtained because of the reversibility of the symmetric inclusion process.

We start by identifying the cheap self-duality function via the following lemma. We denote by $\Gamma(\cdot)$ the Gamma function, i.e., for $t > 0$

$$\Gamma(t) = \int_0^\infty x^{t-1} e^{-x} dx, \tag{5.22}$$

and remind the reader the basic recursion $\Gamma(t+1) = t\Gamma(t)$.

Lemma 5.7 (Reversible Measures and Cheap Self-duality) *For all $\lambda > 0$ the measure*

$$M(\eta) = \prod_{x \in V} \frac{\lambda^{\eta_x}}{\eta_x!} \frac{\Gamma(\eta_x + \alpha)}{\Gamma(\alpha)} \tag{5.23}$$

is a reversible measure for the SIP(α) *on a finite graph V. As a consequence, the function*

$$D^{\mathrm{ch}}(\xi, \eta) = \prod_{x \in V} d^{\mathrm{ch}}(\xi_x, \eta_x), \qquad \textit{with} \qquad d^{\mathrm{ch}}(k, n) = \frac{n!\Gamma(\alpha)}{\Gamma(\alpha + n)} \delta_{k,n} \tag{5.24}$$

is a cheap self-duality function.

Proof This follows from the detailed balance relation (2.43), that is itself a consequence of the fact that

$$\begin{aligned} &\frac{\lambda^n \Gamma(\alpha + n)}{n!\Gamma(\alpha)} \frac{\lambda^m \Gamma(\alpha + m)}{m!\Gamma(\alpha)} n(\alpha + m) \\ &= \frac{\lambda^{n-1}\Gamma(\alpha + n - 1)}{(n-1)!\Gamma(\alpha)} \frac{\lambda^{m+1}\Gamma(\alpha + m + 1)}{(m+1)!\Gamma(\alpha)} (m+1)(\alpha + n - 1), \end{aligned} \tag{5.25}$$

which holds for all $n, m \in \mathbb{N}$. Then (5.24) follows from item 1 of Theorem 2.9. □

We can then act with the symmetries found in Sect. 5.3 in order to construct a triangular self-duality function for SIP(α).

Theorem 5.8 (Self-duality of SIP(α)**)** *The symmetric inclusion process with parameter $\alpha > 0$ on a finite graph V with generator* (5.1) *is self-dual with self-duality function*

$$D(\xi, \eta) = \prod_{x \in V} d(\xi_x, \eta_x) \qquad \textit{with} \qquad d(k, n) = \frac{n!}{(n-k)!} \frac{\Gamma(\alpha)}{\Gamma(\alpha + k)} \mathbb{1}_{\{k \le n\}}. \tag{5.26}$$

Proof We first consider the process on two vertices. We use the symmetry $e^{S^+} = e^{K_1^+ + K_2^+}$ which commutes with L_{12} by Proposition 5.5. As we did in Chap. 3 for independent random walkers, also here we choose to start from an exponential symmetry since we aim at producing a self-duality function in factorized form. In the spirit of item 2 of Theorem 2.9, we work with this symmetry on the cheap self-duality function identified in Lemma 5.7. It thus suffices to show that, for $k \le n$

$$e^{K^+}[d^{\mathrm{ch}}(\cdot, n)](k) = d(k, n). \tag{5.27}$$

For a function $f : \mathbb{N} \to \mathbb{R}$, using $(K^+ f)(k) = (\alpha + k) f(k + 1)$, we have

$$(e^{K^+} f)(k) = \sum_{\ell=0}^{\infty} \frac{\Gamma(\alpha + k + \ell)}{\Gamma(\alpha + k)\ell!} f(k + \ell). \tag{5.28}$$

Then

$$\begin{aligned} e^{K^+}[d^{\mathrm{ch}}(\cdot, n)](k) &= \sum_{\ell=0}^{\infty} \frac{\Gamma(\alpha + k + \ell)}{\Gamma(\alpha + k)\ell!} d^{\mathrm{ch}}(k + \ell, n) \\ &= \sum_{\ell=0}^{\infty} \frac{\Gamma(\alpha + k + \ell)}{\Gamma(\alpha + k)\ell!} \cdot \frac{n!\Gamma(\alpha)}{\Gamma(\alpha + n)} \delta_{\ell, n-k} \\ &= \frac{\Gamma(\alpha + n)}{\Gamma(\alpha + k)(n - k)!} \cdot \frac{n!\Gamma(\alpha)}{\Gamma(\alpha + n)} \mathbb{1}_{\{k \le n\}} \qquad (5.29) \\ &= d(k, n), \end{aligned}$$

which indeed yields (5.27). The general case of a finite graph V follows then because the generator (5.14) is given by the sum $L = \sum_{\{x,y\} \in E} p(x, y) L_{x,y}$ where $L_{x,y}$ is the two-site generator (5.15) working on η_x, η_y. Hence the two-site self-duality result applies for each edge $\{x, y\}$, and then the general result follows from the symmetry of $p(x, y)$. □

Remark 5.9 (Working with S^- Instead of S^+) Notice that working with e^{S^-} on the n-variable of the single-site cheap duality function $d^{\mathrm{ch}}(k, n)$ yields exactly the same single-site self-duality polynomial (5.26).

By properly renormalizing the reversible measures (5.23), we obtain a one-parameter family of reversible probability distributions for the symmetric inclusion process. These are homogeneous product measure whose single-site marginals are discrete-Gamma distributions (α, λ), i.e. with parameters $\lambda \in (0, 1)$ and $\alpha \in (0, \infty)$:

$$\nu_{\lambda,\alpha}(n) = (1 - \lambda)^{\alpha} \frac{\lambda^n}{n!} \frac{\Gamma(\alpha + n)}{\Gamma(\alpha)}, \qquad n \in \mathbb{N}. \tag{5.30}$$

Notice that for $\alpha \in \mathbb{N}$, $\nu_{\lambda,\alpha}$ is a Negative-Binomial (α, λ), and in particular, for $\alpha = 1$ it is a Geometric (λ). With a slight abuse we use the same notation $\nu_{\lambda,\alpha}$ for the homogeneous product measure on $\mathbb{N}^V$ with marginals (5.30).

Remark 5.10 (Invariant Measures) The symmetric inclusion process preserves the total number of particles. As a consequence, homogeneous products of discrete-Gamma are reversible but not ergodic. For finite graphs, to find the ergodic measures one needs to condition on the total number of particles. Let us discuss this with 2 sites and consider two independent random variables X, Y distributed according to (5.30) and define the random variable $U_N := X|X + Y = N$, where $N \in \mathbb{N}$. It is easy to verify that U_N is distributed as a Beta-Binomial with parameters (N, α, α):

$$P(U_N = k) = \mathbb{E}\left(\binom{N}{k} B^k (1 - B)^{N-k}\right), \tag{5.31}$$

where the expectation is taken over a random variable B distributed as a Beta(α, α), i.e., B takes values in the interval $[0, 1]$ and has probability density

$$f_B(x) = \frac{x^{\alpha-1}(1 - x)^{\alpha-1}}{B(\alpha, \alpha)}, \quad 0 \leq x \leq 1. \tag{5.32}$$

Notice that in the particular case $\alpha = 1$, U_N has a discrete-Uniform distribution on the set $\{0, \ldots, N\}$. Let now $\{\eta(t) = (\eta_1(t), \eta_2(t)),\ t \geq 0\}$ be the SIP(α) on two sites. Starting this process from a configuration $\eta = (\eta_1, \eta_2) \in \mathbb{N}^2$ with $\eta_1 + \eta_2 = N$, we have that $\{\eta(t),\ t \geq 0\}$ converges in distribution to $(X, N - X)$ with X distributed as a Beta-Binomial with parameters (N, α, α). This will be used in Chap. 8 when we define the thermalization of the symmetric inclusion process.

As we did in Proposition 3.20 for independent walkers, we prove here a relation involving the expectations of the self-duality polynomials (5.26) of SIP(α) with respect to the reversible probability measures.

Proposition 5.11 (Expectation of the Duality Function in the Reversible Distribution) *We have for all $k \leq n$, $\lambda \in (0, 1)$ and $\alpha \in (0, \infty)$*

$$\sum_{n=0}^{\infty} d(k, n)\nu_{\lambda,\alpha}(n) = \left(\frac{\lambda}{1 - \lambda}\right)^k, \tag{5.33}$$

where $d(k, n)$ is the single-site SIP(α) *self-duality function in the right-hand-side of* (5.26). *As a consequence, for the self-duality function $D(\xi, \eta)$ defined in* (5.26) *we have*

$$\int D(\xi, \eta)\nu_{\lambda,\alpha}(d\eta) = \left(\frac{\lambda}{1 - \lambda}\right)^{|\xi|} = \left(\int D(\delta_x, \eta)\nu_{\lambda,\alpha}(d\eta)\right)^{|\xi|}, \tag{5.34}$$

for all $x \in V$.

Proof The proof is an explicit and elementary computation:

$$\begin{aligned}\sum_{n=0}^{\infty} d(k,n)\nu_{\lambda,\alpha}(n) &= \sum_{n=k}^{\infty} \frac{n!}{(n-k)!}\frac{\Gamma(\alpha)}{\Gamma(\alpha+k)}\frac{\lambda^n}{n!}\frac{\Gamma(\alpha+n)}{\Gamma(\alpha)}(1-\lambda)^{\alpha} \\ &= (1-\lambda)^{\alpha}\lambda^k \sum_{n=0}^{\infty} \frac{\lambda^n}{n!}\frac{\Gamma(\alpha+n+k)}{\Gamma(\alpha+k)} \\ &= (1-\lambda)^{\alpha}\lambda^k \frac{\int_0^\infty \sum_{n=0}^{\infty}\frac{\lambda^n}{n!}x^n x^{\alpha+k-1}e^{-x}\,dx}{\int_0^\infty x^{\alpha+k-1}e^{-x}\,dx} \\ &= (1-\lambda)^{\alpha}\lambda^k \frac{\int_0^\infty x^{\alpha+k-1}e^{\lambda x}e^{-x}\,dx}{\int_0^\infty x^{\alpha+k-1}e^{-x}\,dx} \\ &= \frac{\lambda^k}{(1-\lambda)^k}. \end{aligned} \tag{5.35}$$

This proves (5.33). Then (5.34) follows from (5.33) and the fact that

$$\frac{\lambda}{1-\lambda} = \int D(\delta_x,\eta)\nu_{\lambda,\alpha}(d\eta).$$

for all $x \in V$. □

Remark 5.12 Later on in the book (e.g. in Chap. 10) we will use different parametrizations for the reversible product measure $\nu_{\lambda,\alpha}$ defined (5.30). A possible reparametrization is the one labeled by the so-called weighted particle density parameter $\rho := \frac{\lambda}{1-\lambda}$, that is, for a discrete-Gamma(α,λ) random variable, equal to the expected number of particles divided by α. With this reparametrization the *duality transform* in (5.33) gives exactly ρ^k.

5.5 Self-duality as a Change of Representation

The following theorem shows a deeper connection between the K operators in (5.5) and the cheap self-duality function, provided by the discrete representation of $\mathfrak{su}(1,1)$.

Proposition 5.13 (Cheap Self-duality as a Change of Representation) *Let K^u, $u \in \{-,0,+\}$ be the operators defined in* (5.5), *and let d^{ch} be the single-site cheap duality function defined in* (5.24), *then we have the following duality relations:*

$$\begin{aligned} K^+ &\xrightarrow{d^{\mathrm{ch}}} K^- \\ K^- &\xrightarrow{d^{\mathrm{ch}}} K^+ \\ K^0 &\xrightarrow{d^{\mathrm{ch}}} K^0. \end{aligned} \tag{5.36}$$

As a consequence, the generator of the symmetric inclusion process L defined in (5.14) *is self-dual with self-duality function* D^{ch} (5.24).

Proof In order to prove (5.36) we have to show that

$$\begin{aligned}\left[K^+d^{\text{ch}}(\cdot,n)\right](k) &= \left[K^-d^{\text{ch}}(k,\cdot)\right](n)\\ \left[K^-d^{\text{ch}}(\cdot,n)\right](k) &= \left[K^+d^{\text{ch}}(k,\cdot)\right](n)\\ \left[K^0d^{\text{ch}}(\cdot,n)\right](k) &= \left[K^0d^{\text{ch}}(k,\cdot)\right](n).\end{aligned} \tag{5.37}$$

We have

$$\begin{aligned}[K^+d^{\text{ch}}(\cdot,n)](k) &= n!\frac{\Gamma(\alpha)}{\Gamma(\alpha+n)}(\alpha+k)\,\delta_{k+1,n}\\ &= \frac{n!\Gamma(\alpha)}{\Gamma(\alpha+n-1)}\,\delta_{k,n-1} = [K^-d^{\text{ch}}(k,\cdot)](n),\end{aligned} \tag{5.38}$$

from which follows the first relation. The second identity can be proved in a similar way, whereas the third one is trivial. We now want to use this to prove the generator duality. Thanks to the decomposition of the generator (5.14) in single-edge generators, it is sufficient to prove the result for the two-site system with $V = \{1,2\}$. Let K_x^u, $u \in \{-,0,+\}$, $x \in V$ as in (5.16). Using Theorem 2.40 we obtain:

$$\begin{aligned}K_1^+K_2^- &\xrightarrow{D^{\text{ch}}} K_1^-K_2^+\\ K_1^-K_2^+ &\xrightarrow{D^{\text{ch}}} K_1^+K_2^-\\ K_1^0K_2^0 &\xrightarrow{D^{\text{ch}}} K_1^0K_2^0.\end{aligned} \tag{5.39}$$

Hence the cheap-duality relation between generators

$$L_{12} \xrightarrow{D^{\text{ch}}} L_{12}, \tag{5.40}$$

then follows using the abstract form (5.18). □

Proposition 5.14 (Triangular Self-duality as a Change of Representation) *Consider now the operators* k^+, k^- *and* k^0 *defined on functions* $f : \mathbb{N} \to \mathbb{R}$ *by*

$$\begin{cases}\left(\mathrm{k}^+ f\right)(n) &= (\alpha+n)f(n+1) - 2(\frac{\alpha}{2}+n)f(n) + nf(n-1)\\ \left(\mathrm{k}^- f\right)(n) &= nf(n-1)\\ \left(\mathrm{k}^0 f\right)(n) &= (n+\frac{\alpha}{2})f(n) - nf(n-1).\end{cases} \tag{5.41}$$

They form a representation of the conjugate Lie algebra $\mathfrak{su}(1,1)^*$*. Moreover the operators* k^+, k^- *and* k^0 *are in duality relation with the operators* K^+, K^- *and* K^0 *defined in* (5.5)

$$\begin{aligned} K^+ &\xrightarrow{d} \mathrm{k}^- \\ K^- &\xrightarrow{d} \mathrm{k}^+ \\ K^0 &\xrightarrow{d} \mathrm{k}^0. \end{aligned} \tag{5.42}$$

where d *is the single-site triangular duality relation defined in* (5.26)*. As a consequence, the generator of the symmetric inclusion process* L *defined in* (5.14) *is self-dual with self-duality function* D *defined in* (5.26)*.*

Proof To prove the result we remark that the operators k^u can be rewritten in terms of K^v, where $u, v \in \{-, 0, +\}$. We have indeed

$$\mathrm{k}^+ = K^+ - 2K^0 + K^-, \qquad \mathrm{k}^- = K^-, \qquad \mathrm{k}^0 = K^0 - K^-. \tag{5.43}$$

As a consequence

$$[\mathrm{k}^-, \mathrm{k}^+] = [K^-, K^+] - 2[K^-, K^0] + [K^-, K^-] = -2K^0 + 2K^- = -2\mathrm{k}^0, \tag{5.44}$$

whereas

$$\begin{aligned} [\mathrm{k}^+, \mathrm{k}^0] &= [K^+, K^0] - [K^+, K^-] - 2[K^0, K^0] + 2[K^0, K^-] + [K^-, K^0] \\ &\quad - [K^-, K^-] \\ &= K^+ - 2K^0 + K^- = \mathrm{k}^+, \end{aligned} \tag{5.45}$$

and

$$[\mathrm{k}^-, \mathrm{k}^0] = [K^-, K^0] - [K^-, K^-] = -K^- = -\mathrm{k}^-. \tag{5.46}$$

Then k^+, k^- and k^0 satisfy the commutation relations (5.6) that define the conjugate Lie algebra $\mathfrak{su}(1,1)^*$. In order to prove (5.42) we need to show that

$$[K^+ d(\cdot, n)](k) = [\mathrm{k}^- d(k, \cdot)](n). \tag{5.47}$$

We have

$$\begin{aligned} [\mathrm{k}^- d(k, \cdot)](n) &= n \cdot \frac{(n-1)!}{(n-1-k)!} \cdot \frac{\Gamma(\alpha)}{\Gamma(\alpha+k)} \mathbb{1}_{\{k \le n-1\}} \\ &= (\alpha + k) \cdot \frac{n!}{(n-1-k)!} \cdot \frac{\Gamma(\alpha)}{\Gamma(\alpha+k+1)} \mathbb{1}_{\{k+1 \le n\}} \\ &= [K^+ d(\cdot, n)](k). \end{aligned} \tag{5.48}$$

Moreover

$$[\mathrm{k}^+ d(k,\cdot)](n) = \left\{(\alpha+n)\,\frac{(n+1)!}{(n+1-k)!}\,\mathbb{1}_{\{k\le n+1\}} - (\alpha+2n)\,\frac{n!}{(n-k)!}\mathbb{1}_{\{k\le n\}} \right.$$
$$\left. +n\,\frac{(n-1)!}{(n-1-k)!}\mathbb{1}_{\{k\le n-1\}}\right\}\frac{\Gamma(\alpha)}{\Gamma(\alpha+k)}$$
$$= k\,\frac{n!}{(n-k+1)!}\,\frac{\Gamma(\alpha)}{\Gamma(\alpha+k-1)} = [K^- d(\cdot,n)](k). \tag{5.49}$$

In a similar way it can be proved that $[K^0 d(\cdot,n)](k) = [\mathrm{k}^0 d(k,\cdot)](n)$. Then, defining "copies" of the operators (5.41), i.e. k_x^u, $u \in \{-,0,+\}$, $x \in V$, using Theorem 2.40 we obtain

$$\begin{aligned} K_1^+ K_2^- &\xrightarrow{D} \mathrm{k}_1^- \mathrm{k}_2^+ \\ K_1^- K_2^+ &\xrightarrow{D} \mathrm{k}_1^+ \mathrm{k}_2^- \\ K_1^0 K_2^0 &\xrightarrow{D} \mathrm{k}_1^0 \mathrm{k}_2^0, \end{aligned} \tag{5.50}$$

with D the self-duality function defined in (5.26) for $V = \{1,2\}$. The third statement can be proved by observing that the two-site generator, rewritten in terms of the operators k^+, k^- and k^0, has exactly the same abstract form of as in (5.10):

$$L_{1,2} = \mathrm{k}_1^+ \mathrm{k}_2^- + \mathrm{k}_1^- \mathrm{k}_2^+ - 2\mathrm{k}_1^0 \mathrm{k}_2^0 + \frac{\alpha^2}{2}. \tag{5.51}$$

Indeed we have

$$\begin{aligned} &\mathrm{k}_1^+ \mathrm{k}_2^- + \mathrm{k}_1^- \mathrm{k}_2^+ - 2\mathrm{k}_1^0 \mathrm{k}_2^0 \\ &= \left(K_1^+ - 2K_1^0 + K_1^-\right) K_2^- + K_1^- \left(K_2^+ - 2K_2^0 + K_2^-\right) \\ &\quad -2\left(K_1^0 - K_1^-\right)\left(K_2^0 - K_2^-\right) \\ &= K_1^+ K_2^- - 2K_1^0 K_2^- + K_1^- K_2^- + K_1^- K_2^+ - 2K_1^- K_2^0 + K_1^- K_2^- \\ &\quad -2K_1^0 K_2^0 + 2K_1^- K_2^0 + 2K_1^0 K_2^- - 2K_1^- K_2^- \\ &= K_1^+ K_2^- + K_1^- K_2^+ - 2K_1^0 K_2^0. \end{aligned} \tag{5.52}$$

Hence the self-duality relation for the two-site generator

$$L_{12} \xrightarrow{D} L_{12}, \tag{5.53}$$

follows using again Theorem 2.40. Then the generator self-duality immediately follows from the single-edge decomposition of L given in (5.14). □

5.6 The Abstract Generator Revisited

Here we show how the abstract generator of the symmetric inclusion process can be generated naturally from a co-product applied to a non-trivial central element of the universal enveloping algebra of $\mathfrak{su}(1,1)$, the so-called Casimir element.

We remind the reader the commutation relations of $\mathfrak{su}(1,1)$.

Definition 5.15 We denote by $\mathfrak{su}(1,1)^*$ the conjugate Lie algebra generated by the elements K^+, K^-, K^0 satisfying the commutation relations

$$[K^{\pm}, K^0] = \pm K^{\pm}, \qquad [K^+, K^-] = 2K^0. \tag{5.54}$$

In this section we denote by $U(\mathfrak{su}(1,1)^*)$ the universal enveloping algebra of $\mathfrak{su}(1,1)^*$.

An element of an algebra is called *central* if it commutes with the algebra generators, and, as a consequence, with all the elements of the algebra. We start by identifying a well-known and relevant central element.

Lemma 5.16 *The so-called* Casimir element*:*

$$C = (K^0)^2 - \frac{1}{2}(K^+K^- + K^-K^+) \tag{5.55}$$

is a central element of $U(\mathfrak{su}(1,1)^*)$.

Proof We start by proving $[C, K^0] = 0$. We have

$$\begin{aligned}[C, K^0] &= -\frac{1}{2}\left([K^+, K^0]K^- + K^+[K^-, K^0] + [K^-, K^0]K^+ + K^-[K^+, K^0]\right)\\ &= -\frac{1}{2}\left(K^+K^- - K^+K^- - K^-K^+ + K^-K^+\right) = 0,\end{aligned} \tag{5.56}$$

while, for $[C, K^-]$ we have

$$\begin{aligned}[C, K^-] &= K^0[K^0, K^-] + [K^0, K^-]K^0 - \frac{1}{2}([K^+, K^-]K^- + K^-[K^+, K^-])\\ &= K^0K^- + K^-K^0 - K^0K^- - K^-K^0 = 0.\end{aligned} \tag{5.57}$$

The computation of $[C, K^+]$ is similar and left to the reader. □

We now illustrate how the abstract generator of the symmetric inclusion process is naturally connected to the Casimir element via a co-product. We start with the definition of the co-product.

Definition 5.17 (Coproduct) For $u \in \{+, -, 0\}$, we define

$$\Delta(K^u) = K^u \otimes I + I \otimes K^u = K_1^u + K_2^u, \tag{5.58}$$

and extend the definition of Δ to all elements of $U(\mathfrak{su}(1,1)^*)$ as follows: $\forall\, g, h \in U(\mathfrak{su}(1,1)^*)$

$$\Delta(g+h) = \Delta(g) + \Delta(h) \qquad \text{and} \qquad \Delta(gh) = \Delta(g)\Delta(h). \tag{5.59}$$

Proposition 5.18 *The coproduct in Definition 5.17 is an algebra homomorphism between $U(\mathfrak{su}(1,1)^*)$ and $U(\mathfrak{su}(1,1)^*) \otimes U(\mathfrak{su}(1,1)^*)$.*

Proof As we did in Remark 3.6 for the Heisenberg algebra, the only thing we have to show is that Δ preserves the commutation relations (5.54), i.e., that

$$\Delta[K^u, K^v] = [\Delta(K^u), \Delta(K^v)], \tag{5.60}$$

for $u, v \in \{+, -, 0\}$. We will show this for $u = +$ and $v = -$, leaving the analogous computations for the other cases to the reader. We will use repeatedly the obvious fact that $[K_1^u, K_2^v] = 0$ for $u, v \in \{+, -, 0\}$. We have

$$\Delta(K^+K^-) = (K_1^+ + K_2^+)(K_1^- + K_2^-) = K_1^+K_1^- + K_2^+K_2^- + K_1^+K_2^- + K_2^+K_1^-, \tag{5.61}$$

and

$$\Delta(K^-K^+) = (K_1^- + K_2^-)(K_1^+ + K_2^+) = K_1^-K_1^+ + K_2^-K_2^+ + K_1^-K_2^+ + K_2^-K_1^+. \tag{5.62}$$

As a consequence,

$$\Delta([K^+, K^-]) = [K_1^+, K_1^-] + [K_2^+, K_2^-] = 2K_1^0 + 2K_2^0, \tag{5.63}$$

and, on the other hand,

$$\begin{aligned}[\Delta(K^+), \Delta(K^-)] &= [K_1^+ + K_2^+, K_1^- + K_2^-] = [K_1^+, K_1^-] + [K_2^+, K_2^-] \\ &= 2K_1^0 + 2K_2^0, \end{aligned} \tag{5.64}$$

from which it follows that $\Delta([K^+, K^-]) = [\Delta(K^+), \Delta(K^-)]$. □

Remark 5.19 We have seen in Proposition 5.18 that the definition of the co-product is consistent with the commutation relations. Notice that this is a general fact that does not depend on the particular Lie algebra. Indeed if one defines the co-product as

$$\Delta(g) = g \otimes I + I \otimes g \tag{5.65}$$

for all elements g of the Lie algebra, then automatically one has

$$\begin{aligned}
&[\Delta(g), \Delta(h)] \\
&= \Delta(g)\Delta(h) - \Delta(h)\Delta(g) \\
&= (g \otimes I + I \otimes g)(h \otimes I + I \otimes h) - (h \otimes I + I \otimes h)(g \otimes I + I \otimes g) \\
&= (gh \otimes I + g \otimes h + h \otimes g + I \otimes gh) - (hg \otimes I + g \otimes h + h \otimes g + I \otimes hg) \\
&= gh \otimes I + I \otimes gh - hg \otimes I - I \otimes hg \\
&= \Delta[g, h]
\end{aligned} \tag{5.66}$$

and, as a consequence, the definition of Δ can be extended to the universal enveloping algebra.

The following lemma shows the relation between $\Delta(C)$ and the abstract generator of the symmetric inclusion process.

Lemma 5.20 *Let C denote the Casimir element defined in* (5.55). *Then we have*

$$\Delta(C) = -(K_1^+ K_2^- + K_2^+ K_1^-) + 2K_1^0 K_2^0 + C_1 + C_2, \tag{5.67}$$

with $C_1 = C \otimes I$ and $C_2 = I \otimes C$.

Proof We have

$$\begin{aligned}
\Delta(C) &= \Delta\left((K^0)^2\right) - \frac{1}{2}\Delta\left(K^+K^- + K^-K^+\right) \\
&= (K_1^0 + K_2^0)(K_1^0 + K_2^0) - \frac{1}{2}(K_1^+ + K_2^+)(K_1^- + K_2^-) \\
&\quad - \frac{1}{2}(K_1^- + K_2^-)(K_1^+ + K_2^+) \\
&= -(K_1^+ K_2^- + K_2^+ K_1^-) + 2K_1^0 K_2^0 + C_1 + C_2
\end{aligned} \tag{5.68}$$

that proves (5.67). □

If we now recall the abstract expression of the generator of the symmetric inclusion process on two sites

$$L_{1,2} = K_1^+ K_2^- + K_1^- K_2^+ - 2K_1^0 K_2^0 + \frac{\alpha^2}{2}. \tag{5.69}$$

we see that it coincides with the coproduct of the Casimir modulo addition of a constant, more precisely

$$L_{1,2} = -\Delta(C) + (C_1 + C_2) + \frac{\alpha^2}{2}. \tag{5.70}$$

From Lemma 5.16 we know that the Casimir C is a central element of $U(\mathfrak{su}(1,1)^*)$. I.e., $[C, K^u] = 0$ for $u \in \{+, -, 0\}$. This implies that $\Delta(C)$ commutes with $\Delta(K^u) = K_1^u + K_2^u$, $u \in \{-, 0, +\}$. Thus we have obtained via (5.70) the symmetries of the generator of the symmetric inclusion process on two sites, namely

$$[L_{1,2}, K_1^+ + K_2^+] = 0, \qquad [L_{1,2}, K_1^- + K_2^-] = 0, \qquad [L_{1,2}, K_1^0 + K_2^0] = 0. \tag{5.71}$$

The fact that the operators S^+, S^- and S^0 defined in (5.19) are symmetries of the generator of the symmetric inclusion process on a general lattice V

$$L = \frac{1}{2} \sum_{x,y \in V} p(x,y) L_{x,y} \tag{5.72}$$

can then be deduced in a similar way. The single-edge generator $L_{x,y}$ can indeed be written in terms of $[\Delta(C)]_{xy}$ as follows:

$$L_{x,y} = [\Delta(C)]_{xy} + (C_x + C_y) + \frac{\alpha^2}{2}, \tag{5.73}$$

where C_x and $[\Delta(C)]_{xy}$ are both elements of the algebra $\otimes_{x \in V} U(\mathfrak{su}(1,1)^*)$. More precisely C_x is the Casimir element tensor identities in all sites different from x and $[\Delta(C)]_{xy}$, with $x, y \in V$, is equal to $\Delta(C)$ on the edge $\{x, y\}$ and equal to the identity on all other components. Indeed we have that, for $u \in \{-, 0, +\}$,

$$\begin{aligned} [L, S^u] &= \frac{1}{2} \left[\sum_{x,y \in V} p(x,y) [\Delta(C)]_{xy}, S^u \right] \\ &= \frac{1}{2} \sum_{x,y \in V} p(x,y) \Big[[\Delta(C)]_{xy}, K_x^u + K_y^u \Big] = 0, \end{aligned} \tag{5.74}$$

where in the first equality we used that K_z^u commutes with $[\Delta(C)]_{xy}$ for all $z \notin \{x, y\}$ and in the second equality we used that $[\Delta(C), \Delta(K^u)] = 0$.

5.7 Ergodic Tempered Measures

So far we identified in (5.30) reversible product measures for SIP(α), namely product of discrete-Gamma distributions. These measures are the analogue of the homogeneous Poisson measures for independent random walkers. In Chap. 3, for independent random walkers, in the infinite volume setting $V = \mathbb{Z}^d$ we used self-duality to show the ergodicity of Poisson product measures and the fact that among the set of tempered invariant measures, Poisson product measures are the only ergodic ones. We now want to prove the same for the symmetric inclusion process, i.e., in a general infinite-volume setting, we want to characterize the set of ergodic tempered measures. In doing so, we will actually slightly generalize the context to a process where we have duality and some conditions on the duality functions. Then we define in that general setting the notion of tempered measures and prove an ergodic theorem, i.e., characterize the set of ergodic tempered measures. The advantage of this set up is that it will not only include the symmetric inclusion process but also all the later examples to come, such as the Brownian energy process in Chap. 6 and the symmetric partial exclusion process in Chap. 7.

We follow the road of Chap. 3 to characterize the set of ergodic measures. We formulate an abstract theorem about tempered ergodic measures in a context where one has duality polynomials. We then indicate, without giving full details how this applies in the context of SIP(α) on $V = \mathbb{Z}^d$ with translation invariant p. We refer to the literature for details such as the existence of a successful coupling.

We consider a Markov process $\{\eta(t) : t \geq 0\}$ on a state space of the form $\Omega = E^V$ where in the example of the symmetric inclusion process we have $E = \mathbb{N}$. In later examples one has more general single-site state space, for instance $E = [0, \infty)$ for the Brownian energy process in Chap. 6. Furthermore we assume the existence of a dual process on finite particle configurations, i.e., on the state space

$$\widehat{\Omega}_f = \bigcup_{n \in \mathbb{N}} \widehat{\Omega}_n, \tag{5.75}$$

where

$$\widehat{\Omega}_n = \left\{ \xi \in \mathbb{N}^V : |\xi| = \sum_x \xi_x = n \right\} \tag{5.76}$$

denotes the set of dual configurations with n particles.

We denote the duality function by $\mathcal{D} : \widehat{\Omega}_f \times \Omega \to \mathbb{R}$.

We then assume the following conditions for the duality function.

Definition 5.21 (Regular Duality Functions) We call the duality function regular if the following conditions are met.

1. *The duality functions are homogeneously factorizing*, i.e., they are of the form

$$\mathcal{D}(\xi, \eta) = \prod_{x \in V} d(\xi_x, \eta_x), \tag{5.77}$$

where for all $k \geq 1$ we assume $d(k, \cdot)$ is a non-negative function and $d(0, \cdot) = 1$.
2. *Growth condition.* There exists a sequence $c_n, n \in \mathbb{N}$ such that the following holds. If, for a probability measure μ on Ω we have the bound

$$\sup_{\xi \in \widehat{\Omega}_n} \int \mathcal{D}(\xi, \eta) d\mu(\eta) \leq c_n, \qquad \text{for all } n \in \mathbb{N}, \tag{5.78}$$

then μ is uniquely determined by the expectations

$$\hat{\mu}(\xi) = \int \mathcal{D}(\xi, \eta) d\mu(\eta). \tag{5.79}$$

We call such a sequence $c_n, n \in \mathbb{N}$ an allowed growth sequence.
3. *Exponential growth sequences are allowed.* The sequence $c_n = \theta^n$ is an allowed growth sequence for all $\theta > 0$.
4. *Density condition.* For every μ satisfying (5.78), the linear combinations of $\mathcal{D}(\xi, \cdot)$ are dense in $L^2(\mu)$.

As underlying motivating example we have in mind the self-duality polynomials for symmetric inclusion process (5.26) and the self-duality polynomials for independent random walkers (3.44). Let us give three examples of allowed growth sequences associated to a given single-site duality function d.

1. Monomials. $\Omega = [0, \infty)^V$, $d(k, z) = z^k$, $(c_k)_{k \geq 0}$ a sequence such that the Carleman condition is satisfied, i.e.,

$$\sum_k c_k^{-1/k} < \infty. \tag{5.80}$$

An example is $c_k = k!\theta^k$ for some $\theta > 0$ which corresponds to the moments of an exponential distribution.
2. Falling factorials. $\Omega = \mathbb{N}^V$, $d(k, n) = n!/(n-k)!$, $c_k = k!r^k$, for some $r > 0$.
3. Monomials in the compact setting. $\Omega = \{0, 1, \ldots, \alpha\}^V$, $d(k, n) = n^k$, $c_k = \alpha^k$.

Given the Markov process $\{\eta(t) : t \geq 0\}$, its dual $\{\xi(t) : t \geq 0\}$ and the duality function $\mathcal{D}$ satisfying the assumption of regularity, we define the set of tempered probability measures as follows.

Definition 5.22 (Tempered Measures) Assume that the duality function is regular, with regularity sequence $c_n, n \in \mathbb{N}$. Then we call a probability measure μ tempered if we have

1. *Allowed moment growth.* There exist an allowed growth sequence in the sense of Definition 5.21 such that

$$\sup_{|\xi|=n} \int \mathcal{D}(\xi, \eta) d\mu(\eta) \leq c_n. \tag{5.81}$$

2. *Second moment bounds for the duality function.*

$$\sup_{|\xi|=n, |\xi'|=n} \int \mathcal{D}(\xi, \eta)\mathcal{D}(\xi', \eta) d\mu(\eta) \leq d_n, \tag{5.82}$$

for some sequence d_n of non-negative numbers.

We denote by $\mathscr{P}$ the set of tempered probability measures on Ω.

Definition 5.23 (Measure Associated to a Duality Function) For a given regular duality function $\mathcal{D}$ we define for $\theta \in [0, \infty)$ the measure μ_θ via

$$\int D(\xi, \eta) d\mu_\theta(\eta) = \theta^{|\xi|}. \tag{5.83}$$

We call $\mathscr{R}$ the set of product measures of this type.

We remark that, by definition, μ_θ are tempered product measures.

For the dual process $\{\xi(t) : t \geq 0\}$, we assume the following.

Definition 5.24 (Canonical Dual) The dual process is called canonical if the following three conditions are met.

1. *Conservation law.* The process conserves the number of particles, i.e., for $\xi \in \widehat{\Omega}_n$, we have $\xi(t) \in \widehat{\Omega}_n$ for all $t \geq 0$.
2. *Non-positive recurrence.* For all $\xi, \xi' \in \widehat{\Omega}_f$,

$$\lim_{t\to\infty} p_t(\xi, \xi') = 0. \tag{5.84}$$

Notice that this excludes the case V finite. It is generically satisfied when $V = \mathbb{Z}^d$ and the transition rates of individual particles are translation invariant and finite range.

3. *Bounded harmonic functions are constant.* I.e., if $f : \widehat{\Omega}_n \to \mathbb{R}$ is a bounded function such that

$$f(\xi) = \mathbb{E}_\xi f(\xi(t)), \tag{5.85}$$

then $f(\xi) = g(n)$ for some $g : \mathbb{N} \to \mathbb{R}$.

The following lemma shows the relation between tempered invariant measures and bounded harmonic functions of the dual process, see Theorem 3.23 for the analogous statement and proof for independent walkers.

Lemma 5.25 *Assume that the dual process is canonical in the sense of Definition 5.24. If μ is tempered and invariant, then $\hat{\mu}$ is bounded and harmonic, and thus depends only on the number of dual particle. I.e., for $\xi \in \widehat{\Omega}_n$*

$$\hat{\mu}(\xi) = g(n), \tag{5.86}$$

for some $g : \mathbb{N} \to \mathbb{R}$.

Remark 5.26 The third condition of Definition 5.24 is typically verified via the construction of a successful coupling, i.e., from two different initial conditions $\xi, \xi' \in \widehat{\Omega}_n$ there exists a coupling of the dual processes starting from ξ and ξ' such that in this coupling, the coupling time is finite.

The existence of a successful coupling for the dual is ensured in the following setting.

1. For independent random walks, we discussed the existence of a successful coupling (the coordinate-wise Ornstein coupling) in Chap. 2.
2. For the symmetric exclusion process, the existence of a successful coupling is proved in [70], see also [172].
3. For the symmetric inclusion process, using ideas from both [190], and [70] a successful coupling is constructed for SIP(α) in [158].

Then we have the following general result.

Theorem 5.27 (General Ergodic Theorem) *Let $\{\eta(t), t \geq 0\}$ denote a configuration process, with a canonical dual process, in the sense of Definition 5.24, with regular duality function in the sense of Definition 5.21. Let us denote by $\mathcal{I}$ the set of ergodic invariant measure. Then*

$$\mathcal{I} \cap \mathcal{P} = \mathcal{R} \cap \mathcal{P}. \tag{5.87}$$

In other words the only tempered ergodic measures are the those of Definition 5.23.

Proof We have defined the context in such a way that we can proceed very analogously to how we proceeded for independent walkers, i.e., in the characterization of tempered ergodic measures in Chap. 2. The essential ingredients are the bounds provided by the temperedness, i.e., (5.82), combined with (5.84). We leave the details to the reader. □

5.8 Propagation of Positive Correlations

In this section we see, as an application of self-duality for the SIP, that starting from local-equilibrium product measures, the evolution produces measures with positive correlations. This is another statement that confirms the "bosonic" or "attractive" character of the inclusion process. In fact, as showed by Liggett in [172], under the symmetric exclusion process evolution, product measures evolve towards negatively-correlated distributions. However, for the symmetric exclusion process, which allows at most one particle per site, there is essentially only one type of product measures, namely products of Bernoulli. We will see that for symmetric inclusion process instead, a general initial product measure can lead to both positive and negative correlations, and, in order to have propagation of positive correlations we need to start from a well-chosen product measure, whose marginals are discrete-Gamma, with scale parameters that can be site-dependent.

The way the correlation inequality is proved is exactly along the lines of Liggett's proof. We first need a labeled particle representation for SIP. We fix a countable lattice V and a symmetric irreducible function $p : V \times V \to [0, \infty)$. Then we consider the dynamics of a finite number of SIP particles, say n, and we attach to each of them a label that they will keep forever. The evolution of their positions $X_1(t), \dots X_n(t)$ define a continuous-time Markov chain on V^n with generator

$$L_n f(x_1, \dots, x_n) = \sum_{i=1}^{n} \sum_{j=1}^{n} \sum_{y \in V} p(x_i, y)(\alpha + \mathbb{1}_{\{x_j = y\}})(f(\mathbf{x}(i, y)) - f(\mathbf{x})), \tag{5.88}$$

where $\mathbf{x} = (x_1, \dots, x_n)$ and with $\mathbf{x}(i, y)$ denoting the n-tuple obtained from $\mathbf{x}$ by replacing the i-th component x_i by y. We will also use the notation $\mathbf{x}(i, j; u, v)$ for the vector arising from $\mathbf{x}$ by substituting x_i with u and x_j with v. We denote by $S_n(t) = e^{tL_n}$ semigroup corresponding to the generator (5.88).

As we did in Sect. 5.2 for the configuration process, also for the coordinates process, we isolate the "independent" part of the dynamics. In other words we define a system of n independent random walkers moving with rate $\alpha p(x, y)$ between x and y, and we label them. In the coordinate representation this has generator

$$L_n^{\text{irw}} f(\mathbf{x}) = \sum_{i=1}^{n} \sum_{j=1}^{n} \sum_{y \in V} \alpha p(x_i, y)(f(\mathbf{x}(i, y)) - f(\mathbf{x})), \tag{5.89}$$

and corresponding semigroup $S_n^{\text{irw}}(t) := e^{tL_n^{\text{irw}}}$. So we have

$$(L_n - L_n^{\text{irw}}) f(x) = \sum_{i=1}^{n} \sum_{j=1}^{n} p(x_i, x_j)(f(\mathbf{x}(i, x_j)) - f(\mathbf{x})), \tag{5.90}$$

which, by symmetry of p, can be rewritten as follows for a symmetric function $f : V^n \to \mathbb{R}$,

$$(L_n - L_n^{\text{irw}})f(x) = \frac{1}{2}\sum_{i=1}^{n}\sum_{j=1}^{n} p(x_i, x_j)(f(\mathbf{x}(i, x_j)) + f(\mathbf{x}(j, x_i) - 2f(\mathbf{x})). \tag{5.91}$$

This quantity has positive sign if on top of being symmetric, f is also positive-definite. Let us define this concept now more in detail.

Definition 5.28 A function $f : V^2 \to \mathbb{R}$ is positive-definite if for all $\varphi : V \to \mathbb{R}$

$$\sum_{x,y} \varphi(x)\varphi(y)f(x, y) \geq 0. \tag{5.92}$$

A function $f : V^n \to \mathbb{R}$ is positive-definite if for all i, j and $\mathbf{x} \in V^n$, the map

$$(u, v) \mapsto f(\mathbf{x}(i, j; u, v)) \tag{5.93}$$

is positive-definite. We denote by $\mathfrak{S}_n$ the set of positive-definite symmetric functions $f : V^n \to \mathbb{R}$.

The simplest examples of elements of $\mathfrak{S}_n$ are of the form

$$f(\mathbf{x}) = \prod_{i=1}^{n} \rho(x_i), \tag{5.94}$$

for $\rho : V \to [0, \infty)$. Also functions of the form

$$(x_1, \ldots, x_n) \mapsto \int_W \prod_{i=1}^{n} \rho(x_i, \omega)d\lambda(\omega) \tag{5.95}$$

belong to $\mathfrak{S}_n$ whenever $\rho : V \times W \to [0, \infty)$, with W a measurable space, and λ is a positive finite measure. We also notice that, for $a : V^2 \to [0, \infty)$ and $f \in \mathfrak{S}_n$,

$$\mathbf{x} \mapsto \sum_{y_1,\ldots,y_n} \prod a(x_i, y_i)f(\mathbf{x}) \tag{5.96}$$

is positive-definite (provided it is well-defined) and symmetric. As a consequence, for all $f \in \mathfrak{S}_n$, we have $S_n^{\text{irw}}(t)f \in \mathfrak{S}_n$. We now prove the following.

Theorem 5.29 *For $f \in \mathfrak{S}_n$ we have, for all $\mathbf{x} \in V^n$*

$$S_n(t)f(\mathbf{x}) \geq S_n^{\text{irw}}(t)f(\mathbf{x}). \tag{5.97}$$

Proof Put $A_n := L_n - L_n^{\mathrm{irw}}$, then, using (5.91), we have that for all $f \in \mathfrak{S}_n$,

$$A_n f(\mathbf{x}) = \frac{1}{2}\sum_{i=1}^{n}\sum_{j=1}^{n} p(x_i, x_j)\left(f(\mathbf{x}(i, x_j)) - 2f(\mathbf{x}) + f(\mathbf{x}(j, x_i))\right) \geq 0, \tag{5.98}$$

where we used that for $f : V^2 \to \mathbb{R}$ positive-definite,

$$f(x, x) + f(y, y) - 2f(x, y) = \sum_{u,v\in V} f(u, v)\varphi(u)\varphi(v) \geq 0, \tag{5.99}$$

with $\varphi(u) = \mathbb{1}_{\{u=x\}} - \mathbb{1}_{\{u=y\}}$. Now if $f \in \mathfrak{S}_n$, then for $0 \leq s \leq t$, also $S_n^{\mathrm{irw}}(s)f \in \mathfrak{S}_n$, hence $A_n S_n^{\mathrm{irw}}(s)f \geq 0$, and then, since $S_n(\cdot)$ is a positivity preserving semigroup, we deduce that $S_n(t-s)A_n S_n^{\mathrm{irw}}(s)f \geq 0$. Then, using the variation of constants formula, we conclude that

$$S_n(t)f - S_n^{\mathrm{irw}}(t)f = \int_0^t S_n(t-s)(A_n(S_n^{\mathrm{irw}}(s)f)) \geq 0, \tag{5.100}$$

for all $f \in \mathfrak{S}_n$. □

As a first consequence of this inequality, we show that two SIP particles spend more time together than their independent walker counterparts. This statement can easily be generalized to more than two SIP particles.

Proposition 5.30 *Let us denote by $\mathbb{P}_t^{\mathrm{sip}(\alpha)}(x, y; u, v)$ the transition probabilities of two labeled SIP(α) particles to move in time t from (x, y) to (u, v) and $\mathbb{P}_t^{\mathrm{irw}(\alpha)}(x, y; u, v) = p_t(x, u)p_t(y, v)$ the corresponding independent random walk transition probabilities. Then we have for all $x, y, u \in V$ and $t \geq 0$,*

$$\mathbb{P}^{\mathrm{sip}(\alpha)}(x, y; u, u) \geq \mathbb{P}^{\mathrm{irw}(\alpha)}(x, y; u, u). \tag{5.101}$$

Proof The statement follows immediately from Theorem 5.29 and the fact that the function $f(x, y) = \mathbb{1}_{\{u=x=y\}}$ is positive-definite and symmetric. □

In order to apply Theorem 5.29 to prove the announced result on preservation of positive correlations, we need to introduce some further notations. For a probability measure μ on the configuration space of the (unlabeled) SIP(α), i.e. $\Omega = \mathbb{N}^V$, we define the so-called n-point function:

$$C_\mu(x_1, \ldots, x_n) = \int D\left(\sum_{i=1}^{n} \delta_{x_i}, \eta\right) d\mu(\eta), \tag{5.102}$$

where D is the SIP(α) self-duality function defined in (5.26).

Definition 5.31 Let ρ be a function $\rho : V \to [0, \infty)$ modeling a *density profile* over the lattice V. We denote by μ_ρ the *inhomogeneous product measure associated to* ρ, namely the product measure whose site-x marginal is the discrete-Gamma distribution with parameters $\left(\alpha, \frac{\rho_x}{1+\rho_x}\right)$.

Remark 5.32 Notice that if the density profile ρ is constant, then μ_ρ is one of the reversible measures of the symmetric exclusion process (see (5.30)). In general, i.e. for inhomogeneous density profiles, the measure μ_ρ is only "locally at equilibrium". Notice that the parametrization of the discrete-Gamma marginals is chosen in such a way to make the local particle density ρ_x equal to the expected number of particles at site x divided by α, namely

$$\rho_x = \mathbb{E}_{\mu_\rho}\left[\frac{\eta_x}{\alpha}\right] = \mathbb{E}_{\mu_\rho}[D(\delta_x, \eta)], \tag{5.103}$$

where the second identity follows from the fact that $d(1, n) = \frac{n}{\alpha}$ (see (5.26)).

Remark 5.33 (Characterizing Property of μ_ρ) A characterizing property of the probability measure μ_ρ is the following identity. For $\rho : V \to [0, \infty)$ $\mu = \mu_\rho$ if and only if

$$C_{\mu_\rho}(x_1, \ldots, x_n) = \prod_{i=1}^{n} \rho(x_i). \tag{5.104}$$

More precisely, the n-point functions with respect to a probability measure μ on Ω are of the form (5.104) if and only if μ is equal to μ_ρ, inhomogeneous product measure associated to ρ.

We define the density profile at time $t > 0$ by

$$\rho_t(x) = \int D(\delta_x, \eta) d\mu_\rho(t)(\eta) = \sum p_t(x, y)\rho(y), \tag{5.105}$$

where $p_t(x, y)$ is the transition probability of the continuous-time random walk jumping at rate $\alpha p(x, y)$ from x to y. Finally we define an order relation $\mu \prec \nu$ between two probability measures on Ω if we have

$$\int D(\xi, \eta)\, d\mu(\eta) \le \int D(\xi, \eta)\, d\nu(\eta), \tag{5.106}$$

for all particle configurations ξ in Ω with a finite number of particles. Notice that this partial order is different from the usual stochastic domination partial order defined by all monotone functions. Indeed the latter is stronger, and it can happen that $\mu \prec \nu$ but ν does not stochastically dominate μ. We can now prove the following correlation inequality.

Theorem 5.34 *Let μ_ρ be a local equilibrium distribution for* SIP(α) *with density profile ρ, and let $\mu_\rho(t)$ denote the distribution at time $t > 0$ of* SIP(α) *when started from μ_ρ at time zero. Then we have, for all $x_1, \ldots, x_n \in V$,*

$$C_{\mu_\rho(t)}(x_1, \ldots, x_n) \geq \prod_{i=1}^{n} \rho_t(x_i), \tag{5.107}$$

and, as a consequence, $\mu_{\rho(t)} \prec \mu_\rho(t)$. Moreover, the variables $\eta_{x_1}, \ldots, \eta_{x_n}$ are positively correlated under the measure $\mu_\rho(t)$.

Proof Denote $\rho_n : V^n \to \mathbb{R}$ the function that assigns to a n-tuple $\mathbf{x} = (x_1, \ldots, x_n)$ the product $\rho_n(\mathbf{x}) = \prod_{i=1}^{n} \rho(x_i)$. This function is clearly symmetric and positive-definite. Using self-duality of SIP(α) with self-duality function (5.26) and applying (5.97) to $\rho_n(\mathbf{x})$, we obtain

$$\begin{aligned} C_{\mu_\rho(t)}(x_1, \ldots, x_n) &= (S_n(t)\rho_n)(x_1, \ldots, x_n) \\ &\geq (S_n^{\text{irw}}(t)\rho_n)(x_1, \ldots, x_n) = \prod_{i=1}^{n} \rho_t(x_i). \end{aligned} \tag{5.108}$$

The inequality $\mu_{\rho(t)} \prec \mu_\rho(t)$ follows now by the definition of the order $\prec$. The statement about positiveness of correlations follows immediately from the fact $D(\delta_x, \eta) = \frac{\eta_x}{\alpha}$ with $\alpha > 0$, and the fact that, if $x_1 \neq x_2 \neq \ldots \neq x_n$ then

$$D\left(\sum_{i=1}^{n} \delta_{x_i}, \eta\right) = \prod_{i=1}^{n} D(\delta_{x_i}, \eta). \tag{5.109}$$

This concludes the proof. □

Remark 5.35 Notice that we did not use any particular property of the self-duality functions other than the fact that

$$\int D(\sum_{i=1}^{n} \delta_{x_i}, \eta) d\mu_\rho = \prod_{i=1}^{n} \rho(x_i) = \rho_n(\mathbf{x}). \tag{5.110}$$

In order for the function $\rho_n(\mathbf{x})$ to be positive-definite we need the non-negativity of ρ, but not the specific form of the self-duality functions. This implies that the result in Theorem 5.34 holds true for any choice of the self-duality functions and any choice of the measure μ_ρ satisfying the relation (5.110).

5.9 Inhomogeneous Symmetric Inclusion Process

In this section we will see that it is possible to generalize the self-duality result proved in Sect. 5.4 to a system of inclusion particles in which the parameter $\alpha > 0$ is not constant but site-dependent. To this aim we introduce the vector $\boldsymbol{\alpha} = \{\alpha_x,\ x \in V\}$ with $\alpha_x > 0$. Let $p : V^2 \to [0, \infty)$ be a symmetric transition function, then we define the inhomogeneous inclusion process parametrized with the vector $\boldsymbol{\alpha}$, SIP($\boldsymbol{\alpha}$), as the process $\{\eta(t) : t \geq 0\}$ of interacting particles moving on the lattice V with generator

$$L = \frac{1}{2} \sum_{x,y \in V} p(x, y)\, L_{x,y}, \tag{5.111}$$

$$L_{x,y} f(\eta) = \eta_x(\alpha_y + \eta_y)[f(\eta^{x,y}) - f(\eta)] + \eta_y(\alpha_x + \eta_x)[f(\eta^{y,x}) - f(\eta)], \tag{5.112}$$

and state space $\Omega = \mathbb{N}^V$.

Via a detailed-balance computation, one can see that the inhomogeneous process is still reversible with respect to the product measures of the form

$$M(\eta) = \prod_{x \in V} \frac{\lambda^{\eta_x}}{\eta_x!} \frac{\Gamma(\eta_x + \alpha_x)}{\Gamma(\alpha_x)}, \qquad \text{with} \quad \lambda > 0, \tag{5.113}$$

that is the inhomogeneous version of (5.23). From this reversible measure one can produce a cheap self-duality function. In order to obtain a non-trivial self-duality function, now, it is sufficient to find a triangular symmetry as we did in Sect. 5.3 for the homogeneous case, and act with it on the cheap duality. It is convenient, to this aim, to rewrite the single-edge generator, in terms of the generators of the conjugate algebra $\mathfrak{su}(1,1)^*$. This can be easily done mimicking the homogeneous case. We have indeed:

$$L_{x,y} = K_x^+ K_y^- + K_x^- K_y^+ - 2K_x^0 K_y^0 + \frac{\alpha_x \alpha_y}{2}. \tag{5.114}$$

where the operators K_x^v, $v \in \{0, +\}$, $x \in V$ defined in (5.5) for the homogeneous case, are now site-dependent, since, at each site $x \in V$ corresponds a different parameter α_x, whereas K_x^- has the same action for all x:

$$\begin{aligned} K_x^+ f(n) &= (\alpha_x + n) f(n+1), \\ K_x^- f(n) &= n f(n-1), \\ K_x^0 f(n) &= \left(\tfrac{\alpha_x}{2} + n\right) f(n). \end{aligned} \tag{5.115}$$

The change in the definition of these operators doesn't modify the commutation relations that read again:

$$
\begin{aligned}
[K_x^{\pm}, K_y^0] &= \pm K_x^{\pm}\delta_{x,y}, \\
[K_x^{+}, K_y^{-}] &= 2K_x^0\delta_{x,y}.
\end{aligned}
\tag{5.116}
$$

Due to the fact that the generator in the abstract form (5.114) is the same as in (5.18) modulo addition of a constant, and since the commutation relations remain unperturbed, we can deduce a symmetry result. The inhomogeneous inclusion process generator $L_{x,y}$ commutes with $K_x^v + K_y^v$ for $v \in \{+, -, 0\}$. This can be easily checked, we do it for $v = +$, we have:

$$
\begin{aligned}
&[K_x^+K_y^- + K_x^-K_y^+ - 2K_x^0K_y^0, K_x^+ + K_y^+] \\
&= [K_x^-, K_x^+]K_y^+ - 2[K_x^0, K_x^+]K_y^0 \\
&+ K_x^+[K_y^-, K_y^+] - 2K_x^0[K_y^0, K_y^+] \\
&= -2K_x^0K_y^+ + 2K_x^+K_y^0 - 2K_x^+K_y^0 + 2K_x^0K_y^+ = 0.
\end{aligned}
\tag{5.117}
$$

The result can be easily generalized to the system with more sites and then to the entire generator L that is thus commuting with the $S^v := \sum_{x\in V} K_x^v$. We can combine now, the cheap-duality result and the symmetry statement in order to obtain a triangular self-duality for the inhomogeneous process.

Theorem 5.36 (Self-duality of Inhomogeneous SIP($\boldsymbol{\alpha}$)) *The inhomogeneous symmetric inclusion process parameters* $\boldsymbol{\alpha} = \{\alpha_x,\ x \in V\}$ *on a finite graph* V *with generator* (5.111)-(5.112) *is self-dual with self-duality function*

$$
D(\xi, \eta) = \prod_{x\in V} d_x(\xi_x, \eta_x),
\tag{5.118}
$$

and

$$
d_x(k, n) = \frac{n!}{(n-k)!}\frac{\Gamma(\alpha_x)}{\Gamma(\alpha_x + k)}\mathbb{1}_{\{k\le n\}}.
\tag{5.119}
$$

Proof The result is obtained, similarly to the proof of Theorem 5.8, by acting with the symmetry e^{S^+}, on the cheap duality function $d^{\mathrm{ch}}(\xi, \eta) = \frac{\delta_{\xi,\eta}}{M(\xi)}$, where $M(\cdot)$ is the reversible measure defined in (5.113). □

5.10 Additive Structure on Ladder Graphs

In this section we prove that the SIP($\boldsymbol{\alpha}$) defined on a ladder graph has the remarkable property that, summing over the levels of the ladder, defines a new symmetric inclusion process with parameters obtained by summing the components of $\boldsymbol{\alpha}$ over the ladder. The proof of this property relies on computations involving the generator in its abstract form.

We start by introducing the idea of a particle system on a ladder. Let us denote by V^* a set of sites. We consider a finite *fibred* vertex set over V^*, i.e., a finite set of the form $V = \{(x, \ell) : x \in V^*, \ \ell \in \{1, \dots, L_x\}\}$, and we call such a set a "ladder vertex set". The interpretation is that at each "site" $x \in V^*$ (name which we reserve for the base-points of each fiber) there are L_x levels of a ladder.

Let $p : V \times V \to [0, \infty)$ be an irreducible symmetric function. We call *p level-independent* if $p((x, \ell), (y, s)) = p((x, \ell'), (y, s'))$ for all $\ell, \ell' \in [L_x]$ and $s, s' \in [L_y]$ (where we use the notation $[n] = \{1, \dots, n\}$). In other words, this means that p only depends on the base-points of the fibres, i.e. we can define a symmetric edge-weight function over V^*, $p^* : V^* \times V^* \to [0, \infty)$ such that $p((x, \ell), (y, s)) = p^*(x, y)$, for all $\ell \in [L_x]$, $s \in [L_y]$.

For a configuration $\eta \in \mathbb{N}^V$ we define its contraction as the configuration $\eta^* \in \mathbb{N}^{V^*}$ such that

$$\eta^*_x = \sum_{\ell=1}^{L_x} \eta_{(x,\ell)}. \tag{5.120}$$

We then have the following theorem.

Theorem 5.37 *Consider the* SIP($\boldsymbol{\alpha}$) *on a* fibred *vertex set V over V^*, with edge weights $p : V \times V \to [0, \infty)$ and parameters $\boldsymbol{\alpha} = \{\alpha_{(x,\ell)}, \ x \in V^*, \ \ell \in [L_x]\}$. Assume that p is level-independent. Then the contracted process $\{\eta^*(t) : t \geq 0\}$ is the* SIP($\boldsymbol{\alpha}^*$) *on V^* with*

$$\boldsymbol{\alpha}^* = \{\alpha^*_x, \ x \in V^*\} \qquad \textit{and} \qquad \alpha^*_x = \sum_{\ell=1}^{L_x} \alpha_{(x,\ell)}. \tag{5.121}$$

Proof Let us denote $\psi : \mathbb{N}^V \to \mathbb{N}^{V^*}$ the contraction map such that $\psi(\eta) = \eta^*$. We have to show that, for all $f : \mathbb{N}^{V^*} \to \mathbb{R}$,

$$L^{\mathrm{SIP}(\boldsymbol{\alpha})}(f \circ \psi) = (L^{\mathrm{SIP}(\boldsymbol{\alpha}^*)} f) \circ \psi. \tag{5.122}$$

To this aim we write the generator $L^{\mathrm{SIP}(\boldsymbol{\alpha})}$ of the symmetric inclusion process with parameters $(\boldsymbol{\alpha})$ in its abstract form

$$L^{\mathrm{SIP}(\boldsymbol{\alpha})} = \sum_{x,y\in V} p^*(x,y)$$

$$\sum_{\ell=1}^{L_x}\sum_{s=1}^{L_y}\left(K^+_{(x,\ell)}K^-_{(y,s)} + K^-_{(x,\ell)}K^+_{(y,s)} - 2K^0_{(x,\ell)}K^0_{(y,s)} + \tfrac{1}{2}\,\alpha_{(x,\ell)}\alpha_{(y,s)}\right).$$

Here we denote by $K^-_{(x,\ell)}$ the operators given by

$$\begin{aligned}
K^+_{(x,\ell)} f(n) &= (\alpha_{(x,\ell)} + n) f(n+1),\\
K^-_{(x,\ell)} f(n) &= n f(n-1),\\
K^0_{(x,\ell)} f(n) &= \left(\tfrac{\alpha_{(x,\ell)}}{2} + n\right) f(n).
\end{aligned} \tag{5.123}$$

First of all we notice that, for what concerns the constant term, when we sum over the ladders we have

$$\sum_{\ell=1}^{L_x}\sum_{s=1}^{L_y} \tfrac{1}{2}\,\alpha_{(x,\ell)}\alpha_{(y,s)} = \tfrac{1}{2}\,\alpha^*_x\alpha^*_y. \tag{5.124}$$

Hence, in order to prove our result it is sufficient to show that, for all functions $f : \mathbb{N}^{V^*} \to \mathbb{R}$ only depending on the x-coordinate, one has

$$\sum_{\ell=1}^{L_x} K^v_{(x,\ell)}(f \circ \psi) = (K^v_x f) \circ \psi, \tag{5.125}$$

for $v \in \{+, -, 0\}$, where

$$\begin{aligned}
K^+_x f(n) &= (\alpha^*_x + n) f(n+1),\\
K^-_x f(n) &= n f(n-1),\\
K^0_x f(n) &= \left(\tfrac{\alpha^*_x}{2} + n\right) f(n).
\end{aligned} \tag{5.126}$$

We prove (5.125) only for the case $v = +$, as the proof of the other cases is similar. Let $f : \mathbb{N}^{V^*} \to \mathbb{R}$ be a function only depending on the x-coordinate, and let $\eta \in \mathbb{N}^V$ then we first notice that for all $\ell \in \{1, \dots, L_x\}$ it holds that

$$f \circ \psi(\eta + \delta_{x,\ell}) = f(\eta^* + \delta_x) = f(\psi(\eta) + \delta_x). \tag{5.127}$$

As a consequence

$$\begin{aligned}\left(\sum_{\ell=1}^{L_x} K^{v}_{(x,\ell)}(f\circ\psi)\right)(\eta) &= \sum_{\ell=1}^{L_x}(\alpha_{(x,\ell)}+\eta_{(x,\ell)})f(\eta^*+\delta_x)\\ &= (\alpha^*_x+\eta^*_x)f(\eta^*+\delta_x)\\ &= (K^+_x f)(\eta^*)\\ &= (K^+_x f)\circ\psi(\eta).\end{aligned}$$

□

Remark 5.38 The result in Theorem 5.37 might appear surprising. In the contraction procedure from η to η^* a part of the information is lost. Being $\{\eta(t) : t \geq 0\}$ a Markov process, one might wonder whether $\{\eta^*(t) : t \geq 0\}$ is still a Markov process. This is the case by (5.122). Indeed, because of the level independence of p, the rates satisfy the lumping condition in [61] ensuring that the contraction from η to η^* indeed preserves the Markov property. Notice that this is also true because of the (bi)linearity of the rates in both α and η_i, and hence this provides another proof of the theorem (see [116]).

5.11 Additional Notes

The inclusion process was introduced in [115] as a dual of a model of heat conduction (the Brownian momentum process). Basic properties including the analogues of Liggett's correlation inequalities for the exclusion process were proved in [117]. In the literature on population dynamics, the inclusion process on the complete graph appears as the Moran model (with parent independent mutation rate). The precise connections between the inclusion process and related processes in population dynamics is discussed in [42]. The identification of the $SU(1,1)$ symmetry of the inclusion process and the consequent self-dualities were first described in [116]. Later the connection with the Casimir element was proved and used to construct an appropriate asymmetric version of the inclusion process which has self-duality properties coming from quantum Lie algebra symmetries (see [43]). Concerning the use of duality in the context of population dynamics, the emphasis is more on the coalescent processes, describing the ancestral relations of individual backwards in time. This leads e.g. to moment duality of the Moran model with the Kingman's coalescent block counting process see e.g. [182], [87]. The symmetric inclusion process manifests condensation phenomena for small α, i.e., when $\alpha \to 0$ the process typically forms piles of size $1/\alpha$ which have a limiting dynamics. The self-duality allows to study the coarsening process, i.e., how starting from a homogeneous initial distribution the condensates are formed in the course of time and how they move on an appropriate time scale. The study of condensation in the symmetric inclusion process was initiated in [127] for the condensation properties of the stationary measure, and [128] for the dynamics of the condensate. The condensation is studied with potential theoretic methods and a conjecture of the existence of multiple time scales was formulated in [29], which was proved in [147]. In [59] for symmetric inclusion process on the complete graph, in the condensation limit, a measure-valued diffusion process is obtained for the size-biased and appropriately scaled empirical measures of mass distribution.

Chapter 6
Duality for the Brownian Energy Process

In this chapter we study the Brownian energy process, a diffusion process modelling heat conduction, where energy is diffusively exchanged among sites. We will introduce it as a many-particle limit of the symmetric inclusion process, in the spirit of population models where diffusion processes arise when limits of large population size are taken, and the variables become the ratios of the numbers individuals of different types with respect to the total population size. Naturally, the generator of the Brownian energy process, in its abstract form, reads exactly as the generator of the symmetric inclusion process, in terms of the generator of the $\mathfrak{su}(1,1)$ Lie algebra. This algebraic structure is also at the origin of the duality relation between the two processes. We also introduce the Brownian momentum process, a diffusion process modelling the evolution of interacting momenta on a lattice, also sharing the same algebraic structure.

6.1 The Brownian Energy Process on Two Vertices

Let $\{\eta(t) : t \geq 0\}$ be the inhomogeneous SIP($\boldsymbol{\alpha}$), with $\boldsymbol{\alpha} = (\alpha_1, \alpha_2)$ on the two-site lattice $V = \{1, 2\}$, i.e. the process with generator:

$$L_{1,2} f(\eta) = \eta_1(\eta_2+\alpha_2)[f(\eta^{1,2}) - f(\eta)] + \eta_2(\eta_1+\alpha_1)[f(\eta^{2,1}) - f(\eta)], \qquad (6.1)$$

with $\eta^{x,y} = \eta - \delta_x + \delta_y$. As we did in Sect. 4.1 for the system of independent random walkers, we study the many- particle limit, i.e. the limit as $N \to \infty$ of the process $\{\eta(t)/N : t \geq 0\}$ started with order N particles. We start by scaling the initial conditions as follows:

$$\eta_1^{(N)} = \lfloor N\zeta_1 \rfloor, \qquad \eta_2^{(N)} = \lfloor N\zeta_2 \rfloor, \qquad (6.2)$$

C. Giardinà, F. Redig, *Duality for Markov Processes*, Grundlehren der mathematischen Wissenschaften 365, https://doi.org/10.1007/978-3-032-04099-2_6

where $\zeta_1, \zeta_2 \in [0, \infty)$. Then we have

$$\lim_{N\to\infty} \frac{\eta_x^{(N)}}{N} = \zeta_x, \qquad \text{for} \quad x = 1, 2. \tag{6.3}$$

We then study the behaviour of the limiting process $\{\eta(t)/N : t \geq 0\}$. Let $f : [0, \infty) \to \mathbb{R}$ be a smooth function. We have

$$\begin{aligned} & f\left((\zeta_1, \zeta_2) - \tfrac{\delta_1}{N} + \tfrac{\delta_2}{N}\right) - f(\zeta_1, \zeta_2) \\ &= \frac{1}{N}\left(\frac{\partial}{\partial \zeta_2} - \frac{\partial}{\partial \zeta_1}\right) f(\zeta_1, \zeta_2) + \frac{1}{2N^2}\left(\frac{\partial}{\partial \zeta_2} - \frac{\partial}{\partial \zeta_1}\right)^2 f(\zeta_1, \zeta_2) + O\left(\frac{1}{N^3}\right). \end{aligned} \tag{6.4}$$

Using this and expanding the generator working on a function of the form $(\eta_1, \eta_2) \mapsto f(\frac{\eta_1}{N}, \frac{\eta_2}{N})$ up to second order yields

$$\begin{aligned} & \lim_{N\to\infty} Lf(\tfrac{\eta_1}{N}, \tfrac{\eta_2}{N}) \\ &= (\alpha_2\zeta_1 - \alpha_1\zeta_2)\left(\frac{\partial}{\partial \zeta_2} - \frac{\partial}{\partial \zeta_1}\right) f(\zeta_1, \zeta_2) + \zeta_1\zeta_2\left(\frac{\partial}{\partial \zeta_2} - \frac{\partial}{\partial \zeta_1}\right)^2 f(\zeta_1, \zeta_2). \end{aligned} \tag{6.5}$$

It is then a standard application of the Trotter-Kurtz theorem [159] that the sequence of processes $\{\eta(t)/N : t \geq 0\}$ converges weakly in path-space to the diffusion process with generator

$$\mathcal{L}_{1,2} = (\alpha_2\zeta_1 - \alpha_1\zeta_2)\left(\frac{\partial}{\partial \zeta_2} - \frac{\partial}{\partial \zeta_1}\right) + \zeta_1\zeta_2\left(\frac{\partial}{\partial \zeta_2} - \frac{\partial}{\partial \zeta_1}\right)^2, \tag{6.6}$$

and state space $[0, \infty)^V$, starting from the configuration (ζ_1, ζ_2) at time zero.

In stochastic differential equation language, the process $\{(\zeta_1(t), \zeta_2(t)) : t \geq 0\}$ evolves as

$$\begin{aligned} d\zeta_1(t) &= -(\alpha_2\zeta_1(t) - \alpha_1\zeta_2(t))dt + \sqrt{2\zeta_1(t)\zeta_2(t)}dB_t \\ d\zeta_2(t) &= (\alpha_2\zeta_1(t) - \alpha_1\zeta_2(t))dt - \sqrt{2\zeta_1(t)\zeta_2(t)}dB_t. \end{aligned} \tag{6.7}$$

From this we see that the sum $\zeta_1(t) + \zeta_2(t)$ is a conserved quantity for the Brownian energy process on two sites. In other words, if initially $\zeta_1 + \zeta_2 = z$ then $\{\zeta_1(t) : t \geq 0\}$ evolves as a diffusion process on the interval $[0, z]$ governed by the equation:

$$d\zeta_1(t) = (-\alpha_2\zeta_1(t) + \alpha_1(z - \zeta_1(t)))dt + \sqrt{2\zeta_1(t)(z - \zeta_1(t))}dB_t, \tag{6.8}$$

where $\{B_t : t \geq 0\}$ is the standard one-dimensional Brownian motion. This process is called the (inhomogeneous) Brownian energy process with parameter vector $\boldsymbol{\alpha} = (\alpha_1, \alpha_2)$ on two sites, and abbreviated BEP($\boldsymbol{\alpha}$) on two sites.

6.2 Duality on Two Vertices

We now use the convergence of the rescaled inclusion process $\{\eta(t)/N : t \geq 0\}$ to the Brownian energy process, to obtain a duality relation between BEP($\boldsymbol{\alpha}$) and SIP($\boldsymbol{\alpha}$) from the self-duality of the symmetric inclusion process. Notice that this result is analogous to the result proven in Theorem 4.3 about duality between independent random walkers and the corresponding deterministic system.

Theorem 6.1 (Symmetric Inclusion Process Is Dual to Brownian Energy Process) *Let* $V = \{1, 2\}$ *and define*

$$D(\xi, \zeta) = \prod_{x=1}^{2} d_{\alpha_x}(\xi_x, \zeta_x) \qquad \textit{with} \quad d_\alpha(n, z) = z^n \frac{\Gamma(\alpha)}{\Gamma(\alpha + n)}, \tag{6.9}$$

for $\zeta = (\zeta_1, \zeta_2) \in [0, \infty)^V$ *and* $\xi = (\xi_1, \xi_2) \in \mathbb{N}^V$. *Then we have*

$$\mathbb{E}_\xi^{\mathrm{SIP}(\boldsymbol{\alpha})}\left(D(\xi(t), \zeta)\right) = \mathbb{E}_\zeta^{\mathrm{BEP}(\boldsymbol{\alpha})}\left(D(\xi, \zeta(t))\right). \tag{6.10}$$

Proof Notice that, for $\zeta_1 > 0$, we have

$$\lim_{N\to\infty} \frac{\lfloor N\zeta_1 \rfloor!}{(\lfloor N\zeta_1 \rfloor - k)!} = \zeta_1^k. \tag{6.11}$$

Start from self-duality of the symmetric inclusion process, and take η_1, η_2 as in (6.2). Then we have

$$\mathbb{E}_\xi^{\mathrm{SIP}(\boldsymbol{\alpha})}(D(\xi(t), \eta)) = \mathbb{E}_\zeta^{\mathrm{SIP}(\boldsymbol{\alpha})}(D(\xi, \eta(t))). \tag{6.12}$$

Divide this identity by $N^{\xi_1+\xi_2}$, take the limit $N \to \infty$ and use the weak convergence of $\{\eta(t)/N : t \geq 0\}$ to $\{\zeta(t) : t > 0\}$ and the dominated convergence theorem to conclude (6.10). □

6.3 Abstract Form of the Generator

In order to understand better the duality properties of the Brownian energy process, in particular how to naturally generate the duality function (6.9), we study the abstract form of the generator. As we did in Sect. 4.2 for the duality between independent random walkers and the deterministic system, we would like to recognize the duality result obtained in Theorem 6.1 as the outcome of a change of representation of a (Lie) algebra. In Sect. 4.2 we had "discrete" and "continuous" representations $a, a^\dagger$ and $A, A^\dagger$ of the generators of the (conjugate) Heisenberg algebra. In a similar way, here we have a representation of the generators

$\mathscr{K}_x^+, \mathscr{K}_x^-, \mathscr{K}_x^0$ of the algebra $\mathfrak{su}(1,1)$ in terms of differential operators. These operators work on smooth functions $f : [0, \infty) \to \mathbb{R}$ as follows:

$$\begin{aligned} \mathscr{K}_x^+ f(z) &= z f(z) \\ \mathscr{K}_x^- f(z) &= z f''(z) + \alpha_x f'(z) \\ \mathscr{K}_x^0 f(z) &= z f'(z) + \tfrac{\alpha_x}{2} f(z), \end{aligned} \tag{6.13}$$

and satisfy the commutation relations of the algebra $\mathfrak{su}(1,1)$

$$[\mathscr{K}_x^+, \mathscr{K}_y^-] = -2\mathscr{K}_x^0 \delta_{x,y}, \qquad [\mathscr{K}_x^0, \mathscr{K}_y^\pm] = \pm \mathscr{K}_x^\pm \delta_{x,y}, \tag{6.14}$$

i.e., they form a right representation of the algebra $\mathfrak{su}(1,1)$. It is easy to verify that the generator of the BEP($\boldsymbol{\alpha}$) on two sites can be written in terms of these operators in the following way:

$$\mathcal{L}_{1,2} = \mathscr{K}_1^+ \mathscr{K}_2^- + \mathscr{K}_1^- \mathscr{K}_2^+ - 2\mathscr{K}_1^0 \mathscr{K}_2^0 + \tfrac{\alpha_1 \alpha_2}{2}\,. \tag{6.15}$$

Now we recognize that the generator of the symmetric inclusion process SIP($\boldsymbol{\alpha}$) and the generator of Brownian energy process BEP($\boldsymbol{\alpha}$) have the same abstract form (compare (5.18) to (6.15)). The generators of the two processes arise from two different representations: a "discrete" representation for SIP($\boldsymbol{\alpha}$) and a "continuous" representation for BEP($\boldsymbol{\alpha}$)) of the same abstract object. Then duality between the two processes follows as a consequence of the duality between these two representations. This is the content of the theorem below.

Theorem 6.2 (Duality as Change of Representation) *Let $d_\alpha(n,z)$ be defined as in* (6.9). *For $x \in \{1,2\}$ we have the dualities:*

$$\begin{aligned} K_x^+ &\xrightarrow{d_{\alpha x}} \mathscr{K}_x^+ \\ K_x^- &\xrightarrow{d_{\alpha x}} \mathscr{K}_x^- \\ K_x^0 &\xrightarrow{d_{\alpha x}} \mathscr{K}_x^0. \end{aligned} \tag{6.16}$$

As a consequence we have the duality between the Brownian energy process on two sites and the symmetric inclusion process on two sites

$$L_{1,2} \xrightarrow{D} \mathcal{L}_{1,2}, \tag{6.17}$$

with duality function D defined in (6.9).

Proof To see (6.16) we compute

$$\begin{aligned}
[\mathscr{K}_x^+ d_{\alpha_x}(n,\cdot)](z) &= z\frac{z^n\Gamma(\alpha_x)}{\Gamma(\alpha_x+n)} \\
&= (\alpha_x+n)z^{n+1}\frac{\Gamma(\alpha_x)}{\Gamma(\alpha_x+n+1)} \\
&= (\alpha_x+n)d_{\alpha_x}(n+1,z) \\
&= [K^+ d_{\alpha_x}(\cdot,z)](n), \qquad (6.18)
\end{aligned}$$

$$\begin{aligned}
[\mathscr{K}_x^- d_{\alpha_x}(n,\cdot)](z) &= z\frac{n(n-1)z^{n-2}\Gamma(\alpha_x)}{\Gamma(\alpha_x+n)} + \alpha_x\frac{nz^{n-1}\Gamma(\alpha_x)}{\Gamma(\alpha_x+n)} \\
&= \frac{n(n+\alpha_x-1)z^{n-1}\Gamma(\alpha_x)}{\Gamma(\alpha_x+n)} = nd_{\alpha_x}(n-1,z) \\
&= [K^- d_{\alpha_x}(\cdot,z)](n), \qquad (6.19)
\end{aligned}$$

$$\begin{aligned}
[\mathscr{K}_x^0 d_{\alpha_x}(n,\cdot)](z) &= z\frac{nz^{n-1}\Gamma(\alpha_x)}{\Gamma(\alpha_x+n)} + \frac{\alpha_x}{2}\frac{z^n\Gamma(\alpha_x)}{\Gamma(\alpha_x+n)} \\
&= \left(\frac{\alpha_x}{2}+n\right)d_{\alpha_x}(n,z) \\
&= [K^0 d_{\alpha_x}(\cdot,z)](n). \qquad (6.20)
\end{aligned}$$

This proves the first statement. The consequence (6.17) follows via Proposition 2.40. □

Once we have obtained the duality between the Brownian energy process on two sites and the symmetric inclusion process on two sites, we can straightforwardly extend it to a general finite graph. This will be discussed in the next section.

6.4 Duality of Inhomogeneous BEP($\boldsymbol{\alpha}$) and SIP($\boldsymbol{\alpha}$)

We start from a finite set V, a symmetric irreducible function $p : V \times V \to \mathbb{R}$, and a profile of α's: $\boldsymbol{\alpha} : V \to (0,\infty)$. We define the symmetric inclusion process and Brownian energy process associated to these parameters as follows.

Definition 6.3 (Inhomogeneous SIP and BEP) We define the following:

(a) The SIP($\boldsymbol{\alpha}$) with edge rates p is the Markov process on $\mathbb{N}^V$ with generator

$$Lf(\eta) = \frac{1}{2} \sum_{x,y \in V} p(x, y) \left[\eta_x(\alpha_y + \eta_y)(f(\eta^{x,y}) - f(\eta)) \right.$$
$$\left. + \eta_y(\alpha_x + \eta_x)(f(\eta^{y,x}) - f(\eta)) \right]. \tag{6.21}$$

(b) The BEP($\boldsymbol{\alpha}$) with edge rates p is the Markov process on $[0, \infty)^V$ with generator acting on smooth functions as follows

$$\mathcal{L}f(\zeta) = \frac{1}{2} \sum_{x,y \in V} p(x, y) \left[(\alpha_y \zeta_x - \alpha_x \zeta_y) \left(\frac{\partial}{\partial \zeta_y} - \frac{\partial}{\partial \zeta_x} \right) \right.$$
$$\left. + \zeta_x \zeta_y \left(\frac{\partial}{\partial \zeta_x} - \frac{\partial}{\partial \zeta_y} \right)^2 \right] f(\zeta). \tag{6.22}$$

Because the generators of both processes defined in Definition 6.3 are sums of generators working on two vertices, from the duality Theorem 6.2 we obtain the results described below.

Theorem 6.4 (Duality) *The following holds:*

(a) The SIP($\boldsymbol{\alpha}$) *with edge rates p is dual to the* BEP($\boldsymbol{\alpha}$) *with edge rates p with duality function*

$$D_{\boldsymbol{\alpha}}(\xi, \zeta) = \prod_{x \in V} d_{\alpha_x}(\xi_x, \zeta_x), \tag{6.23}$$

with d_α defined by

$$d_\alpha(n, z) = z^n \frac{\Gamma(\alpha)}{\Gamma(\alpha + n)}. \tag{6.24}$$

(b) The BEP($\boldsymbol{\alpha}$) *with edge rates p has stationary and reversible measures given by products of Gamma distributions with shape parameters given by $\boldsymbol{\alpha}$, and constant scale parameter θ, i.e., the measures*

$$\mu_{\boldsymbol{\alpha},\theta}(d\zeta) = \prod_{x \in V} \frac{1}{\theta^{\alpha_x}} \frac{\zeta_x^{\alpha_x - 1}}{\Gamma(\alpha_x)} e^{-\zeta_x/\theta} \, d\zeta_x. \tag{6.25}$$

(c) When $\sum_{x\in V}\zeta_x = z$ *then the conditional measure*

$$\nu^M(d\zeta) := \mu_{\boldsymbol{\alpha},\theta}\Big(d\zeta \;\Big|\; \sum_{x\in V}\zeta_x = z\Big) \tag{6.26}$$

is equal to the distribution of $z\mathscr{X}$ *where* $\mathscr{X}$ *is a random vector with Dirichlet distribution with parameters* $\boldsymbol{\alpha}$*, i.e., it has the probability density function*

$$f_{\mathscr{X}}(\zeta) = C_{\boldsymbol{\alpha}} \prod_{x\in V} \zeta_x^{\alpha_x - 1} \cdot \mathbb{1}_{\{\sum_{x\in V}\zeta_x = 1\}} \tag{6.27}$$

where $C_{\boldsymbol{\alpha}}$ *is the normalizing constant which is given by*

$$C_{\boldsymbol{\alpha}} = \frac{\Gamma\left(\sum_{x\in V}\alpha_x\right)}{\prod_{x\in V}\Gamma(\alpha_x)}. \tag{6.28}$$

This canonical measure is also reversible for the BEP($\boldsymbol{\alpha}$) *with edge rates* p*.*

(d) The canonical measures in item (c) are ergodic for the BEP($\boldsymbol{\alpha}$) *with edge rates* p*.*

Proof Item (a) follows from the duality of the two-site system, as proven in Theorem 6.1. For item (b), we start with the relation between the duality functions and the measures $\mu_{\boldsymbol{\alpha},\theta}$:

$$\int D(\xi,\zeta)\mu_{\boldsymbol{\alpha},\theta}(d\zeta) = \theta^{|\xi|}, \tag{6.29}$$

with $|\xi| = \sum_{x\in V}\xi_x$. The result then follows by duality and the fact that in the SIP($\boldsymbol{\alpha}$) the number of particles is conserved. The reversibility can be obtained either by direct verification, i.e., by showing that the generator of BEP($\boldsymbol{\alpha}$) is selfadjoint on $L^2(\mu_{\boldsymbol{\alpha},\theta})$, or by realizing that BEP($\boldsymbol{\alpha}$) is the many-particle limit of SIP($\boldsymbol{\alpha}$), for which products of discrete-Gamma distributions are reversible measures. Then, taking the scaling limits of these measures one obtains exactly the measures $\mu_{\boldsymbol{\alpha},\theta}$. Item (c) then follows immediately because, conditioning a reversible measure on a time-invariant event, yields a reversible measure. For item (d) finally, we proceed as follows. For a finite configuration ξ we have

$$\lim_{t\to\infty}\mathbb{E}_\zeta\left(D(\xi,\zeta(t))\right) = \lim_{t\to\infty}\mathbb{E}_\xi\left(D(\xi(t),\zeta)\right). \tag{6.30}$$

The symmetric inclusion process started from a configuration ξ is an irreducible Markov chain on the finite state space $\{\eta\in\mathbb{N}^V : |\eta| = |\xi|\}$. As a consequence it converges to its unique stationary distribution, and then the limit in the r.h.s. of (6.30) depends only on $|\zeta|$ and is equal to the expectation of $D(\xi,\zeta)$ w.r.t. the canonical measure, i.e. the one with density (6.27). The convergence (6.30) holds for all initial conditions ζ with fixed sum. As a consequence we have that the limit is ergodic. Indeed we obtain from (6.30) that the set of invariant measures on ζ with fixed sum is a singleton and therefore an extreme point. □

Remark 6.5 (Propagation of Positive Correlations) For the Brownian anergy process, it is easy to infer the analogous results of Theorem (5.34) for the symmetric inclusion process. Namely, via duality of BEP($\boldsymbol{\alpha}$) and SIP($\boldsymbol{\alpha}$), one proves that starting the Brownian energy process from the an inhomogeneous product measures with marginals Gamma distribution, the evolution produces measures with positive correlations We leave this as a useful exercise to the reader.

6.5 Generating Functions

In this section we discuss the use of generating functions for the family of models with $\mathfrak{su}(1,1)$ symmetry. This parallels what was done in Sect. 4.3 for the models with Heisenberg algebra symmetry.

As a first result we show that, with the generating function method, the self-duality of the inclusion process is equivalent to the duality between Brownian energy process and inclusion process. We can already see a signature of this by considering the single-site self-duality polynomial of the symmetric inclusion process,

$$d_x(k,n) = \frac{n!}{(n-k)!} \frac{\Gamma(\alpha_x)}{\Gamma(\alpha_x + k)} \mathbb{1}_{\{n \geq k\}}, \tag{6.31}$$

for a fixed $x \in V$, and defining the generating function

$$g_x(k,z) =: \sum_{n=0}^{\infty} d_x(k,n) \frac{z^n}{n!} = \frac{\Gamma(\alpha_x)}{\Gamma(\alpha_x + k)} z^k e^z. \tag{6.32}$$

We immediately notice that (6.32) coincides with the single-site duality function given in (6.9) for the duality between the Brownian energy process and the symmetric inclusion process (up to the factor e^z that will be inessential from the duality point of view because of the conservation law).

Theorem 6.6 (Duality and Generating Function for SIP($\boldsymbol{\alpha}$) and BEP($\boldsymbol{\alpha}$), Part 1) *For a given set V, fix a profile $\boldsymbol{\alpha} : V \to (0, \infty)$ and define the functions $D : \mathbb{N}^V \times \mathbb{N}^V \to \mathbb{R}$ and $G : \mathbb{N}^V \times [0,\infty)^V \to \mathbb{R}$*

$$D(\xi,\eta) = \prod_{x \in V} d_x(\xi_x, \eta_x) \qquad \text{and} \qquad G(\xi,\zeta) = \prod_{x \in V} g_x(\xi_x, \zeta_x), \tag{6.33}$$

with

$$g_x(k,z) =: \sum_{n=0}^{\infty} d_x(k,n)\frac{z^n}{n!}. \tag{6.34}$$

Let L be the generator of symmetric inclusion process defined in (6.21) *and $\mathcal{L}$ be the generator of the Brownian energy process defined in* (6.22)*. Then the duality property between BEP($\boldsymbol{\alpha}$) and SIP($\boldsymbol{\alpha}$) with duality function G*

$$(\mathcal{L}G(\xi,\cdot))(\zeta) = (LG(\cdot,\zeta))(\xi) \qquad \textit{for all} \quad \xi \in \mathbb{N}^V,\ \zeta \in [0,\infty)^V, \tag{6.35}$$

is equivalent to the self-duality property of SIP($\boldsymbol{\alpha}$) with self-duality function D

$$(LD(\xi,\cdot))(\eta) = (LD(\cdot,\eta))(\xi) \qquad \textit{for all} \quad \xi,\eta \in \mathbb{N}^V. \tag{6.36}$$

Proof Due to the symmetry of $p : V \times V \to \mathbb{R}$, the generator L can be rewritten as the sum

$$L = \frac{1}{2}\sum_{x,y\in V} p(x,y)L_{x,y}, \tag{6.37}$$

where $L_{x,y}$ is the single-edge generator working on η_x, η_y (and not changing the other η_z's for $z \neq x, y$). Clearly, because of (6.37) and the product nature of the duality functions involved in the theorem, imposing the duality relation on a set V is equivalent to imposing it for each couple of sites, so, it is enough to prove the statement of the theorem for the single-edge generators. For functions $f : \mathbb{N}^2 \to \mathbb{R}$, we recall the definition of $L_{1,2}$

$$\begin{aligned} L_{1,2}f(n_1,n_2) &= n_1(n_2+\alpha_2)(f(n_1-1,n_2+1) - f(n_1,n_2)) \\ &\quad + n_2(n_1+\alpha_1)(f(n_1+1,n_2-1) - f(n_1,n_2)), \end{aligned} \tag{6.38}$$

and for smooth functions $f : [0,\infty)^2 \to \mathbb{R}$ we recall the definition of $\mathcal{L}_{1,2}$

$$(\mathcal{L}_{12}f)(z_1,z_2) = \left(z_1z_2\left(\frac{\partial}{\partial z_1} - \frac{\partial}{\partial z_2}\right)^2 - (\alpha_2 z_1 - \alpha_1 z_2)\left(\frac{\partial}{\partial z_1} - \frac{\partial}{\partial z_2}\right)\right) f(z_1,z_2). \tag{6.39}$$

We prove the "if" part of the theorem. We thus assume that $d(k_1,n_1)d(k_2,n_2)$ is a self-duality function for $L_{1,2}$, i.e. for all natural numbers k_1, k_2 and n_1, n_2,

$$\big(L_{1,2}\, d_1(k_1,\cdot)d_2(k_2,\cdot)\big)(n_1,n_2) = \big(L_{1,2}\, d_1(\cdot,n_1)d_2(\cdot,n_2)\big)(k_1,k_2), \tag{6.40}$$

and would like to prove that, for all $z_1, z_2 \in \mathbb{R}$ and for all $k_1, k_2 \in \mathbb{N}$,

$$\big(\mathcal{L}_{1,2}\, g_1(\cdot, n_1) g_2(\cdot, n_2)\big)(z_1, z_2) = \big(L_{1,2}\, g_1(\cdot, n_1) g_2(\cdot, n_2)\big)(k_1, k_2). \tag{6.41}$$

Using the definition of the generating function g_x given in (6.34) and using the assumed self-duality (6.40), the right hand side of (6.41) reads

$$\begin{aligned}\big(L_{1,2}\, g_1(\cdot, z_1) g_2(\cdot, z_2)\big)(k_1, k_2) &= \sum_{n_1=0}^{\infty}\sum_{n_2=0}^{\infty} \big(L_{1,2}\, d_1(\cdot, n_1) d_2(\cdot, n_2)\big)(k_1, k_2) \frac{z_1^{n_1}}{n_1!}\frac{z_2^{n_2}}{n_2!} \\ &= \sum_{n_1=0}^{\infty}\sum_{n_2=0}^{\infty} \big(L_{1,2}\, d_1(k_1, \cdot) d_2(k_2, \cdot)\big)(n_1, n_2) \frac{z_1^{n_1}}{n_1!}\frac{z_2^{n_2}}{n_2!}.\end{aligned} \tag{6.42}$$

We now show this is equal to the left hand side of (6.41). We first compute

$$\sum_{n=0}^{\infty} n\, d_1(k_1, n-1)\frac{z_1^n}{n!} = z_1 \sum_{n=1}^{\infty} d_1(k_1, n-1)\frac{z_1^{n-1}}{(n-1)!} = z_1 g_1(k_1, z_1). \tag{6.43}$$

Moreover we compute

$$\sum_{n=0}^{\infty} d_2(k_2, n+1)\frac{z_2^n}{n!} = \sum_{m=1}^{\infty} d_2(k_2, m)\frac{z_2^{m-1}}{(m-1)!} = \frac{\partial}{\partial z_2} g_2(k_2, z_2), \tag{6.44}$$

and

$$\begin{aligned}\sum_{n=0}^{\infty} n\, d_2(k_2, n+1)\frac{z_2^n}{n!} &= z_2 \sum_{n=1}^{\infty} d_2(k_2, n+1)\frac{z_2^{n-1}}{(n-1)!} \\ &= z_2 \frac{\partial^2}{\partial z_2^2} \sum_{n=0}^{\infty} d_2(k_2, n+1)\frac{z_2^{n+1}}{(n+1)!} \\ &= z_2 \frac{\partial^2}{\partial z_2^2} g_2(k_2, z_2),\end{aligned} \tag{6.45}$$

which implies

$$\sum_{n=0}^{\infty} (\alpha_2 + n) d_2(k_2, n+1)\frac{z_2^n}{n!} = \left(\alpha_2 \frac{\partial}{\partial z_2} + z_2 \frac{\partial^2}{\partial z_2^2}\right) g_2(k_2, z_2). \tag{6.46}$$

As a consequence, we may write

$$\sum_{n_1,n_2=0}^{\infty} n_1(\alpha_2+n_2)[d_1(k_1,n_1-1)d_2(k_2,n_2+1)-d_1(k_1,n_1)d_2(k_2,n_2)]\frac{z_1^{n_1}z_2^{n_2}}{n_1!n_2!}$$
$$=\left(\alpha_2 z_1\frac{\partial}{\partial z_2}+z_1z_2\frac{\partial^2}{\partial z_2^2}-\alpha_2 z_1\frac{\partial}{\partial z_1}-z_1z_2\frac{\partial}{\partial z_1}\frac{\partial}{\partial z_2}\right)(g_1(k_1,z_1)g_2(k_2,z_2)). \tag{6.47}$$

Similarly, we have

$$\sum_{n_1,n_2=0}^{\infty} n_2(\alpha_1+n_1)[d_1(k_1,n_1+1)d_2(k_2,n_2-1)-d_1(k_1,n_1)d_2(k_2,n_2)]\frac{z_1^{n_1}z_2^{n_2}}{n_1!n_2!}$$
$$=\left(\alpha_1 z_2\frac{\partial}{\partial z_1}+z_1z_2\frac{\partial^2}{\partial z_1^2}-\alpha_1 z_2\frac{\partial}{\partial z_2}-z_1z_2\frac{\partial}{\partial z_1}\frac{\partial}{\partial z_2}\right)(g_1(k_1,z_1)g_2(k_2,z_2)). \tag{6.48}$$

Adding up (side-by-side) the previous two equations we obtain

$$\begin{aligned}&\big(L_{1,2}\,g_1(\cdot,z_1)g_2(\cdot,z_2)\big)(k_1,k_2)\\ &=\left(z_1z_2\left(\frac{\partial}{\partial z_1}-\frac{\partial}{\partial z_2}\right)^2-(\alpha_2 z_1-\alpha_1 z_2)\left(\frac{\partial}{\partial z_1}-\frac{\partial}{\partial z_2}\right)\right)\\ &\quad\times(g_1(k_1,z_1)g_2(k_2,z_2))\\ &=\big(\mathcal{L}_{1,2}\,g_1(\cdot,n_1)g_2(\cdot,n_2)\big)(z_1,z_2)\end{aligned} \tag{6.49}$$

This proves that (6.40) implies (6.41). The implication in the other direction follows from a similar reasoning. □

As a second application of the generating function method, we now prove a self-duality property for the Brownian energy process. Since we are in a continuum context, the method based on applying a symmetry to a cheap duality function is not natural to implement. Indeed, while in the discrete context the cheap duality function is a diagonal operator of the form $\frac{1}{M(\xi)}\delta_{\xi,\eta}$, with M a reversible measure, it is not clear how to define an analogous object in the continuous context. The generating function method, on the other hand, seems to be the most natural approach to lift duality from the discrete to the continuous setting. Discrete self-dualities can be "lifted" to continuous-discrete dualities by applying a generating function once. Moreover, iterating this procedure, continuous-continuous self-dualities can be obtained by applying the generating function twice.

Theorem 6.7 (Duality and Generating Functions for SIP(α) and BEP(α), Part 2) *The following three statements are equivalent:*

1. $D(\xi, \eta) = \prod_{x \in V} d_x(\xi_x, \eta_x)$ *is a self-duality function for the symmetric inclusion process with generator* L *in* (6.21).
2. $G(\xi, \zeta) = \prod_{x \in V} g_x(\xi_x, \zeta_x)$ *with*

$$g_x(k, z) = \sum_{n=0}^{\infty} d_x(k, n) \tfrac{z^n}{n!}, \tag{6.50}$$

is a duality function between symmetric inclusion process and the Brownian energy process with generator $\mathcal{L}$ *in* (6.22).

3. $H(v, \zeta) = \prod_{x \in V} h_x(v_x, \zeta_x)$ *with*

$$h_x(v, z) = \sum_{k=0}^{\infty} \sum_{n=0}^{\infty} d_x(k, n) \tfrac{v^k z^n}{n! k!} = \sum_{k=0}^{\infty} g_x(k, z) \tfrac{v^k}{k!}, \tag{6.51}$$

is a self-duality function for the Brownian energy process with generator $\mathcal{L}$ *in* (6.22).

Proof The equivalence between 1 and 2 has already been proven in Theorem 6.6. The equivalence between 2 and 3 can be proved similarly. Namely, we start from

$$\mathcal{L}_{1,2}\big(g_1(k_1, \cdot) g_2(k_2, \cdot)\big)(z_1, z_2) = L_{1,2}\big(g_1(\cdot, z_1) g_2(\cdot, z_2)\big)(k_1, k_2), \tag{6.52}$$

and write

$$\begin{aligned} &\sum_{k_1, k_2 = 0}^{\infty} \mathcal{L}_{1,2}\big(g_1(k_1, \cdot) g_2(k_2, \cdot)\big)(z_1, z_2) \frac{v_1^{k_1}}{k_1!} \frac{v_2^{k_2}}{k_2!} \\ &= \sum_{k_1, k_2 = 0}^{\infty} L_{1,2}\big(g_1(\cdot, z_1) g_2(\cdot, z_2)\big)(k_1, k_2) \frac{v_1^{k_1}}{k_1!} \frac{v_2^{k_2}}{k_2!}. \end{aligned} \tag{6.53}$$

On the left hand side we recognize the definition of the generating function, thus obtaining

$$\sum_{k_1, k_2 = 0}^{\infty} \mathcal{L}_{1,2}\big(g_1(k_1, \cdot) g_2(k_2, \cdot)\big)(z_1, z_2) \frac{v_1^{k_1}}{k_1!} \frac{v_2^{k_2}}{k_2!} = \mathcal{L}_{1,2}\big(h_1(v_1, \cdot) h_2(v_2, \cdot)\big)(z_1, z_2). \tag{6.54}$$

By an explicit computation that just uses the definitions of L and $\mathcal{L}$, one can check that

$$\sum_{k_1,k_2=0}^{\infty}\Big(L_{1,2}\big(g_1(\cdot,z_1)g_2(\cdot,z_2)\big)(k_1,k_2)\Big)\frac{v_1^{k_1}}{k_1!}\frac{v_2^{k_2}}{k_2!}=\mathcal{L}_{1,2}\big(h_1(\cdot,z_1)h_2(\cdot,z_2)\big)(v_1,v_2). \tag{6.55}$$

Thus the equivalence between item 2 and item 3 follows by combining (6.52), (6.54) and (6.55). □

Example If we consider the single-site duality function between BEP($\boldsymbol{\alpha}$) and SIP($\boldsymbol{\alpha}$)

$$g_x(n,z)=z^n\frac{\Gamma(\alpha_x)}{\Gamma(\alpha_x+n)} \tag{6.56}$$

and compute its generating function the we get a single-site self-duality function for BEP($\boldsymbol{\alpha}$) in terms of modified Bessel function $I_\alpha(y)$, i.e.

$$\begin{aligned}h_x(v,z)&=\sum_{n=0}^{\infty}g_x(n,z)\frac{v^n}{n!}\\&=\sum_{n=0}^{\infty}\frac{(zv)^n}{n!}\frac{\Gamma(\alpha_x)}{\Gamma(\alpha_x+n)}\\&=\Gamma(\alpha_x)(\sqrt{zv})^{1-\alpha_x}I_{\alpha_x-1}(2\sqrt{zv}),\end{aligned} \tag{6.57}$$

where we recall that

$$I_\alpha(y)=\sum_{k=0}^{\infty}\frac{(\frac{y}{2})^{2k+\alpha}}{k!\Gamma(\alpha+1+k)}. \tag{6.58}$$

The self-duality function of BEP($\boldsymbol{\alpha}$) is then the product

$$H(v,\zeta)=\prod_x h_x(v_x,\zeta_x). \tag{6.59}$$

6.6 Intertwining

We know that from duality we can get intertwining (see Theorem 2.29). In this section we provide the description of the intertwining for the family of models with $\mathfrak{su}(1,1)$ symmetry. This parallels what was done in Sect. 4.4 for the models with

Heisenberg algebra symmetry. Here the situation is richer, since both the symmetric inclusion process and the Brownian energy process admit a reversible measure and thus we have intertwining in both directions.

6.6.1 *Intertwining Between* BEP(α) *and* SIP(α)

The starting point is provided by the single-site duality function between SIP($\boldsymbol{\alpha}$) and BEP($\boldsymbol{\alpha}$), namely

$$d(n, z) = z^n \frac{\Gamma(\alpha)}{\Gamma(\alpha + n)}, \tag{6.60}$$

and the single-site reversible weight for SIP($\boldsymbol{\alpha}$)

$$M(n) = \frac{\Gamma(\alpha + n)}{n!\Gamma(\alpha)}. \tag{6.61}$$

This then leads to the following single-site intertwiner (cf. (2.106)) between BEP($\boldsymbol{\alpha}$) and SIP($\boldsymbol{\alpha}$)

$$\begin{aligned} \Lambda f(z) &= \sum_{n=0}^{\infty} d(n, z) M(n) f(n) \\ &= \sum_{n=0}^{\infty} \frac{z^n}{n!} f(n). \end{aligned} \tag{6.62}$$

Here we implicitly assume that $f(\cdot)$ is such that the r.h.s. of (6.62) is absolutely convergent for z at least in an open neighborhood of the origin. In practice, we will only consider at most polynomially growing f (such as e.g. self-duality functions of SIP). Then we have the following inertwining result.

Theorem 6.8 (Intertwining Between BEP($\boldsymbol{\alpha}$) **and** SIP($\boldsymbol{\alpha}$)**)** *For a function* $f : \mathbb{N}^V \to \mathbb{R}$*, let* $\Lambda f : [0, \infty)^V \to \mathbb{R}$ *be defined by*

$$(\Lambda f)(\zeta) = \sum_{\eta \in \mathbb{N}^V} f(\eta) \prod_{x \in V} \frac{\zeta_x^{\eta_x}}{\eta_x!} e^{-\zeta_x}. \tag{6.63}$$

Then we have the intertwining

$$\mathcal{L}\Lambda = \Lambda L, \tag{6.64}$$

where $\mathcal{L}$ is the generator of the Brownian energy process with profile $\boldsymbol{\alpha}$ defined in (6.22) *and L denotes the generator of the inclusion process generator with profile $\boldsymbol{\alpha}$ defined in* (6.21).

Proof It is enough to consider the two-site generators, that we recall hereafter:

$$\mathcal{L}_{12} f(\zeta_1, \zeta_2) = \left(\zeta_1\zeta_2\left(\frac{\partial}{\partial \zeta_1} - \frac{\partial}{\partial \zeta_2}\right)^2 - (\alpha_2\zeta_1 - \alpha_1\zeta_2)\left(\frac{\partial}{\partial \zeta_1} - \frac{\partial}{\partial \zeta_2}\right)\right) f(\zeta_1, \zeta_2), \tag{6.65}$$

$$\begin{aligned} L_{12} f(\eta_1, \eta_2) &= \eta_1(\alpha + \eta_2)(f(\eta_1 - 1, \eta_2 + 1) - f(\eta_1, \eta_2)) \\ &\quad + \eta_2(\alpha + \eta_1)(f(\eta_1 + 1, \eta_2 - 1) - f(\eta_1, \eta_2)). \end{aligned} \tag{6.66}$$

We start from

$$(\mathcal{L}_{12}\Lambda f)(\zeta_1, \zeta_2) = \mathcal{L}_{12} \sum_{\eta_1, \eta_2 \in \mathbb{N}^2} f(\eta_1, \eta_2) \frac{\zeta_1^{\eta_1}}{\eta_1!} \frac{\zeta_2^{\eta_2}}{\eta_2!} e^{-\zeta_1 - \zeta_2}. \tag{6.67}$$

Acting with $\mathcal{L}_{12}$ on the (ζ_1, ζ_2) variables, after rearranging terms one finds

$$\begin{aligned} (\mathcal{L}_{12}\Lambda f)(\zeta_1, \zeta_2) = &\sum_{\eta_1, \eta_2 \in \mathbb{N}^2} \frac{f(\eta_1, \eta_2)}{\eta_1! \eta_2!} e^{-\zeta_1 - \zeta_2} \\ &\Big(\eta_1(\eta_1 - 1 + \alpha_1)\zeta_1^{\eta_1 - 1}\zeta_2^{\eta_2 + 1} - \eta_1(\alpha_2 + \eta_2)\zeta_1^{\eta_1}\zeta_2^{\eta_2} \\ &+ \eta_2(\eta_2 - 1 + \alpha_2)\zeta_1^{\eta_1 + 1}\zeta_2^{\eta_2 - 1} - \eta_2(\alpha_1 + \eta_1)\zeta_1^{\eta_1}\zeta_2^{\eta_2}\Big). \end{aligned} \tag{6.68}$$

By appropriate shifting of the discrete variables η_1 and η_2, one then recognizes, on the r.h.s., the action of the symmetric inclusion process generator, which yields

$$(\mathcal{L}_{12}\Lambda f)(\zeta_1, \zeta_2) = \sum_{\eta_1, \eta_2 \in \mathbb{N}^2} L_{12} f(\eta_1, \eta_2) \frac{\zeta_1^{\eta_1}}{\eta_1!} \frac{\zeta_2^{\eta_2}}{\eta_2!} e^{-\zeta_1 - \zeta_2}. \tag{6.69}$$

This proves (6.64). □

As already remarked in Sect. 4.3, the intertwiner Λ has the probabilistic interpretation of averaging over a (inhomogeneous) product Poisson distribution, i.e.,

$$\Lambda f(\zeta) = \int f(\eta) \nu_\zeta(d\eta), \tag{6.70}$$

where ν_ζ is the product Poisson measure on $\mathbb{N}^V$ with parameter ζ_x at $x \in V$. Then we have the following result.

Corollary 6.9 (Consequences of Intertwining Between BEP($\boldsymbol{\alpha}$) and SIP($\boldsymbol{\alpha}$)) *Denote by $\mathcal{S}(t)$ the semigroup of the Brownian energy process with profile $\boldsymbol{\alpha}$ and by $S(t)$ the semigroup of the symmetric inclusion process with profile $\boldsymbol{\alpha}$. Then the following holds true:*

(a) Λ is an intertwiner between the semigroups $\mathcal{S}(t)$ and $S(t)$, i.e. for all $t > 0$

$$\mathcal{S}(t)\Lambda = \Lambda S(t). \tag{6.71}$$

(b) As a consequence, we have the following propagation of inhomogeneous Poisson product measures for the symmetric inclusion process $\{\eta(t), t \geq 0\}$: for all $\zeta \in [0, \infty)^V$ and $f : \mathbb{N}^V \to \mathbb{R}$ bounded:

$$\int \Big(\mathbb{E}_\eta f(\eta(t))\Big)\nu_\zeta(d\eta) = \mathbb{E}_\zeta\Big(\int f(\eta)\nu_{\zeta(t)}(d\eta)\Big), \tag{6.72}$$

where $\{\zeta(t), t \geq 0\}$ is the Brownian energy process.

Proof The first statement is a consequence of the lifting of the analogous relations for generators. To prove the second statement, we have

$$\begin{aligned}\int \Big(\mathbb{E}_\eta f(\eta(t))\Big)\nu_\zeta(d\eta) &= \int \Big(S(t)f(\eta)\Big)\nu_\zeta(d\eta)\\ &= \Lambda(S(t)f)(\zeta)\\ &= \mathcal{S}(t)(\Lambda f)(\zeta)\\ &= \mathbb{E}_\zeta(\Lambda f(\zeta(t)))\\ &= \mathbb{E}_\zeta\Big(\int f(\eta)\nu_{\zeta(t)}(d\eta)\Big).\end{aligned} \tag{6.73}$$

This concludes the proof. □

Remark 6.10 (Interpretation of Intertwining) The intertwining relation (6.72) can be rewritten as

$$\int \Big(\mathbb{E}_\eta f(\eta(t))\Big)\nu_\zeta(d\eta) = \int d\zeta'\Big(\int f(\eta)\nu_{\zeta'}(d\eta)\Big)p_t(\zeta,\zeta'), \tag{6.74}$$

where $p_t(\zeta, \zeta')$ denotes the transition probability density in the Brownian energy process. From this rewriting, the probabilistic meaning becomes more transparent: starting the symmetric inclusion process from an inhomogeneous Poisson product measure ν_ζ with parameter ζ and evolving it at time t, has the same distribution as a mixture of Poisson product measures, where the mixture is provided by the Brownian energy process initialized at ζ and then evolved at time t. This is to be compared with the Doob's theorem (Theorem 4.14) for independent walkers, for which Poisson product measure are mapped to Poisson product measures in the course of the evolution where the profile of the Poisson parameters evolves according to the *deterministic* dual dynamics. Here the BEP dynamics is stochastic, and therefore, starting from a Poisson product measures at time $t > 0$ we obtain a mixture of Poisson product measures.

Remark 6.11 (Mixtures of Products of Poisson Distributions Are Closed for the Symmetric Inclusion Process) Equivalently, another way to rewrite (6.72) is

$$\left(\otimes_x \nu_{\zeta_x}\right) S(t) = \mathbb{E}_{\zeta}^{\mathrm{BEP}(\boldsymbol{\alpha})}\left[\otimes_x \nu_{\zeta_x(t)}\right]. \tag{6.75}$$

From this relation it follows that the set of mixtures of products of Poisson measures is closed under the SIP dynamics. Indeed, if we start from an initial measure of the type:

$$\int \lambda(d\zeta)\left(\otimes_x \nu_{\zeta_x}\right), \tag{6.76}$$

then evolving under the SIP dynamics, at time $t > 0$ we find the measure

$$\left(\int \lambda(d\zeta)\left(\otimes_x \nu_{\zeta_x}\right)\right) S(t) = \int \lambda_t(d\zeta)\left(\otimes_x \nu_{\zeta_x}\right), \tag{6.77}$$

where $\lambda_t(d\zeta)$ is the distribution of BEP at time $t > 0$ started from $\lambda(d\zeta)$ at time 0.

Remark 6.12 (Stationarity of Products of Gamma Distribution for BEP($\boldsymbol{\alpha}$) **Implies Stationarity of Product of Discrete-Gamma for** SIP($\boldsymbol{\alpha}$)**)** A particular instance is when the mixing measure $\lambda(d\zeta)$ is a product of Gamma distributions with parameters (α_x, θ_x) at site x:

$$\lambda(d\zeta) = \prod_x \zeta_x^{\alpha_x - 1} e^{-\zeta_x/\theta_x} \frac{1}{\theta_x^{\alpha_x}\Gamma(\alpha_x)} d\zeta_x. \tag{6.78}$$

Then the mixture of Poisson product measures with such mixing measure turns out to be the product measure with marginals given by the discrete-Gamma distribution with parameters $(\alpha_x, \theta_x/(1+\theta_x))$:

$$\int \lambda(d\zeta)\left(\otimes_x \nu_{\zeta_x}(d\eta_x)\right) = \prod_x \left(\frac{1}{1+\theta_x}\right)^{\alpha_x} \frac{\left(\frac{\theta_x}{1+\theta_x}\right)^{\eta_x}}{\eta_x!} \frac{\Gamma(\eta_x + \alpha_x)}{\Gamma(\alpha_x)} d\eta_x, \tag{6.79}$$

which is indeed stationary for SIP, as we would find from (6.77), because inhomogeneous product of Gamma distributions with parameters (α_x, θ_x) is stationary for BEP($\boldsymbol{\alpha}$).

6.6.2 *Intertwining Between* SIP(α) *and* BEP(α)

So far we have been thinking the symmetric inclusion process as the dual of the Brownian energy process. The duality relation is a symmetric relation and thus we can exchange the role of the two processes. In this way we can construct the intertwiner in the other direction.

Theorem 6.13 (Intertwining Between SIP($\boldsymbol{\alpha}$) **and** BEP($\boldsymbol{\alpha}$)**)** *For a function* $f : [0,\infty)^V \to \mathbb{R}$*, let* $Uf : \mathbb{N}^V \to \mathbb{R}$ *be defined by*

$$(Uf)(\eta) = \int d\zeta\, f(\zeta) \prod_{x\in V} \frac{\zeta_x^{\alpha_x+\eta_x-1}}{\Gamma(\alpha_x+\eta_x)} e^{-\zeta_x}. \tag{6.80}$$

Then we have the intertwining

$$LU = U\mathcal{L}, \tag{6.81}$$

where $\mathcal{L}$ *is the generator of the Brownian energy process with profile* $\boldsymbol{\alpha}$ *defined in* (6.22) *and* L *denotes the generator of the inclusion process with profile* $\boldsymbol{\alpha}$ *defined in* (6.21).

Proof It is enough to consider the two-site generators. On one hand we have

$$(Uf)(\eta_1,\eta_2) = \int d\zeta_1\, d\zeta_2\, f(\zeta_1,\zeta_2) \frac{\zeta_1^{\alpha_1+\eta_1-1}}{\Gamma(\alpha_1+\eta_1)} \frac{\zeta_2^{\alpha_2+\eta_2-1}}{\Gamma(\alpha_2+\eta_2)} e^{-\zeta_1-\zeta_2}, \tag{6.82}$$

and acting with L_{12} on the (η_1,η_2) variables one finds

$$\begin{aligned}
&(L_{12}Uf)(\eta_1,\eta_2) \\
&= \int d\zeta_1\, d\zeta_2\, f(\zeta_1,\zeta_2) e^{-\zeta_1-\zeta_2} \\
&\left[\eta_1(\eta_2+\alpha_2)\left(\frac{\zeta_1^{\alpha_1+\eta_1-2}}{\Gamma(\alpha_1+\eta_1-2)}\frac{\zeta_2^{\alpha_2+\eta_2}}{\Gamma(\alpha_2+\eta_2)} - \frac{\zeta_1^{\alpha_1+\eta_1-1}}{\Gamma(\alpha_1+\eta_1)}\frac{\zeta_2^{\alpha_2+\eta_2-1}}{\Gamma(\alpha_2+\eta_2)}\right)\right. \\
&\left. +\eta_2(\eta_1+\alpha_1)\left(\frac{\zeta_1^{\alpha_1+\eta_1}}{\Gamma(\alpha_1+\eta_1)}\frac{\zeta_2^{\alpha_2+\eta_2-2}}{\Gamma(\alpha_2+\eta_2-2)} - \frac{\zeta_1^{\alpha_1+\eta_1-1}}{\Gamma(\alpha_1+\eta_1)}\frac{\zeta_2^{\alpha_2+\eta_2-1}}{\Gamma(\alpha_2+\eta_2)}\right)\right].
\end{aligned} \tag{6.83}$$

On the other hand we have

$$(\mathcal{L}_{12}f)(\zeta_1,\zeta_2) = \left(\zeta_1\zeta_2\Big(\frac{\partial}{\partial\zeta_1} - \frac{\partial}{\partial\zeta_2}\Big)^2 - (\alpha_2\zeta_1-\alpha_1\zeta_2)\Big(\frac{\partial}{\partial\zeta_1} - \frac{\partial}{\partial\zeta_2}\Big)\right) f(\zeta_1,\zeta_2), \tag{6.84}$$

and, acting with U, we obtain

$$
\begin{aligned}
(U\mathcal{L}_{12}f)(\eta_1,\eta_2) = \int d\zeta_1\, d\zeta_2\; f(\zeta_1,\zeta_2)e^{-\zeta_1-\zeta_2} \\
\Bigg[\frac{\zeta_1^{\alpha_1+\eta_1}}{\Gamma(\alpha_1+\eta_1)}\frac{\zeta_2^{\alpha_2+\eta_2}}{\Gamma(\alpha_2+\eta_2)}\Big(\frac{\partial}{\partial\zeta_1}-\frac{\partial}{\partial\zeta_2}\Big)^2 f(\zeta_1,\zeta_2) \\
-\alpha_2\frac{\zeta_1^{\alpha_1+\eta_1}}{\Gamma(\alpha_1+\eta_1)}\frac{\zeta_2^{\alpha_2+\eta_2-1}}{\Gamma(\alpha_2+\eta_2)}\Big(\frac{\partial}{\partial\zeta_1}-\frac{\partial}{\partial\zeta_2}\Big) f(\zeta_1,\zeta_2) \\
+\alpha_1\frac{\zeta_1^{\alpha_1+\eta_1-1}}{\Gamma(\alpha_1+\eta_1)}\frac{\zeta_2^{\alpha_2+\eta_2}}{\Gamma(\alpha_2+\eta_2)}\Big(\frac{\partial}{\partial\zeta_1}-\frac{\partial}{\partial\zeta_2}\Big) f(\zeta_1,\zeta_2)\Bigg].
\end{aligned}
\tag{6.85}
$$

By integrating by parts this expression, one shows that (6.85) equals (6.83), which proves (6.81). □

The intertwiner U has the probabilistic interpretation of averaging over an (inhomogeneous) product Gamma distribution, i.e.,

$$(Uf)(\eta) = \int f(\zeta)\mu_\eta(d\zeta), \tag{6.86}$$

where μ_η is the product Gamma measure on $[0,\infty)^V$ with shape parameter $\alpha_x + \eta_x$ at site $x \in V$:

$$\mu_\eta(d\zeta) = \prod_x \frac{1}{\Gamma(\alpha_x+\eta_x)}\zeta_x^{\alpha_x+\eta_x-1}e^{-\zeta_x}\, d\zeta_x. \tag{6.87}$$

Then we have the following result.

Corollary 6.14 (Consequences of Intertwining Between SIP($\boldsymbol{\alpha}$) **and** BEP($\boldsymbol{\alpha}$)**)** *Denoting by $\mathcal{S}(t)$ the semigroup of the Brownian energy process with profile $\boldsymbol{\alpha}$ and by $S(t)$ the semigroup of the inclusion process with profile $\boldsymbol{\alpha}$, the following holds true:*

(a) U is an intertwiner between the semigroups $S(t)$ and $\mathcal{S}(t)$, i.e. for all $t > 0$

$$S(t)U = U\mathcal{S}(t). \tag{6.88}$$

(b) As a consequence, we have the following propagation of inhomogeneous Gamma product measures for the Brownian energy process $\{\zeta(t), t \geq 0\}$: for

all $\eta \in \mathbb{N}^V$ *and* $f : [0,\infty)^V \to \mathbb{R}$ *bounded:*

$$\int \Big(\mathbb{E}_\zeta f(\zeta(t))\Big)\mu_\eta(d\zeta) = \mathbb{E}_\eta\Big(\int f(\zeta)\mu_{\eta(t)}(d\zeta)\Big), \tag{6.89}$$

where $\{\eta(t), t \geq 0\}$ *is the symmetric inclusion process.*

Proof The first statement is a consequence of the lifting of the analogous relations for generators. To prove the second statement, we have

$$\begin{aligned}\int \Big(\mathbb{E}_\zeta f(\zeta(t))\Big)\mu_\eta(d\zeta) &= \int \Big(\mathcal{S}(t)f(\zeta)\Big)\mu_\eta(d\zeta)\\ &= U(\mathcal{S}(t)f)(\eta)\\ &= S(t)(Uf)(\eta)\\ &= \mathbb{E}_\eta(Uf(\eta(t)))\\ &= \mathbb{E}_\eta\Big(\int f(\zeta)\mu_{\eta(t)}(d\zeta)\Big).\end{aligned} \tag{6.90}$$

□

Remark 6.15 (Interpretation of Intertwining) The intertwining relation (6.89) can be rewritten as

$$\int \Big(\mathbb{E}_\zeta f(\zeta(t))\Big)\mu_\eta(d\zeta) = \sum_{\eta'}\Big(\int f(\zeta)\mu_{\eta'}(d\zeta)\Big)p_t(\eta,\eta'), \tag{6.91}$$

where $p_t(\eta,\eta')$ denotes the transition probability of the symmetric inclusion process. From this rewriting, the probabilistic meaning becomes more transparent: starting the Brownian energy process from an inhomogeneous Gamma product measure μ_η with shape parameter $\eta+\alpha$ and evolving at time t, has the same distribution as a mixture of Gamma product measures, where the mixture is provided by the transition probability of the symmetric inclusion process initialized at η and then evolved at time t.

Remark 6.16 (Mixtures of Products of Gamma Distributions Are Closed for Brownian Energy Process) From the previous remark it follows that initial Gamma product measures are not exactly reproduced by the Brownian energy process but turned instead into convex combinations of Gamma product measures with random weights provided by the inclusion process. Namely

$$\left(\otimes_x \mu_{\eta_x}\right)\mathcal{S}(t) = \mathbb{E}_\eta^{\mathrm{SIP}}\left[\otimes_x \mu_{\eta_x(t)}\right]. \tag{6.92}$$

From this it follows that the set of mixtures of products of Gamma measures is closed under the BEP dynamics. Indeed, if we start from an initial measure

$$\int \lambda(d\eta)\left(\otimes_x \mu_{\eta_x}\right), \tag{6.93}$$

and evolve under the BEP dynamics, at time $t > 0$ we find again a mixed measure

$$\left(\int \lambda(d\eta)\left(\otimes_x \mu_{\eta_x}\right)\right) \mathcal{S}(t) = \int \lambda_t(d\eta)\left(\otimes_x \mu_{\eta_x}\right), \tag{6.94}$$

where $\lambda_t(d\eta)$ is the distribution of SIP(α) at time $t > 0$ started from $\lambda(d\eta)$ at time 0.

Remark 6.17 (Stationarity of Products of Negative Binomials Distributions for SIP($\boldsymbol{\alpha}$) **Implies Stationarity of Product of Gamma Distributions for** BEP($\boldsymbol{\alpha}$)) If the mixing measure $\lambda(d\zeta)$ is a product of discrete-Gamma distributions with parameters (α_x, θ_x) at site x:

$$\lambda(d\eta) = \prod_x (1-\theta_x)^{\alpha_x} \frac{\theta_x^n}{n!} \frac{\Gamma(\alpha_x + \eta_x)}{\Gamma(\alpha_x)} d\eta_x, \tag{6.95}$$

then the result of the mixture (6.93) is a product of Gamma distributions with shape parameter α_x and scale parameter $\frac{\theta_x}{1-\theta_x}$:

$$\int \lambda(d\eta)\left(\otimes_x \mu_{\eta_x}(d\zeta_x)\right) = \prod_x \left(\frac{1-\theta_x}{\theta_x}\right)^{\alpha_x} \frac{\zeta_x^{\alpha_x - 1} e^{-\left(\frac{1-\theta_x}{\theta_x}\right)\zeta_x}}{\Gamma(\alpha_x)} d\zeta_x, \tag{6.96}$$

which is stationary for BEP($\boldsymbol{\alpha}$). This is consistent with the formula (6.94), and the fact that products of discrete-Gamma distributions (6.95) are invariant for SIP($\boldsymbol{\alpha}$).

Remark 6.18 The essential point of the proof of the previous theorem lies in the fact that in the action of K's operators (as in (5.115)) is linearly dependent, both on the η_x variables and on the representation parameters α_x. Therefore these operators have natural behavior with respect to contraction of both α and η. The same linearity property holds for the $\mathscr{K}$s operators (6.13) appearing in the abstract form of the generator of the Brownian energy process. As a consequence, the same result can be proven for BEP($\boldsymbol{\alpha}$).

6.7 The Brownian Momentum Process

The Brownian momentum process (sometimes we will use the abbreviation BMP process) is another diffusion process which is similar to the Brownian energy process. However the BMP process on a set V takes values in $\mathbb{R}^V$ and the variables are interpreted as momenta. The generator of BMP has the same abstract form of the generator of the Brownian energy process in terms of the generators of the non-compact $\mathfrak{su}(1,1)$ Lie algebra. More precisely, the K's operators are written again in a representation in term of differential operators, but this is different than the one used for the construction of Brownian energy process.

We start by defining the process for the simplest case where the process parameters α_x are equal to 1 for all sites $x \in V$. Moreover we restrict to the two-site case $V = \{1, 2\}$. Consider the following operator acting on smooth compactly supported functions $f : \mathbb{R}^2 \to \mathbb{R}$ (this can be extended to smooth at most polynomially growing functions, in particular polynomials as in Proposition 6.19 below).

$$\mathcal{L}_{1,2} = \frac{1}{4}\left(\mathfrak{z}_1 \frac{\partial}{\partial \mathfrak{z}_2} - \mathfrak{z}_2 \frac{\partial}{\partial \mathfrak{z}_1}\right)^2. \tag{6.97}$$

If we use polar coordinates (θ, R) defined by

$$\mathfrak{z}_1^2 + \mathfrak{z}_2^2 = R^2 \qquad\qquad \theta = \arctan(\mathfrak{z}_2/\mathfrak{z}_1), \tag{6.98}$$

then the operator simply reads

$$\mathcal{L}_{1,2} = \frac{1}{4}\frac{\partial^2}{\partial \theta^2}. \tag{6.99}$$

The change to polar coordinates also indicates a clear interpretation: the radial coordinate $R(t) = \mathfrak{z}_1^2(t) + \mathfrak{z}_2^2(t)$ is a constant of the dynamics, whereas the angular coordinate $\theta(t)$ undergoes a Brownian motion on the circle. We will now see how the generator $\mathcal{L}_{1,2}$ derives from a representation of the $\mathfrak{su}(1,1)$ Lie algebra.

Proposition 6.19 *Consider the following operators working on smooth compactly supported functions $f : \mathbb{R} \to \mathbb{R}$*

$$\begin{aligned}
\mathbb{K}^+ f(z) &= \frac{1}{2} z^2 f(z), \\
\mathbb{K}^- f(z) &= \frac{1}{2} f''(z), \\
\mathbb{K}^0 f(z) &= \frac{1}{4}(2z f'(z) + f(z)).
\end{aligned} \tag{6.100}$$

Then these operators form a representation of the Lie algebra $\mathfrak{su}(1,1)$, i.e., they satisfy the commutation relations

$$[\mathbb{K}^0, \mathbb{K}^\pm] = \pm \mathbb{K}^\pm, \qquad [\mathbb{K}^-, \mathbb{K}^+] = 2\mathbb{K}^0. \tag{6.101}$$

Moreover, the generator (6.97) *reads*

$$\mathcal{L}_{1,2} = \mathbb{K}_1^+\mathbb{K}_2^- + \mathbb{K}_2^+\mathbb{K}_1^- - 2\mathbb{K}_1^0\mathbb{K}_2^0 + \frac{1}{8}. \tag{6.102}$$

Proof We verify $[\mathbb{K}^-, \mathbb{K}^+] = 2\mathbb{K}^0$, leaving the computation of the other commutation relations to the reader. We have

$$[\mathbb{K}^-\mathbb{K}^+]f(z) = -\frac{1}{4}\left(z^2 f''(z) - (z^2 f)''\right) = \frac{1}{4}(4zf'(z) + 2f(z)) = 2\mathbb{K}^0 f(z). \tag{6.103}$$

Next

$$\begin{aligned}
&\mathbb{K}_1^+\mathbb{K}_2^- + \mathbb{K}_2^+\mathbb{K}_1^- - 2\mathbb{K}_1^0\mathbb{K}_2^0 + \frac{1}{2} \\
&= \frac{1}{4}\mathfrak{z}_1^2\frac{\partial}{\partial \mathfrak{z}_2}^2 + \frac{1}{4}\mathfrak{z}_2\frac{\partial}{\partial \mathfrak{z}_1}^2 - \frac{1}{2}\mathfrak{z}_1\frac{\partial}{\partial \mathfrak{z}_2}\mathfrak{z}_2\frac{\partial}{\partial \mathfrak{z}_2} - \frac{1}{4}\mathfrak{z}_1\frac{\partial}{\partial \mathfrak{z}_1}\frac{1}{4}\mathfrak{z}_2\frac{\partial}{\partial \mathfrak{z}_2} \\
&= \frac{1}{4}(\mathfrak{z}_1\frac{\partial}{\partial \mathfrak{z}_2} - \mathfrak{z}_2\frac{\partial}{\partial \mathfrak{z}_1})^2.
\end{aligned} \tag{6.104}$$

□

Notice that the generator $\mathcal{L}_{1,2}$ in (6.102) has the same abstract form of the generator of the homogeneous SIP(α) with $\alpha = 1/2$. In view of proving a duality relation between the Brownian momentum process and the symmetric inclusion process with parameter $\alpha = 1/2$, we show in the following proposition how to intertwine this new continuous representation with the discrete representation used for the construction of the inclusion process.

Proposition 6.20 *Consider the function* $d : \mathbb{N} \times \mathbb{R} \to \mathbb{R}$ *given by*

$$d(n, z) = \frac{z^{2n}}{(2n-1)!!}, \tag{6.105}$$

where $(2n-1)!! = \prod_{k=1}^n (2k-1)$. *Then we have*

$$K^u \xrightarrow{d} \mathbb{K}^u, \tag{6.106}$$

for $u \in \{+, -, 0\}$, *where* K^u *are the operators defined in* (5.5), *with* $\alpha = \frac{1}{2}$.

Proof This follows from explicit computation:

$$[\mathbb{K}^+ d(n,\cdot)](z) = \frac{1}{2}z^2\frac{z^{2n}}{(2n-1)!!} = (n+\tfrac{1}{2})\frac{z^{2(n+1)}}{(2n+1)!!} = [K^+ d(\cdot, z)](n), \tag{6.107}$$

$$[\mathbb{K}^- d(n,\cdot)](z) = \frac{1}{2}\frac{2n(2n-1)z^{2n-2}}{(2n-1)!!} = n\frac{z^{2(n-1)}}{(2(n-1)-1)!!} = [K^- d(\cdot, z)](n), \tag{6.108}$$

$$[\mathbb{K}^0 d(n,\cdot)](z) = \frac{1}{2}\frac{2nz^{2n}}{(2n-1)!!} + \frac{1}{4}\frac{z^{2n}}{(2n-1)!!} = (n+\tfrac{1}{4})d(n,z) = [K^0 d(\cdot, z)](n). \tag{6.109}$$

□

Theorem 6.21 *The Brownian momentum on two sites is dual to the symmetric inclusion process* SIP(1/2) *on two sites, with duality functions*

$$D(\xi, \mathfrak{z}) = \prod_{x=1}^{2} d(\xi_x, \mathfrak{z}_x), \tag{6.110}$$

where d is defined in (6.105)*. As a consequence, the Gaussian product measure*

$$\mu_\sigma(d\mathfrak{z}_1 d\mathfrak{z}_2) = \frac{1}{2\pi\sigma^2} e^{-\frac{1}{2\sigma^2}(\mathfrak{z}_1^2+\mathfrak{z}_2^2)} d\mathfrak{z}_1 d\mathfrak{z}_2 \tag{6.111}$$

is invariant and reversible for all $\sigma^2 > 0$.

Proof The duality property immediately follows from Proposition 6.20, together with the fact that the abstract form of the generator of BMP on one hand and the abstract form of the generator of the SIP($\frac{1}{2}$) on the other hand are the same. In order to prove the second statement, we notice that the relation between $\mu_\sigma(d\mathfrak{z}_1 d\mathfrak{z}_2)$ and the duality function (6.110) is given by

$$\int D(\xi, \mathfrak{z})\mu_\sigma(d\mathfrak{z}_1 d\mathfrak{z}_2) = \sigma^{2(\xi_1+\xi_2)}. \tag{6.112}$$

Then invariance of these measures follows from conservation of particles for the SIP. Reversibility is a consequence of a direct computation, showing that the operator $\mathcal{L}_{1,2}$ is symmetric, i.e., for f, g smooth compactly supported functions

$$\langle f, Lg\rangle = \langle Lf, g\rangle, \tag{6.113}$$

with $\langle\cdot,\cdot\rangle$ the $L^2(\mu_\sigma(d\mathfrak{z}_1 d\mathfrak{z}_2))$ inner product. □

We conclude this section by discussing the relation between the Brownian energy process and the Brownian momentum process.

Lemma 6.22 *Let $\{(\mathfrak{z}_1(t), \mathfrak{z}_2(t)) : t \geq 0\}$ denote the Brownian Momentum process. Define $(\zeta_1(t), \zeta_2(t)) = (\mathfrak{z}_1^2(t), \mathfrak{z}_2^2(t))$, then $\{(\zeta_1(t), \zeta_2(t)) : t \geq 0\}$ is* BEP($\frac{1}{2}$).

Proof Let $\varphi : [0,\infty)^2 \to \mathbb{R}$ be a smooth and compactly supported function, and define $f(\mathfrak{z}_1,\mathfrak{z}_2) := \varphi(\mathfrak{z}_1^2,\mathfrak{z}_2^2)$. Using the chain rule, we compute

$$\left(\mathfrak{z}_1\frac{\partial}{\partial \mathfrak{z}_2} - \mathfrak{z}_2\frac{\partial}{\partial \mathfrak{z}_1}\right) f(\mathfrak{z}_1,\mathfrak{z}_2) = \frac{1}{2}(\mathfrak{z}_1^2 - \mathfrak{z}_2^2)((\frac{\partial}{\partial \mathfrak{z}_1} - \frac{\partial}{\partial \mathfrak{z}_2})\varphi)(\mathfrak{z}_1^2,\mathfrak{z}_2^2)$$
$$+ \mathfrak{z}_1^2\mathfrak{z}_2^2((\frac{\partial}{\partial \mathfrak{z}_1} - \frac{\partial}{\partial \mathfrak{z}_2})^2\varphi)(\mathfrak{z}_1^2,\mathfrak{z}_2^2), \qquad (6.114)$$

which means

$$\mathcal{L}_{1,2}^{\mathrm{BMP}} f(\mathfrak{z}_1,\mathfrak{z}_2) = \mathcal{L}_{1,2}^{\mathrm{BEP}(\frac{1}{2})}\varphi(\zeta_1,\zeta_2). \qquad (6.115)$$

Since both $\mathcal{L}_{1,2}^{\mathrm{BMP}}$ and $\mathcal{L}_{1,2}^{\mathrm{BEP}(\frac{1}{2})}$ generate well-defined Markov processes, the lemma is proved. □

We can now generalize the Brownian momentum process, by considering $\alpha \in \mathbb{N}$ momenta per site, i.e., consider the generator

$$\mathcal{L}_{1,2}^{\mathrm{BMP}(\alpha)} = \sum_{\ell,s=1}^{\alpha} (\mathfrak{z}_{1,\ell}\partial_{2,s} - \mathfrak{z}_{1,s}\partial_{1,\ell})^2, \qquad (6.116)$$

where we used the notation BMP(α) for the Brownian Momentum process of parameter $\alpha \in \mathbb{N}$. This can be viewed as the BMP on a ladder graph with α levels, i.e., the underlying vertex set is $\{1,2\} \times \{1,\ldots,\alpha\}$. We then have the following straightforward consequences of Lemma 6.22, together with Theorem 5.37.

Theorem 6.23 *Consider the process* BMP(α), *with generator* (6.116). *Define for* $x = 1, 2$

$$\zeta_x = \sum_{\ell=1}^{\alpha} \mathfrak{z}_{(x,\ell)}^2, \qquad (6.117)$$

then $\{(\zeta_1(t),\zeta_2(t)) : t \geq 0\}$ *evolves according to* BEP($\frac{\alpha}{2}$).

Proof For $\alpha = 1$ the theorem is true because of Lemma 6.22. For general α, it follows that $\{(\mathfrak{z}_{(1,\ell)}^2(t), \mathfrak{z}_{(2,\ell)}^2(t)),\ \ell = 1,\ldots,\alpha;\ t \geq 0\}$ evolves according to the BEP($\frac{1}{2}$) on the ladder graph $\{1,2\} \times \{1,\ldots,\alpha\}$. Therefore, by Theorem 5.37, $\{(\zeta_1(t),\zeta_2(t)), t \geq 0\}$ evolves according to BEP($\frac{\alpha}{2}$) □

6.8 Additional Notes

The Brownian energy process and its duality properties were studied first in [116], where also the connection with the continuous representation of $SU(1, 1)$ is given. In population dynamics, the Wright-Fisher diffusion with parent independent mutation rate is a related diffusion process, for which a dual was found in [88]. The intertwining between the Brownian energy process and the inclusion process is from [198]. The equivalent formulation in terms of the evolution of a product of Poisson measures is natural analogue of Doob's theorem in this setting. In [179] several new intertwinings between diffusion processes and discrete processes of birth and death type are derived. The intertwining between the inclusion process and the Brownian energy process is new. We believe that these intertwinings are related to the recent developments in characterizing non-equilibrium steady states as mixtures of product measures see [51], [69], [50], but understanding this connection is at present an open problem. The Brownian momentum process was introduced in [114], and its duality properties were first formulated in [115], then generalized in [116].

To construct a process where both energy and momentum are conserved and such that there is still duality is at present an open problem. Models of heat conduction with momentum conservation are studied e.g. in [11], see also [97].

Chapter 7
Duality for the Symmetric Partial Exclusion Process

In this chapter we introduce the symmetric partial exclusion process, which generalizes the standard symmetric exclusion process, by allowing a maximal particle number per site that can vary from site to site. Because there is an extensive literature on the exclusion process, both in the area of probability theory as well as in the area of quantum spin chains, we give here a concise discussion focusing on duality and intertwining from the algebraic point of view. We construct the single edge generator from the $\mathfrak{su}(2)$ algebra generators, in a discrete representation labeled by the maximal number of particles. We show that the single edge generator of this process is related to the coproduct of the Casimir in a manner which is completely analogous to the computation for the symmetric inclusion process, but now in the setting of the $\mathfrak{su}(2)$ algebra. Next we show that symmetric partial exclusion process has an additive structure given by copies of symmetric exclusion processes on a layered graph, similarly to what we proved in the setting of the symmetric inclusion process. This leads to intertwining of symmetric partial exclusion processes with different maximal occupation numbers. Finally, we consider a scaling limit where the maximal number of particles diverges, and find the deterministic process associated to independent random walkers.

7.1 Introduction

The exclusion process is extensively studied in the literature, it is the prototype model of an interacting particle system. This is true also for duality: it is in the context of the exclusion process that a proper definition of duality of a Markov process emerged [172, 217] and was proven to be useful. In particular Chapter 8 of [172] uses substantially duality to obtain a complete ergodic theory of the symmetric exclusion process on $\mathbb{Z}^d$. Because the exclusion process is so much studied in the literature, this chapter will be more condensed with respect to other chapters, and

C. Giardinà, F. Redig, *Duality for Markov Processes*, Grundlehren der mathematischen Wissenschaften 365, https://doi.org/10.1007/978-3-032-04099-2_7

focuses mainly on the aspects that are more relevant to the spirit of the book, such as the Lie algebraic structure of the process and the intertwining properties. The algebraic approach in the context of the symmetric exclusion process was pioneered in [213], and later extended to the asymmetric setting [210] using the q-deformation of the $\mathfrak{su}(2)$ Lie algebra.

The algebraic approach leads in particular to a natural generalization of the standard symmetric exclusion process, called *symmetric partial exclusion process* and denoted by SEP(α). The parameter α of the symmetric partial exclusion represents the maximal number of particles per site. As a consequence, α is an integer and in the simplest version of the process ($\alpha = 1$), particles are forbidden to be at the same vertex. In the SEP(α) process the rate for a particle to jump decreases linearly in the occupation of the arrival site, and increases linearly in the occupation of the departure site. For $\alpha = 1$, the symmetric exclusion process is recovered and for generic $\alpha \in \mathbb{N} \setminus \{1\}$ particles are *discouraged* to go to vertices which are already occupied, as opposed to the SIP(α) process where particles are *encouraged* to join the same vertex. The symmetric partial exclusion process is thus a particle system which is the "fermionic" companion of the symmetric inclusion process, or equivalently, the SIP(α) process is the "bosonic" companion of the SEP(α) process. The generalization which consists of having a maximum number of particles that is even allowed to depend on the site, emerges naturally from the abstract generator of the standard symmetric exclusion process, by choosing a representation which can even be site dependent. The representation parameter corresponds to the maximal number of particles at each site.

The fact that the symmetric partial exclusion process is a close relative of the symmetric inclusion process is also seen at the level of the Lie algebra underlying the process. We shall see that the underlying algebra responsible for the self-duality properties of SEP(α) is the $\mathfrak{su}(2)$ Lie algebra, and the abstract generator will be very familiar to the reader who is already acquainted with the abstract generator of the SIP(α) process. In particular, the generator has exactly the same relation with the Casimir element of $\mathfrak{su}(2)$, i.e., up to trivial central elements it equals the co-product of the Casimir, yielding the Heisenberg XXX spin chain [1]. The process with parameter α is obtained by considering an $(\alpha + 1)$-dimensional representation of $\mathfrak{su}(2)$.

Being associated to the algebra $\mathfrak{su}(2)$, for the SEP(α) process we do not have a companion diffusion process arising in the many-particle scaling, such as the Brownian energy process that was found starting from the SIP(α) process in the setting of the non compact $\mathfrak{su}(1, 1)$ Lie algebra. The only possible scaling limit corresponds to letting the maximal number of particles at each site go to infinity, and this leads or to a deterministic system in the continuum or to a system of independent random walkers. The reason that "the many particle limit" of the previous chapter, which produces the BEP(α) from the SIP(α) is not possible when starting from the SEP(α) is that it would produce a non-positive definite diffusion matrix, i.e., this limit does not lead to the generator of a Markov diffusion process.

7.2 Process Definition and Connection with $\mathfrak{su}(2)$

We provide straightaway the definition of the inhomogeneous version of the process, where site $x \in V$ has a maximum number of particles $\alpha_x \in \mathbb{N}$. We start with two sites.

Definition 7.1 (Symmetric Partial Exclusion Process, Two Sites) The symmetric partial exclusion process SEP(α_1, α_2) on two vertices $1, 2$ with parameters $\alpha_1, \alpha_2 \in \mathbb{N} \setminus \{0\}$ is the process on $\{0, \ldots, \alpha_1\} \times \{0, \ldots, \alpha_2\}$ with generator

$$\begin{aligned} L_{1,2} f(\eta_1, \eta_2) &= \eta_1(\alpha_2 - \eta_2)(f(\eta_1 - 1, \eta_2 + 1) - f(\eta)) \\ &\quad + \eta_2(\alpha_1 - \eta_1)(f(\eta_1 + 1, \eta_2 - 1) - f(\eta)). \end{aligned} \tag{7.1}$$

The algebraic description of the process will use the $\mathfrak{su}(2)$ Lie algebra.

Definition 7.2 ($\mathfrak{su}(2)$ Lie Algebra) The Lie algebra $\mathfrak{su}(2)$ is defined by the generators J^+, J^-, J^0 which satisfy the commutation relations

$$[J^0, J^\pm] = \pm J^\pm, \qquad [J^-, J^+] = -2J^0. \tag{7.2}$$

Representation theory of $\mathfrak{su}(2)$ is well studied. The next lemma provides $(\alpha+1)$-dimensional irreducible representations.

Lemma 7.3 (Representation) *A representation of the conjugate algebra of $\mathfrak{su}(2)$ indexed by $\alpha \in \mathbb{N}$ is given by the following operators working on $f : \{0, 1, \ldots, \alpha\} \to \mathbb{R}$*

$$\begin{aligned} J^{\alpha,+} f(n) &= (\alpha - n) f(n+1) \\ J^{\alpha,-} f(n) &= n f(n-1) \\ J^{\alpha,0} f(n) &= (-\tfrac{\alpha}{2} + n) f(n), \end{aligned} \tag{7.3}$$

with the convention $f(-1) := 0$. The operators in (7.3) satisfy the commutation relations with opposite sign of (7.2).

Proof The proof is an explicit computation, we verify $[J^{\alpha,+}, J^{\alpha,-}] = -2J^0$, leaving the other commutators as an exercise. We have

$$\begin{aligned} J^{\alpha,+} J^{\alpha,-} f(n) &= (\alpha - n) J^{\alpha,-} f(n+1) = (\alpha - n)(n+1) f(n), \\ J^{\alpha,-} J^{\alpha,+} f(n) &= n J^{\alpha,+} f(n-1) = n(\alpha - n + 1) f(n), \end{aligned} \tag{7.4}$$

so that

$$J^{\alpha,+}J^{\alpha,-}f(n) - J^{\alpha,-}J^{\alpha,+}f(n) = (\alpha - 2n)f(n) = -2J^{\alpha,0}f(n). \tag{7.5}$$

□

Lemma 7.4 (Algebraic Form of the Generator) *The generator of* SEP(α_1, α_2) *is given by*

$$L_{1,2} = J_1^{\alpha_2,+}J_2^{\alpha_1,-} + J_1^{\alpha_2,-}J_2^{\alpha_1,+} + 2J_1^{\alpha_1,0}J_2^{\alpha_2,0} - \tfrac{\alpha_1\alpha_2}{2}. \tag{7.6}$$

Proof The proof is a by explicit computation, using (7.3). Because the computation is completely analogous to the similar computation explicitly performed in Lemma (5.3), we leave this to the reader. □

Given the algebraic description of the process, it is now immediate to deduce duality properties.

Theorem 7.5 (Self-Duality, Two Sites) *The following holds:*

(a) The generator $L_{1,2}$ *commutes with* $J_1^{\alpha_1,u} + J_2^{\alpha_2,u}$ *for* $u \in \{+, -, 0\}$.
(b) A reversible measure for the generator $L_{1,2}$ *is given by*

$$M(\eta_1, \eta_2) = \binom{\alpha_1}{\eta_1}\binom{\alpha_2}{\eta_2}.$$

(c) The SEP(α_1, α_2) *is self-dual with self-duality function given by*

$$D(\xi_1, \xi_2; \eta_1, \eta_2) = d_{\alpha_1}(\xi_1, \eta_1)d_{\alpha_2}(\xi_2, \eta_2), \tag{7.7}$$

with the single-site self-duality function given by

$$d_\alpha(k, n) = \frac{\binom{n}{k}}{\binom{\alpha}{k}}\mathbb{1}_{\{k \le n\}}. \tag{7.8}$$

(d) The SEP(α_1, α_2) *has reversible product probability measures given by the product of binomials parametrized by the success probability* $\rho \in (0, 1)$ *and given by*

$$\nu_\rho^{\alpha_1,\alpha_2} = \binom{\alpha_1}{\eta_1}\binom{\alpha_1}{\eta_2}\rho^{\eta_1+\eta_2}(1-\rho)^{\alpha_1+\alpha_2-\eta_1-\eta_2}. \tag{7.9}$$

(e) The relation between $\nu_\rho^{\alpha_1,\alpha_2}$ *and the self-duality function* $D(\xi_1, \xi_2; \eta_1, \eta_2)$ *reads*

$$\int D(\xi_1, \xi_2; \eta_1, \eta_2)\, \nu_\rho^{\alpha_1,\alpha_2}(d\eta_1 d\eta_2) = \rho^{\xi_1+\xi_2}.$$

Proof The proof of item (a) is analogous to the proof of Lemma 5.5. The proof of (b) follows from the detailed balance relation

$$M(\eta_1, \eta_2)\eta_1(\alpha_2 - \eta_2) = M(\eta_1 - 1, \eta_2 + 1)(\eta_2 + 1)(\alpha_1 - \eta_1 + 1). \tag{7.10}$$

By item (b),

$$D_{\text{cheap}}(\xi_1, \xi_2; \eta_1, \eta_2) = \frac{1}{M(\eta_1, \eta_2)}\delta_{\xi_1,\eta_1}\delta_{\xi_2,\eta_2} \tag{7.11}$$

is a cheap self-duality function. Now work with the symmetry $e^{J^{\alpha_1,+}+J^{\alpha_2,+}}$ on the (ξ_1, ξ_2) variables of this cheap duality function. Use here that for $k \le n$

$$\begin{aligned}
\frac{1}{\binom{\alpha}{n}}\left(e^{J^{\alpha,+}}\delta_{\cdot,n}\right)(k) &= \binom{\alpha}{n}^{-1}\frac{(J^{\alpha,+})^{n-k}\delta_{\cdot,n}(k)}{(n-k)!} \\
&= \frac{(\alpha-k)(\alpha-k-1)\dots(\alpha-n+1)}{(n-k)!}\binom{\alpha}{n}^{-1} \\
&= \frac{1}{(n-k)!}\left(\frac{\alpha!}{(\alpha-k)!n!}\right)^{-1} \\
&= \frac{\binom{n}{k}}{\binom{\alpha}{k}}.
\end{aligned} \tag{7.12}$$

This proves item (c). Item (d) follows from item (b), and item (e) is a simple computation left to the reader. □

The two vertex self-duality immediately implies self-duality for a general graph, where along each edge $\{x, y\}$, $x, y \in V$ we copy the two-site generator (with parameters α_x, α_y). We first define the process and the state the general self-duality results on the next theorem.

Definition 7.6 (Symmetric Partial Exclusion Process on a Graph) Let V be a finite set and let $p : V \times V \to [0, \infty)$ be a positive symmetric and irreducible function. Let $\boldsymbol{\alpha} : V \to \mathbb{N} \setminus \{0\}$. Then the symmetric partial exclusion process SEP($\boldsymbol{\alpha}$) is defined as the process with generator

$$\begin{aligned}
L = \sum_{x,y\in V} p(x, y)\big[&\eta_x(\alpha_y - \eta_y)(f(\eta^{x,y}) - f(\eta)) \\
&+\eta_y(\alpha_x - \eta_x)(f(\eta^{y,x}) - f(\eta))\big],
\end{aligned} \tag{7.13}$$

where, as usual, $\eta^{x,y} = \eta - \delta_x + \delta_y$ is the configuration which arises from η by the jump of a particle from x to y.

Theorem 7.7 (Self-duality) *The following holds:*

(a) The process SEP($\boldsymbol{\alpha}$) *is self-dual with self-duality functions*

$$D(\xi, \eta) = \prod_{x \in V} d_{\alpha_x}(\xi_x, \eta_x), \tag{7.14}$$

with d_α given in (7.8).

(b) For all $\rho \in [0, 1]$ the product of binomial distributions with parameters α_i, ρ given by

$$\nu_\rho^{\boldsymbol{\alpha}} = \otimes_{x \in V} \nu_\rho^{\alpha_x}, \tag{7.15}$$

with

$$\nu_\rho^{\alpha_x}(n) = \binom{\alpha_x}{n} \rho^n (1-\rho)^{\alpha_x - n}, \tag{7.16}$$

is reversible for SEP($\boldsymbol{\alpha}$).

Proof This is an immediate consequence of the corresponding results for the two vertex case contained in Theorem 7.5. □

Remark 7.8 If $\alpha_x = 1$ for all $i \in V$, then every site carries at most one particle, and the self-duality function (7.14) becomes simpler:

$$D(\xi, \eta) = \prod_{x \in V : \xi_x = 1} \eta_x = \mathbb{1}_{\{\xi \leq \eta\}}, \tag{7.17}$$

where $\leq$ is the point-wise order between configurations. This duality function is the Siegmund duality function which we already met in the introduction and which is the basic duality function used in the book of Liggett, for the symmetric exclusion process and for various other models such as spin systems, the voter model and the contact process [172]. Also [225] considers generalized dualities in the setting of $\mathfrak{su}(2)$ with spin 1/2, i.e., $\alpha = 1$ for all $x \in V$.

7.3 The Abstract Generator

In what follows we show that, up to constants and central elements, the abstract generator is the co-product of the Casimir, i.e., the analogue of Sect. 5.6, now in the context of the universal enveloping algebra $\mathcal{U}(\mathfrak{su}(2))$.

Definition 7.9 (Co-product) The co-product $\Delta : \mathfrak{su}(2) \to \mathfrak{su}(2) \otimes \mathfrak{su}(2)$ is given by (analogously to (5.58))

$$\Delta(J^v) = J_1^v + J_2^v, \tag{7.18}$$

for $v \in \{+, -, 0\}$.

Similarly, we introduce the distinguished central element.

Definition 7.10 (Casimir) The Casimir element of $\mathfrak{su}(2)$ is given by

$$C = \frac{1}{2}(J^+J^- + J^-J^+) + (J^0)^2. \tag{7.19}$$

This element C is central, i.e., it commutes with the three generators J^+, J^-, J^0 and, as a consequence, with any other element of the universal enveloping algebra. The following lemma shows the connection between the abstract generator of SEP and the coproduct of the Casimir. We omit its proof, because it is exactly as the proof of Lemma 5.20, replacing K-operators by J-operators.

Lemma 7.11 (Co-product of the Casimir) *The co-product of the Casimir is given by*

$$\Delta(C) = (J_1^+J_2^- + J_2^+J_1^- - 2J_1^0J_2^0) + C_1 + C_2. \tag{7.20}$$

As a consequence, the abstract generator of SEP commutes with $\Delta(J^v)$ *for* $v \in \{+, -, 0\}$.

7.4 Additive Structure

The fact that α appears linearly in the representation (7.3) leads to an additive structure analogous to what we encountered in Sect. 5.10 in the study of SIP. We state here the analogous results without proofs, which are exact copies of the proofs in Sect. 5.10 replacing K operators by J operators.

Given a base set V^* and a sequence of positive integers L_x for $x \in V^*$, we consider a "vertex set with ladders", i.e., $V = \{(x, s) : x \in V^*, s = 1, \dots, L_x\}$, where we interpret the s-coordinate as the "ladder height" at x. We then consider a positive symmetric irreducible transition function $p((x, s), (y, r))$ which does not depend on the "ladder-level" (i.e., the second coordinate)

$$p((x, s), (y, r)) = p^*(x, y) \tag{7.21}$$

We then have the following analogue of Theorem 5.37. Because the proof is identical to that of Theorem 5.37 with obvious adaptations, we omit it.

Theorem 7.12 (Process on a Ladder Graph) *Let V be a vertex set with ladders, with base set V^* and transition function $p(\cdot,\cdot)$ as in (7.21). Let $\{\eta(t) : t \geq 0\}$ denote the* SEP($\boldsymbol{\alpha}$) *process on V with parameters $\boldsymbol{\alpha} : V \to \mathbb{N}$. Then the contracted process $\{\eta^*(t), t \geq 0\}$ defined via*

$$(\eta^*(t))_x = \sum_{s=1}^{L_x} (\eta(t))_{x,s}, \quad x \in V^*, \tag{7.22}$$

is a SEP($\boldsymbol{\alpha}^*$) *process on V^* with transition function $p^*(\cdot,\cdot)$ and parameters $\boldsymbol{\alpha}^* : V^* \to \mathbb{N}$ defined by*

$$\alpha_x^* = \sum_{s=1}^{L_x} \alpha_{(x,s)}. \tag{7.23}$$

7.5 Intertwining Between Different α's

The contraction in Theorem 7.12 can be interpreted as an intertwining. Indeed, let us denote by $L^{\mathrm{SEP}(\boldsymbol{\alpha})}$ and $L^{\mathrm{SEP}(\boldsymbol{\alpha}^*)}$ the generators of the $\eta(t)$ process, resp. $\eta^*(t)$ process, of Theorem 7.12, and let us denote by $\Omega_{\boldsymbol{\alpha}}$, resp. $\Omega_{\boldsymbol{\alpha}^*}$, their respective state spaces. Then the map $T : \eta \to \eta^*$ defined by equation (7.22) yields an intertwining between "ladder SEP" and "contracted ladder SEP" via the relation

$$L^{\mathrm{SEP}(\boldsymbol{\alpha})} \Lambda = \Lambda L^{\mathrm{SEP}(\boldsymbol{\alpha}^*)}, \tag{7.24}$$

where Λ maps functions f^* on the state space $\Omega_{\boldsymbol{\alpha}^*}$ to functions Λf^* on the state space $\Omega_{\boldsymbol{\alpha}}$ via

$$(\Lambda f^*)(\eta) = f^*(T\eta). \tag{7.25}$$

This intertwining exists also in the "opposite direction". We will illustrate this for ladder with at most one particle per vertex. It will give us that the self-duality functions for SEP($\boldsymbol{\alpha}^*$) can be obtained from the self-duality functions of the ladder SEP($\mathbf{1}$) combined with intertwining. Because the self-duality functions for SEP($\mathbf{1}$) are well-known and easy, this provides an easy way to obtain a family of self-duality functions for SEP($\boldsymbol{\alpha}^*$), via intertwining, starting from the self-duality functions of SEP($\mathbf{1}$) which are simple and known.

So we now assume that on V, the vertex set with ladders, $\alpha_{(x,s)} = 1$ for all $x \in V^*, s = 1, \ldots, L_x$. As a consequence $\alpha_x^* = L_x$ for all $x \in V^*$. For $\eta^* \in \Omega_{\boldsymbol{\alpha}^*}$, denote by $C(\eta^*)$ the set of compatible ladder SEP(**1**) configurations. In particular the cardinality of this set is:

$$|C(\eta^*)| = \prod_{x \in V^*} \binom{L_x}{\eta_x^*}. \tag{7.26}$$

Indeed, for each x and η^*, we have to choose η_x^* ladder places at x to put the particles. For $f : \Omega_{\boldsymbol{\alpha}} \to \mathbb{R}$, define $\Lambda^* f : \Omega_{\boldsymbol{\alpha}^*} \to \mathbb{R}$ via

$$\Lambda^* f(\eta^*) = \sum_{\eta \in C(\eta^*)} \frac{1}{|C(\eta^*)|} f(\eta). \tag{7.27}$$

We have the following "inverse" (w.r.t. (7.24)) intertwining relation:

$$L^{\mathrm{SEP}(\boldsymbol{\alpha}^*)} \Lambda^* = \Lambda^* L^{\mathrm{SEP}(\mathbf{1})}. \tag{7.28}$$

The proof is a simple explicit computation left to the reader. This relation means in words the following. Fix $\eta^* \in \Omega_{\boldsymbol{\alpha}^*}$. Choose uniformly a compatible ladder SEP(**1**) configuration $\eta \in C(\eta^*)$ and evolve it for a time $t > 0$, according to the ladder SEP(**1**); alternatively, evolve η^* according to the SEP($\boldsymbol{\alpha}^*$) that is obtained by "contracting" the SEP(**1**), and next choose a uniform element compatible with the evolved configuration. The intertwining result is that both evolutions lead to the same configuration (in distribution).

Because we have chosen $\boldsymbol{\alpha} = 1$ for the ladder SEP, as we have seen before, the self-duality functions are simple and of the form

$$D(\xi, \eta) = \prod_{(x,s) \in V} (a + b\xi_{(x,s)} + c\eta_{(x,s)} + d\eta_{(x,s)}\xi_{(x,s)}), \tag{7.29}$$

where $a, d, c, d \in \mathbb{R}$. Combining this with the intertwining yields that if we work with Λ^* both on the ξ and η variables, we find all self-duality functions of the contracted SEP($\boldsymbol{\alpha}^*$).

Theorem 7.13 (Self-duality on Ladder Graph) *The functions*

$$\begin{aligned} &D(\xi^*, \eta^*) \\ &= \frac{1}{|C(\eta^*)||C(\xi^*)|} \sum_{\xi \in C(\xi^*)} \sum_{\eta \in C(\eta^*)} \prod_{(x,s) \in V} (a + b\xi_{(x,s)} + c\eta_{(x,s)} + d\eta_{(x,s)}\xi_{(x,s)}) \end{aligned} \tag{7.30}$$

are self-duality functions of SEP($\boldsymbol{\alpha}^*$). *In particular*

$$D(\xi^*, \eta^*) = \prod_{x \in V^*} \frac{\binom{\eta_x^*}{\xi_x^*}}{\binom{\alpha_x^*}{\xi_x^*}} \tag{7.31}$$

is a self-duality function for SEP($\boldsymbol{\alpha}^*$).

Proof The first statement follows from the intertwining (7.28), combined with the fact that the for the SEP(**1**), (7.29) are self-duality functions.

To find the self-duality with (7.31) we show that it arises by acting with the intertwiner Λ^* on the left and on the right on the Siegmund self-duality function

$$D(\xi, \eta) = \mathbb{1}_{\{\xi \le \eta\}}. \tag{7.32}$$

Acting with Λ^* both on the η^* and ξ^* variables gives

$$(\Lambda^*_{\text{left}} \Lambda^*_{\text{right}} D)(\xi^*, \eta^*) = \frac{1}{|C(\xi^*)|} \frac{1}{|C(\eta^*)|} \sum_{\xi \in C(\xi^*)} \sum_{\eta \in C(\eta^*)} \mathbb{1}_{\{\xi \le \eta\}}, \tag{7.33}$$

where Λ^*_{left} denotes Λ^* working on the ξ variable and Λ^*_{right} denotes Λ^* working on the η variable. This expression can now be interpreted as the probability that two independent uniformly chosen configurations $\xi \in C(\xi^*), \eta \in C(\eta^*)$ satisfy the inequality $\xi \le \eta$, i.e., where there are particles in ξ, there are also particles in η. This probability equals

$$\prod_{x \in V^*} \frac{\binom{\eta^*_x}{\xi^*_x}}{\binom{\alpha^*_x}{\xi^*_x}}, \tag{7.34}$$

because given η^*, and $x \in V^*$, we have to choose occupied places in $\eta \in C(\eta^*)$ to place the ξ^*_x particles of ξ, of which there are exactly η^*_x at $x \in V^*$, and at every site $x \in V^*$ there are α^*_x available places. □

7.6 Scaling Limits

At first sight, one might think that taking a "many particle limit" for the symmetric partial exclusion process produces a new Markov process in the continuum, with a finite maximal value on each site that would be the analogous of the maximum occupancy α_x at site x in the discrete setting. Indeed, we recall that the scaling limit obtained by considering a large number of particles N and then studying the scaled process $\eta(t)/N$ as $N \to \infty$ was already studied in Sect. 4.1 for the independent random walkers process, yielding a deterministic system of ODE's, and in Sect. 6.1 for the symmetric inclusion process, yielding the Brownian energy process.

However, for the SEP(α), this procedure does not lead to a Markov diffusion process as we will see now. A first problem is that on a finite graph the total number of particles is upper bounded by $\sum_{x \in V} \alpha_x$ and thus is not possible to achieve a limit of infinite population without also scaling the α_x. Even when we ignore this problem, and proceed formally, via a Taylor expansion of the generator of the scaled process $\eta(t)/N$, we start from

$$L^{(N)} f\left(\tfrac{\eta}{N}\right) = \sum_{x,y\in V} p(x,y)\,\eta_x(\alpha_y - \eta_y)\left(f\left(\tfrac{\eta}{N} - \tfrac{1}{N}\delta_x + \tfrac{1}{N}\delta_y\right) - f(\tfrac{\eta}{N})\right). \tag{7.35}$$

Using then the symmetry of p, we get $L^{(N)} \to \mathcal{L}$ as $N \to \infty$ where

$$\widetilde{\mathcal{L}} = \frac{1}{2}\sum_{x,y\in V} p(x,y)\left((\alpha_x\zeta_y - \alpha_y\zeta_x)\left(\frac{\partial}{\partial\zeta_x} - \frac{\partial}{\partial\zeta_y}\right) - \zeta_x\zeta_y\left(\frac{\partial}{\partial\zeta_x} - \frac{\partial}{\partial\zeta_y}\right)^2\right). \tag{7.36}$$

However $\widetilde{\mathcal{L}}$ is a differential operator which is *not* the generator of a Markov diffusion process. Indeed, if we rewrite the second order derivatives like

$$a_{x,y}(\zeta)\frac{\partial^2}{\partial\zeta_x\partial\zeta_y}, \tag{7.37}$$

then the quadratic form generated by the 2×2 matrix withe elements $a_{x,y}(\zeta) = -\zeta_x\zeta_y$ is not positive definite, which is impossible for the generator of a diffusion process. Note the similarity to the many particle limit of the symmetric inclusion process that instead produced the positive definite "diffusivity matrix" $+\zeta_x\zeta_y$ and thus allowed for the definition of the Brownian energy process.

In order to obtain a process that is a proper scaling limit of the symmetric partial exclusion process we thus need to change strategy. Having identified, in the formal application of the many particle limit, the source of problems in the term with second order derivatives, the idea is to rescale time to eliminate the problematic term. We illustrate this (for the sake of simplicity) in the homogeneous setting, i.e., $\alpha_x = \alpha \in \mathbb{N}$ for all $x \in V$.

Define a sequence of initial configurations $\eta^{(\alpha)}$, $\alpha \in \mathbb{N}$ with a number of particles of order α, i.e. for all $x \in V$

$$\eta_x^{(\alpha)} = \lfloor \alpha\zeta_x \rfloor, \tag{7.38}$$

for some configuration $\zeta : V \to [0,1]^V$. This implies that $\eta_x^{(\alpha)}/\alpha \to \zeta_x$ as $\alpha \to \infty$.

Theorem 7.14 (Scaling Limit of SEP(α)) *Let $\{\eta^{(\alpha)}(t), t \geq 0\}$ be the symmetric partial exclusion process with parameters $\alpha_x = \alpha$, and symmetric transition function $p : V \times V \to \mathbb{R}$, initialized from $\eta^{(\alpha)}$ defined in (7.38). The process $\{\zeta^{(\alpha)}(t), t \geq 0\}$ defined by*

$$\zeta_x^{(\alpha)}(t) = \eta_x^{(\alpha)}(\alpha t)/\alpha, \tag{7.39}$$

converges weakly in path space as $\alpha \to \infty$, *to the deterministic process* $\{\zeta(t), t \geq 0\}$ *on* $[0, 1]^V$ *which is the solution of the system of ODE's given by*

$$\frac{d\zeta_x(t)}{dt} = \sum_{y \in V} p(x, y)(\zeta_y(t) - \zeta_x(t)). \tag{7.40}$$

Proof The proof is a consequence of the Trotter-Kurtz theorem. It is enough to show that the generator $L^{(\alpha)}$ of the process $\zeta^{(\alpha)}(t)$ converges as $\alpha \to \infty$ to

$$\mathcal{L}f(\zeta) = \sum_{x,y \in V} p(x, y)(\zeta_y - \zeta_x)\frac{\partial f(\zeta)}{\partial \zeta_x}. \tag{7.41}$$

which is the generator of the deterministic system (7.40) For α fixed, we have

$$L^{(\alpha)} f(\zeta) = \alpha \sum_{x,y \in V} p(x, y)\, \zeta_x(1 - \zeta_y) \left(f\left(\zeta - \tfrac{1}{\alpha}\delta_x + \tfrac{1}{\alpha}\delta_y\right) - f(\zeta)\right). \tag{7.42}$$

Assuming now that $f : [0, 1]^V \to \mathbb{R}$ is smooth, by Taylor expansion, we find

$$\lim_{\alpha \to \infty} L^{(\alpha)} f = \mathcal{L}f, \tag{7.43}$$

where the convergence is uniform. Because such smooth f are a core of the generator $\mathcal{L}$, we conclude that $\{\zeta^{(\alpha)}(t), t \geq 0\} \to \{\zeta(t) : t \geq 0\}$ as $\alpha \to \infty$, where the convergence is weak convergence in the Skorohod topology. □

Remark 7.15 Notice that the limiting system of ODE's (7.40) is the same as the one we found for independent random walkers. The only difference is that the initial condition is taken from $\zeta(0) \in [0, 1]^V$, and thus remains in this set $[0, 1]^V$, because the evolution $\zeta(t)$ is a contraction in the supremum norm and preserves positivity. In other words, in this scaling limit, we do no longer see the effect of exclusion in the evolution.

Remark 7.16 How does the self-duality of symmetric partial exclusion process "propagate" to the scaling limit of Theorem 7.14? Let $\{\eta(t), \xi(t), t \geq 0\}$ be two copies of the SEP(α) process on V starting, respectively, from $\eta, \xi \in \{0, 1, \dots, \alpha\}^{|V|}$. Then we recall the self-duality relation

$$\mathbb{E}_\eta D(\xi, \eta(\alpha t)) = \mathbb{E}_\xi D(\xi(\alpha t), \eta), \tag{7.44}$$

with D as in Theorem 7.7, namely

$$D(\xi, \eta) = \prod_{x \in V} \frac{\eta_x(\eta_x - 1)\dots(\eta_x - \xi_x + 1)}{\alpha(\alpha - 1)\dots(\alpha - \xi_x + 1)}. \tag{7.45}$$

We put $\eta = \lfloor \alpha\zeta^{(\alpha)} \rfloor$, $\eta(\alpha t) = \lfloor \alpha\zeta^{(\alpha)}(t) \rfloor$ and take the limit as $\alpha \to \infty$. On one hand we find, using the convergence $\{\zeta^{(\alpha)}(t), t \geq 0\} \to \{\zeta(t) : t \geq 0\}$, that

$$D(\xi, \eta(\alpha t)) \to \prod_{x \in V} \zeta_x(t)^{\xi_x}. \tag{7.46}$$

On the other hand we easily find that $\xi(\alpha t) \to \sigma(t)$, where $\{\sigma(t), t \geq 0\}$ denotes the independent random walk process and thus

$$D(\xi(\alpha t), \eta^{(\alpha)}) \to \prod_{x \in V} \zeta_x^{\sigma_x(t)}. \tag{7.47}$$

As a consequence, the self-duality relation (7.44) implies that

$$\mathbb{E}_\zeta \mathcal{D}(\xi, \zeta(t)) = \mathbb{E}_\xi \mathcal{D}(\sigma(t), \zeta), \tag{7.48}$$

with

$$\mathcal{D}(\xi, \zeta) = \prod_{x \in V} \zeta_x^{\xi_x}. \tag{7.49}$$

We recovered in this way the duality relation between the linear system of ODE's and the independent random walker process.

7.7 Additional Notes

Self-duality property for the standard symmetric exclusion process (SEP(1) in this book) dates back to Spitzer [217] who introduced it to characterize the stationary distribution. Later on, Ligget [172] gave a systematic treatment of duality for spin systems and its applications in the characterization of ergodic properties. From these very first results duality techniques for SEP were extensively used. The most common applications of duality are related to the derivation of scaling limits, such as the hydrodynamic equation [71].

Also the relation between the symmetric exclusion process and the XXX quantum spin chain with spin $1/2$ has been known for a long time in the theoretical physics literature [191]. The symmetric partial exclusion process duality was formalized for the first time by Sandow and Schütz in [213], where they used the $\mathfrak{su}(2)$ symmetry of the spin chain to identify the self-duality function. In the mathematical literature, the symmetric partial exclusion process was also considered in [38] in relation to the spectral gap. More recent works include [115, 116] and [117], where duality was used to find correlation inequalities. In [49] the authors find explicit formulas for the Laplace transform of the two-particle dynamics transition probabilities, and then, via duality the time-dependent covariances of the process with an arbitrary number of particles. We remark that the partial exclusion process is not the only generalization of SEP allowing for a given number of particles per site. Another process of this type had already been considered in [144] (where it is referred to as to K-exclusion process). The latter though, does not have a simple algebraic structure, thus the SEP(α) emerges as the most natural generalization of SEP showing a factorized self-duality property.

The extension of the self-duality property to the asymmetric exclusion process (ASEP) is due to Schütz [210]. This result immediately found a vast number of applications, allowing, among the other things, to compute the current fluctuations [140] and several properties of the transition probabilities [139]. In the case of the ASEP, several applications of duality rely on Bethe ansatz techniques, which allow to solve the dual dynamics [211]. These methods are applicable for a whole class of models that are referred as to integrable stochastic systems (see e.g. [205] for a review on integrable models). The self-duality function of ASEP is structurally very different from the one of SEP. It exhibits indeed a nested product structure reminding the Gartner transform [112]. Rather than correlations (as in the symmetric case), expectations of self-duality functions allow to compute suitable q-exponential moments of the current (where q is the parameter tuning the asymmetry). Thanks to its structure, this self-duality function has played an important role in the proof of convergence to the KPZ equation of WASEP (weakly asymmetric exclusion process) and, more generally, to the identification of models belonging to the KPZ-universality class (see e.g. [22, 33, 62, 63, 141]).

The asymmetric version of SEP(α) was introduced in [41] where the authors called it the ASEP(q, α) process, q being the asymmetry parameter. The generator is constructed in such a way to have a duality property from a ($\alpha + 1$)-dimensional representation of a quantum Hamiltonian with $\mathcal{U}_q(\mathfrak{sl}_2)$ invariance. The self-duality relation for ASEP(q, α) follows then as a consequence of the algebraic structure. The self-duality function has again a nested-product structure that allows to compute the q-exponential moment of the current for suitable initial conditions by means of a single dual particle. The model is not integrable, and thus the dynamic of n dual particle could not be solved. A generalization of ASEP(q, α) is proposed in [176]. The process allows for multiple jumps of particles between neighbouring sites and its generator is constructed using the Temperley-Lieb algebra. Finally we mention [185] for another processes with $\mathcal{U}_q(\mathfrak{sl}_2)$ symmetry.

The duality functions that emerge in all the aforementioned cases are standard dualities, i.e. they show a triangular structure. Orthogonal polynomial duality functions for SEP(α) were introduced more recently in a series of papers [45, 98, 99, 198] for the SEP(α). In [48] the authors find a q-orthogonal polynomial duality functions for the asymmetric process ASEP(q, α). In the last few years self-duality techniques for SEP(α) have been broadly used in the realm of scaling-limits investigations [7, 57, 93, 201], among these we stress the innovative role played by orthogonal dualities for the definition of higher order density fields [7, 57]. Among the applications of self-duality and algebraic approach in the context of asymmetric exclusion processes, we mention the key role played in the study of shocks. We mention e.g. [17] for an analysis of microscopic shock dynamics for ASEP(1), [16] for its multispecies version and [212] for the process conditioned to low current.

Several steps forward have been done, in the last few years, in the direction of finding duality relations for multispecies exclusion systems. The first result was obtained by Kuan in [156] where a duality function is found for an ASEP(1) with particles of two types. This is achieved by using symmetries of the quantum groups $\mathcal{U}_q(\mathfrak{gl}_3)$ and $\mathcal{U}_q(\mathfrak{sp}_4)$. In the already mentioned [16] the authors define a multi-species version of ASEP(1), find a duality relation and use it to study multiple shocks dynamics. In [157] the author defines a general multi-species version of ASEP(q, α) and finds duality relations.

A different version of asymmetric partial exclusion process, in its multispecies version, is introduced in [56, 58]. The authors refer to this model as mASEP and find duality functions exploiting the mathematical structure provided by the deformed quantum Knizhnik-Zamolodchikov equation.

Chapter 8
Duality for Other Models

In this chapter we consider additional models with duality properties. We start with a class of models of mass transport, both discrete and continuous, inspired by the well-known KMP (Kipnis-Marchioro-Presutti) model. We obtain these models from a "thermalization procedure" applied to the basic models of the previous chapters. The thermalization of a model preserves symmetries and duality properties. As a consequence, all these thermalized models automatically satisfy duality properties. This yields a one parameter family of discrete and continuous models of KMP type with duality properties. Other models which we obtain via thermalization include the Kac model and the Aldous averaging model. Next we study two additional models and their duality properties using the algebraic approach for the Heisenberg algebra. The first is the Ginzburg-Landau model with quadratic potential which we show to be dual to independent random walkers. The second is the Wright-Fisher diffusion with mutation (and its finite population companion the Moran model), where we have the well-known dualities of population genetics, namely duality with the coalescent. In both cases the dualities can be understood from a change of representation in the Heisenberg algebra.

8.1 Introduction

In this chapter we describe the dualities of several other models of interacting stochastic systems. Many of these models are well-known in the literature and have played a very important role in the context where they have been proposed. We will see here how these models are naturally related to the models we have already studied so far, and fit well in the algebraic approach. In this sense they can be considered as "derived models" where dualities can be obtained straightforwardly from the dualities that we have studied up to now. We restrict to five models:

C. Giardinà, F. Redig, *Duality for Markov Processes*, Grundlehren der mathematischen Wissenschaften 365, https://doi.org/10.1007/978-3-032-04099-2_8

1. The Aldous averaging process [2], introduced as an interacting particle system approach to social dynamics.
2. The Kac model [143], introduced as a model of particle collisions in kinetic theory.
3. The Kipnis-Marchioro-Presutti model [152], that provided one of the first exactly solvable stochastic models of heat conduction (Fourier law).
4. The stochastic Ginzburg-Landau model [129], that played a crucial role in the development of the entropy method for the derivation of the hydrodynamic limit [129].
5. The Wright-Fisher diffusion process and the Moran model, which are the prototype models of mathematical population genetics [87].

It is our aim in this chapter to show that for all these models, the existence of a dual process can be understood using the algebraic approach described in this book.

For instance, some of these models are obtained from the models that we discussed in previous chapters by the so-called procedure of "*thermalization*". We shall explain this procedure in general terms in Sect. 8.2 and then apply it to discuss the duality of the Aldous averaging process. We then continue to discuss the duality of the Kipnis-Marchioro-Presutti model in Sect. 8.3 and the duality of the Kac model in Sect. 8.4. Since the thermalization procedure conserves the duality property of the original models, we will identify the dual processes as the thermalized version of the duals of the original models.

We shall discuss the duality of the stochastic Ginzburg-Landau model with quadratic potential in Sect. 8.5 and the duality of the Wright-Fisher diffusion and Moran model in Sect. 8.6. Those dualities can be understood as a consequence of a change of representation of a Lie algebra.

8.2 Thermalization: Definition and a First Example

So far, for the processes which we have studied we start from the "single-edge generator" which acts on the variables associated to the two vertices of an edge. Then we prove duality for this generator via symmetries acting on the cheap duality associated to a reversible measure. The thus obtained duality functions are in factorized form (i.e., a product over the vertex variables). This allows to extend directly to a model on a general graph where along the edges the "single edge generator" is copied. In the previous chapters, new processes were obtained by taking the many particle limit, which leads for the symmetric inclusion process to the Brownian energy process, and from independent symmetric random walkers to a deterministic system of ordinary differential equations.

In this chapter, we describe yet another method, which we call "thermalization", to generate processes which satisfy duality properties by construction. These models will be of the type "mass redistribution models" where at each edge, at fixed rate, mass is exchanged between the vertices of the edge, in a way which conserves the total mass of the edge upon each transition.

To explain this procedure, let us start with the example of the single edge generator of independent random walkers. We remind that this generator is given by

$$L_{1,2}f(\eta) = \eta_1(f(\eta^{1,2}) - f(\eta)) + \eta_2(f(\eta^{2,1}) - f(\eta)), \tag{8.1}$$

with $\eta = (\eta_1, \eta_2)$ and $\eta^{1,2} = (\eta_1 - 1, \eta_2 + 1)$.

Starting from a given configuration (η_1, η_2) with $\eta_1 + \eta_2 = N$, the process generated by $L_{1,2}$ converges to a unique stationary measure $\mu^N_{1,2}$ which is given by

$$\mu^N_{1,2}(\eta_1 = k, \eta_2 = N - k) = \binom{N}{k}\left(\frac{1}{2}\right)^N. \tag{8.2}$$

I.e., η_1 is Bin$(N, 1/2)$ and $\eta_2 = N - \eta_1$.

This Binomial distribution can be obtained by conditioning the reversible measure ν_ρ which is a product of Poisson with vertex-independent parameter ρ to the event $\eta_1 + \eta_2 = N$:

$$\begin{aligned}\nu_\rho(\eta_1 = k, \eta_2 = N - k|\eta_1 + \eta_2 = N) &= \frac{\rho^k \rho^{N-k} e^{-2\rho}}{k!(N-k)! \sum_{l=0}^N \frac{\rho^l \rho^{N-l} e^{-2\rho}}{l!(N-l)!}} \\ &= \binom{N}{k}\left(\frac{1}{2}\right)^N. \end{aligned} \tag{8.3}$$

Alternatively, one can check directly by detailed balance that $\mu_{1,2}$ is reversible. The reversible measure $\mu_{1,2}$ is called the "canonical" reversible measure with total particle number N, corresponding to the "grand canonical" reversible measure ν_ρ which is the product of Poisson, where the number of particles is not fixed.

We then introduce a new process as follows: at rate 1, η_1, η_2 changes to η'_1, η'_2 where η'_1 is Bin$(\eta_1 + \eta_2, 1/2)$ and $\eta'_2 = \eta_1 + \eta_2 - \eta'_1$. We call this process the thermalization of the process with generator $L_{1,2}$. It has generator

$$\begin{aligned}L^{\mathrm{th}}_{1,2}f(\eta_1, \eta_2) &= \sum_{k=0}^{\eta_1+\eta_2} \binom{\eta_1 + \eta_2}{k}\left(\frac{1}{2}\right)^{\eta_1+\eta_2} (f(k, \eta_1 + \eta_2 - k) - f(\eta_1, \eta_2)) \\ &= \mathbb{E}\left(f(Z, \eta_1 + \eta_2 - Z) - f(\eta_1, \eta_2)\right), \end{aligned} \tag{8.4}$$

where Z is Bin$(\eta_1 + \eta_2, 1/2)$. Alternatively, this can be written as

$$L^{\mathrm{th}}_{1,2}f(\eta_1, \eta_2) = \lim_{t\to\infty} (e^{tL_{1,2}} - I)f(\eta_1, \eta_2), \tag{8.5}$$

where $e^{tL_{1,2}}$ denotes as usual the semigroup of the process $(\eta_1(t), \eta_2(t))$.

The equality (8.5) explains the name "thermalization": at rate 1, the configuration of the edge $\{1, 2\}$ is updated according to the unique (canonical) stationary measure of that edge, obtained by running the process with generator $L_{1,2}$ for "infinite time".

As we have seen before in Chap. 3, the process of independent random walkers is dual to the deterministic process with generator

$$\mathcal{L}_{1,2} = -(\zeta_1 - \zeta_2)\left(\frac{\partial}{\partial \zeta_1} - \frac{\partial}{\partial \zeta_2}\right). \tag{8.6}$$

This system, when initiated from a configuration $(\zeta_1, \zeta_2) \in [0, \infty)^2$ evolves in the course of time as

$$(\zeta_1(t), \zeta_2(t)) = \left(\frac{\zeta_1 + \zeta_2}{2} + \frac{\zeta_1 - \zeta_2}{2} e^{-2t}, \frac{\zeta_1 + \zeta_2}{2} - \frac{\zeta_1 - \zeta_2}{2} e^{-2t}\right), \tag{8.7}$$

which converges exponentially fast to the unique fixed point

$$\left(\frac{\zeta_1 + \zeta_2}{2}, \frac{\zeta_1 + \zeta_2}{2}\right). \tag{8.8}$$

Therefore, for this system, we define its thermalization

$$\mathcal{L}^{\text{th}}_{1,2} f(\zeta_1, \zeta_2) = \lim_{t\to\infty} (e^{t\mathcal{L}_{1,2}} - I) f(\eta_1, \eta_2) = f\left(\frac{\zeta_1 + \zeta_2}{2}, \frac{\zeta_1 + \zeta_2}{2}\right) - f(\zeta_1, \zeta_2). \tag{8.9}$$

I.e., at rate one we redistribute the mass of the edge according to the unique fixed point (which is the analogue here of the unique canonical reversible measure). This process is called the "Aldous averaging model" [2] on two vertices. Using (8.5), (8.9) we immediately obtain the following dualities.

1. Self-duality of thermalized independent walkers. I.e., Self-duality of the process with generator $L^{\text{th}}_{1,2}$ with self-duality function

$$D(\xi, \eta) = \frac{\eta_1!}{(\eta_1 - \xi_1)!} \frac{\eta_2!}{(\eta_2 - \xi_2)!}. \tag{8.10}$$

   This follows from the self-duality of $L_{1,2}$ with this self-duality function.
2. Duality between thermalized independent walkers with generator $L^{\text{th}}_{1,2}$ and the thermalized deterministic process (Aldous averaging process) with generator $\mathcal{L}^{\text{th}}_{1,2}$ with duality function

$$D(\xi, \zeta) = \zeta_1^{\xi_1} \zeta_2^{\xi_2}. \tag{8.11}$$

3. Self-duality of the thermalized deterministic process (Aldous averaging process) with generator $\mathcal{L}^{\text{th}}_{1,2}$ with self-duality function

$$D(u, \zeta) = e^{u_1\zeta_1 + u_2\zeta_2}. \tag{8.12}$$

We can then copy the thermalized processes along the edges of a graph $G = (V, E)$ and have the analogue of the above mentioned dualities for these processes.

Definition 8.1 (Thermalized Random Walkers and Aldous Averaging Process) Let V be a finite set, and $p : V \times V \to [0, \infty)$ a symmetric irreducible transition function.

(1) *Thermalized random walkers.* We define the process of thermalized random walkers via the generator

$$L^{\text{th}} f(\eta) = \sum_{x,y \in V} p(x, y)\mathbb{E}\left(f(T(x, y, \eta, Z^{[\eta_x+\eta_y]}) - f(\eta)\right), \tag{8.13}$$

where $Z^{[\eta_x+\eta_y]}$ is a random variable with $\text{Bin}(\eta_x + \eta_y, 1/2)$ distribution, $\mathbb{E}$ is expectation w.r.t. $Z^{[\eta_x+\eta_y]}$ and

$$\left(T(x, y, \eta, Z^{[\eta_x+\eta_y]})\right)_z = \begin{cases} \eta_z \text{ for } z \notin \{x, y\} \\ Z^{[\eta_x+\eta_y]} \text{ for } z = x \\ \eta_x + \eta_y - Z^{[\eta_x+\eta_y]} \text{ for } z = y. \end{cases} \tag{8.14}$$

(2) *Aldous averaging process.*
We define the Aldous averaging process via the generator

$$\mathcal{L}^{\text{th}} f(\zeta) = \sum_{x,y \in V} p(x, y) \left(f(\mathcal{T}^{xy}(\zeta)) - f(\zeta)\right), \tag{8.15}$$

where

$$(\mathcal{T}^{xy}(\zeta))_z = \begin{cases} \zeta_z \text{ for } z \notin \{x, y\} \\ \frac{\zeta_x+\zeta_y}{2} \text{ for } z \in \{x, y\}. \end{cases} \tag{8.16}$$

We then have the following duality theorem, which follows immediately from the corresponding single edge duality results discussed above.

Theorem 8.2

1. *The process of thermalized random walkers is self-dual with self-duality function given by*

$$D(\xi, \eta) = \prod_{x \in V} \frac{\eta_x!}{(\eta_x - \xi_x)!}. \tag{8.17}$$

2. *The Aldous averaging process is dual to the process of thermalized random walkers with duality function given by*

$$D(\xi, \zeta) = \prod_{x \in V} \zeta_x^{\xi_x}. \tag{8.18}$$

3. *The Aldous averaging process is self-dual with self-duality function given by*

$$D(u, \zeta) = \prod_{x \in V} e^{u_x \zeta_x}. \tag{8.19}$$

8.3 Continuous and Discrete KMP Models

8.3.1 Some Background on the KMP Model

The Kipnis-Marchioro-Presutti model was introduced in [152] as a model of heat conduction. It has been the very first (stochastic) model for which one could prove Fourier's law. In [152] the model was defined by considering a one-dimensional system that is coupled to two "reservoirs" at the left and right boundaries. We here focus on the evolution in the bulk (which we allow to be a finite graph $G = (V, E)$). The analysis of processes with boundary reservoirs is dealt with in Chap. 11.

In the KMP process, we consider a graph $G = (V, E)$, and a configuration $\zeta \in \Omega_V = [0, \infty)^V$ which is interpreted as associating to vertices $x \in V$ an "energy" $\zeta_x \geq 0$. These energies evolve as follows. Each edge has a Poisson clock of rate 1. When the clock of the edge $\{x, y\}$ rings, the energy of the edge is redistributed uniformly, conserving the total energy, i.e., according to the rule

$$(\zeta_x, \zeta_y) \to ((\zeta_x + \zeta_y)U, (\zeta_x + \zeta_y)(1 - U)), \tag{8.20}$$

where U is a uniform random variable in $[0, 1]$.

This simple redistribution model is a caricature of a realistic Hamiltonian model of heat conduction such as an interacting system of (non-harmonic) oscillators where energy and momentum are exchanged in the course of the Hamiltonian evolution. Of course, in the KMP model only energy is conserved, whereas in a Hamiltonian model also momentum would be conserved.

Because the KMP model satisfies Fourier's law, i.e., the macroscopic equation for the time-evolution of the energy is the heat equation, and is to some degree "exactly solvable" (though not integrable), it became one of the paradigmatic models in the literature of interacting particle systems and non-equilibrium statistical physics. Already in the original paper [152], duality played a fundamental role to compute the energy profile in a linear chain coupled to reservoirs at left and right ends, and in the proof of local equilibrium (see [111]).

Clearly this model has the flavour of a "thermalized model". In fact we will show that is corresponds to the thermalization of the Brownian energy process with parameter $\alpha = 1$. This hidden structure makes it possible to reveal a one-parameter family of KMP models, as well as a one-parameter family of discrete KMP models, which correspond to thermalization of the symmetric inclusion process (which is dual to Brownian energy process). For all these models we will automatically have duality and self-duality properties, inherited from the corresponding dualities between the symmetric inclusion process and the Brownian energy process.

8.3.2 Thermalization of Symmetric Inclusion Process and of Brownian Energy Process

As we have seen in the first section where duality of the Aldous averaging process was discussed, one can obtain new models via thermalization and inherits automatically dualities from the models one starts from. In this section we focus on thermalization in the context of the symmetric inclusion process SIP(α) and the Brownian energy process BEP(α). The analogue of the Binomial distribution which we encountered for independent random walkers now becomes the Beta Binomial distribution for SIP, and the analogue of the "fixed point" for the continuous deterministic system derived from independent random walks now becomes a redistribution according to a Beta distributed random variable.

The thermalization of SIP(α) will therefore be a discrete mass redistribution model based on the Beta Binomial distribution, whereas the thermalization of BEP(α) will be a continuous mass redistribution model based on the Beta distribution. As we already mentioned above, the latter, when $\alpha = 1$ is the famous KMP model [152].

We start with recalling some elementary facts of the Beta Binomial and Beta distributions. We recall that the Beta function is defined via

$$B(\alpha, \beta) = \int_0^1 x^{\alpha-1}(1-x)^{\beta-1}dx = \frac{\Gamma(\alpha)\Gamma(\beta)}{\Gamma(\alpha+\beta)}. \tag{8.21}$$

Definition 8.3 (Beta Binomial Distribution) A random variable X taking values in $\{0, 1, \dots, N\}$ is called Beta Binomial distributed with parameters (N, α, β) when its probability mass function is given by

$$\mathbb{P}(X = k) = \binom{N}{k} \frac{B(\alpha + k, N - k + \beta)}{B(\alpha, \beta)}. \tag{8.22}$$

A random variable X taking values in $[0, 1]$ is called Beta distributed with parameters (α, β) if its probability density is given by

$$f_X(x) = \frac{x^{\alpha-1}(1 - x)^{\beta-1}}{B(\alpha, \beta)} \mathbb{1}_{\{0 \le x \le 1\}}. \tag{8.23}$$

Remark 8.4 The following is easily verified:

(1) One can obtain the Beta Binomial distribution by considering a binomial distribution with parameters N, p where p is random and has a Beta distribution with parameters (α, β), i.e.,

$$\binom{N}{k} \frac{B(\alpha + k, N - k + \beta)}{B(\alpha, \beta)} = \frac{1}{B(\alpha, \beta)} \int_0^1 \left(\binom{N}{k} p^k (1 - p)^{N-k} \right) p^{\alpha-1}(1 - p)^{\beta-1}\, dp. \tag{8.24}$$

(2) The Beta binomial distribution with parameters $(N, 1, 1)$ coincides with the (discrete) uniform distribution on the set $\{0, 1, \ldots, N\}$.

We also recall for the convenience of the reader the related discrete Gamma and continuous Gamma distributions.

Definition 8.5 (Discrete Gamma and Continuous Gamma Distribution)

1. A random variable X taking values in $\mathbb{N}$ is called discrete Gamma with scale parameter $\lambda \in (0, 1)$ and shape parameter $\alpha > 0$ if

$$\mathbb{P}(X = n) = \frac{\lambda^n \Gamma(\alpha + n)}{n! \Gamma(\alpha)} (1 - \lambda)^\alpha. \tag{8.25}$$

2. A random variable X taking values in $[0, \infty)$ is called Gamma with scale parameter $\theta \in (0, \infty)$ and shape parameter $\alpha > 0$ if its probability density is given by

$$f_X(x) = \frac{\theta^\alpha x^{\alpha-1} e^{-\theta x}}{\Gamma(\alpha)} \mathbb{1}_{\{x>0\}}. \tag{8.26}$$

We have the following elementary lemma which relates the reversible product measures of SIP(α), resp. BEP(α) to Beta Binomial and Beta distributions.

Lemma 8.6

1. *Let (η_1, η_2) be distributed as a product of two independent discrete Gamma distributions with identical scale parameters and with respective shape parameter α and β, i.e.,*

$$\mathbb{P}(\eta_1 = k, \eta_2 = l) = \frac{\lambda^{k+l}}{l!k!}\frac{\Gamma(\alpha+k)\Gamma(\beta+l)}{\Gamma(\alpha)\Gamma(\beta)}(1-\lambda)^{\alpha+\beta}.$$

Then conditional on $\eta_1+\eta_2 = N$, η_1 is Beta Binomial with parameters (N, α, β).

2. *Let (ζ_1, ζ_2) be distributed as a product of two independent Gamma distributions with identical scale parameters and respective shape parameters α and β, i.e., with joint probability density*

$$f_{(\zeta_1,\zeta_2)}(z_1, z_2) = \theta^{\alpha+\beta}\frac{z_1^{\alpha-1} z_2^{\beta-1}}{\Gamma(\alpha)\Gamma(\beta)}e^{-\theta(z_1+z_2)}\mathbb{1}_{\{0\le z_1\}}\mathbb{1}_{\{0\le z_2\}}. \tag{8.27}$$

Then conditional on $\zeta_1 + \zeta_2 = x$, η_1 is distributed as xB where B is Beta(α, β) distributed.

Proof This follows from simple explicit computation. □

We then have the following explicit form of the thermalizations of SIP(α) and BEP(α).

Proposition 8.7 *The following holds.*

1. Thermalization of SIP($\boldsymbol{\alpha}$). *Let $L_{1,2}$ denote the generator of SIP($\boldsymbol{\alpha}$) defined on two sites, i.e.,*

$$L_{1,2}f(\eta_1, \eta_2) = \eta_1(\alpha_2 + \eta_2)(f(\eta^{1,2}) - f(\eta)) + \eta_2(\alpha_1 + \eta_1)(f(\eta^{2,1}) - f(\eta)). \tag{8.28}$$

Let $L^{\text{th}}_{1,2}$ denote the corresponding thermalized process, i.e.,

$$L^{\text{th}}_{1,2}f(\eta_1, \eta_2) = \lim_{t\to\infty}(e^{tL_{1,2}} - I)f(\eta_1, \eta_2). \tag{8.29}$$

Then we have

$$L^{\text{th}}_{1,2}f(\eta_1, \eta_2) = \mathbb{E}(f(X, \eta_1 + \eta_2 - X) - f(\eta_1, \eta_2)), \tag{8.30}$$

where the expectation is w.r.t. X, a Beta Binomial random variable with parameters $(\eta_1 + \eta_2, \alpha_1, \alpha_2)$.

2. Thermalization of BEP *($\boldsymbol{\alpha}$). Let $\mathcal{L}_{1,2}$ denote the generator of BEP($\boldsymbol{\alpha}$) defined on two sites, i.e.,*

$$\mathcal{L}_{1,2} f(\zeta_1, \zeta_2) = \Big(\zeta_1\zeta_2\Big(\frac{\partial}{\partial\zeta_1} - \frac{\partial}{\partial\zeta_2}\Big)^2 - (\alpha_2\zeta_1 - \alpha_1\zeta_2)\Big(\frac{\partial}{\partial\zeta_1} - \frac{\partial}{\partial\zeta_2}\Big)\Big) f(\zeta_1, \zeta_2). \tag{8.31}$$

Let $\mathcal{L}^{\text{th}}_{1,2}$ denote the corresponding thermalized process, i.e.,

$$\mathcal{L}^{\text{th}}_{1,2} f(\eta_1, \eta_2) = \lim_{t\to\infty} (e^{t\mathcal{L}_{1,2}} - I) f(\eta_1, \eta_2). \tag{8.32}$$

Then we have

$$\mathcal{L}^{\text{th}}_{1,2} f(\eta_1, \eta_2) = \mathbb{E}(f(X(\zeta_1 + \zeta_2), (1 - X)(\zeta_1 + \zeta_2)) - f(\zeta_1, \zeta_2)), \tag{8.33}$$

where the expectation is w.r.t. X, a Beta random variable with parameters (α_1, α_2).

Proof The first item follows from the fact that the process with generator $L_{1,2}$, when started from (η_1, η_2), converges to its unique reversible distribution which is the distribution of $(X, \eta_1 + \eta_2 - X)$ where X is a Beta Binomial random variable with parameters $(\eta_1 + \eta_2, \alpha_1, \alpha_2)$. This in turn follows from the reversibility of the product of discrete Gamma distribution with identical scale parameters and shape parameters (α_1, α_2), together with Lemma 8.6. The second item follows in the same spirit from the fact that the process with generator $\mathcal{L}_{1,2}$, when started from (ζ_1, ζ_2), converges to its unique reversible distribution which is the distribution of $(X(\zeta_1 + \zeta_2), (1 - X)(\zeta_1 + \zeta_2))$, where X is a Beta random variable with parameters $(\eta_1 + \eta_2, \alpha_1, \alpha_2)$. □

8.3.3 *Generalized Discrete and Continuous KMP Processes*

We can now, by copying the thermalized generators along the edges of a graph define the generalized KMP model, the generalized discrete KMP model and obtain immediate duality relations between these models.

Definition 8.8 (Thermalized SIP or Discrete Generalized KMP Process) Let $G = (V, E)$ be a finite graph, and let $p : E \to [0, \infty)$ denote an irreducible symmetric transition function. Let $\boldsymbol{\alpha} : V \to (0, \infty)$. Then the discrete generalized KMP($\boldsymbol{\alpha}$) process or thermalized SIP($\boldsymbol{\alpha}$) is defined as the process on the state space $\Omega_V = \mathbb{N}^V$ via its generator

$$Lf(\eta) = \sum_{xy\in E} p(x, y)\mathbb{E}(f(T(X_{xy}, \eta)) - f(\eta)), \tag{8.34}$$

where

$$(T(X_{xy},\eta))_z = \begin{cases} \eta_z \text{ if } z \notin \{x,y\} \\ X_{xy} \text{ if } z = x \\ \eta_x + \eta_y - X_{xy} \text{ if } z = y, \end{cases} \tag{8.35}$$

where X_{xy} is Beta Binomial with parameters $(\eta_x + \eta_y, \alpha_x, \alpha_y)$ and where the expectation in (8.34) is w.r.t. this variable X_{xy}.

Definition 8.9 (Thermalized BEP or Generalized KMP Process) Let $G = (V, E)$ be a finite graph, and let $p : E \to [0, \infty)$ denote an irreducible symmetric transition function. Let $\boldsymbol{\alpha} : V \to (0, \infty)$. Then the generalized KMP($\boldsymbol{\alpha}$) process or thermalized BEP($\boldsymbol{\alpha}$) is defined as the process on the state space $\Omega_V = [0, \infty)^V$ by its generator

$$\mathcal{L} f(\zeta) = \sum_{xy\in E} p(x, y)\mathbb{E}(f(\mathcal{T}(X_{xy}, \eta)) - f(\eta)), \tag{8.36}$$

where

$$(\mathcal{T}(X_{xy},\eta))_z = \begin{cases} \zeta_z & \text{if } z \notin \{x,y\} \\ X_{xy}(\zeta_x + \zeta_y) & \text{if } z = x \\ (\zeta_x + \zeta_y)(1 - X_{xy}) & \text{if } z = y, \end{cases} \tag{8.37}$$

where X_{xy} is a random variable which is Beta distributed with parameters (α_x, α_y) and where the expectation in (8.36) is w.r.t. this random variable X_{xy}.

Remark 8.10 We now see that $\boldsymbol{\alpha} = 1$ reduces to the original Kipnis-Marchioro-Presutti process discussed in the introduction to this chapter. The redistribution rule is thus uniform on every edge for this model. The discrete case with $\boldsymbol{\alpha} = 1$ corresponds to a natural analogous discrete uniform redistribution rule.

Combining Lemma 8.6 and Proposition 8.7 we obtain the following dualities and relation between the discrete and continuous generalized KMP processes.

Theorem 8.11 (Duality and Basic Properties of Generalized KMP Models) *Let $G = (V, E)$ be a finite graph, and let $p : E \to [0, \infty)$ denote an irreducible symmetric transition function. Let $\boldsymbol{\alpha} : V \to (0, \infty)$.*

1. Self-duality of discrete KMP. *The discrete KMP($\boldsymbol{\alpha}$) process is self-dual with self-duality function*

$$D(\xi, \eta) = \prod_{x\in V} d_{\alpha(x)}(\xi_x, \eta_x), \tag{8.38}$$

where $d_\alpha(k, n) = \frac{n!\Gamma(\alpha)}{(n-k)!\Gamma(\alpha+k)}$ is the single-site self-duality function of SIP($\boldsymbol{\alpha}$).

2. Duality between continuous and discrete KMP. *The continuous and discrete KMP process are dual with duality function*

$$D(\xi, \zeta) = \prod_{x \in V} d_{\alpha(x)}(\xi_x, \zeta_x), \tag{8.39}$$

where $d_\alpha(k, z) = \frac{z^k \Gamma(\alpha)}{\Gamma(\alpha+k)}$ *is the single-site self-duality function for duality between SIP($\boldsymbol{\alpha}$) and BEP($\boldsymbol{\alpha}$).*

3. Many particle limit of discrete KMP gives continuous KMP. *The continuous KMP process can be obtained as a "many-particle limit" of the discrete KMP process as follows. Let* $\zeta \in [0, \infty)^V$ *be given and let* $\eta^{(N)} \in \mathbb{N}^V$ *be a sequence of configurations such that* $\eta^{(N)}/N \to \zeta$ *as* $N \to \infty$. *Let* $\eta^{(N)}(t)$ *denote the discrete KMP process starting from* $\eta^{(N)}$, *and let* $\zeta(t)$ *denote the continuous KMP process starting from* ζ. *Then* $\{\eta^{(N)}(t) : t \geq 0\}$ *converges weakly in pathspace to* $\{\zeta(t) : t \geq 0\}$
4. Reversible product measures of discrete and continuous KMP. *The discrete KMP process has reversible measures given by the product of discrete Gamma distributions, with constant scale parameter* $\lambda \in (0, 1)$ *and with shape parameter* α_x *at site* $x \in V$. *The continuous KMP process has reversible measures the product of Gamma distributions with constant scale parameter* $\theta > 0$ *and with shape parameter* α_x *at site* $x \in V$.

Proof The dualities follow immediately because the models are thermalized models of SIP, resp. BEP, and therefore the self-duality of SIP, and the duality between SIP and BEP carries over to the thermalized models.

To see the convergence $\{\eta^{(N)}(t) : t \geq 0\}$ to $\{\zeta(t) : t \geq 0\}$ notice that if X_N is Beta Binomial with parameters (N, α, β) then

$$\frac{X_N}{N} \to X, \qquad \text{as } N \to \infty. \tag{8.40}$$

where X is Beta distributed with parameters (α, β), and where the convergence is in distribution. As a consequence we have generator convergence of the process $\eta^{(N)}(t)/N$ to the generator of the process $\zeta(t)$.

Finally, to prove item 4), notice that thermalization does not change the reversible measures. □

Remark 8.12 To appreciate better the special role of the Beta distribution in the context of the KMP model, one can consider a generalization by considering the model (8.33) with a general distribution for the random variable X. It can then be shown that this model has reversible product measures if and only if X is Beta distributed [60]. Since we want duality functions in product form, it is important to be able to start from the cheap duality function coming from a reversible measure. Therefore, we expect that only for X beta distributed we can have duality functions in factorized form.

Remark 8.13 In the KMP model the total "energy" $\zeta_1 + \zeta_2$ is redistributed over the sites constituting the edge. In chapter XII we will introduce a very similar model where instead a fraction of the energy of a *vertex* is moved to the *other vertex*. As we will see this model, though having very similar duality properties is exactly solvable in the setting of a chain geometry with left and right boundary reservoirs, which is not the case for KMP.

8.4 The Kac Model

The Kac model was introduced by Mark Kac [143], in the context of kinetic theory, as a lattice model of particles with stochastically evolving velocities. The aim was to have a simple microscopic model from which in a kinitic limit the linear Boltzmann equation can be derived rigorously. The time evolution for the probability distribution of velocities is a linear master equation, called the Kac master equation. The model has been studied rigorously in several works, see for instance [32, 52, 177].

In the Kac model the basic dynamic rule yielding interaction of particles by collisions is the following. The vertices of a finite graph $G = (V, E)$ label the velocities of particles that evolve as a continuous-time Markov chain $\{\zeta(t),\ t \geq 0\}$ with $\zeta(t) \in \Omega_V = \mathbb{R}^V$. An irreducible symmetric transition rate function $p : V \times V \to \mathbb{R}_+$ is given. After waiting a random time that is exponentially distributed with parameter $p(x, y) > 0$, the edge $\{x, y\}$ is selected and we let particle x and particle y collide with a random scattering angle. At collision times the velocities (ζ_x, ζ_y) of particles associated to the edge xy are updated according to the random rotation rule

$$\left(\zeta_x, \zeta_y\right) \to \left(\zeta_x \cos\Theta_{x,y} + \zeta_y \sin\Theta_{x,y}, -\zeta_x \sin\Theta_{x,y} + \zeta_y \cos\Theta_{x,y}\right), \tag{8.41}$$

where $\Theta_{x,y}$ is a random variable that is uniformly distributed on $[0, 2\pi]$. More explictly, the Kac model $\{\zeta(t),\ t \geq 0\}$ is then the Markov process with state space $\mathbb{R}^V$ and generator

$$\mathcal{L}^{Kac} = \sum_{(x,y)\in E} p(x, y)\frac{1}{2\pi}\int_0^{2\pi} d\theta_{x,y}[f(\mathcal{T}(\zeta, \theta_{x,y}) - f(\zeta)], \tag{8.42}$$

with

$$\mathcal{T}(\zeta, \theta_{x,y}) = \begin{cases} \zeta_z \text{ for } z \notin \{x, y\} \\ \zeta_x \cos\theta_{x,y} + \zeta_y \sin\theta_{x,y} \text{ for } z = x \\ -\zeta_x \sin\theta_{x,y} + \zeta_y \cos\theta_{x,y} \text{ for } z = y. \end{cases} \tag{8.43}$$

Our first aim is to show that the velocity process $\{\zeta(t),\ t \geq 0\}$ of the Kac model can be obtained as the thermalization of the Brownian momentum process of Sect. 6.7. We recall that the Brownian momentum process on the graph $G = (V, E)$ is the diffusion process $\{\zeta(t),\ t \geq 0\}$ taking values in $\mathbb{R}^V$ defined by the generator

$$\mathcal{L} = \sum_{(x,y)\in E} p(x, y)\mathcal{L}_{x,y}, \tag{8.44}$$

with

$$\mathcal{L}_{x,y} = \frac{1}{4}\Big(\zeta_x \frac{\partial}{\partial \zeta_y} - \zeta_y \frac{\partial}{\partial \zeta_x}\Big)^2. \tag{8.45}$$

The single-edge generator $\mathcal{L}_{x,y}$ evolves only the variables on the vertices x and y, leaving all other variables unchanged. To obtain the thermalization of $\{\zeta(t),\ t \geq 0\}$ we use that the reversible product measures of the Brownian momentum process are products of centered Normal distributions with identical variance as we have seen before in Sect. 6.7.

As a consequence, when the process with generator $\mathcal{L}_{x,y}$ is started from (ζ_x, ζ_y) then it converges to its unique stationary distribution which is the product of two centered Normal distributions (ζ'_x, ζ'_y) with identical variance and expectation zero conditioned to $(\zeta'_x)^2+(\zeta'_y)^2 = \zeta_x^2+\zeta_y^2$. This is because $\mathcal{L}_{x,y}$ conserves the Euclidean length of the vector (ζ_x, ζ_y), and this is the unique conserved quantity. An easy computation shows that this marginal distribution μ_{ζ_x,ζ_y} is given by the uniform distribution on the circle

$$C_{\zeta_x,\zeta_y} = \{(\zeta'_x, \zeta'_y) \in \mathbb{R}^2 \, : \, (\zeta'_x)^2 + (\zeta'_y)^2 = \zeta_x^2 + \zeta_y^2\}. \tag{8.46}$$

This allows us to obtain the thermalization of the BMP process, which is formulated in the following proposition.

Proposition 8.14 *Parametrizing the points of* C_{ζ_x,ζ_y} *as*

$$\begin{aligned}\zeta'_x &= \zeta_x \cos\theta_{x,y} + \zeta_y \sin\theta_{x,y},\\ \zeta'_y &= -\zeta_x \sin\theta_{x,y} + \zeta_y \cos\theta_{x,y},\end{aligned} \tag{8.47}$$

with the angle $\theta_{x,y} \in [0, 2\pi)$, *we have that the thermalized single-edge generator corresponding to the generator* (8.45) *reads*

$$\begin{aligned}\mathcal{L}^{\mathrm{th}}_{x,y} f(\zeta) &= \lim_{t\to\infty} (e^{t\mathcal{L}_{x,y}} - I) f(\zeta)\\ &= \mathbb{E}[f(\mathcal{T}(\zeta, \Theta_{x,y})] - f(\zeta).\end{aligned} \tag{8.48}$$

where $\mathcal{T}(\zeta, \Theta_{x,y})$ *is as in* (8.43), *with* $\Theta_{x,y}$ *uniformly distributed in* $[0, 2\pi)$ *and where* $\mathbb{E}$ *denotes expectation with respect to this random variable* $\Theta_{x,y}$.

Because the BMP process is dual to SIP with $\alpha = 1/2$ by Theorem 6.21, we obtain immediately that the Kac process is dual to the thermalized SIP with parameter $\alpha = 1/2$.

This is formulated in the next theorem.

Theorem 8.15 (Duality for the Kac Process) *The Kac process* $\{\zeta(t),\ t \geq 0\}$ *with state space* $\mathbb{R}^V$ *and generator* $\mathcal{L}^{Kac}$ *in* (8.42) *is dual to the thermalized inclusion process with parameter* $\alpha = \frac{1}{2}$, *i.e., the process with generator* (8.34) *with* $\boldsymbol{\alpha} = \frac{1}{2}$. *The duality function is given by*

$$D(\xi, \zeta) = \prod_{x \in V} \frac{\zeta_x^{2\xi_x}}{(2\xi_x - 1)!!}. \tag{8.49}$$

Proof This follows from the duality between BMP and SIP with $\alpha = 1/2$ and the fact that the Kac process is the thermalization of BMP. □

8.5 The Ginzburg-Landau Model

The stochastic Ginzburg Landau model [129] is a Markov diffusion process on the state space $\Omega_V = \mathbb{R}^V$ which describes the time evolution of real variables $\varphi_x(t)$, $x \in V$ which represent the amount of "charge" at time $t \geq 0$ at vertex $x \in V$ of a graph $G = (V, E)$. Charges are randomly redistributed between vertices connected by an edge as a diffusion process that conserves the total charge and depends on a Hamiltonian function $\mathcal{H} : \mathbb{R}^V \to \mathbb{R}$. Notice that we use the letter φ to describe the configuration, rather than the earlier used ζ, in order to distinguish this process from the earlier introduced diffusion processes such as the Brownian energy process or the Brownian momentum process.

The generator of the stochastic Ginzburg Landau model is acting on smooth compactly supported functions as follows:

$$\mathcal{L} = \sum_{(x,y) \in E} \mathcal{L}_{x,y}, \tag{8.50}$$

where the single-edge term $\mathcal{L}_{x,y}$ is given by:

$$\mathcal{L}_{x,y} = -\left(\frac{\partial \mathcal{H}}{\partial \varphi_x} - \frac{\partial \mathcal{H}}{\partial \varphi_y}\right)\left(\frac{\partial}{\partial \varphi_x} - \frac{\partial}{\partial \varphi_y}\right) + \left(\frac{\partial}{\partial \varphi_x} - \frac{\partial}{\partial \varphi_y}\right)^2. \tag{8.51}$$

The single edge generator $\mathcal{L}_{1,2}$ generates a diffusion process on $(\varphi_1(t), \varphi_2(t))$ which is described by the stochastic differential equation

$$d\varphi_1(t) = -\left(\frac{\partial \mathcal{H}}{\partial \varphi_1} - \frac{\partial \mathcal{H}}{\partial \varphi_2}\right) dt + \sqrt{2} dB_{1,2}(t) = -d\varphi_2(t).$$

where $B_{1,2}(t)$ is a standard Brownian motion. The charge conservation rule is expressed by $d(\varphi_1(t) + \varphi_2(t)) = 0$.

On the graph $G = (V, E)$, the evolution is governed by the system of stochastic differential equations

$$d\varphi_x(t) = - \sum_{y \in V, y \sim x} \left\{\frac{\partial \mathcal{H}}{\partial \varphi_x}(\varphi(s)) - \frac{\partial \mathcal{H}}{\partial \varphi_y}(\varphi(s))\right\} dt + \sqrt{2} \sum_{y \in V, y \sim x} dB_{x,y}(t), \tag{8.52}$$

where $x \in V$, the notation $y \sim x$ denotes that y is a neighbor of x in the graph (V, E), and where for $(x, y) \in V \times V$ such that $\{x, y\} \in E$ (also denoted $x \sim y$)

$$B_{x,y}(t) = -B_{y,x}(t) = W_{\{x,y\}}(t), \tag{8.53}$$

where $W_e(t)$ with $e \in E$ is a collection of independent standard Brownian motions.

As a consequence of the antisymmetry property (8.53) (notice that also the drift term in (8.52) is antisymmetric) one has $d(\sum_x \varphi_x(t)) = 0$ and therefore

$$\sum_{x \in V} \varphi_x(t) = \sum_{x \in V} \varphi_x(0), \tag{8.54}$$

i.e., the total "charge" is conserved.

The diffusion process has a drift term that is dictated by the Hamiltonian $\mathcal{H} : \mathbb{R}^V \to \mathbb{R}$ and a "noise" term associated to edges which acts with opposite sign on both ends of the edge. For the dynamics to exist and to be well-defined, sufficient regularity properties on the Hamiltonian $\mathcal{H}$ must be specified (even for a finite graph G). We do not discuss this issue and refer to [46], where standard assumptions are specified in the context of lattice models.

For any Hamiltonian $\mathcal{H}$ for which the diffusion process associated to the generator (8.50) exists, the Boltzmann-Gibbs measure with Hamiltonian $\mathcal{H}$ is reversible, provided $e^{-\mathcal{H}}$ is integrable. More precisely, if we define the Boltzmann-Gibbs measure as

$$\mu(d\varphi) = \frac{1}{\mathcal{Z}} \cdot e^{-\mathcal{H}(\varphi)} d\varphi, \tag{8.55}$$

where the normalizing partition function is

$$\mathcal{Z} = \int e^{-\mathcal{H}(\varphi)}\, d\varphi, \tag{8.56}$$

then we have the following result.

Theorem 8.16 *The Boltzmann-Gibbs measure in* (8.55) *is a reversible measure for the Ginzburg-Landau dynamics.*

Proof It suffices to show that for f, g smooth compactly supported functions we have

$$\int (\mathcal{L} f) g d\mu = \int g(\mathcal{L} f) d\mu. \tag{8.57}$$

So let us fix f, g, two smooth compactly supported functions. Remark that the single edge generator can be rewritten as follows

$$\mathcal{L}_{x,y} = e^{\mathcal{H}} \nabla_{x,y}(e^{-\mathcal{H}} \nabla_{x,y}), \tag{8.58}$$

where

$$\nabla_{x,y} = \frac{\partial}{\partial \varphi_x} - \frac{\partial}{\partial \varphi_y}. \tag{8.59}$$

As a consequence, using partial integration,

$$\int (\mathcal{L}_{x,y} f) g d\mu = \frac{1}{\mathcal{Z}} \int g \nabla_{x,y}(e^{-\mathcal{H}} \nabla_{x,y} f) d\varphi = -\int (\nabla_{x,y} g)(\nabla_{x,y} f) d\mu. \tag{8.60}$$

The expression $-\int (\nabla_{x,y} g)(\nabla_{x,y} f) d\mu$ is clearly symmetric in f and g, which implies that μ is reversible. □

Duality is not expected to hold for the general Ginzburg-Landau model, i.e., with a generic Hamiltonian $\mathcal{H}$. However, as we will show, we have duality for the quadratic Hamiltonian

$$\mathcal{H}(\varphi) = \frac{1}{2} \sum_{x \in V} \varphi_x^2, \tag{8.61}$$

see [46, 126].

In this case the dual process is the system of independent random walkers of Chap. 3. More precisely we have the following theorem.

Theorem 8.17 (Duality for Quadratic Ginzburg-Landau Process) *The Ginzburg-Landau process $\{\varphi(t)\,,\ t\geq 0\}$ on a graph $G=(V,E)$ with generator*

$$\mathcal{L}=\sum_{(x,y)\in E}\Big[-\big(\varphi_x-\varphi_y\big)\Big(\frac{\partial}{\partial\varphi_x}-\frac{\partial}{\partial\varphi_y}\Big)+\Big(\frac{\partial}{\partial\varphi_x}-\frac{\partial}{\partial\varphi_y}\Big)^2\Big], \tag{8.62}$$

is dual to the independent random walk process $\{\eta(t)\,,\ t\geq 0\}$ on a graph $G=(V,E)$ with generator

$$Lf(\eta)=\sum_{(x,y)\in E}\Big[\eta_x(f(\eta^{x,y})-f(\eta))+\eta_y(f(\eta^{y,x})-f(\eta))\Big], \tag{8.63}$$

where $\eta^{x,y}$, as usual, denotes the configuration obtained from η by moving a particle from vertex x to vertex y. The duality function is

$$D(\eta,\varphi)=\prod_{x\in V}H_{\eta_x}(\varphi_x). \tag{8.64}$$

where $H_\eta(\varphi_x)$ is the Hermite polynomial of degree η_x.

Proof We provide two independent proofs. The first one is based on explicit computations using identities for the Hermite polynomials and the second one is based on the algebraic approach described in this book.

First Proof: Direct Computation To alleviate notation we do not write the argument of the polynomials. We start by writing out the action of the single-edge Ginzburg-Landau generator with quadratic Hamiltonian:

$$\begin{aligned}\mathcal{L}_{x,y}D(\eta,\cdot)(\varphi)=\Big[\prod_{z\in V,z\neq x,y}H_{\eta_z}\Big]\Big[&H''_{\eta_x}H_{\eta_y}+H_{\eta_x}H''_{\eta_y}-2H'_{\eta_x}H'_{\eta_y}\\&-\varphi_xH'_{\eta_x}H_{\eta_y}-\varphi_yH_{\eta_x}H'_{\eta_y}+\varphi_xH_{\eta_x}H'_{\eta_y}+\varphi_yH'_{\eta_x}H_{\eta_y}\Big].\end{aligned} \tag{8.65}$$

We regroup terms as follows

$$\begin{aligned}\mathcal{L}_{x,y}D(\eta,\cdot)(\varphi)=\Big[\prod_{z\in V,z\neq x,y}H_{\eta_z}\Big]&\\ \Big[H'_{\eta_x}(\varphi_yH_{\eta_y}-H'_{\eta_y})&+(H''_{\eta_x}-\varphi_xH'_{\eta_x})H_{\eta_y}\\+(\varphi_xH_{\eta_x}-H'_{\eta_x})H'_{\eta_y}&+H_{\eta_x}(H''_{\eta_y}-\varphi_yH'_{\eta_y})\Big],\end{aligned} \tag{8.66}$$

and then use the following identities for Hermite polynomials

$$H'_{\eta_x}(\varphi_x) = \eta_x \, H_{\eta_x - 1}(\varphi_x), \tag{8.67}$$

$$\varphi_x H_{\eta_x}(\varphi_x) - H'_{\eta_x}(\varphi_x) = H_{\eta_x + 1}(\varphi_x), \tag{8.68}$$

$$H''_{\eta_x}(\varphi_x) - \varphi_x H'_{\eta_x}(\varphi_x) = -\eta_x H_{\eta_x}(\varphi_x), \tag{8.69}$$

to find

$$\begin{aligned}\mathcal{L}_{x,y} D(\eta, \cdot)(\varphi) &= \Big[\prod_{z \in V, z \neq x,y} H_{\eta_z} \Big] \Big[\eta_x (H_{\eta_x - 1} H_{\eta_y + 1} - H_{\eta_x} H_{\eta_y}) \\ &\qquad + \eta_y (H_{\eta_x + 1} H_{\eta_y - 1} - H_{\eta_x} H_{\eta_y}) \Big] \\ &= L_{x,y} D(\cdot, \varphi)(\eta),\end{aligned} \tag{8.70}$$

so that we have recovered the action of the single-edge independent random walkers generator.

Second Proof: Two Representations of the Heisenberg Algebra We introduce the operators $A_x^\dagger$, A_x working on smooth functions of a real variable

$$\begin{aligned}A_x^\dagger f(\varphi_x) &= \varphi_x f(\varphi_x) - \frac{\partial f}{\partial \varphi_x}(\varphi_x), \\ A_x f(\varphi_x) &= \frac{\partial f}{\partial \varphi_x}(\varphi_x).\end{aligned} \tag{8.71}$$

On the tensor product space, these operators satisfy the "canonical commutation relation"

$$\left[A_x, A_y^\dagger\right] = \left[\frac{\partial}{\partial \varphi_x}, \varphi_y - \frac{\partial}{\partial \varphi_y}\right] = \left[\frac{\partial}{\partial \varphi_x}, \varphi_y\right] = I \delta_{x,y}, \tag{8.72}$$

where I denotes the identity operator and $\delta_{x,y}$ is the Kronecker delta function. Next, we consider the operators working on functions $f : \mathbb{N} \to \mathbb{R}$

$$\begin{aligned}a_x^\dagger f(\eta_x) &= f(\eta_x + 1), \\ a_x f(\eta_x) &= \eta_x f(\eta_x - 1), \qquad\qquad a_x f(0) = 0.\end{aligned} \tag{8.73}$$

As we have seen before, on the tensor product space, these operators satisfy the "dual canonical commutation relations"

$$[a_x, a_y^\dagger] = -I \delta_{x,y}. \tag{8.74}$$

The operators A_x, $A_x^\dagger$ and a_x, $a_x^\dagger$ satisfy a duality relation via the Hermite orthogonal polynomials $D(\eta_x, \varphi_x) = H_{\eta_x}(\varphi_x)$:

$$\big(AD(\eta_x, \cdot)\big)(\varphi_x) = \big(aD(\cdot, \varphi_x)\big)(\eta_x), \tag{8.75}$$

$$\big(A^\dagger D(\eta_x, \cdot)\big)(\varphi_x) = \big(a^\dagger D(\cdot, \varphi_x)\big)(\eta_x). \tag{8.76}$$

In fact, the dualities (8.75) and (8.76) are an equivalent formulation of the identities (8.67) and (8.68), as we discussed in the last example of Sect. 2.5. In terms of the A_x, $A_x^\dagger$ operators the generator (8.62) of the Ginzburg-Landau diffusion process reads

$$\mathcal{L}_{x,y} = -(A_x^\dagger - A_y^\dagger)(A_x - A_y). \tag{8.77}$$

Similarly, we know from Chap. 3 that, in terms of the a_x, $a_x^\dagger$ operators, the generator of the independent random walkers process is

$$L_{x,y} = -(a_x - a_y)(a_x^\dagger - a_y^\dagger). \tag{8.78}$$

We see that the duality between $\mathcal{L}$ and L is a consequence of the basic dualities (8.75) and (8.76) and of the general principle that says "replace A_x by a_x, replace $A_x^\dagger$ by $a_x^\dagger$ and write product of these operators in reversed order".

□

Remark 8.18 In the proof of the duality, we have only proved "generator" duality. To some extent even that proof is formal because the generator is defined as the closure of the differential operator (8.62) on smooth compactly supported functions $\mathcal{C}_c^\infty(\Omega_V)$, and the Hermite polynomials are of course not in that class. However, once the existence of the diffusion process (8.52), is established, one can use standard multi-variate Itô calculus to show the semigroup duality from the formal generator duality, i.e., via Itô calculus one obtains that for the Hermite polynomials one has

$$H_\eta(\varphi(t)) - H_\eta(\varphi(0)) - \int_0^t LH.(\varphi(0))(\eta(s))ds = M(t), \tag{8.79}$$

where L is the generator of independent random walkers acting on the η variable, and where $\{M(t) : t \geq 0\}$ is a martingale.

8.6 The Wright-Fisher Diffusion and the Moran Model

Duality is a crucially important concept in the context of mathematical population genetics, see e.g. [87], [66]. Usually, duality in this context means relating the forward-in-time models of allele frequency evolution with the backwards-in-time genealogical models. In Sect. 2.5 we discussed the basic example of such a duality: duality between the Wright-Fisher diffusion for genetic drift and its genealogical counterpart, the block counting process of the Kingman coalescence. In this section we add some material that expands this basic example in several directions and shows that these dualities all fit in the context of the Heisenberg algebra. We restrict our discussion to populations of two types only, for the generalization to multi-type populations see [42, 125].

Notice that our aim in this section is not to add much to the extensive literature for the models of population dynamics. Rather, we want to illustrate how also these examples fit naturally in the algebraic framcwork, by considering two representations of the Heisenberg algebra (different from the ones we have encountered so far).

8.6.1 Coalescent: Duality with a Pure Death Process

The first generalization we consider is the addition of mutation to the Wright diffusion with two-types. We call $\{X(t)\,,\, t\,\geq\,0\}$ the fraction of individuals of type 1 in an infinitely large population. Besides "random genetic drift" we include mutations: thus each individual of type 2 mutates at rate $\frac{\alpha_1}{2}$ into type 1 and, conversely, individuals of type 1 mutate at rate $\frac{\alpha_2}{2}$ into type 2. The Markov process $\{X(t)\,,\, t \geq 0\}$ is a diffusion process with generator

$$\mathcal{L} = \frac{1}{2}x(1-x)\frac{d^2}{dx^2} + \frac{1}{2}\Big(\alpha_1(1-x) - \alpha_2 x\Big)\frac{d}{dx}. \tag{8.80}$$

The action of this generator on the monomial x^n does not yield in general a Markov process in the n-variable (except in the case $\alpha_1 = 1$ and $\alpha_2 = 0$). To have a dual Markov process we need to scale the monomial with an appropriate function of n. This implies that the dual death process provided by the block counting process of Kingman's coalescencent gets an extra rate describing mutations.

Theorem 8.19 *The diffusion process* $\{X(t)\,,\, t \geq 0\}$ *with generator* (8.80) *is dual to the death process* $\{N(t)\,,\, t \geq 0\}$ *with generator*

$$Lf(n) = \frac{n(n-1+\alpha_1+\alpha_2)}{2}\Big(f(n-1) - f(n)\Big), \tag{8.81}$$

with duality function $D : \mathbb{N} \times [0,1] \to \mathbb{R}$ *given by*

$$D(n,x) = x^n \frac{\Gamma(\alpha_1)}{\Gamma(\alpha_1+n)} \frac{\Gamma(\alpha_1+\alpha_2+n)}{\Gamma(\alpha_1+\alpha_2)}. \tag{8.82}$$

Proof The statement of the theorem can be checked with an elementary computation. To understand the origin of the duality function we provide instead an algebraic proof, and see how the duality arises from two representations of the Heisenberg algebra. Using the standard representation of the Heisenberg algebra

$$A^\dagger = x, \qquad A = \frac{d}{dx}, \tag{8.83}$$

yielding $[A, A^\dagger] = I$, the generator (8.80) can be rewritten in the abstract form

$$\mathcal{L} = \frac{1}{2}A^\dagger(1 - A^\dagger)A^2 + \frac{1}{2}(\alpha_1(1 - A^\dagger) - \alpha_2 A^\dagger)A. \tag{8.84}$$

Similarly, consider the discrete representation of the dual Heisenberg algebra given by

$$a^\dagger f(n) = \frac{\alpha_1 + n}{\alpha_1 + \alpha_2 + n} f(n+1),$$

$$af(n) = n\frac{\alpha_1 + \alpha_2 + n - 1}{\alpha_1 + n - 1} f(n-1), \tag{8.85}$$

where as usual we make the convention $af(0) = 0$. The operators $a, a^\dagger$ form a discrete representation of the dual Heisenberg algebra, i.e., $[a, a^\dagger] = -I$. Using this representation one can write the generator (8.81) as

$$L = \frac{1}{2}a^2(1 - a^\dagger)a^\dagger + \frac{1}{2}a(\alpha_1(1 - a^\dagger) - \alpha_2 a^\dagger). \tag{8.86}$$

Thus the duality between the process $\{X(t), t \geq 0\}$ with generator (8.80) and the process $\{N(t), t \geq 0\}$ with generator (8.81) follows from the general composition rule of dualities if one can prove that

$$\big(A^\dagger D(n, \cdot)\big)(x) = \big(a^\dagger D(\cdot, x)\big)(n), \tag{8.87}$$

$$\big(AD(n, \cdot)\big)(x) = \big(aD(\cdot, x)\big)(n), \tag{8.88}$$

where D is given by (8.82). We verify the first of these dualities and leave to the reader the verification of the second. The l.h.s. of (8.87) reads

$$\big(A^\dagger D(n, \cdot)\big)(x) = xD(n, x) = x^{n+1}\frac{\Gamma(\alpha_1)}{\Gamma(\alpha_1 + n)}\frac{\Gamma(\alpha_1 + \alpha_2 + n)}{\Gamma(\alpha_1 + \alpha_2)}, \tag{8.89}$$

whereas the r.h.s. is given by

$$
\begin{aligned}
\big(a^{\dagger}D(\cdot,x)\big)(n) &= \frac{\alpha_1+n}{\alpha_1+\alpha_2+n}D(n+1,x) \\
&= \frac{\alpha_1+n}{\alpha_1+\alpha_2+n}x^{n+1}\frac{\Gamma(\alpha_1)}{\Gamma(\alpha_1+n+1)}\frac{\Gamma(\alpha_1+\alpha_2+n+1)}{\Gamma(\alpha_1+\alpha_2)} \\
&= x^{n+1}\frac{\Gamma(\alpha_1)}{\Gamma(\alpha_1+n)}\frac{\Gamma(\alpha_1+\alpha_2+n)}{\Gamma(\alpha_1+\alpha_2)}, \qquad (8.90)
\end{aligned}
$$

where we used the property of the Gamma function $\Gamma(n+1) = n\Gamma(n)$. □

8.6.2 Conservative Duals: Birth and Death Processes

In the previous example, the dual process used to infer properties of Wright-Fisher diffusion is given by a death process describing the dynamics of the former in reverse time. By using the algebraic approach it has been observed that duality exists also keeping the forward direction of time for the dual process. More precisely, for the Wright-Fisher diffusion with mutation one has duality with the Moran process with mutation. Thus the dual is now a birth and death process.

Theorem 8.20 *The diffusion process* $\{X(t)\,,\ t \geq 0\}$ *with generator* (8.80) *is dual to the birth and death process* $\{N(t)\,,\ t \geq 0\}$ *with generator*

$$
\begin{aligned}
Lf(n) &= \frac{n(N-n+\alpha_2)}{4}\Big(f(n-1)-f(n)\Big) \\
&+ \frac{(N-n)(n+\alpha_1)}{4}\Big(f(n+1)-f(n)\Big), \qquad (8.91)
\end{aligned}
$$

with duality function $D : \{0, 1, \ldots, N\} \times [0, 1] \to \mathbb{R}$ *given by*

$$
D(n,x) = x^n(1-x)^{N-n}\frac{\Gamma(\alpha_1)}{\Gamma(\alpha_1+n)}\frac{\Gamma(\alpha_2)}{\Gamma(\alpha_2+N-n)}. \qquad (8.92)
$$

Proof The duality between Wright-Fisher diffusion and Moran process is just an instance of the more general duality between the Brownian Energy Process and the Symmetric Inclusion Process. Thus the algebraic structure behind this duality is the one of the $\mathfrak{su}(1,1)$ algebra.

Consider indeed the generator of the two-sites BEP process with inhomogeneities α_1 and α_2

$$
\mathcal{L}^{BEP} = x_1x_2\left(\frac{\partial}{\partial x_1}-\frac{\partial}{\partial x_2}\right)^2 - (\alpha_2x_1-\alpha_1x_2)\left(\frac{\partial}{\partial x_1}-\frac{\partial}{\partial x_2}\right), \qquad (8.93)
$$

restricted to the simplex $x_1 + x_2 = 1$. Then defining $x_1 = x$ and $x_2 = 1 - x$ one sees that

$$\mathcal{L}^{BEP} = 4\mathcal{L}, \tag{8.94}$$

where $\mathcal{L}$ is the generator in (8.80). Similarly, start from the generator of the two-sites SIP process with inhomogeneities α_1 and α_2

$$\begin{aligned} L^{SIP} f(n_1, n_2) &= n_1(n_2 + \alpha_2)(f(n_1 - 1, n_2 + 1) - f(n_1, n_2)) \\ &+ n_2(n_1 + \alpha_1)(f(n_1 + 1, n_2 - 1) - f(n_1, n_2)). \end{aligned} \tag{8.95}$$

Restricting to the simplex $n_1 + n_2 = N$ and defining $n_1 = n$ and $n_2 = N - n$ one sees that

$$\mathcal{L}^{SIP} = 4L, \tag{8.96}$$

where L is the generator in (8.91). □

8.6.3 *Dualities for Finite Size Populations*

We close this section by observing that dualities for population of finite sizes can also be put in the framework of the algebraic approach described in this book. We show this by considering the simple example of the neutral Moran process with N individuals. Let $K(t)$ denotes the number of individuals of type 1 at time $t \geq 0$.

Theorem 8.21 *The neutral Moran process $\{K(t)\,,\, t \geq 0\}$ with generator*

$$L_N f(k) = \frac{1}{2} k(N - k)\Big(f(k-1) - 2f(k) + f(k+1)\Big), \tag{8.97}$$

is dual to the death process $\{N(t)\,,\, t \geq 0\}$ with generator

$$Lf(n) = \frac{n(n-1)}{2}\Big(f(n-1) - f(n)\Big), \tag{8.98}$$

with duality function $D : \mathbb{N} \times \{0, 1, \dots, N\} \to \mathbb{R}$ given by

$$D(n, k) = \frac{\binom{k}{n}}{\binom{N}{n}} = \frac{k(k-1)\cdots(k-(n-1))}{N(N-1)\cdots(N-(n-1))}. \tag{8.99}$$

Proof The proof follows by combining the following observations:

(i) The generator (8.97) is rewritten as

$$L_N = a_N^\dagger(1 - a_N^\dagger)a_N^2, \tag{8.100}$$

where the ladder operators defined by

$$a_N^\dagger f(k) = \sum_{r=0}^{k-1}(-1)^{k-1-r}\frac{\binom{N}{r}}{\binom{N}{k}}f(r),$$

$$a_N f(k) = (N-k)f(k+1) + (2k-N)f(k) - kf(k-1). \tag{8.101}$$

give a representation of the Heisenberg algebra $[a_N, a_N^\dagger] = I$.

(ii) The generator (8.98) is rewritten as

$$L = a^2(1 - a^\dagger)a^\dagger, \tag{8.102}$$

where the ladder operators

$$\begin{aligned} a^\dagger f(n) &= f(n+1), \\ af(n) &= nf(n-1), \end{aligned} \tag{8.103}$$

give a representation of the dual Heisenberg algebra $[a, a^\dagger] = -I$.

(iii) The duality function (8.99) is the intertwiner between these two representations:

$$\big(a_N^\dagger D(n,\cdot)\big)(k) = \big(a^\dagger D(\cdot,k)\big)(n), \tag{8.104}$$

$$\big(a_N D(n,\cdot)\big)(k) = \big(aD(\cdot,k)\big)(n). \tag{8.105}$$

□

8.7 Additional Notes

The KMP model on a chain $\{1,\ldots,N\}$ with boundary reservoirs at left and right end of the chain was introduced in [152] (see also [111]) as a stochastic model of heat conduction which satisfies Fourier law. The absorbing dual was discovered in that paper and used to prove properties of the non-equilibrium steady state such as local equilibrium. Macroscopic properties of the model have been further analyzed using the "macroscopic fluctuation theory" in [24], [27]. In the recent work [69] new properties of KMP process and its non-equilibrium steady state are proved using a "hidden temperature model". The identification of the KMP model as a thermalization of the Brownian energy process is from [116]. As a consequence, a one-parameter family of KMP models with similar duality properties were introduced. Other models of mass redistribution arise in the context of agent based models of wealth distribution see e.g. [55], [238] for more background on the context and various models of this type. In [60] and [231], [197] such models are analyzed with duality techniques.

The Aldous averaging process is studied in [2], whereas the Kac ring model served as one of the first kinetic models where the Boltzmann equation can be derived rigorously. This model is a toy model for many problems in statistical physics which turn around the micro-to-macro problem, such as the irreversibility paradox, see e.g. [35]. To our knowledge the dualities for both of these models are new. In [195] a class of discrete averaging processes is analyzed and mixing properties are proved.

The stochastic Ginzburg-Landau model is a special case of a large class of models of interacting diffusions, of which the hydrodynamic limit was studied first in [109]. Large deviations for the density profile in such models of this type lead to the Guo-Papanicolao-Varadhan method, see [129]. In [46] the Ginzburg-Landau model of the current chapter was used to prove existence of stationary states with non-zero current in infinite-volume systems. This result is strongly based on the duality with independent walkers discussed here.

The Moran model and the Wright Fisher diffusion are extensively studied in the literature of stochastic models of population dynamics. The main aim of the section on this subject is to relate the known dualities to representations of creation and annihilation operators. More connections between dualities from population dynamics and the algebraic approach can be found in [42].

Chapter 9
Orthogonal Dualities

In this chapter we consider orthogonal dualities, i.e., we search for duality functions that satisfy an orthogonality condition in the Hilbert space weighted by the reversible measure of the process. As the standard duality functions are associated to the kernel of an intertwiner between two Lie algebra representations, here we prove that orthogonal dualities are associated to *unitary* intertwiners. The orthogonal dualities for the processes studied in the previous chapters turn out to be classical orthogonal polynomials which can indeed be obtained by Gram-Schmidt orthogonalization of the basic dualities.

9.1 Introduction

In the previous chapters, we have discussed several dualities for various processes. In this chapter we ask if and when these dualities can be turned into *orthogonal dualities*. By this we mean that, if we have a process $\{\eta(t) : t \geq 0\}$ with state space Ω and reversible measure μ, then we search for a duality function that is orthogonal in the Hilbert space $L^2(\Omega, \mu)$. More precisely, considering the Hilbert space, with inner product

$$\langle f, g \rangle = \int_\Omega f(\eta) g(\eta) \mu(d\eta), \tag{9.1}$$

consisting of all measurable functions $f : \Omega \to \mathbb{R}$ with $\langle f, f \rangle = ||f||^2 < \infty$, we search for a duality function $D : \widehat{\Omega} \times \Omega \to \mathbb{R}$ such that the sequence of functions $D(\xi, \cdot)$ labeleled by $\xi \in \widehat{\Omega}$ satisfies the orthogonality condition

$$\langle D(\xi, \cdot), D(\xi', \cdot) \rangle = \delta_{\xi,\xi'} a(\xi), \tag{9.2}$$

C. Giardinà, F. Redig, *Duality for Markov Processes*, Grundlehren der mathematischen Wissenschaften 365, https://doi.org/10.1007/978-3-032-04099-2_9

where $\delta_{\xi,\xi'}$ denotes the Kronecker delta function between two configurations in the state space of the dual process and a is the norm squared:

$$a(\xi) = ||D(\xi, \cdot)||^2. \tag{9.3}$$

Here, and throughout this chapter, we assume that the state space of the dual process is countable.

9.1.1 Usefulness of Orthogonal Dualities

There are several reasons to be interested in orthogonal dualities. Clearly, whenever the duality function (viewed as a sequence of functions labeled by the configurations of the dual process) form a basis of the Hilbert space $L^2(\Omega, \mu)$, then one can in principle study the statistical dynamical properties of any observable by expanding it in this base and following the evolution of the coefficients. If the base happens to be an orthogonal one, then one needs to study fewer coefficients, since many terms in the expansion vanish as a consequence of the orthogonality condition.

Consider for instance time-dependent expectations with respect to the stationary measure of an observable $f \in L^2(\Omega, \mu)$ with expansion

$$f(\eta) = \sum_{\xi \in \widehat{\Omega}} c_f(\xi) D(\xi, \eta), \tag{9.4}$$

and coefficients

$$c_f(\xi) = \frac{1}{a(\xi)} \langle f, D(\xi, \cdot) \rangle. \tag{9.5}$$

By reversibility of μ we have stationarity of μ, i.e., $\int (S_t f) d\mu = \int f d\mu$ for all $t \geq 0$. Assuming $\int f d\mu = 0$, we will show in Proposition 9.2 that the variance at time $t > 0$ can be written as

$$\mathbb{V}ar[S_t f] = \sum_{\xi, \xi'} c_f(\xi) c_f(\xi')\, a(\xi)\, p_{2t}(\xi', \xi), \tag{9.6}$$

where $p_t(\xi, \xi')$ is the transition probability of the dual process, i.e.

$$p_t(\xi, \xi') = \mathbb{P}(\xi(t) = \xi' \,|\, \xi(0) = \xi). \tag{9.7}$$

Thus orthogonal dualities can be used to estimate the speed of convergence to equilibrium in $L^2(\Omega, \mu)$. More generally, if we consider another observable $g \in L^2(\Omega, \mu)$ with expansion

$$g(\eta) = \sum_{\xi} c_g(\xi) D(\xi, \eta), \tag{9.8}$$

then, considering two times $t, t' > 0$, in Proposition 9.2 we prove the following formula for the time-dependent covariance

$$\mathbb{C}ov[S_t f, S_{t'} g] = \int (S_t f)(S_{t'} g) d\mu = \sum_{\xi, \xi'} c_f(\xi)\, c_g(\xi')\, a(\xi)\, p_{t+t'}(\xi', \xi). \tag{9.9}$$

As a consequence, having orthogonal dualities can be used to identify functions with positive correlations. A third application of orthogonal dualities will be given in the context of fluctuation fields and the Boltzmann-Gibbs principle in the context of hydrodynamic limits. This is explained later in Chap. 12, where we study the fluctuation fields associated to orthogonal duality functions under diffusive space-time rescaling.

9.1.2 Orthogonal Polynomials and Gram-Schmidt Orthogonalization

Our next question is thus how to identify orthogonal duality functions. For processes that we already know to admit a duality function in the form of a monomial or a polynomial, the natural guess in the quest of orthogonal dualities are the *orthogonal polynomials* in $L^2(\Omega, \mu)$. The original process will be associated with the variables of the polynomials, whereas the dual process will be associated with the degrees of the polynomial. We shall see indeed that this guess is correct for our examples, where we will encounter some of the *classical orthogonal polynomials*. This covers in particular the self-dualities of discrete processes and the dualities between discrete and continuous processes. To achieve self-dualities of continuous process, other special functions are needed, such as the *Bessel functions*.

It turns out that the dualities with product of orthogonal polynomials as duality function can be derived from the dualities of the previous chapters by applying the Gram-Schmidt orthogonalization procedure. The reason for this is that the linear transformation corresponding to the Gram-Schmidt orthogonalization procedure is actually a symmetry of the generator, as we will explain in detail in Sect. 9.8.

9.1.3 A Remark on the Norm of the Orthogonal Duality Function

Before concluding this introductory section, let us explain a relation between the norm of orthogonal duality functions and a reversible measure for the dual process in the setting where the dual process has a countable state space (this is the case for particle systems where particles move on a finite set of vertices V, in which case the configuration space is a subset of $\mathbb{N}^V$) and a reversible measure $\widehat{\mu}$. We will see that the orthogonality condition with respect to the reversible measure μ of the original process fixes the squared norm $a(\xi) = ||D(\xi, \cdot)||^2$ of the orthogonal duality function to

$$a(\xi) = \frac{c}{\widehat{\mu}(\xi)}, \tag{9.10}$$

where $c > 0$ is a constant and $\widehat{\mu}$ is the reversible measure of the dual process. Indeed, the reversibility with respect to the measure μ implies that the semigroup S_t is self-adjoint in $L^2(\Omega, \mu)$, i.e.,

$$\langle S_t D(\xi, \cdot), D(\xi', \cdot)\rangle = \langle D(\xi, \cdot), S_t D(\xi', \cdot)\rangle. \tag{9.11}$$

Now, considering the left hand side of the above equation we have by duality

$$\begin{aligned}\langle S_t D(\xi, \cdot), D(\xi', \cdot)\rangle &= \int \mu(d\eta)[S_t D(\xi, \cdot)](\eta) D(\xi', \eta) \\ &= \int \mu(d\eta)[\widehat{S}_t D(\cdot, \eta)](\xi) D(\xi', \eta),\end{aligned} \tag{9.12}$$

and the orthogonality condition (9.2) yields

$$\langle S_t D(\xi, \cdot), D(\xi', \cdot)\rangle = ||D(\xi', \cdot)||^2 \; p_t(\xi, \xi'), \tag{9.13}$$

where we recall that $p_t(\xi, \xi')$, defined in (9.7), denotes the transition probability of the dual process to go in a time t from the configuration ξ to the configuration ξ'.

Similarly, for the right hand side of (9.11) we have

$$\langle D(\xi, \cdot), S_t D(\xi', \cdot)\rangle = ||D(\xi, \cdot)||^2 \; p_t(\xi', \xi). \tag{9.14}$$

Combining (9.13) and (9.14) with (9.11) we deduce that the squared norm of the duality function has to be (up to a constant) the inverse of the reversible measure of the dual process.

9.1.4 Chapter Organization

The content of the rest of this chapter is organized as follows. We start by considering the simplest possible setting. In Sect. 9.2 we shall prove that the product of Charlier polynomials is an orthogonal self-duality function for independent random walkers. This is followed, in Sect. 9.3, by the illustration of how orthogonal dualities can be used for general systems to quantify the speed of relaxation to the equilibrium by estimating the decay of the time-dependent variance. In particular, for independent random walkers on $\mathbb{Z}^d$, whose equilibrium is given by a product of Poisson, we will see that orthogonal duality with Charlier polynomials yields a bound on the time-dependent variance of the type $t^{-d/2}$ for large times t.

In Sect. 9.4, we present a complete list of orthogonal duality relations for the processes studied so far in this book. The list includes orthogonal self-dualities of discrete processes, orthogonal dualities between discrete and continuos processes and self-dualities of continuous processes. In the subsequent chapters, we provide multiple proofs for those orthogonal duality relations. Each proof reveals a particular aspect of the approach presented in this book.

The first option to prove orthogonal dualities is to use *structural properties* of classical orthogonal polynomials and other special functions. This is described in Sect. 9.5. It amounts to a sequence of algebraic manipulations that show how duality is a consequence of three important properties of hypergeometric functions: their recurrence relations, their difference/differential equations and the introduction of a "creation" operator that raises the dual variable by one unit.

From the algebraic perspective it is natural to ask how orthogonal dualities are connected to the representation theory of the Lie algebras. We therefore show in Sect. 9.6 that they arise from *unitary equivalent representations* of Lie algebras.

In the case of self-dualities for Markov processes with countable state space, we know that the self-duality functions can be obtained by acting with a symmetry of the generator on the cheap self-duality function. We show in Sect. 9.7 that the orthogonal self-duality functions can be obtained by using a *unitary symmetry*.

Finally, in Sect. 9.8, we prove that the dualities with product of orthogonal polynomials as duality function can be derived from the "basic" dualities by applying the Gram-Schmidt orthogonalization procedure.

9.2 Orthogonal Self-duality for Independent Random Walkers

Independent random walkers have the homogeneous Poisson product measure ν_ρ, with expectation $\rho > 0$, as reversible measure. Then a natural conjecture is that an homogeneous product of Charlier polynomials (which are the orthogonal polynomials of the Poisson measure) is an orthogonal self-duality function. In this section we show that this conjecture is correct.

We first recall some basic properties of the univariate Charlier polynomials. Let $C_n(x)$ denote the nth-order Charlier polynomial in the variable x (later we will give their explicit form, see (9.76)). They satisfy the orthogonality relation

$$\sum_{x=0}^{\infty} \frac{\rho^x}{x!} e^{-\rho} C_n(x) C_m(x) = \frac{n!}{\rho^n} \delta_{m,n}, \tag{9.15}$$

where $n, m \in \mathbb{N}$ are the degrees of two polynomials. We see that the squared norm is

$$||C_n||^2 = \frac{n!}{\rho^n}, \tag{9.16}$$

and thus they are correctly normalized for being a candidate orthogonal duality function. Indeed, since we aim for self-duality, we know from (9.10) that the norm of the self-duality function must be (up to constants) the inverse of Poisson measure with parameter ρ.

As all other orthogonal polynomials, the Charlier polynomials satisfy a three-term recurrence relation, given by

$$xC_n(x) = -\rho C_{n+1}(x) + (n + \rho) C_n(x) - nC_{n-1}(x), \tag{9.17}$$

where it is understood that $C_{-1}(x) = 0$. Furthermore, they belong to the classical orthogonal polynomials and thus they satisfy a second-order difference equation, given by

$$\begin{aligned} &x[C_n(x+1) - 2C_n(x) + C_n(x-1)] + (\rho - x)[C_n(x+1) - C_n(x)] \\ &\quad + nC_n(x) = 0. \end{aligned} \tag{9.18}$$

Finally, the Charlier polynomials can also be represented using the Rodrigues formula

$$C_n(x) = \frac{x!}{\rho^x} (\nabla_\ell)^n \left[\frac{\rho^x}{x!} \right], \tag{9.19}$$

where ∇_ℓ is the discrete left derivative

$$\nabla_\ell f(x) = f(x) - f(x-1). \tag{9.20}$$

From this one can check the identity

$$\rho C_n(x) - xC_n(x-1) = \rho C_{n+1}(x), \tag{9.21}$$

which can be interpreted as a raising relation. Namely, the operator R defined by

$$Rf(x) = \rho f(x) - xf(x-1), \tag{9.22}$$

increases by one the degree of the polynomial, i.e.

$$RC_n(x) = \rho C_{n+1}(x). \tag{9.23}$$

Theorem 9.1 (Self-duality of Independent Random Walkers and Charlier Polynomials) *The independent random walkers process on a set V with generator L in* (3.5) *has a family (parametrize by $\rho > 0$) of orthogonal self-duality function given by*

$$D_\rho(\xi, \eta) = \prod_{x \in V} C_{\xi_x}(\eta_x), \tag{9.24}$$

where $C_{\xi_x}(\eta_x)$ are the Charlier polynomials with orthogonality relation (9.15).

Proof As we did several times, given the product structure of the duality function it is enough to consider two sites. We thus need to verify that for all $x_1, x_2, n_1, n_2 \in \mathbb{N}$ it holds

$$L_{12}D_{12}(n_1, n_2; \cdot, \cdot)(x_1, x_2) = L_{12}D_{12}(\cdot, \cdot; x_1, x_2)(n_1, n_2), \tag{9.25}$$

where

$$\begin{aligned} L_{12}D_{12}(n_1, n_2; \cdot, \cdot)(x_1, x_2) &= x_1 \left[C_{n_1}(x_1 - 1)C_{n_2}(x_2 + 1) - C_{n_1}(x_1)C_{n_2}(x_2)\right] \\ &\quad + x_2 \left[C_{n_1}(x_1 + 1)C_{n_2}(x_2 - 1) - C_{n_1}(x_1)C_{n_2}(x_2)\right], \end{aligned} \tag{9.26}$$

and

$$\begin{aligned} L_{12}D_{12}(\cdot, \cdot; x_1, x_2)(n_1, n_2) &= n_1 \left[C_{n_1-1}(x_1)C_{n_2+1}(x_2) - C_{n_1}(x_1)C_{n_2}(x_2)\right] \\ &\quad + n_2 \left[C_{n_1+1}(x_1)C_{n_2-1}(x_2) - C_{n_1}(x_1)C_{n_2}(x_2)\right]. \end{aligned} \tag{9.27}$$

The left hand side of (9.25) can be re-written as

$$\begin{aligned} L_{12}D_{12}(n_1, n_2; \cdot, \cdot)(x_1, x_2) &= [x_1 C_{n_1}(x_1 - 1)]C_{n_2}(x_2 + 1) - [x_1 C_{n_1}(x_1)]C_{n_2}(x_2) \\ &\quad + C_{n_1}(x_1 + 1)[x_2 C_{n_2}(x_2 - 1)] - C_{n_1}(x_1)[x_2 C_{n_2}(x_2)]. \end{aligned} \tag{9.28}$$

From this we see that to prove self-duality we need to express the terms

$$xC_n(x-1)\,, \qquad xC_n(x)\,, \qquad C_n(x+1), \tag{9.29}$$

using $C_n(x)$, $C_{n-1}(x)$, $C_{n+1}(x)$. To get those expressions, we use the properties of the Charlier polynomials. The expression for the first term is simply obtained from the raising operator relation (9.21) which gives

$$xC_n(x-1) = \rho C_n(x) - \rho C_{n+1}(x). \tag{9.30}$$

The expression for the second term is provided by the three-term recurrence relation (9.17) which gives

$$xC_n(x) = -\rho C_{n+1}(x) + (n+\rho)C_n(x) - nC_{n-1}(x). \tag{9.31}$$

For the third term, its expression is obtained inserting (9.30) and (9.31) into the difference Eq. (9.18). This gives

$$C_n(x+1) = C_n(x) - \frac{n}{\rho}C_{n-1}(x). \tag{9.32}$$

We now plug the expressions that we found in the generator:

$$\begin{aligned}
&L_{12}D_{12}(n_1,n_2;\cdot,\cdot)(x_1,x_2)\\
&= \Big[\rho C_{n_1}(x_1) - \rho C_{n_1+1}(x_1)\Big]\Big[C_{n_2}(x_2) - \frac{n_2}{\rho}C_{n_2-1}(x_2)\Big]\\
&- \Big[-\rho C_{n_1+1}(x_1) + (n_1+\rho)C_{n_1}(x_1) - n_1C_{n_1-1}(x_1)\Big]C_{n_2}(x_2)\\
&+ \Big[C_{n_1}(x_1) - \frac{n_1}{\rho}C_{n_1-1}(x_1)\Big]\Big[\rho C_{n_2}(x_2) - \rho C_{n_2+1}(x_2)\Big]\\
&- C_{n_1}(x_1)\Big[-\rho C_{n_2+1}(x_2) + (n_2+\rho)C_{n_2}(x_2) - n_2C_{n_2-1}(x_2)\Big].
\end{aligned} \tag{9.33}$$

Expanding the products above, only polynomials having degree n_1+n_2 survive. In particular, after simplifications, one is left with

$$\begin{aligned}
L_{12}D_{12}(n_1,n_2;\cdot,\cdot)(x_1,x_2) &= n_1\left[C_{n_1-1}(x_1)C_{n_2+1}(x_2) - C_{n_1}(x_1)C_{n_2}(x_2)\right]\\
&\quad + n_2\left[C_{n_1+1}(x_1)C_{n_2-1}(x_2) - C_{n_1}(x_1)C_{n_2}(x_2)\right],
\end{aligned} \tag{9.34}$$

and thus the self-duality relation (9.25) is proved. □

9.3 Orthogonal Duality and Relaxation to Equilibrium

In this section we show some applications of orthogonal dualities, in particular how they can be used to estimate the speed of convergence to equilibrium in the L^2 sense. We first prove the expressions for the variance (9.6) and covariance (9.9) of functions claimed in the introductory section. We then consider the case of independent random walkers in $\mathbb{Z}^d$ and prove, exploiting independence, that the variance of a function in L_2 vanishes as $t^{-d/2}$.

9.3.1 *Variance, Covariance and Orthogonal Polynomials*

We start with the following simple proposition.

Proposition 9.2 (Variance, Covariance and Orthogonal Polynomials) *Consider* $\{\eta(t)\,,\, t \geq 0\}$, *a Markov process with state space* Ω *and reversible measure* μ. *Let* $\{\xi(t)\,,\, t \geq 0\}$ *be a dual Markov process with a discrete state space* $\widehat{\Omega}$ *and let* $D : \widehat{\Omega} \times \Omega \to \mathbb{R}$ *be a duality function. Assume that* $D(\xi, \cdot)$ *form an orthogonal basis of* $L_2(\Omega, \mu)$ *with*

$$\langle D(\xi, \cdot), D(\xi', \cdot)\rangle = \int_\Omega D(\xi, \eta)D(\xi', \eta)\mu(d\eta) = a(\xi)\delta_{\xi,\xi'}. \tag{9.35}$$

Then we have:

(a) for $t > 0$ *and* $\xi, \xi' \in \widehat{\Omega}$,

$$\int [S_t D(\xi, \cdot)](\eta)\, D(\xi', \eta)\, \mu(d\eta) = a(\xi')\, p_t(\xi, \xi'), \tag{9.36}$$

where $p_t(\xi, \xi')$ *is the transition probability of the dual process:*

$$p_t(\xi, \xi') = \mathbb{P}(\xi(t) = \xi' \,|\, \xi(0) = \xi); \tag{9.37}$$

(b) for $f \in L^2(\Omega, \mu)$ *with* $\mathbb{E}[f] = \int f(\eta)\mu(d\eta) = 0$,

$$\mathbb{V}ar[S_t f] = \sum_{\xi,\xi'} p_{2t}(\xi', \xi)c_f(\xi)c_f(\xi')a(\xi), \tag{9.38}$$

where

$$c_f(\xi) = \frac{1}{a(\xi)}\langle f, D(\xi, \cdot)\rangle; \tag{9.39}$$

(c) for $f, g \in L^2(\Omega, \mu)$ such that $\int f(\eta)\mu(d\eta) = \int g(\eta)\mu(d\eta) = 0$

$$\mathbb{C}ov[S_t f, S_{t'} g] = \sum_{\xi,\xi'} c_f(\xi) c_g(\xi')\, a(\xi)\, p_{t+t'}(\xi', \xi). \tag{9.40}$$

Proof Item (a): using duality we have

$$[S_t D(\xi, \cdot)](\eta) = [\widehat{S}_t D(\cdot, \eta)](\xi) = \sum_{\xi''} p_t(\xi, \xi'') D(\xi'', \eta), \tag{9.41}$$

and then

$$\begin{aligned} \int [S_t D(\xi, \cdot)](\eta)\, D(\xi', \eta)\, \mu(d\eta) &= \int \sum_{\xi''} p_t(\xi, \xi'') D(\xi'', \eta)\, D(\xi', \eta)\, \mu(d\eta) \\ &= a(\xi')\, p_t(\xi, \xi'), \end{aligned} \tag{9.42}$$

where in the last step we used the orthogonality (9.35). This proves (9.36).

Item (b): by reversibility we have that

$$\mathbb{V}ar[S_t f] = \langle S_t f, S_t f \rangle = \langle S_{2t} f, f \rangle. \tag{9.43}$$

Hence, decomposing

$$f = \sum_{\xi} c_f(\xi) D(\xi, \cdot), \tag{9.44}$$

one gets

$$\mathbb{V}ar[S_t f] = \sum_{\xi,\xi'} c_f(\xi) c_f(\xi') \langle S_{2t} D(\xi, \cdot), D(\xi', \cdot) \rangle. \tag{9.45}$$

Applying item (a) (9.38) follows. The proof of item (c) is similar and left to the reader. □

9.3.2 Application to Independent Random Walkers

We shall now discuss in more detail how to apply the results of Proposition 9.2 to the case of independent random walkers. As we saw in Theorem 9.1, this process satisfies self-duality with an orthogonal self-duality function given by product of Charlier polynomials. We start from the homogeneous Poisson product measure

ν_ρ, with expectation ρ, and recall the notation $D_\rho(\xi, \eta)$ for the corresponding orthogonal self-duality polynomials:

$$D_\rho(\xi, \eta) = \prod_{x \in V} C_{\xi_x}(\eta_x), \tag{9.46}$$

with

$$\int D_\rho(\xi, \eta) D_\rho(\xi', \eta) \nu_\rho(d\eta) = a(\xi')\delta_{\xi,\xi'}, \tag{9.47}$$

where

$$a(\xi) = \|D_\rho(\xi, \cdot)\|^2 = \prod_{x \in V} \frac{\xi_x!}{\rho^{\xi_x}}. \tag{9.48}$$

We will always work on $V = \mathbb{Z}^d$ with translation invariant and symmetric underlying random walk, and denote

$$p_t(x, y) = \pi_t(y - x), \qquad x, y \in \mathbb{Z}^d, \tag{9.49}$$

the corresponding symmetric and translation invariant transition probability for the position of each single particle (notice that, with a slight abuse, we use the same notation for the transition probability in the configuration state-space, $p_t(\xi, \xi')$, $\xi, \xi' \in \Omega$). Additionally we assume finite

$$\sum_{v \in \mathbb{Z}^d} \pi_t(v)\|v\|^2 < \infty. \tag{9.50}$$

Under (9.50) we have the estimate

$$p_t(0, x) \le p_t(0, 0) \le Ct^{-d/2}. \tag{9.51}$$

For any *allowed* configuration $\eta \in \Omega_{\text{alw}} \subseteq \mathbb{N}^V$ (see Definition 3.14) we denote by $\{\eta(t)\,,\, t \ge 0\}$ the process of independent walkers starting from η.

We can then state the main result of this section, which estimates the variance of $S_t f$ in terms of random walk return probabilities and a norm of Sobolev type of f. We make use of the following notation. Let $n \in \mathbb{N}$ and denote by $\mathbf{x} \in \mathbb{Z}^{nd}$ the coordinates vector $\mathbf{x} := (x_1, \ldots, x_n)$, with $x_i \in \mathbb{Z}^d$, $i = 1, \ldots, n$. We denote by $\xi(\mathbf{x})$ the configuration associated to $\mathbf{x}$, i.e. $\xi_x(\mathbf{x}) = \sum_{i=1}^n \mathbb{1}_{x=x_i}$. We define $|\mathbf{x}| := |\xi(\mathbf{x})| = n$. Here x_i is the position of the i-th particle, where particles are labeled in such a way that the dynamics is symmetric. Then, if $|\xi| = n$, the number of labeled configurations corresponding to ξ is $c(n, \xi) := \frac{n!}{\prod_x \xi_x!}$.

Theorem 9.3 (Speed of Relaxation to Equilibrium for Independent Random Walkers) *Let f be in $L^2(\Omega, \nu_\rho)$ such that for some $0 < z < 1$ the weighted norm*

$$\|f\|_z^2 := \sum_{n=1}^{\infty} z^n \sum_{\xi:|\xi|=n} |\hat{f}(\xi)| \sum_{\xi':|\xi'|=n} a(\xi')|\hat{f}(\xi')|, \tag{9.52}$$

is finite. Then, denoting $p_{2t}(0,0)$ the return probability for a single random walk on $\mathbb{Z}^d$, we have the estimate

$$Var_{\nu_\rho}(S_t f) \le \|f\|^2_{p_{2t}(0,0)}. \tag{9.53}$$

As a consequence, if (9.50) *holds, then for all t large enough such that $p_{2t}(0,0) < z$,*

$$Var_{\nu_\rho}(S_t f) \le \frac{c}{z} t^{-d/2} \|f\|_z^2. \tag{9.54}$$

Proof First we will pass to labeled configurations, and rewrite

$$\begin{aligned} Var_{\nu_\rho}(S_t f) &= \sum_{\xi,\xi'} a(\xi) p_{2t}(\xi,\xi') \hat{f}(\xi)\hat{f}(\xi') \\ &= \sum_{n=1}^{\infty} \sum_{\mathbf{x},\mathbf{x}'\in\mathbb{Z}^{dn}} \frac{a(\mathbf{x})}{c(n,\xi(\mathbf{x}))c(n,\xi(\mathbf{x}'))} \cdot p_{2t}(\mathbf{x},\mathbf{x}')\hat{f}(\mathbf{x})\hat{f}(\mathbf{x}'), \end{aligned} \tag{9.55}$$

where, with a slight abuse of notation, we use the same symbol for $\hat{f}(\xi)$ and $\hat{f}(\mathbf{x})$. Notice that the sum in the rhs of (9.55) starts at $n = 1$, because f has zero expectation, so its projection on the polynomial of order zero is zero. The essential point now is that $p_t(\mathbf{x},\mathbf{x}') = \pi_t(\mathbf{x}' - \mathbf{x})$ is translation invariant, and hence the sum over $\mathbf{x}'$ acts as a convolution. Let us further abbreviate

$$\hat{f}_n(\mathbf{x}) := \frac{\hat{f}(\mathbf{x})}{c(n,\xi(\mathbf{x}))}. \tag{9.56}$$

Then we can rewrite (9.55) as

$$\begin{aligned} &\sum_{n=1}^{\infty} \sum_{\mathbf{x},\mathbf{x}'\in\mathbb{Z}^{dn}} a(\mathbf{x}) p_{2t}(\mathbf{x},\mathbf{x}')\hat{f}_n(\mathbf{x})\hat{f}_n(\mathbf{x}') \\ &= \sum_{n=1}^{\infty} \sum_{\mathbf{x},\mathbf{x}'\in\mathbb{Z}^{dn}} a(\mathbf{x}) \pi_{2t}(\mathbf{x}' - \mathbf{x})\hat{f}_n(\mathbf{x})\hat{f}_n(\mathbf{x}') \end{aligned}$$

$$= \sum_{n=1}^{\infty} \sum_{\mathbf{x} \in \mathbb{Z}^{dn}} a(\mathbf{x}) \hat{f}_n(\mathbf{x}) \cdot \pi_{2t} * \hat{f}_n(\mathbf{x})$$

$$= \sum_{n=1}^{\infty} \langle a \hat{f}_n, \pi_{2t} * \hat{f}_n \rangle_n, \tag{9.57}$$

where $\langle g, h \rangle_n = \sum_{\mathbf{x} \in Z^{dn}} g(x)h(x)$ denotes the innerproduct in $l_2(\mathbb{Z}^{dn})$ and where $*$ denotes convolution. Now notice that the convolution with π_{2t} is a self-adjoint semigroup (by symmetry of $p_t(\mathbf{x}, \mathbf{x}')$ in $\mathbf{x}, \mathbf{x}'$) in $l_2(\mathbb{Z}^{dn})$ and therefore, using Cauchy-Schwarz and Young's inequality, we rewrite

$$\begin{aligned} \langle a \hat{f}_n, \pi_{2t} * \hat{f}_n \rangle_n &= \langle \pi_t * (a \hat{f}_n), \pi_t * \hat{f}_n \rangle_n \\ &\leq \|\pi_t * (a \hat{f}_n)\|_2 \|\pi_t * (\hat{f}_n)\|_2 \\ &\leq \|\pi_t\|_2^2 \|a \hat{f}_n\|_1 \|\hat{f}_n\|_1. \end{aligned} \tag{9.58}$$

Now notice that

$$\|a \hat{f}_n\|_1 = \sum_{\mathbf{x} \in \mathbb{Z}^{dn}} \frac{a(\mathbf{x})}{c(n, \xi(\mathbf{x}))} |f(\mathbf{x})| = \sum_{\xi : |\xi| = n} a(\xi) |\hat{f}(\xi)|, \tag{9.59}$$

and similarly,

$$\|\hat{f}_n\|_1 = \sum_{\xi : |\xi| = n} |\hat{f}(\xi)|. \tag{9.60}$$

On the other hand,

$$\|\pi_t\|_2^2 = \sum_{\mathbf{x}} p_t(\mathbf{0}, \mathbf{x}) p_t(\mathbf{x}, \mathbf{0}) = p_{2t}(0, 0)^n. \tag{9.61}$$

Inserting these into (9.57) we arrive to

$$Var_{\nu_\rho}(S_t f) \leq \|f\|^2_{p_{2t}(0,0)}. \tag{9.62}$$

From the definition of the norm (9.52) we can further write, for all t large enough such that $p_{2t}(0, 0) < z$,

$$Var_{\nu_\rho}(S_t f) \leq \frac{p_{2t}(0, 0)}{z} \|f\|_z^2. \tag{9.63}$$

Finally using the asymptotics

$$p_{2t}(0, 0) \leq c t^{-d/2}, \tag{9.64}$$

with some constant $c > 0$, it follows

$$Var_{\nu_\rho}(S_t f) \leq \frac{c}{z} t^{-d/2} \| f \|_z^2. \tag{9.65}$$

□

9.4 Overview of Orthogonal Dualities

In this section we classify orthogonal duality relations for the Markov processes considered in previous chapters. After a preliminary paragraph in which we recall the definitions of some hypergeometric functions, we present three theorems containing, respectively: (1) the orthogonal self-dualities of discrete processes; (2) the orthogonal dualities between discrete and continuos processes; (3) the self-dualities of continuous processes. As for all duality relations, the "generating function" method applies to orthogonal dualities as well, and reveals that orthogonal discrete self-dualities, orthogonal dualities discrete-continuos and continuos self-dualities are essentially equivalent (in the sense that one of them implies the others, recall e.g. Theorem 4.6).

9.4.1 Preliminaries: Hypergeometric Polynomials

Hypergeometric Series

The *hypergeometric function* ${}_rF_s$ is defined by the hypergeometric series

$${}_rF_s\left(\begin{matrix} a_1, \ldots, a_r \\ b_1, \ldots, b_s \end{matrix}; z\right) = \sum_{k=0}^{\infty} \frac{(a_1)_k \cdots (a_r)_k}{(b_1)_k \cdots (b_s)_k} \frac{z^k}{k!}, \tag{9.66}$$

where $(a)_k$ denotes the Pochhammer symbol

$$(a)_k = a(a+1)\cdots(a+k-1) = \frac{\Gamma(a+k)}{\Gamma(a)}. \tag{9.67}$$

This series is absolutely convergent for all z if $r \leq s$ and for $|z| < 1$ if $r = s+1$. It is divergent for all $z \neq 0$ if $r > s+1$, unless the series is finite. This follows directly from the ratio test applied to the series. Indeed calling

$$c_k = \frac{(a_1)_k \cdots (a_r)_k}{(b_1)_k \cdots (b_s)_k} \frac{z^k}{k!}, \tag{9.68}$$

one has

$$\frac{c_{k+1}}{c_k} = \frac{(k+a_1)(k+a_2)\cdots(k+a_r)z}{(k+b_1)(k+b_2)\cdots(k+b_s)(k+1)}, \tag{9.69}$$

and therefore

$$\lim_{k\to\infty}\left|\frac{c_{k+1}}{c_k}\right| = \begin{cases} 0 \text{ if } r < s+1, \\ |z| \text{ if } r = s+1, \\ \infty \text{ if } r > s+1. \end{cases} \tag{9.70}$$

Whenever one of the numerator parameter a_j is a negative integer $-n$, the hypergeometric function ${}_rF_s$ is a finite sum up to n, i.e. a polynomial of degree n. This can be seen by observing that $(-n)_k = 0$ when $k > n$, since in this case the product $(-n)(-n+1)\cdots(-n+k-1)$ contains a zero. We also remark the identity

$$(-n)_k = \frac{n!}{(n-k)!}(-1)^k \,. \tag{9.71}$$

Classical Orthogonal Polynomials

We shall now review several classical orthogonal polynomials, both in discrete and continuous variables. Notably, in the first set we will encounter Meixner polynomials, Krawtchouk polynomials and Charlier polynomials. Following [153] we recall their definitions in terms of the hypergeometric series.

Meixner Polynomials

They are the orthogonal polynomials of the Negative Binomial distribution $\mathrm{NB}(\alpha, p)$ with parameters $0 < p < 1$ and $\alpha > 0$:

$$M_n(x;\alpha,p) = {}_2F_1\left(\begin{matrix}-n,-x\\ \alpha\end{matrix}; 1-\frac{1}{p}\right), \qquad x \in \mathbb{N}, \tag{9.72}$$

with orthogonality relation

$$\sum_{x=0}^{\infty}\frac{\Gamma(\alpha+x)}{\Gamma(\alpha)}\frac{p^x}{x!}(1-p)^{\alpha}M_n(x;\alpha,p)M_m(x;\alpha,p) = \frac{n!}{p^n}\frac{\Gamma(\alpha)}{\Gamma(\alpha+n)}\delta_{m,n}. \tag{9.73}$$

Krawtchouk Polynomials

They are the orthogonal polynomials of the Binomial distribution $B(\alpha, p)$ with parameters $0 < p < 1$ and $\alpha \in \mathbb{N}$:

$$K_n(x; \alpha, p) = {}_2F_1\left(\begin{matrix} -n, -x \\ -\alpha \end{matrix}; \frac{1}{p}\right), \qquad x \in \{0, 1, \dots, \alpha\}, \tag{9.74}$$

with orthogonality relation

$$\sum_{x=0}^{\alpha} \binom{\alpha}{x} p^x (1-p)^{\alpha - x} K_n(x; \alpha, p) K_m(x; \alpha, p) = n! \frac{(-1)^n}{(-\alpha)_n} \left(\frac{1-p}{p}\right)^n \delta_{m,n}. \tag{9.75}$$

Charlier Polynomials

They are the orthogonal polynomials of the Poisson distribution $\mathrm{Poi}(\lambda)$ with parameter $\lambda > 0$:

$$C_n(x; \lambda) = {}_2F_0\left(\begin{matrix} -n, -x \\ - \end{matrix}; -\frac{1}{\lambda}\right), \qquad x \in \mathbb{N}, \tag{9.76}$$

with orthogonality relation

$$\sum_{x=0}^{\infty} \frac{\lambda^x}{x!} e^{-\lambda} C_n(x; \lambda) C_m(x; \lambda) = \frac{n!}{\lambda^n} \delta_{m,n}. \tag{9.77}$$

Remark 9.4 (Duality of Polynomials) We observe that the Meixner polynomials, Krawtchouk polynomials and Charlier polynomials, as it is clear from inspection of the hypergeometric function defining them, are symmetric if one exchanges the variable and the degree, i.e.

$$p_n(x) = p_x(n). \tag{9.78}$$

In the orthogonal polynomial literature this property is sometimes called "duality", however it should not be confused with the Markov duality that is the topic of this book.

Next, we recall the definitions [153] of the Hermite polynomials and of the Laguerre polynomials. They belong to the classical orthogonal polynomials of a continuous variable.

Hermite Polynomials

They are the orthogonal polynomials of the Gaussian distribution $\mathcal{N}(0, 1)$ with zero mean and unit variance:

$$H_n(x) = (2x)^n {}_2F_0\left(\begin{matrix} -n/2, -(n-1)/2 \\ - \end{matrix} \middle| -\frac{1}{x^2} \right), \qquad x \in \mathbb{R}, \tag{9.79}$$

with orthogonality relation

$$\int_{-\infty}^{\infty} \frac{1}{\sqrt{\pi}} e^{-x^2} H_n(x) H_m(x) dx = 2^n n! \delta_{m,n}. \tag{9.80}$$

Laguerre Polynomials

They are the orthogonal polynomials of the Gamma$(\alpha + 1, 1)$ distribution, with shape parameter $\alpha + 1 > 0$ and unit scale parameter:

$$L_n(x; \alpha) = \frac{\Gamma(\alpha + 1 + n)}{\Gamma(\alpha + 1)} \frac{1}{n!} {}_1F_1\left(\begin{matrix} -n \\ \alpha + 1 \end{matrix} ; x \right), \qquad x \in \mathbb{R}_+, \tag{9.81}$$

with orthogonality relation

$$\int_0^{\infty} \frac{x^\alpha e^{-x}}{\Gamma(\alpha + 1)} L_n(x; \alpha) L_m(x; \alpha) dx = \frac{1}{n!} \frac{\Gamma(n + \alpha + 1)}{\Gamma(\alpha + 1)} \delta_{m,n}. \tag{9.82}$$

Bessel Functions

Finally, we shall need other special functions of hypergeometric nature such as the Bessel function. They are the canonical solutions of the Bessel equation:

$$J_\alpha(x) = \frac{(x/2)^\alpha}{\Gamma(\alpha + 1)} {}_0F_1\left(\begin{matrix} - \\ \alpha + 1 \end{matrix} ; -\frac{x^2}{4} \right), \qquad x \in \mathbb{R}_+, \quad \alpha > 0, \tag{9.83}$$

with orthogonality relation

$$\int_0^1 x J_\alpha\left(x u_{\alpha,m}\right) J_\alpha\left(x u_{\alpha,n}\right) dx = \frac{\delta_{m,n}}{2} \left[J_{\alpha+1}\left(u_{\alpha,m}\right)\right]^2,$$

where $u_{\alpha,m}$ is the mth zero of $J_\alpha(x)$.

9.4.2 Orthogonal Self-duality of Discrete Processes

The next theorem and its corollary summarize the orthogonal self-dualities of the processes with discrete state space studied in previous chapters.

Theorem 9.5 (Orthogonal Self-dualities of the Main Discrete Processes) *For a given set V, the following holds:*

1. Self-duality of SIP with Meixner polynomials. *Consider the family of functions (parametrize by $0 < p < 1$)*
$$D_p(\xi, \eta) = \prod_{x \in V} M_{\xi_x}(\eta_x; \alpha, p), \tag{9.84}$$
where $\xi, \eta \in \mathbb{N}^V$ and $M_{\xi_x}(\eta_x; \alpha, p)$ are the Meixner polynomials defined in (9.72). *Then D_p is an orthogonal self-duality function of the symmetric inclusion process* SIP(α).
2. Self-duality of SEP with Krawtchouk polynomials. *Consider the family of functions (parametrize by $0 < p < 1$)*
$$D_p(\xi, \eta) = \prod_{x \in V} K_{\xi_x}(\eta_x; \alpha, p), \tag{9.85}$$
where $\xi, \eta \in \{0, 1, \ldots, \alpha\}^V$ and $K_{\xi_x}(\eta_x; \alpha, p)$ are the Krawtchouk polynomials defined in (9.74). *Then D_p is an orthogonal self-duality function of the symmetric partial exclusion process* SEP(α).
3. Self-duality of IRW with Charlier polynomials. *Consider the family of functions (parametrize by $\rho > 0$)*
$$D_\rho(\xi, \eta) = \prod_{x \in V} C_{\xi_x}(\eta_x; \rho), \tag{9.86}$$
where $\xi, \eta \in \mathbb{N}^V$ and $C_{\xi_x}(\eta_x; \rho)$ are the Charlier polynomials defined in (9.76). *Then D_ρ is an orthogonal self-duality function of the independent random walkers process.*

Proof See Sect. 9.5. □

Since the thermalization procedure described in Chap. 8 conserves the duality properties, the following corollary immediately follows.

Corollary 9.6 (Orthogonal Self-dualities of Discrete KMP and Thermalized Independent Walkers)

(1) The function D_p in (9.84) *is an orthogonal self-duality function of the thermalized* SIP(α) *process (or discrete generalized KMP process) introduced in Definition 8.8.*

(2) *Similarly, the function D_ρ in (9.86) is an orthogonal self-duality function of thermalized independent walkers introduced in Definition 8.1.*

9.4.3 Orthogonal Duality Between Continuous and Discrete Processes

The next theorem and its corollary concern orthogonal dualities between continuous and discrete processes.

Theorem 9.7 (Orthogonal Dualities Between Continuous and Discrete Processes) *For a given set V, the following holds.*

1. Orthogonal duality between SIP(1/2) and BMP with Hermite polynomials. *Consider the function*

$$D(\xi, z) = \prod_{x \in V} \frac{H_{2\xi_x}(z_x)}{(2\xi_x - 1)!!} = \prod_{x \in V} \left(-\frac{1}{2}\right)^{\xi_x} {}_1F_1\left(\begin{matrix} -\xi_x \\ \frac{1}{2} \end{matrix} \middle| z_x^2\right), \tag{9.87}$$

where $\xi \in \mathbb{N}^V$, $z \in \mathbb{R}^V$ and $H_{\xi_x}(z_x)$ are the Hermite polynomials defined in (9.79). Then D is an orthogonal duality function between the Brownian momentum process and the symmetric inclusion process with parameter $\alpha = 1/2$.

2. Orthogonal duality between SIP and BEP with Laguerre polynomials. *Consider the function*

$$D(\xi, z) = \prod_{x \in V} \frac{\xi_x! \, \Gamma(\alpha)}{\Gamma(\alpha + \xi_x)} L_{\xi_x}(z_x;\, \alpha - 1) = \prod_{x \in V} {}_1F_1\left(\begin{matrix} -\xi_x \\ \alpha \end{matrix} ; z_x\right), \tag{9.88}$$

where $\xi \in \mathbb{N}^V$, $z \in \mathbb{R}_+^V$ and $L_{\xi_x}(z_x; \alpha - 1)$ are the Laguerre polynomials defined in (9.81). Then D is an orthogonal duality function between the Brownian energy process BEP(α) *and the symmetric inclusion process* SIP(α).

Proof See Sect. 9.5. □

Remark 9.8 The expression (9.87) in terms of the ${}_1F_1$ hypergeometric function is obtained by using the two identities

$$H_{2n}(x) = (-1)^n \frac{(2n)!}{n!} \, {}_1F_1\left(\begin{matrix} -n \\ \frac{1}{2} \end{matrix} ; x^2\right)$$

and

$$(2n - 1)!! = \frac{(2n)!}{2^n n!}$$

Actually, the expression can be further simplified using that the inclusion process conserves the number of particles. Namely, we could just consider $\prod_{x\in V} {}_1F_1\left(\begin{smallmatrix}-\xi_x \\ \frac{1}{2}\end{smallmatrix}\middle| z_x^2\right)$ as a duality function between Brownian momentum process and the symmetric inclusion process with parameter $1/2$, as duality functions are always defined up to functions that are constant for the dual dynamics.

Since the thermalization procedure described in Chap. 8 conserves the duality properties, the following corollary immediately follows.

Corollary 9.9 (Orthogonal Dualities of the KMP Process and of the Kac Process)

1. *The function D in* (9.87) *is an orthogonal duality function between the Kac model with generator* (8.42) *and the thermalized* SIP($\frac{1}{2}$) *(or generalized discrete KMP process with parameter* $1/2$*) introduced in Definition 8.8.*
2. *The function D in* (9.88) *is an orthogonal duality function between the thermalized* BEP(α) *(or generalized KMP process) of Definition 8.9 and the thermalized* SIP(α) *(or generalized discrete KMP process) introduced in Definition 8.8.*

9.4.4 Self-duality of Continuous Processes

The next theorem and its corollary present self-dualities of processes with continuous state spaces.

Theorem 9.10 (Self-duality of Continuous Processes) *For a given set V, the following holds true.*

1. Self-duality of BMP with trigoniometric functions. *Consider the function*

$$D(v,z) = \prod_{x\in V} \cos(z_x v_x) = \prod_{x\in V} {}_0F_1\left(\begin{matrix}- \\ \frac{1}{2}\end{matrix} ; -\frac{v_x^2 z_x^2}{4}\right), \tag{9.89}$$

where $v, z \in \mathbb{R}^V$. *Then D is a self-duality function for the Brownian momentum process.*

2. Self-duality of BEP with Bessel functions. *Consider the function*

$$D(v,z) = \prod_{x\in V} \Gamma(\alpha)\left(\frac{v_x z_x}{2}\right)^{-\frac{\alpha}{2}+\frac{1}{2}} J_{\alpha-1}(\sqrt{v_x z_x}) = \prod_{x\in V} {}_0F_1\left(\begin{matrix}- \\ \alpha\end{matrix} ; -\frac{v_x z_x}{4}\right), \tag{9.90}$$

where $v, z \in \mathbb{R}_+^V$ *and* $J_{\alpha-1}(\sqrt{v_x z_x})$ *are the Bessel functions defined in* (9.83). *Then D is a self-duality function for the Brownian energy process BEP(*α*).*

3. Laplace duality for the deterministic process of Chapter III. *Consider the function*

$$D(v, z) = \prod_{x\in V} \exp(v_x z_x) = \prod_{x\in V} {}_0F_0\left(\begin{matrix} - \\ - \end{matrix} ; v_x z_x \right), \tag{9.91}$$

where $v, z \in \mathbb{R}_+^V$. *Then* D *is a self-duality function for the deterministic process considered in Chap. 4.*

Proof See Sect. 9.5. □

Remark 9.11 Recall that the Brownian energy process with $\alpha = 1/2$ can be identified (component-wise) with the square of the Brownian momentum process. This is consistent with the fact that (9.89) can be obtained from (9.90) by specializing it to $\alpha = 1/2$ and using the identity

$$\sqrt{\frac{|xy|}{2}} J_{-1/2}(xy) = \sqrt{\frac{1}{\pi}} \cos(xy), \tag{9.92}$$

Since the thermalization procedure described in Chap. 8 conserves the duality properties, the following corollary immediately follows.

Corollary 9.12 (Orthogonal Self-dualities of the KMP Model and of the Kac Model)

1. *The function* D *in* (9.89) *is a self-duality function for the Kac model with generator* (8.42).
2. *The function* D *in* (9.90) *is a self-duality function thermalized* BEP(α) *(or generalized KMP process) of Definition 8.9*
3. *The function* D *in* (9.91) *is a self-duality function for the averaging process of Definition 8.1.*

9.5 Structural Relations of Hypergeometric Functions

The first method we present to prove the orthogonal dualities stated in Sect. 9.2 is based on explicit computations. The proof relies on the *hypergeometric structure* of the duality functions. Similarly to what we did in Sect. 9.2 for the Charlier polynomials, the idea here is to use three structural properties of classical orthogonal polynomials: the first is the differential or difference hypergeometric equation they satisfy; the second is the three-term recurrence relation; the third is the expression for the raising operator, that can be derived by using Rodriguez formula. These three properties provide three identities that can be directly used to express the action of the generator and verify the duality relation.

We denote the discrete left and right derivatives as

$$(\nabla_l)p_n(x) = p_n(x) - p_n(x-1),$$
$$(\nabla_r)p_n(x) = p_n(x+1) - p_n(x), \tag{9.93}$$

and the discrete Laplacian as

$$\Delta p_n(x) = p_n(x+1) - 2p_n(x) + p_n(x-1). \tag{9.94}$$

Then the three properties of classical (hypergeometric) orthogonal polynomials can be described in general terms as follows:

1. A *difference equation* for discrete variables

$$\sigma_n(x)\Delta p_n(x) + \tau_n(x)(\nabla_r)p_n(x) + \lambda_n p_n(x) = 0, \tag{9.95}$$

   or a *differential equation* for continuous variables

$$\sigma_n(x)\frac{d^2}{dx^2}p_n(x) + \tau_n(x)\frac{d}{dx}p_n(x) + \lambda_n p_n(x) = 0, \tag{9.96}$$

   where the sequences $\sigma_n(x)$ and $\tau_n(x)$ are polynomials in the x variable of at most second and first degree, respectively, and λ_n is a sequence of constants.
2. A *three-term recurrence relation* of the form

$$xp_n(x) = \alpha_n p_{n+1}(x) + \beta_n p_n(x) + \gamma_n p_{n-1}(x), \tag{9.97}$$

   for a sequence of coefficients α_n, β_n, γ_n.
3. A *raising operator*

$$R_n p_n(x) = r_n p_{n+1}(x), \tag{9.98}$$

   where the operator R_n on the left side acts on the x variable. This operator is defined from the Rodrigues formula, that yields a relation for the first difference operator of polynomials $p_n(x)$ in terms of the polynomials themselves.

In the remaining of this section we verify the self-dualities and dualities described in Theorems 9.5, 9.7, and 9.10.

9.5.1 Proof of Theorem 9.5

The independent random walkers case was already proved in Sect. 9.2. We provide here full details for the symmetric inclusion process orthogonal self-duality, whereas for the symmetric partial exclusion process we only write the basic identities used in the proof. As usual, because of the product form of the self-duality and duality functions, it is enough to consider a two-site system.

Orthogonal Self-duality of the Symmetric Inclusion Process, SIP(α)

We recall from [153] some identities for the Meixner polynomials defined in (9.72) (we shorthand here $M_n(x)$ for $M_n(x; \alpha, p)$. They satisfy the difference equation

$$x\,[M_n(x+1) - 2M_n(x) + M_n(x-1)] \\ + (\alpha p - x + xp)\,[M_n(x+1) - M_n(x)] + n(1-p)M_n(x) = 0\,, \tag{9.99}$$

and their three-term recurrence relation read

$$xM_n(x) = \frac{p(n+\alpha)}{p-1} M_{n+1}(x) - \frac{n + p(n+\alpha)}{p-1} M_n(x) + \frac{n}{p-1} M_{n-1}(x)\,. \tag{9.100}$$

From the Rodrigues formula

$$M_n(x) = \frac{x!}{(\alpha)_x\, p^x} (\nabla_l)^n \left[\frac{(\alpha+n)_x\, p^x}{x!}\right], \tag{9.101}$$

one can extract the raising operator

$$[p(n+\alpha+x)]M_n(x) - xM_n(x-1) = p(n+\alpha)M_{n+1}(x)\,. \tag{9.102}$$

The action of the generator Symmetric Inclusion Process SIP(α) working on the self-duality function for two sites is given by

$$\begin{aligned} &L_{12}D_{12}(n_1, n_2; \cdot, \cdot)(x_1, x_2) \\ &= x_1(\alpha + x_2)\left[M_{n_1}(x_1 - 1)M_{n_2}(x_2 + 1) - M_{n_1}(x_1)M_{n_2}(x_2)\right] \\ &+ (\alpha + x_1)x_2\left[M_{n_1}(x_1 + 1)M_{n_2}(x_2 - 1) - M_{n_1}(x_1)M_{n_2}(x_2)\right]. \end{aligned} \tag{9.103}$$

We see that we need an expression for the following terms:

$$xM_n(x)\,, \qquad xM_n(x-1)\,, \qquad (\alpha + x)M_n(x+1)\,. \tag{9.104}$$

in terms of $M_{n-1}(x)$, $M_n(x)$, $M_{n+1}(x)$. To get those, we simply perform algebraic manipulations on Eqs. (9.99), (9.100), (9.102). We then arrive at

$$\begin{aligned} xM_n(x) &= \frac{p(\alpha+n)}{p-1}M_{n+1}(x) - \frac{n+p(n+\alpha)}{p-1}M_n(x) \\ &\quad + \frac{n}{p-1}M_{n-1}(x), \\ xM_n(x-1) &= \frac{p(\alpha+n)}{p-1}M_{n+1}(x) - \frac{p(\alpha+2n)}{p-1}M_n(x) \\ &\quad + \frac{p}{p-1}nM_{n-1}(x), \\ (\alpha+x)M_n(x+1) &= \frac{p}{p-1}(\alpha+n)M_{n+1}(x) - \frac{\alpha+2n}{p-1}M_n(x) + \frac{1}{p-1}nM_{n-1}(x). \end{aligned} \tag{9.105}$$

We notice that on the left hand side we have operators that act on the x variable, whereas on the right hand side we have operators acting on the n variable. These relations allow us to expand the SIP(α) generator in Eq. (9.103) as

$$\begin{aligned} &L_{12}D_{12}(n_1, n_2; \cdot, \cdot)(x_1, x_2) \\ &= \left[\frac{p(\alpha+n_1)}{p-1}M_{n_1+1}(x_1) - \frac{p(\alpha+2n_1)}{p-1}M_{n_1}(x_1) + \frac{pn_1}{p-1}M_{n_1-1}(x_1)\right] \times \\ &\left[\frac{p(\alpha+n_2)}{p-1}M_{n_2+1}(x_2) - \frac{\alpha+2n_2}{p-1}M_{n_2}(x_2) + \frac{n_2}{p-1}M_{n_2-1}(x_2)\right] \\ &- \left[\frac{p(\alpha+n_1)}{p-1}M_{n_1+1}(x_1) - \frac{n_1+p(n_1+\alpha)}{p-1}M_{n_1}(x_1) + \frac{n_1}{p-1}M_{n_1-1}(x_1)\right] \times \\ &\left[\frac{p(\alpha+n_2)}{p-1}M_{n_2+1}(x_2) - \frac{n_2+p(n_2+\alpha)}{p-1}M_{n_2}(x_2)\right. \\ &\left. + \frac{n_2}{p-1}M_{n_2-1}(x_2) + \alpha M_{n_2}(x_2)\right] \\ &+ \left[\frac{p(\alpha+n_2)}{p-1}M_{n_2+1}(x_2) - \frac{p(\alpha+2n_2)}{p-1}M_{n_2}(x_2) + \frac{pn_2}{p-1}M_{n_2-1}(x_2)\right] \times \\ &\left[\frac{p(\alpha+n_1)}{p-1}M_{n_1+1}(x_1) - \frac{\alpha+2n_1}{p-1}M_{n_1}(x_1) + \frac{n_1}{p-1}M_{n_1-1}(x_1)\right] \\ &- \left[\frac{p(\alpha+n_2)}{p-1}M_{n_2+1}(x_2) - \frac{n_2+p(n_2+\alpha)}{p-1}M_{n_2}(x_2) + \frac{n_2}{p-1}M_{n_2-1}(x_2)\right] \times \end{aligned}$$

$$\left[\frac{p(\alpha+n_1)}{p-1}M_{n_1+1}(x_1)-\frac{n_1+p(n_1+\alpha)}{p-1}M_{n_1}(x_1)\right.$$
$$\left.+\frac{n_1}{p-1}M_{n_1-1}(x_1)+\alpha M_{n_1}(x_1)\right] \tag{9.106}$$

Simplifying further this expression, a long but straightforward computation shows that only products of polynomials with degree $n_1 + n_2$ survive in the expression above. More precisely, after simplifications, one is left with

$$\begin{aligned}
&L_{12}D_{12}(n_1, n_2; \cdot, \cdot)(x_1, x_2)\\
&= n_1(\alpha+n_2)\left[M_{n_1-1}(x_1)M_{n_2+1}(x_2)-M_{n_1}(x_1)M_{n_2}(x_2)\right]\\
&+ (\alpha+n_1)n_2\left[M_{n_1+1}(x_1)M_{n_2-1}(x_2)-M_{n_1}(x_1)M_{n_2}(x_2)\right]\\
&= L_{12}D_{12}(\cdot, \cdot, x_1, x_2)(n_1, n_2),
\end{aligned} \tag{9.107}$$

which is exactly the self-duality relation with product of Meixner polynomials as orthogonal self-duality function.

Orthogonal Self-duality of Symmetric Partial Exclusion Process, SEP(α)

The structural identities of the Krawtchouk polynomials $K_n(x)$ give [153]

$$\begin{aligned}
xK_n(x) &= -p(\alpha-n)K_{n+1}(x)+(n+2pj-2pn)K_n(x)\\
&\quad -n(1-p)K_{n-1}(x)\\
xK_n(x-1) &= -p(\alpha-n)K_{n+1}(x)+p(\alpha-2n)K_n(x)\\
&\quad +npK_{n-1}(x)\\
(\alpha-x)K_n(x+1) &= p(\alpha-n)K_{n+1}(x)+(1-p)(\alpha-2n)K_n(x)\\
&\quad -\frac{n(1-p)^2}{p}K_{n-1}(x).
\end{aligned} \tag{9.108}$$

Inserting these expressions into the generator it is possible to verify the self-duality of the symmetric partial exclusion process by proceeding similarly to what has been done for the self-duality of the symmetric inclusion process. □

9.5.2 *Proof of Theorem 9.7*

Orthogonal Duality Between the Brownian Momentum Process and the Symmetric Inclusion Process with Parameter $\alpha = \frac{1}{2}$

From [153] we know that the Hermite polynomials $H_n(x)$ satisfy the differential equation

$$H_n''(x) - 2xH_n'(x) + 2nH_n(x) = 0, \tag{9.109}$$

and the recurrence relation

$$H_{n+1}(x) - 2xH_n(x) + 2nH_{n-1}(x) = 0\ . \tag{9.110}$$

It is also possible to define a raising operator

$$2xH_n(x) - H_n'(x) = H_{n+1}(x). \tag{9.111}$$

It is convenient to introduce the single site duality function

$$d_n(x) = \frac{1}{(2n-1)!!} H_{2n}(x). \tag{9.112}$$

The action of the 2-site BMP generator on the duality function then reads

$$L_{12}D_{12}(n_1, n_2; \cdot, \cdot)(x_1, x_2) = \frac{1}{4}\left(x_1\frac{\partial}{\partial x_2} - x_2\frac{\partial}{\partial x_1}\right)^2 d_{n_1}(x_1)d_{n_2}(x_2). \tag{9.113}$$

This can be expanded into

$$\begin{aligned} 4L_{12}D_{12}(n_1, n_2; \cdot, \cdot)(x_1, x_2) &= x_1^2 d_{n_1}(x_1)d_{n_2}''(x_2) + d_{n_1}''(x_1)x_2^2 d_{n_2}(x_2) \\ &\quad - x_1 d_{n_1}'(x_1)d_{n_2}(x_2) - d_{n_1}(x_1)x_2 d_{n_2}'(x_2) \\ &\quad - 2x_1 d_{n_1}'(x_1)x_2 d_{n_2}'(x_2), \end{aligned} \tag{9.114}$$

where the prime denotes derivative with respect to the argument. We now need the identities appropriately rewritten in term of the single site duality function $d_n(x)$ in order to get suitable expression for

$$x^2 d_n(x), \qquad d_n''(x), \qquad x d_n'(x). \tag{9.115}$$

From (9.109) (9.110) (9.111) we find

$$
\begin{aligned}
x^2 d_n(x) &= \frac{1}{4}(2n+1)d_{n+1}(x) + \left(2n+\frac{1}{2}\right)d_n(x) + 2nd_{n-1}(x), \\
d_n''(x) &= 8nd_{n-1}(x), \\
xd_n'(x) &= 2nd_n(x) + 4nd_{n-1}(x).
\end{aligned}
\tag{9.116}
$$

Proceeding with the substitution into the generator we find, after appropriate simplification of the terms whose degree is different from $n_1 + n_2$,

$$
\begin{aligned}
&L_{12}D_{12}(n_1, n_2; \cdot, \cdot)(x_1, x_2) \\
&= n_1\left(n_2+\frac{1}{2}\right)\left[d_{n_1-1}(x_1)d_{n_2+1}(x_2) - d_{2n_1}(x_1)d_{n_2}(x_2)\right] \\
&+ n_2\left(n_1+\frac{1}{2}\right)\left[d_{n_1+1}(x_1)d_{n_2-1}(x_2) - d_{n_1}(x_1)d_{n_2}(x_2)\right],
\end{aligned}
\tag{9.117}
$$

which proves the claimed duality.

Orthogonal Duality Between the Brownian Energy Process and the Symmetric Inclusion Process

From [153] we know that the generalized Laguerre polynomials $L_n^{(\alpha-1)}(x)$ satisfy the differential equation

$$
x\frac{d^2}{dx^2}L_n^{(\alpha-1)}(x) + (\alpha - x)\frac{d}{dx}L_n^{(\alpha-1)}(x) + nL_n^{(\alpha-1)}(x) = 0, \tag{9.118}
$$

and the recurrence relation

$$
xL_n^{(\alpha-1)}(x) = -(n+1)L_{n+1}^{(\alpha-1)}(x) + (2n+\alpha)L_n^{(\alpha-1)}(x) - (n+\alpha-1)L_{n-1}^{(\alpha-1)}(x). \tag{9.119}
$$

Furthermore one has a raising operator given by

$$
(\alpha - x + n)L_n^{(\alpha-1)}(x) + x\frac{d}{dx}L_n^{(\alpha-1)}(x) = (n+1)L_{n+1}^{(\alpha-1)}(x)\,. \tag{9.120}
$$

We define the single site duality function

$$d_n(x) = \frac{n!\,\Gamma(\alpha)}{\Gamma(\alpha+n)} L_n^{(\alpha-1)}(x), \tag{9.121}$$

so that, after simple manipulation

$$\begin{aligned} x d_n''(x) + (\alpha - x) d_n'(x) + n d_n(x) &= 0, \\ x d_n(x) = -(n+\alpha) d_{n+1}(x) + (2n+\alpha) d_n(x) &- n d_{n-1}(x), \\ x d_n'(x) = n d_n(x) &- n d_{n-1}(x) \,. \end{aligned} \tag{9.122}$$

Using these identities in the writing of generator L of the Brownian Energy Process one finds

$$\begin{aligned} L_{12} D_{12}(n_1, n_2; \cdot, \cdot)(x_1, x_2) = {} & n_1\,(n_2+\alpha)\left[d_{n_1-1}(x_1) d_{n_2+1}(x_2) - d_{2n_1}(x_1) d_{n_2}(x_2)\right] \\ & + n_2\,(n_1+\alpha)\left[d_{n_1+1}(x_1) d_{n_2-1}(x_2) - d_{n_1}(x_1) d_{n_2}(x_2)\right], \end{aligned} \tag{9.123}$$

which proves the claimed duality. □

9.5.3 *Proof of Theorem 9.10*

Self-duality of the Brownian Energy Process

We define the single site duality function

$$d_\alpha(v, z) = {}_0F_1\left(\begin{matrix} - \\ \alpha \end{matrix}\,; -\frac{vz}{4}\right) = \Gamma(\alpha) \left(\frac{vz}{2}\right)^{-\frac{\alpha}{2}+\frac{1}{2}} J_{\alpha-1}(\sqrt{vz}). \tag{9.124}$$

We have an identity for the partial derivatives

$$\frac{\partial}{\partial z} d_\alpha(v, z) = -\frac{1}{4}\frac{v}{\alpha} d_{\alpha+1}(v, z), \qquad \frac{\partial}{\partial v} d_\alpha(v, z) = -\frac{1}{4}\frac{z}{\alpha} d_{\alpha+1}(v, z), \tag{9.125}$$

and a recurrence identity

$$d_{\alpha+1}(v, z) = d_\alpha(v, z) + \frac{1}{4}\frac{vz}{\alpha(\alpha+1)} d_{\alpha+2}(v, z). \tag{9.126}$$

By using these two identities one can verify that

$$\left\{-\alpha(z_1 - z_2)\left(\frac{\partial}{\partial z_1} - \frac{\partial}{\partial z_2}\right) + z_1 z_2 \left(\frac{\partial}{\partial z_1} - \frac{\partial}{\partial z_2}\right)^2\right\} d_\alpha(v_1, z_1) d_\alpha(v_2, z_2)$$

$$= \left\{-\alpha(v_1 - v_2)\left(\frac{\partial}{\partial v_1} - \frac{\partial}{\partial v_2}\right) + v_1 v_2 \left(\frac{\partial}{\partial v_1} - \frac{\partial}{\partial v_2}\right)^2\right\} d_\alpha(v_1, z_1) d_\alpha(v_2, z_2).$$

Self-duality of the Brownian Momentum Process

This case can be treated similarly to the previous one by considering the single site duality function specialized to $\alpha = 1/2$ with $z = x^2$ and $v = y^2$, which gives

$$d_\alpha(y, x) = {}_0F_1\left(\begin{matrix} - \\ \frac{1}{2} \end{matrix} ; -\frac{y^2 x^2}{4} \right) = \sqrt{|xy|} J(xy; -\frac{1}{2}) = \sqrt{\frac{2}{\pi}} \cos(xy). \qquad (9.127)$$

Self-duality of the Deterministic Process

This self-duality of the deterministic process, with single site self-duality function $d(v, z) = e^{vz}$ can be immediately verified via a simple direct computation. □

9.6 Orthogonal Duality as a Change of Representation of the Lie Algebra

In this section we show that the single-site orthogonal duality are emerge from two representation of a Lie algebra. We collect these dualities in a sequence of propositions, for the Heisenberg algebra, the $\mathfrak{su}(1, 1)$ Lie algebra and the $\mathfrak{su}(2)$ Lie algebra. From these fundamental dualities one can recover the dualities of the Markov processes described above.

We will provide a full proof of this duality at the level of the algebra for the Heisenberg case, with Charlier polynomials as orthogonal duality functions, and thus obtain the self-duality of independent random walkers with the products of Charlier polynomials as self-duality function. The proofs for the other algebras with the appropriate orthogonal polynomials are completely analogous, based on recurrence relations of the polynomials combined with the composition of dualities, and are therefore left to the reader.

Proposition 9.13 (Charlier Polynomials Self-duality of Independent Random Walkers as a Change of Representation) *Consider a representation of the conjugate Heisenberg algebra given by operators $a, a^\dagger$ defined on functions f :*

$\mathbb{N} \to \mathbb{R}$ by

$$\left(a^\dagger f\right)(x) = f(x+1),$$
$$(af)(x) = xf(x-1), \tag{9.128}$$

where $f(-1) = 0$, satisfying $[a, a^\dagger] = -I$. Consider a representation of the Heisenberg algebra given by operators A, $A^\dagger$ defined on functions $f : \mathbb{N} \to \mathbb{R}$ by

$$\left(A^\dagger f\right)(n) = f(n) - \frac{n}{\lambda} f(n-1),$$
$$(Af)(n) = \lambda f(n) - \lambda f(n+1), \tag{9.129}$$

where $f(-1) = 0$, satisfying $[A, A^\dagger] = I$. Then the operators in (9.128) *and in* (9.129) *are in duality relation*

$$a \xrightarrow{d} A,\ a^\dagger \xrightarrow{d} A^\dagger, \tag{9.130}$$

via

$$d(x, n) = e^\lambda C_n(x; \lambda)\ , \tag{9.131}$$

where $C_n(x; \lambda)$ are the Charlier polynomials. As a consequence independent random walkers are self-dual, with self-duality function

$$D(\xi, \eta) = \prod_{x \in V} d(\xi_x, \eta_x). \tag{9.132}$$

Proof The proof of the dualities $a \xrightarrow{d} A,\ a^\dagger \xrightarrow{d} A^\dagger$ is a consequence of the properties of Charlier polynomials $C_n(x; \lambda)$. More precisely $a \xrightarrow{d} A$ follows from (9.30) and $a^\dagger \xrightarrow{d} A^\dagger$ follows from (9.32).

It then suffices to show the self-duality of independent random walks for the two-site case, which in turn follows by composition of dualities. Indeed, observe that

$$A = -\lambda a^\dagger + \lambda I, \tag{9.133}$$

$$A^\dagger = I - \frac{1}{\lambda} a. \tag{9.134}$$

This implies

$$(A_1^\dagger - A_2^\dagger)(A_1 - A_2) = (a_1 - a_2)(a_1^\dagger - a_2^\dagger). \tag{9.135}$$

On the other hand we know that

$$L_{1,2} = -(a_1 - a_2)(a_1^\dagger - a_2^\dagger). \tag{9.136}$$

Therefore, if we define

$$D((x_1, x_2), (n_1, n_2)) = d(x_1, n_1)d(x_2, n_2), \tag{9.137}$$

then by the general composition rule for dualities (see Theorem 2.23), we obtain

$$L_{1,2} = -(a_1 - a_2)(a_1^\dagger - a_2^\dagger) \xrightarrow{D} -(A_1^\dagger - A_2^\dagger)(A_1 - A_2) = L_{1,2}. \tag{9.138}$$

□

Proposition 9.14 (Meixner Polynomials Self-duality of Symmetric Inclusion Process as a Change of Representation) *Consider a representation of the conjugate $\mathfrak{su}(1,1)$ Lie algebra given by operators K^+, K^-, K^0 defined on functions $f : \mathbb{N} \to \mathbb{R}$*

$$\begin{cases} \left(K^+ f\right)(x) &= (\alpha + x) f(x+1), \\ \left(K^- f\right)(x) &= x f(x-1), \\ \left(K^0 f\right)(x) &= (x + \frac{\alpha}{2}) f(x), \end{cases} \tag{9.139}$$

where $f(-1) = 0$, satisfying

$$[K^0, K^\pm] = \mp K^\pm \qquad \text{and} \qquad [K^+, K^-] = 2K^0 \,. \tag{9.140}$$

Consider a representation of the $\mathfrak{su}(1,1)$ Lie algebra given by the operators $\mathscr{K}^+, \mathscr{K}^-, \mathscr{K}^0$ defined on functions $f : \mathbb{N} \to \mathbb{R}$ by

$$\begin{cases} \left(\mathscr{K}^+ f\right)(n) &= \frac{p}{p-1}(\alpha + n) f(n+1) - \frac{1}{p-1}(\alpha + 2n) f(n) \\ &\quad + \frac{1}{p-1} n f(n-1), \\ \left(\mathscr{K}^- f\right)(n) &= \frac{p}{p-1}(\alpha + n) f(n+1) - \frac{p}{p-1}(\alpha + 2n) f(n) \\ &\quad + \frac{p}{p-1} n f(n-1), \\ \left(\mathscr{K}^0 f\right)(n) &= \frac{p}{p-1}(\alpha + n) f(n+1) - \frac{(1+p)(n + \frac{\alpha}{2})}{p-1} f(n) \\ &\quad + \frac{n}{p-1} f(n-1), \end{cases} \tag{9.141}$$

where $f(-1) = 0$*, satisfying*

$$[\mathcal{K}^0, \mathcal{K}^\pm] = \pm\mathcal{K}^\pm \qquad \text{and} \qquad [\mathcal{K}^+, \mathcal{K}^-] = -2\mathcal{K}^0 \,. \tag{9.142}$$

Then the operators in (9.139) *and in* (9.141) *are in duality relation*

$$K^+ \xrightarrow{d} \mathcal{K}^+, \quad K^- \xrightarrow{d} \mathcal{K}^-, \quad K^0 \xrightarrow{d} \mathcal{K}^0, \tag{9.143}$$

via

$$d(n, x) = (1-p)^{-\alpha} M_n(x; \alpha, p), \tag{9.144}$$

where $M_n(x; \alpha, p)$ *are the Meixner polynomials. As a consequence the symmetric inclusion process is self-dual, with self-duality function*

$$D(\xi, \eta) = \prod_{x \in V} d(\xi_x, \eta_x). \tag{9.145}$$

Proof The proof of the dualities (9.143) is a consequence of the properties of Meixener polynomials $M_n(x; \alpha, p)$. The self-duality of the symmetric inclusion process follows by composition of dualities. □

Proposition 9.15 (Krawtchouk Polynomials Self-duality of Symmetric Partial Exclusion Process as a Change of Representation) *Consider a representation of the conjugate* $\mathfrak{su}(2)$ *Lie algebra given by operators* J^+, J^-, J^0 *defined on functions* $f : \mathbb{N} \to \mathbb{R}$

$$\begin{cases} \left(J^+ f\right)(x) & = (\alpha - x) f(x+1), \\ \left(J^- f\right)(x) & = x f(x-1), \\ \left(J^0 f\right)(x) & = (x - \frac{\alpha}{2}) f(x), \end{cases} \tag{9.146}$$

where $f(-1) = f(\alpha+1) = 0$*, satisfying*

$$[J^0, J^\pm] = \mp J^\pm \qquad \text{and} \qquad [J^+, J^-] = -2J^0. \tag{9.147}$$

Consider a representation of the $\mathfrak{su}(2)$ *Lie algebra given by operators* $\mathcal{J}^+, \mathcal{J}^-, \mathcal{J}^0$ *defined on functions* $f : \mathbb{N} \to \mathbb{R}$ *by*

$$\begin{cases} \left(\mathcal{J}^+ f\right)(n) & = p(\alpha - n) f(n+1) + (1-p)(\alpha - 2n) f(n) \\ & \quad -\dfrac{n}{p}(1-p)^2 f(n-1), \\ \left(\mathcal{J}^- f\right)(n) & = p(\alpha - n) f(n+1) + p(\alpha - 2n) f(n) + npf(n-1), \\ \left(\mathcal{J}^0 f\right)(n) & = -p(\alpha - n) f(n+1) + (n - \frac{\alpha}{2})(1-2p) f(n) \\ & \quad -n(1-p) f(n-1), \end{cases} \tag{9.148}$$

where $f(-1) = f(\alpha + 1) = 0$*, satisfying*

$$[\mathcal{J}^0, \mathcal{J}^{\pm}] = \pm\mathcal{J}^{\pm} \qquad \text{and} \qquad [\mathcal{J}^+, \mathcal{J}^-] = 2\mathcal{J}^0. \tag{9.149}$$

Then the operators in (9.146) *and in* (9.148) *are in duality relation*

$$J^+ \xrightarrow{d} \mathcal{J}^+, \quad J^- \xrightarrow{d} \mathcal{J}^-, \quad J^0 \xrightarrow{d} \mathcal{J}^0, \tag{9.150}$$

via

$$d(n, x) = (1 - p)^{\alpha} K_n(x; \alpha, p), \tag{9.151}$$

where $K_n(x; \alpha, p)$ *is the Krawtchouk polynomials. As a consequence the symmetric inclusion process is self-dual, with self-duality function*

$$D(\xi, \eta) = \prod_{x \in V} d(\xi_x, \eta_x). \tag{9.152}$$

Proof The proof of the dualities (9.150) is a consequence of the properties of Meixener polynomials $K_n(x; \alpha, p)$. The self-duality of the symmetric partial exclusion process follows by composition of dualities. □

Proposition 9.16 (Laguerre Polynomials Duality Between BEP(α) **and** SIP(α) **as a Change of Representation)** *Consider a representation of the conjugate* $\mathfrak{su}(1,1)$ *Lie algebra given by operators* K^+, K^-, K^0 *defined on functions* $f : \mathbb{N} \to \mathbb{R}$

$$\begin{cases} (K^+ f)(n) &= -(n+\alpha) f(n+1) + 2(n + \frac{\alpha}{2}) f(n) - n f(n-1), \\ (K^- f)(n) &= -n f(n-1), \\ (K^0 f)(n) &= (n + \frac{\alpha}{2}) f(n) - n f(n-1), \end{cases} \tag{9.153}$$

where $f(-1) = 0$*, satisfying*

$$[K^0, K^{\pm}] = \mp K^{\pm} \qquad \text{and} \qquad [K^+, K^-] = 2K^0. \tag{9.154}$$

Consider a representation of the $\mathfrak{su}(1,1)$ *Lie algebra given by the operators* $\mathscr{K}^+, \mathscr{K}^-, \mathscr{K}^0$ *defined on smooth functions* $f : (0, \infty) \to \mathbb{R}$ *by*

$$\begin{cases} \mathscr{K}^+ &= z, \\ \mathscr{K}^- &= z\frac{\partial^2}{\partial z^2} + \alpha\frac{\partial}{\partial z}, \\ \mathscr{K}^0 &= z\frac{\partial}{\partial z} + \frac{\alpha}{2}, \end{cases} \tag{9.155}$$

satisfying

$$[\mathscr{K}^0, \mathscr{K}^{\pm}] = \pm\mathscr{K}^{\pm} \qquad \text{and} \qquad [\mathscr{K}^+, \mathscr{K}^-] = -2\mathscr{K}^0. \tag{9.156}$$

Then the operators in (9.153) *and in* (9.155) *are in duality relation*

$$K^+ \xrightarrow{d} \mathscr{K}^+, \quad K^- \xrightarrow{d} \mathscr{K}^-, \quad K^0 \xrightarrow{d} \mathscr{K}^0, \tag{9.157}$$

via

$$d(n, z) = \frac{n!\,\Gamma(\alpha)}{\Gamma(\alpha+n)} L_n(z; \alpha - 1) = {}_1F_1\left(\begin{matrix} -n \\ \alpha \end{matrix} \middle| z\right), \tag{9.158}$$

where $L_n(z; \alpha)$ are the Laguerre polynomials. As a consequence there is duality between the Brownian energy process and the symmetric inclusion process, with duality function

$$D(\xi, \zeta) = \prod_{x\in V} d(\xi_x, \zeta_x). \tag{9.159}$$

Proof The proof of the dualities (9.157) is a consequence of the properties of Laguerre polynomials $L_n(z; \alpha)$. The duality between the Brownian energy process and the symmetric inclusion process follows by composition of dualities. □

Proposition 9.17 (Hermite Polynomials Duality Between BMP **and** SIP($\frac{1}{2}$) **as a Change of Representation)** *Consider a representation of the conjugate $\mathfrak{su}(1,1)$ Lie algebra given by operators K^+, K^-, K^0 defined on functions $f : \mathbb{N} \to \mathbb{R}$*

$$\begin{cases} \left(K^+ f\right)(n) &= \frac{2n+1}{8} f(n+1) + \left(n + \frac{1}{4}\right) f(n) - nf(n-1, \\ \left(K^- f\right)(n) &= 4nf(n-1), \\ \left(K^0 f\right)(n) &= (n + \frac{1}{4}) f(n) + 2nf(n-1), \end{cases} \tag{9.160}$$

where $f(-1) = 0$, satisfying

$$[K^0, K^{\pm}] = \mp K^{\pm} \qquad \text{and} \qquad [K^+, K^-] = 2K^0. \tag{9.161}$$

Consider a representation of the $\mathfrak{su}(1,1)$ Lie algebra given by the operators $\mathscr{K}^+, \mathscr{K}^-, \mathscr{K}^0$ defined on smooth functions $f : (0, \infty) \to \mathbb{R}$ by

$$\begin{cases} \mathscr{K}^+ &= \frac{1}{2} z^2, \\ \mathscr{K}^- &= \frac{1}{2} \frac{\partial^2}{\partial z^2}, \\ \mathscr{K}^0 &= \frac{1}{2} z \frac{\partial}{\partial z} + \frac{1}{4}, \end{cases} \tag{9.162}$$

satisfying

$$[\mathscr{K}^0, \mathscr{K}^\pm] = \pm\mathscr{K}^\pm \qquad \textit{and} \qquad [\mathscr{K}^+, \mathscr{K}^-] = -2\mathscr{K}^0. \tag{9.163}$$

Then the operators in (9.160) *and in* (9.162) *are in duality relation*

$$K^+ \xrightarrow{d} \mathscr{K}^+, \quad K^- \xrightarrow{d} \mathscr{K}^-, \quad K^0 \xrightarrow{d} \mathscr{K}^0, \tag{9.164}$$

via

$$d(n, z) = \frac{H_{2n}(z)}{(2n-1)!!}, \tag{9.165}$$

where $H_n(z)$ *are the Hermite polynomials. As a consequence there is duality between the Brownian momentum process and the symmetric inclusion process with parameter* $\frac{1}{2}$, *with duality function*

$$D(\xi, \zeta) = \prod_{x \in V} d(\xi_x, \zeta_x). \tag{9.166}$$

Proof The proof of the dualities (9.164) is a consequence of the properties of Hermite polynomials $H_n(z)$. The duality between the Brownian momentum process and the symmetric inclusion process follows by composition of dualities. □

9.7 Orthogonal Duality and Unitary Intertwining

In Chap. 2, we have seen that the notion of duality between two operators on an L_2 space is intimately connected to that of intertwining in kernel form. In this section we shall see that an orthogonal duality is associated to a *unitary intertwiner*.

In the context of representation of algebras, where the two operators arise as representations of an element of the algebra, an invertible intertwiner defines the notion of equivalent representations. Thus, when the intertwiner constructed from orthogonal duality is invertible, then we have that orthogonal duality is associated to *unitary equivalent representations*.

Let us start by recalling the relation between duality of bounded operators, intertwining and equivalent representations. We have seen in Theorem 2.29 that if $\widehat{A} : \widehat{\mathscr{H}} \to \widehat{\mathscr{H}}$ and $A : \mathscr{H} \to \mathscr{H}$ are two operators on two Hilbert spaces $\widehat{\mathscr{H}} = L^2(\widehat{\Omega}, \widehat{\mu})$ and $\mathscr{H} = L^2(\Omega, \mu)$ such that the operators $(\widehat{A})^*$ (= adjoint of $\widehat{A}$ in $\widehat{\mathscr{H}}$) and A satisfy a duality relation, i.e.

$$\big((\widehat{A})^* D(\cdot, x)\big)(y) = \big(A D(y, \cdot)\big)(x) \qquad \forall (y, x) \in \widehat{\Omega} \times \Omega, \tag{9.167}$$

then the integral operator $\Lambda : \widehat{\mathcal{H}} \to \mathcal{H}$ defined by

$$(\Lambda f)(x) = \int_{\widehat{\Omega}} D(y, x) f(y) d\widehat{\mu}(y), \tag{9.168}$$

is an intertwiner between the two operators A and $\widehat{A}$, i.e.

$$(\Lambda \widehat{A} f)(x) = (A \Lambda f)(x). \tag{9.169}$$

Suppose now that a Lie algebra $\mathfrak{g}$ is given and consider two representations $\widehat{\rho}$ and ρ of $\mathfrak{g}$, so that to each element $X \in \mathfrak{g}$ there are associated linear operators $\widehat{\rho}(X) : \widehat{\mathcal{H}} \to \widehat{\mathcal{H}}$ and $\rho(X) : \mathcal{H} \to \mathcal{H}$. We recall that two representation are said to be *equivalent* if there exists an invertible intertwiner $\Lambda : \widehat{\mathcal{H}} \to \mathcal{H}$ such that for all $X \in \mathfrak{g}$

$$\rho(X)\Lambda = \Lambda \widehat{\rho}(X). \tag{9.170}$$

Thus, if the duality relation (9.167) is satisfied, with the same duality function $D(\cdot, \cdot)$, for all couples of operators $(\widehat{\rho}(X))^*$ and $\rho(X)$ with X in the set of the generators of the Lie algebra $\mathfrak{g}$, then, when the intertwiner in definition (9.168) is invertible, we conclude that $\widehat{\rho}$ and ρ are equivalent representations of the Lie algebra $\mathfrak{g}$.

The novel observation of this section, which is contained in the theorem that follows, is that if furthermore the duality function satisfies an orthogonality relation then we get a unitary intertwiner. If this is invertible, we then get unitary equivalent representations of the Lie algebra $\mathfrak{g}$.

Theorem 9.18 (Orthogonal Dualities and Unitary Intertwiner) *Suppose that $\widehat{A} : \widehat{\mathcal{H}} \to \widehat{\mathcal{H}}$ and $A : \mathcal{H} \to \mathcal{H}$ are two operators on two Hilbert spaces $\widehat{\mathcal{H}} = L^2(\widehat{\Omega}, \widehat{\mu})$ and $\mathcal{H} = L^2(\Omega, \mu)$. We assume $\widehat{\Omega}$ is discrete. Then the following two statements are equivalent:*

1. *A and $\widehat{A}$ have a* unitary *intertwiner with a kernel $D : \widehat{\Omega} \times \Omega \to \mathbb{R}$, i.e.*

$$\Lambda \widehat{A} = A \Lambda, \tag{9.171}$$

and

$$(\Lambda f)(x) = \sum_{y \in \widehat{\Omega}} D(y, x) f(y) \widehat{\mu}(y). \tag{9.172}$$

2. *$(\widehat{A})^*$ and A are in duality relation with* orthogonal *duality function $D : \widehat{\Omega} \times \Omega \to \mathbb{R}$, i.e.*

$$((\hat{A})^* \otimes I) D = (I \otimes A) D, \tag{9.173}$$

and

$$\int_{\Omega} D(y,x)D(y',x)\mu(dx) = \delta_{y,y'} \frac{1}{\widehat{\mu}(y)}. \tag{9.174}$$

Proof We already know from Theorem 2.29 the equivalence between intertwining in kernel form and duality. Thus we just need to check that the intertwining is unitary if and only if the duality function is orthogonal. For this we consider two function $f, g \in \widetilde{\mathcal{H}}$ and look at their scalar product. We have:

$$\begin{aligned}\langle \Lambda f, \Lambda g \rangle_{\mathcal{H}} &= \int_{\Omega} \mu(dx)\, \Lambda f(x)\, \Lambda g(x) \\ &= \int_{\Omega} \mu(dx) \left(\sum_{y \in \widehat{\Omega}} \widehat{\mu}(y) D(y,x) f(y) \right) \left(\sum_{y' \in \widehat{\Omega}} \widehat{\mu}(y') D(y',x) g(y) \right) \\ &= \sum_{y \in \widehat{\Omega}} \widehat{\mu}(y) f(y) \sum_{y' \in \widehat{\Omega}} \widehat{\mu}(y') g(y') \int_{\Omega} \mu(dx)\, D(y,x) D(y',x).\end{aligned} \tag{9.175}$$

Thus we see that the intertwiner Λ is unitary, i.e.

$$\langle \Lambda f, \Lambda g \rangle_{\mathcal{H}} = \langle f, g \rangle_{\widetilde{\mathcal{H}}}, \tag{9.176}$$

if and only if the orthogonality condition holds

$$\int_{\Omega} D(y,x)D(y',x)d\mu(x) = \delta_{y,y'} \frac{1}{\widehat{\mu}(y)}. \tag{9.177}$$

□

Combining together Theorem 2.11 (which states that in the case of self-dualities the intertwiner is a symmetry of the generator) with Theorem 9.18 (which states that the intertwiner constructed from orthogonal dualities is a unitary operator) we deduce that symmetries leading to orthogonal self-dualities have to be *unitary*. In what follows we single out the general expression those symmetries have for the three particle systems that we often studied in this book, namely inclusion, generalized exclusion and independent particles.

We start by recalling the form of the cheap (i.e. diagonal) self-dualities for our three processes. Notice that, up to negligible factors, the elements on the diagonal are the inverse of the weights of the reversible measure:

$$d^{\mathrm{ch}}(x,y) = \begin{cases} \dfrac{y!\Gamma(\alpha)}{\Gamma(\alpha+y)} p^{-y}\delta_{x,y} & \text{for the SIP}(\alpha), \\ \dfrac{(\alpha-y)!y!}{\alpha!}\left(\dfrac{1-p}{p}\right)^{y}\delta_{x,y} & \text{for the SEP}(\alpha), \\ y!\lambda^{-y}\delta_{x,y} & \text{for the IRW.} \end{cases} \tag{9.178}$$

Recall that a linear operator in $L_2(\Omega,\mu)$ is called unitary if $UU^* = U^*U = I$, where U^* is the adjoint of U. In the next theorem we provide the expression for the most general unitary symmetry that will then yield orthogonal duality functions. We also identify the special values of the parameters appearing in these symmetries for which the duality functions reduce to the orthogonal polynomials.

Theorem 9.19 (Orthogonal Self-duality Functions and Unitary Symmetries) *The following results holds.*

1. *For the* SIP(α) *process consider*

$$\begin{aligned} K^+ f(n) &= (\alpha+n) f(\eta+1), \\ K^- f(n) &= n f(n-1), \\ K^0 f(n) &= \left(\tfrac{\alpha}{2}+n\right) f(n), \end{aligned} \tag{9.179}$$

working on $f:\mathbb{N}\to\mathbb{R}$. *Then we have that:*

(a) *the symmetry*

$$S_{\beta,\gamma} = \exp\left(\beta\left(-K^+ + \frac{1}{p}K^-\right)\right)\exp\left(i\gamma K^0\right), \tag{9.180}$$

is unitary for every choice of $\beta,\gamma\in\mathbb{R}$. *As a consequence* $S_{\beta,\gamma}\left(d^{\mathrm{ch}}(x,\cdot)\right)(y)$ *are orthogonal (single site) self-duality functions;*

(b) *choosing* $\beta = \hat\beta := \sqrt{p}\,\mathrm{arctanh}\left(\sqrt{p}\right)$ *and* $\gamma = \hat\gamma := \pi$ *we get the Meixner polynomials up to a constant:* $S_{\hat\beta,\hat\gamma}\left(d^{\mathrm{ch}}(x,\cdot)\right)(y) = (p-1)^{\frac{\alpha}{2}} M(x,y;p)$.

2. *For the* SEP(α) *process consider*

$$\begin{aligned} J^+ f(n) &= (\alpha-n) f(n+1), \\ J^- f(n) &= n f(n-1), \\ J^0 f(n) &= \left(-\tfrac{\alpha}{2}+n\right) f(n), \end{aligned} \tag{9.181}$$

working on $f : \{0, 1, \ldots, \alpha\} \to \mathbb{R}$. Then we have that:

(a) *the symmetry*

$$S_{\beta,\gamma} = \exp\left(\beta\left(-J^+ + \frac{1-p}{p}J^-\right)\right)\exp\left(i\gamma J^0\right), \tag{9.182}$$

is unitary for every choice of $\beta, \gamma \in \mathbb{R}$. As a consequence $S_{\beta,\gamma}\left(d^{\mathrm{ch}}(x,\cdot)\right)(y)$ are orthogonal (single site) self-duality functions;

(b) *choosing $\beta = \hat{\beta} := \sqrt{\frac{p}{1-p}}\arctan\left(\sqrt{\frac{p}{1-p}}\right)$ and $\gamma = \hat{\gamma} := \pi$ we get the Krawtchouk polynomials up to a constant: $S_{\hat{\beta},\hat{\gamma}}\left(d^{\mathrm{ch}}(x,\cdot)\right)(y) = (p-1)^{\frac{\alpha}{2}}K(x,y;p)$.*

3. *For independent random walkers process consider*

$$\begin{aligned} a^\dagger f(n) &= f(n+1), \\ a^f(n) &= nf(n-1), \end{aligned} \tag{9.183}$$

working on $f : \mathbb{N} \to \mathbb{R}$. Then we have that:

(a) *the symmetry*

$$S_{\beta,\gamma} = \exp\left(\beta\left(-pa^\dagger + a\right)\right)\exp\left(i\gamma\, aa^\dagger\right), \tag{9.184}$$

is unitary for every choice of $\beta, \gamma \in \mathbb{R}$. As a consequence $S_{\beta,\gamma}\left(d^{\mathrm{ch}}(x,\cdot)\right)(y)$ are orthogonal (single site) self-duality functions;

(b) *Choosing $\beta = \hat{\beta} := 1$ and $\gamma = \hat{\gamma} := \pi$ we get the Charlier polynomials up to a constant: $S_{\hat{\beta},\hat{\gamma}}\left(d^{\mathrm{ch}}(x,\cdot)\right)(y) = e^{-\frac{\lambda}{2}}C(x,y;\lambda)$.*

Proof We only consider the $\mathfrak{su}(1,1)$ algebra and the SIP(α) process, for the other two processes the proof is similar.

The proof of item 1.(a) regarding the unitarity of $S_{\beta,\gamma}$ amounts to show that $S^*_{\beta,\gamma} = \left(S_{\beta,\gamma}\right)^{-1}$, where $S^*_{\beta,\gamma}$ is the adjoint of $S_{\beta,\gamma}$ in the Hilbert space $L_2(\mathbb{N},\mu)$ with $\mu(x) = \dfrac{\Gamma(\alpha+x)}{x!\Gamma(\alpha)}p^x$. One can check that

$$(K^0)^* = K^0, \qquad (K^+)^* = \frac{1}{p}K^-, \qquad (K^-)^* = pK^+. \tag{9.185}$$

This implies that

$$\left(S_{\beta,\gamma}\right)^* = \exp\left(-i\gamma K^0\right)\exp\left(\beta\left(-\frac{1}{p}K^- + K^+\right)\right) = \left(S_{\beta,\gamma}\right)^{-1}, \tag{9.186}$$

and so unitarity of $S_{\beta,\gamma}$ is proved.

The proof of item 1.(b) can be obtained using generating functions. For details we refer to [45]. □

Remark 9.20 A different expression for the three unitary symmetries $S_{\hat{\beta},\hat{\gamma}}$ of Theorem 9.19 can be given: it is a factorized expression for function of the algebra generators that one can show to be connected to the previous expression via the Baker-Campbel-Hausdorff formula. More precisely, we have that:

1. the $S_{\hat{\beta},\hat{\gamma}}$ in Eq. (9.180) can be rewritten as

$$S_{\hat{\beta},\hat{\gamma}} = e^{K^-} e^{\log(p-1)K^0} e^{pK^+}; \tag{9.187}$$

2. the $S_{\hat{\beta},\hat{\gamma}}$ in Eq. (9.182) can be rewritten as

$$S_{\hat{\beta},\hat{\gamma}} = e^{J^-} e^{\log\left(\frac{1}{p-1}\right)J^0} e^{\frac{p}{1-p}J^+}; \tag{9.188}$$

3. the $S_{\hat{\beta}}$ in Eq. (9.184) can be rewritten as

$$S_{\hat{\beta},\hat{\gamma}} = e^{a} e^{-\lambda/2+i\pi aa^\dagger} e^{\lambda a^\dagger}. \tag{9.189}$$

For more details, see [45].

9.8 Gram-Schmidt Orthogonalization

In this section we prove that the Gram-Schmidt procedure can be used to produce orthogonal duality functions starting from the "basic" (i.e. triangular) duality functions. In particular, for the self-dualities of the main examples (independent walkers, symmetric inclusion process, symmetric partial exclusion process) the Gram-Schmidt procedure applied to the triangular self-duality functions produces the orthogonal polynomials self-dualities.

We focus here on the case of independent random walkers on a finite set V. From the proof it will be clear that the extension to the other models is immediate, replacing the Poisson distribution by the suitable reversible product measure, and the single-site triangular self-duality function of independent walkers by the suitable single-site triangular self-duality function.

The triangular self-duality functions of independent random walkers are given by

$$D(\xi, \eta) = \prod_{x\in V} d(\xi_x, \eta_x), \tag{9.190}$$

where the single-site triangular duality polynomial equals

$$d(k,n) = \tfrac{n!}{(n-k)!}\mathbb{1}_{\{k\le n\}}. \tag{9.191}$$

Let us now fix a reversible product measure ν_ρ, which is a product of Poisson distributions with parameter ρ. Let us also call $d_\rho(k,n)$ the orthogonal polynomials obtained by Gramm-Schmidt orthogonalization of the single-site triangular self-duality polynomial, i.e., denoting $\langle\cdot,\cdot\rangle$ the inner product in $L^2(\nu_\rho)$ we define

$$d_\rho(0,n) = 1 \tag{9.192}$$

$$d_\rho(k,n) = d(k,n) - \sum_{j=0}^{n-1} \frac{\langle d(k,\cdot), d_\rho(j,\cdot)\rangle}{\langle d_\rho(j,\cdot), d_\rho(j,\cdot)\rangle} d_\rho(j,n), \tag{9.193}$$

for $k \ge 1$. We then call $\prod_{x\in V} d_\rho(\xi_x,\eta_x)$ the factorized orthogonal polynomials.

We consider the following subspaces of $L^2(\nu_\rho)$

$$\mathcal{V}_n = \overline{\mathrm{vct}\{D(\xi,\cdot) : |\xi| \le n\}}, \tag{9.194}$$

where the notation vct means the vector space spanned by, and $\overline{\mathcal{V}}$ refers to closure of $\mathcal{V}$ in $L^2(\nu_\rho)$. The subspaces $\mathcal{V}_n$ are increasing, i.e.,

$$\mathcal{V}_n \subset \mathcal{V}_{n+1}. \tag{9.195}$$

Let us denote $P_{\mathcal{V}_n}$ the orthogonal projection in $L^2(\nu_\rho)$ on $\mathcal{V}_n$. We have the following lemma showing that projections on $\mathcal{V}_n$ or $\mathcal{V}_n^\perp$ commute with the semigroup.

Lemma 9.21 *Let $S(t)$ denote the semigroup of the configuration process $\{\eta(t) : t \ge 0\}$ on $L^2(\nu_\rho)$. Then we have the following.*

1. *Invariance of the n-particle spaces:*

$$S(t)\mathcal{V}_n \subset \mathcal{V}_n. \tag{9.196}$$

2. *Invariance of the orthogonal complement of the n-particle spaces:*

$$S(t)\mathcal{V}_n^\perp \subset \mathcal{V}_n^\perp. \tag{9.197}$$

3. *Commutation of the semigroup with $P_{\mathcal{V}_n}$ and $P_{\mathcal{V}_n^\perp}$:*

$$[S(t), P_{\mathcal{V}_n}] = [S(t), P_{\mathcal{V}_n^\perp}] = 0. \tag{9.198}$$

Proof For item 1, by self duality with self-duality function D we have that for $\xi \in \Omega = \mathbb{N}^V$ with $|\xi| \le n$

$$[S(t)D(\xi, \cdot)](\eta) = \sum_{\xi' \in \Omega : |\xi'| = |\xi|} p_t(\xi, \xi') D(\xi', \eta) \in \mathcal{V}_n, \tag{9.199}$$

because the sum in the rhs is convergent in $L^2(\nu_\rho)$ and thus is a limit of elements of $\mathcal{V}_n$.

For item 2, assume that $f \in \mathcal{V}_n^\perp$ and $g \in \mathcal{V}_n$. Then using the self-adjointness of the semigroup in $L^2(\nu_\rho)$ we get (remember we denote $\langle \cdot, \cdot \rangle$ the inner product in $L^2(\nu_\rho)$)

$$\langle S(t) f, g \rangle = \langle f, S(t) g \rangle = 0, \tag{9.200}$$

where in the last step we used $S(t)g \in \mathcal{V}_n$ which follows from item 1.

For item 3, considering an $f \in L^2(\nu_\rho)$ and writing $f = P_{\mathcal{V}_n} f + P_{\mathcal{V}_n^\perp} f$, we obtain

$$P_{\mathcal{V}_n} S(t) f = P_{\mathcal{V}_n} S(t)[P_{\mathcal{V}_n} f + P_{\mathcal{V}_n^\perp} f] = P_{\mathcal{V}_n} S(t) P_{\mathcal{V}_n} f, \tag{9.201}$$

where the last step follows from item 2, i.e., $S(t) P_{\mathcal{V}_n^\perp} f \in \mathcal{V}_n^\perp$.

Similarly, because $S(t)$ leaves $\mathcal{V}_n$ invariant we have

$$S(t) P_{\mathcal{V}_n} f = [P_{\mathcal{V}_n} + P_{\mathcal{V}_n^\perp}] S(t) P_{\mathcal{V}_n} f = P_{\mathcal{V}_n} S(t) P_{\mathcal{V}_n} f. \tag{9.202}$$

We conclude, combining (9.201), (9.202), that

$$P_{\mathcal{V}_n} S(t) = S(t) P_{\mathcal{V}_n}. \tag{9.203}$$

The proof of commutation of $S(t)$ with $P_{\mathcal{V}_n^\perp}$ is completely analogous. □

Then we have the following theorem showing that the orthogonal duality polynomials can be obtained from Gram-Schmidt orthogonalization of the triangular duality polynomials.

Theorem 9.22 *Define, for* $\xi \in \Omega = \mathbb{N}^V$ *with* $|\xi| \le n$

$$D_\rho(\xi, \cdot) = D(\xi, \cdot) - P_{\mathcal{V}_{n-1}} D(\xi, \cdot). \tag{9.204}$$

Then we have

1. $D_\rho(\xi, \eta)$ *is a self-duality function for the process of independent random walkers.*
2. $D_\rho(\xi, \eta) = \prod d_\rho(\xi_x, \eta)$ *where* $d_\rho(k, \cdot)$ *are orthogonal polynomials w.r.t. the Poisson measure, defined via* (9.192).

Proof Item 1 follows because in Lemma 9.21 we proved that $S(t)$ commutes with $P_{\mathcal{V}_{n-1}}$, and therefore also with $I - P_{\mathcal{V}_{n-1}}$, and therefore $(I - P_{\mathcal{V}_{n-1}})D(\xi, \cdot)(\eta)$ is a new self-duality function (a symmetry applied to a self-duality function yields another self-duality function).

To prove item 2, i.e., the fact that $D_\rho(\xi, \eta)$ is the factorized orthogonal polynomial, let us call $\tilde{D}_\rho(\xi, \eta)$ the factorized orthogonal polynomial. Then we remark first that for ξ a configuration with n particles, we have

$$\tilde{D}_\rho(\xi, \cdot) \perp \mathcal{V}_{n-1}. \tag{9.205}$$

This is because $\mathcal{V}_{n-1}$ also equals the closure of the vector space generated by $\tilde{D}_\rho(\xi, \cdot)$ with $|\xi| \leq n - 1$. The latter can be understood from the fact that the triangular duality polynomials can be written as linear combinations of the factorized orthogonal polynomials, combined with the fact that $\tilde{D}_\rho(\xi, \cdot)$ is orthogonal on every polynomial $\tilde{D}_\rho(\xi', \cdot)$ with $|\xi'| \leq n - 1$.

Next we remark that, by construction, $\tilde{D}_\rho(\xi, \cdot)$ is necessarily of the form

$$\tilde{D}_\rho(\xi, \cdot) = D(\xi, \cdot) - F(\xi, \cdot), \tag{9.206}$$

with $F(\xi, \cdot) \in \mathcal{V}_{n-1}$. As a consequence,

$$0 = P_{\mathcal{V}_{n-1}} \tilde{D}_\rho(\xi, \cdot) = P_{\mathcal{V}_{n-1}} D(\xi, \cdot) - P_{\mathcal{V}_{n-1}} F(\xi, \cdot) = P_{\mathcal{V}_{n-1}} D(\xi, \cdot) - F(\xi, \cdot). \tag{9.207}$$

So we conclude

$$F(\xi, \cdot) = P_{\mathcal{V}_{n-1}}(D(\xi, \cdot)), \tag{9.208}$$

which shows that D_ρ and $\tilde{D}_\rho$ coincide. □

9.9 Additional Notes

Orthogonal dualities were introduced simultaneously in [98] and [198] using different methodology (three term recurrence relation resp. generating function method). The proof based on the Gramm-Schmidt method is from [95]. In that paper orthogonal dualities are also obtained for particle systems in the continuum. In [126], [99] orthogonal polynomial duality was obtained from Lie algebra representation theory. In [45] orthogonal dualities were obtained via unitary symmetries, the results from section 7 of the current chapter are based on this paper.

Orthogonal polynomials play an important role in the study of diffusion processes see [8], and for examples of diffusion processes in the context of population dynamics see e.g. [124] for Jacobi polynomials, and [148] for Gegenbauer polynomials. Connection between orthogonal polynomials and Markov intertwinings are studied e.g. in [181]. In [209] orthogonal polynomials are used to define generalized Wiener chaos expansions which in the context of the Wiener process is related to Hermite polynomials. These generalized chaos decompositions are also a way to obtain orthogonal polynomial duality in the continuum, see e.g. [95], [233].

Orthogonal polynomial duality is useful in the study of relaxation to equilibrium (as we saw in this chapter), but also in the study of macroscopic fields as we will see in chapter 11 where we use orthogonal polynomial duality to obtain a quantitative Boltzmann-Gibbs principle, and in [7] where orthogonal polynomial dualities are used to define higher order fluctuation fields, corresponding to "Wick powers" of the infinite-dimensional Ornstein Uhlenbeck process. Finally, orthogonal dualities can be used in the study of properties of cumulants of non-equilibrium steady states, see [94]. In the context of asymmetric processes, representation theory of quantum Lie algebras can be used to obtain orthogonal dualities see e.g. [48], and also [30], [239] for examples of multi-type particle system corresponding to a higher rank quantum Lie algebra.

Chapter 10
Consistency

In this chapter we introduce the notion of consistency for interacting particle systems. A system is consistent if the action of removing at random a particle commutes with the evolution. Beyond considering the problem at the level of particle configurations we will analyze the notion of consistency from the point of view of (labeled) particle positions. We will show that consistent particle systems satisfy a set of recursion relations for joint factorial moments of particle occupancies. We discuss the relation between consistency and self-duality. We show that adding absorbing sites conserves consistency, which is important in the study of boundary driven systems where the dual is an absorbing system.

10.1 Introduction

So far we have considered three basic particle systems—the partial exclusion process (SEP), the inclusion process (SIP) and independent random walkers (IRW)—and showed how their self-duality properties are related to symmetries of the generator. By taking many particle limits and thermalizations, we have linked these particle systems to other interacting particle systems in continuous variables such as the Brownian energy process (BEP) and the Kipnis-Marchioro-Presutti (KMP) model. We have also shown how dualities are related to intertwinings.

In the literature on interacting particle systems [172], often duality is proved using a graphical construction due to Harris, see [132] for the first appearance of this connection between duality, graphical construction, coupling and monotonicity. In the standard symmetric simple exclusion process, the configuration at time $t > 0$ can be constructed from the configuration at time $t = 0$ from a collection of independent rate 1 Poisson processes. More precisely, one associates to each nearest neighbour edge a Poisson process of rate 1, where these processes are independent for different edges. On the event times of the Poisson process associated to an edge

C. Giardinà, F. Redig, *Duality for Markov Processes*, Grundlehren der mathematischen Wissenschaften 365, https://doi.org/10.1007/978-3-032-04099-2_10

xy, we place a double arrow on this edge. On the event times of the Poisson process of the edge xy, the occupancies of the configuration at sites x and y are exchanged, which of course has only an effect if these occupancies are different, which then corresponds to a jump of the particle at the occupied site x to the empty site y or vice-versa.

This graphical construction provides also a coupling of the symmetric exclusion process starting from an arbitrary initial configuration, indeed the particles just follow the arrows associated to the Poisson processes. I.e., we can view the Poisson arrows as a stochastic flow, which, once a realization is fixed, fixes the evolution of an initial configuration deterministically. From this graphical representation, we see that if we consider k particles out of a configuration of $n \geq k$ particles, then the positions of these k particles at a later time $t > 0$ are exactly the same as the graphical construction applied to these k particles, i.e., the same as the position at time $t > 0$ of these k particles in the symmetric exclusion process. In particular, for $k = 1$, this implies that a single particle evolves as a symmetric random walk. The property that k out of n particles move exactly as the original system with k particles is of course satisfied for independent particles, but for an interacting system it is a remarkable property that is certainly not satisfied generically.

In this chapter, we show that this property, which we call *consistency*, can be reformulated as commutation of the semigroup of the particle system with the so-called particle removal operator, or equivalently, intertwining between a system of n particles with a system of $n - 1$ particles. We also show that from this property we can recover duality results and recursion formulas for moments.

As we saw before, the self-duality of the three basic particle systems SEP, IRW, SIP is a consequence of the commutation of the process generator L with respectively $\sum_x J_x^-$, (SEP), $\sum_x K_x^-$, (SIP), $\sum_x a_x$, (IRW). The action of these three operators is however identical and given by

$$S^- f(\eta) = \sum_x \eta_x f(\eta - \delta_x), \tag{10.1}$$

which can be interpreted as (the unnormalized version of) "removing a randomly chosen particle". This provides us with a more probabilistic interpretation of the self-duality property, namely, self-duality following from the fact that the Markov semigroup commutes with the operation of randomly removing a particle. In other words, for the three basic systems, the self-duality is equivalent with the intertwining (or symmetry) which is S^-. The commutation of the Markovian dynamics with the operation of removing a random particle makes sense in a much broader context than discrete configuration processes and includes e.g. independent diffusion processes and certain classes of interaction Brownian motions see [168], which is a setting where self-duality with self-duality functions factorizing over sites does no longer make sense. Therefore, via the notion of consistency it is possible to study self-duality properties of such processes in the continuum, see e.g. [95].

10.2 Definition of Consistency

Consider a Markov process $\{\eta(t),\ t \geq 0\}$ modelling interacting particles moving on a set of vertices V (which we will assume to be finite in the whole of this chapter). As usual, a configuration η is a collection of single-site occupancies $\eta_x,\ x \in V$ and the state space Ω is a subset of $\mathbb{N}^V$. We will assume, moreover, that the process conserves the total number of particles, i.e. $|\eta(t)| = |\eta|$ for all $t \geq 0$. We will denote by L the infinitesimal generator of the process and by $\{\mathcal{S}(t),\ t \geq 0\}$ the associated semigroup.

In order to define the notion of consistency for this class of models we recall the definition of removal operator, which in this chapter we will denote by S^-. More precisely,

$$S^- f(\eta) = \sum_{x \in V} a_x f(\eta), \qquad f : \Omega \to \mathbb{R}, \tag{10.2}$$

with

$$a_x f(\eta) = \begin{cases} \eta_x f(\eta - \delta_x) & \text{if } \eta_x \geq 1 \\ 0 & \text{otherwise.} \end{cases} \tag{10.3}$$

Definition 10.1 (Consistency) A particle system $\{\eta(t) : t \geq 0\}$ on the lattice V with state space Ω is said to be *consistent* if its generator L commutes with the removal operator:

$$[L, S^-] = 0, \tag{10.4}$$

or, equivalently, if the semigroup of the process commutes with the removal operator:

$$\left(\mathcal{S}(t)(S^- f)\right)(\eta) = \left(S^-(\mathcal{S}(t) f)\right)(\eta). \tag{10.5}$$

Remark 10.2 We notice immediately that the main particle systems studied in the first chapters of this book, (IRW, SIP, SEP) satisfy (10.4). For all them, indeed, the generator commutes with the removal operator S^-. Indeed, S^- corresponds $\sum_x a_x$ in Chap. 3 for the IRW, with $\sum_x K_x^-$ for the SIP in Chap. 5 and with $\sum_x J_x^-$ for SEP in Chap. 7 (included the space-inhomogeneous cases). The fact that the removal operators are the same for all these processes (this is not the case for the sum of creation operators $a_x^\dagger$, K_x^+ and J_x^+) suggests the existence of a common structure. In Sect. 10.5 we will come back to these processes and we will see how, in these cases, consistency is related to the self-duality property.

The commutation relation between the semigroup and the removal operator can be rewritten as follows. For all functions $f : \Omega \to \mathbb{R}$, we have

$$\sum_{x \in V} \mathbb{E}_\eta[\eta_x(t) f(\eta(t) - \delta_x)] = \sum_{x \in V} \eta_x \mathbb{E}_{\eta - \delta_x}[f(\eta(t))]. \tag{10.6}$$

In the left hand side we first let evolve the dynamics for a time t, and then, at this time, we remove a particle chosen uniformly at random. On the right hand side instead, we remove a particle at random from the initial configuration η, and then let evolve the dynamics for a time t. Thus a particle system is consistent if the operation of removing a particle uniformly at random commutes with the dynamics.

In order to better understand this probabilistic interpretation of consistency it is convenient to switch to the coordinate description of the process were, instead of particle numbers, the variables are labeled particle positions. This allows to distinguish between the particles, which is not possible if one describes the particle evolution on the level of the configuration, in which the particles are indistinguishable. A configuration $\eta \in \Omega$ with n particles, can be written in terms of the positions of its particles $x_1, \ldots, x_n$ as follows:

$$\eta = \sum_{i=1}^{n} \delta_{x_i}. \tag{10.7}$$

For

$$\mathbf{x} := (x_1, \ldots, x_n) \in V^n \tag{10.8}$$

we denote by $\varphi(\mathbf{x}) = \sum_{i=1}^{n} \delta_{x_i}$ this map from labeled particle configurations to unlabeled particle configurations.

If we denote by $X(t) = (X_1(t), \ldots, X_n(t))$ the evolution at time t of the positions of the n particles starting, at time 0, from $\mathbf{x} := (x_1, \ldots, x_n) \in V^n$, we can write that

$$\eta(t) = \sum_{i=1}^{n} \delta_{X_i(t)}. \tag{10.9}$$

In this coordinate representation the action of the removal operator on f can be read as

$$S^- f(\eta) = S^- f\Big(\sum_{j=1}^{n} \delta_{x_j}\Big) = \sum_{i=1}^{n} f\Big(\sum_{j \neq i} \delta_{x_j}\Big). \tag{10.10}$$

Let $\mathbf{x}^i \in V^{n-1}$ denote the coordinate vector obtained from $\mathbf{x}$ by removing a particle at site x_i. We can then rewrite the consistency relation for the semigroup given in (10.5) as a condition on the process $\{X(t),\ t \geq 0\}$ as follows

$$\sum_{i=1}^{n} \mathbf{E}_{\mathbf{x}} \left[f\Big(\sum_{j \neq i} \delta_{X_j(t)} \Big) \right] = \sum_{i=1}^{n} \mathbf{E}_{\mathbf{x}^i} \left[f\Big(\sum_{j \neq i} \delta_{X_j(t)} \Big) \right]. \tag{10.11}$$

The identity (10.11) can be further simplified by reinterpreting the function f on the configuration space Ω as a function $g : V^n \to \mathbb{R}$ on the coordinate space V^n, with the relation between f and g being the following

$$g(\mathbf{x}) = g(x_1, \ldots, x_n) = f\left(\sum_{i=1}^{n} \delta_{x_i} \right). \tag{10.12}$$

Then (10.11) becomes, for any $g : V^{n-1} \to \mathbb{R}$

$$\sum_{i=1}^{n} \mathbf{E}_{\mathbf{x}} \left[g\big(X_1(t), \ldots, X_{i-1}(t), X_{i+1}(t), \ldots, X_n(t)\big) \right]$$
$$= \sum_{i=1}^{n} \mathbf{E}_{\mathbf{x}^i} \left[g\big(X_1(t), \ldots, X_{i-1}(t), X_{i+1}(t), \ldots, X_n(t)\big) \right], \tag{10.13}$$

from which the probabilistic interpretation of consistency becomes more transparent.

The passage from the configuration description to the coordinate description involves the introduction of a particle labeling. While in the configuration representation particles are undistinguishable, in the coordinate representation particles are given a label that make them distinguishable. The choice of the labeling mechanism is not unique, i.e., for a given configuration process $\{\eta(t),\ t \geq 0\}$ it is possible to define several coordinate processes $\{X(t),\ t \geq 0\}$ compatible with it. If the coordinate process is consistent in the sense of (10.13) then the corresponding configuration process is also consistent in the sense of Definition 10.1.

In the rest of this section we will see how the consistency relation (10.13) can be regarded as a weakening of *strong consistency* which is explained below in the context of independent random walker.

10.2.1 Strong Consistency

Let $\{X(t),\ t \geq 0\}$ an n-dimensional coordinate process on the lattice V^n, i.e. $X(t) = (X_1(t), \ldots, X_n(t))$, with $X_i(t) \in V$ being the position of the i-th particle at time t.

Then the process is said *strongly consistent* if, for all functions $g : V^{n-1} \to \mathbb{R}$ and for all $i \in \{1, \dots, n\}$,

$$\begin{aligned}&\mathbf{E}_{\mathbf{x}}\left[g\big(X_1(t), \dots, X_{i-1}(t), X_{i+1}(t), \dots, X_n(t)\big)\right] \\ &= \mathbf{E}_{\mathbf{x}^i}\left[g\big(X_1(t), \dots, X_{i-1}(t), X_{i+1}(t), \dots, X_n(t)\big)\right].\end{aligned} \tag{10.14}$$

This means that, initializing the system with n particles at position $\mathbf{x}$ and looking at time t at the $n-1$-dimensional marginal law obtained by removing the i-th coordinate, is the same as initialising the system with $n-1$ particles (all particles except the i-th one) and looking at the evolution of these at time t. Strong consistency implies the identity (10.13), and then consistency for the corresponding configuration process. At first sight, it is not obvious that models of interacting particles can satisfy(10.14). However, as explained in the introduction, the symmetric exclusion process with the stirring construction is an example where (10.14) is an obvious consequence of the graphical representation. Many other examples exist, which are constructed from stochastic flows [167, 168].

Of course, if particles are independent from each other, identity (10.14) is automatically satisfied. More precisely, we call $\{X(t),\ t \ge 0\}$ the process on V^n as the collection of n independent copies of the same process $\{X_i(t),\ t \ge 0\}$, $i = 1, \dots, n$ on V. For instance we suppose that these are all random walkers on the lattice V moving at a certain given rate. Then, if we denote by $p_t(\mathbf{x}, \mathbf{y})$ the transition at time t from $\mathbf{x}$ to $\mathbf{y}$, we have that

$$p_t(\mathbf{x}, \mathbf{y}) = \prod_{i=1}^{n} p_t(x_i, y_i), \tag{10.15}$$

where, with a slight abuse of notation we denote by $p_t(x, y)$ the transition probability at time t from site x to site y for the single random walker. As a consequence we have that

$$\begin{aligned}&\sum_{\mathbf{y}} \prod_{j=1}^{n} p_t(x_j, y_j) \cdot g\big(y_1, \dots, y_{i-1}, y_{i+1}, \dots, y_n\big) \\ &= \sum\nolimits_{\mathbf{y^i}} \prod_{\substack{j=1,\\ j\neq i}}^{n} p_t(x_j, y_j) \cdot g\big(y_1, \dots, y_{i-1}, y_{i+1}, \dots, y_n\big),\end{aligned} \tag{10.16}$$

that implies (10.14), i.e. strong consistency.

The rest of the chapter is organized as follows. In Sect. 10.3 we will prove a many-to-few relation in terms of factorial moments, and we will use this to prove a relation between consistency, reversibility and self-duality. In Sect. 10.4 we will give the probabilistic interpretation of consistency in terms of particles removal, whose understanding requires the introduction of a coordinate description of the dynamics.

More precisely we will define the so-called *compatible coordinate processes* for which the variables are the positions of particles at a given time. In Sect. 10.5 we will come back to the main processes studied in the previous chapters. We will reconsider their duality properties in light of the notion of consistency. Finally, in the last two sections, we will provide some examples of consistent processes that are not reversible with respect to a product reversible measure. More precisely in Sect. 10.7 we will give an example of a consistent process with asymmetric interaction, whereas, in Sect. 10.6 we will consider the case of consistent processes with absorbing sites.

10.3 Recurrence Relations for Factorial Moments

In this section we will see that one of the main consequences of consistency is the possibility of going *from many to few* variables. In this context, this means that it is possible to write specific factorial moments of the system with many particles in terms of transition probability functions of the system with fewer particles (see Theorem 10.5 below). This is exactly the same advantage given by duality with the difference that this holds now for a broader class of processes.

10.3.1 Example: Expected Number of Particles

For the sake of clarity we start by considering the simplest problem of finding the expectations of the number of particles in a given site $v \in V$ at time $t \geq 0$. In other words, let $\{\eta(t) \,:\, t \geq 0\}$ be a consistent process on a finite lattice V, we want to compute the expectation $\mathbb{E}_\eta[\eta_v(t)]$, for a fixed $\eta \in \Omega$. In order to do this we analyze the consequences of consistency at the level of the dynamics.

Start from (10.6) with $f(\eta) := \eta_v$. We then obtain

$$\sum_{x \in V} \mathbb{E}_\eta(\eta_x(t)(\eta_v(t) - \delta_x)) = \sum_{x \in V} \eta_x \mathbb{E}_{\eta - \delta_x}(\eta_v(t)). \tag{10.17}$$

Suppose $|\eta| := \sum_{x \in V} \eta_x = n$, then in the left hand side of (10.17) we obtain, using the conservation of the total number of particles,

$$\begin{aligned}
&\left(\sum_{x \in V, x \neq v} \mathbb{E}_\eta[\eta_x(t)\eta_v(t)] \right) + \mathbb{E}_\eta[\eta_v(t)(\eta_v(t) - 1)] \\
&= \mathbb{E}_\eta[(n - \eta_v(t))\eta_v(t)] + \mathbb{E}_\eta[\eta_v(t)(\eta_v(t) - 1)] \\
&= (n-1)\mathbb{E}_\eta[\eta_v(t)]. \qquad (10.18)
\end{aligned}$$

Thus, from (10.17) we get

$$\mathbb{E}_\eta[\eta_v(t)] = \frac{1}{n-1} \sum_{x \in V} \eta_x \mathbb{E}_{\eta-\delta_x}[\eta_v(t)]. \tag{10.19}$$

These are recurrence relations for the expectation of $\eta_v(t)$, connecting the system with n particles to the systems with $n-1$ particles. One can then iterate (10.19) $n-1$ times in order to obtain an expression for $\mathbb{E}_\eta(\eta_v(t))$ in terms of the one-particle dynamics, i.e. in terms of the dynamics of a single random walker. We will explain this in detail in Corollary 10.7 below.

10.3.2 General Case: Weighted Factorial Moments

Here we investigate to which extent consistency can give information about higher order correlation functions and, more generally, about all the moments of the process. We may wonder, for instance, whether it is possible to find closed recurrence relations of the type (10.19) for the k-th moment $\mathbb{E}[\eta_x^k(t)]$, and, to this aim, what is the correct function f to plug in (10.6). It turns out that the most suitable collection of functions is $\{F(\xi, \cdot),\ \xi \in \Omega\}$ with

$$F(\xi, \eta) := \prod_{y \in V} \binom{\eta_y}{\xi_y}, \qquad \eta \in \Omega. \tag{10.20}$$

Plugging these functions in (10.6) produces indeed a closed system of equations. The product of binomial coefficients (10.20) can be obtained from the action of the operator e^{S^-} on the Kronecker delta functions $\delta_\xi(\cdot)$, $\xi \in \Omega$, more precisely, we have

$$(e^{S^-}\delta_\xi)(\eta) = F(\xi, \eta). \tag{10.21}$$

Indeed, recalling the definition $af(n) = nf(n-1)$ for a function $f : \mathbb{N} \to \mathbb{R}$, we get

$$e^a f(n) = \sum_{r=0}^{n} \binom{n}{r} f(n-r), \tag{10.22}$$

and, as a consequence, choosing $f(n) = \delta_k(n)$

$$e^a \delta_k(n) = \binom{n}{k}. \tag{10.23}$$

Therefore, we obtain (10.21) (cf. also (3.52)).

Definition 10.3 (Weighted Factorial Moments) Let $\eta, \xi \in \Omega$ and let $\{\eta(t) : t \geq 0\}$ be a particle system. Let $F(\xi, \cdot)$ be the function in (10.20). We call the expectation $\mathbb{E}_\eta [F(\xi, \eta(t))]$ the *weighted factorial moment* of order $\xi \in \Omega$ at time $t \geq 0$ for the process starting from η at time 0.

Lemma 10.4 (Going from an n-particle System to an $(n-1)$-particle System) *Let $\{\eta(t), t \geq 0\}$ be a consistent particle system on a lattice V. For all $\eta, \xi \in \Omega$ such that $1 \leq |\xi| < |\eta|$, the weighted factorial moments fulfill for all $t \geq 0$*

$$\mathbb{E}_\eta [F(\xi, \eta(t))] = \frac{1}{(|\eta| - |\xi|)} \cdot \sum_{x \in V} \eta_x \mathbb{E}_{\eta - \delta_x} [F(\xi, \eta(t))] . \tag{10.24}$$

Proof We fix $\eta, \xi \in \Omega$ such that $|\xi| \in \{1, \ldots, |\eta| - 1\}$ and apply (10.6) to the function $f = F(\xi, \cdot)$. We get

$$\sum_{x \in V} \eta_x \mathbb{E}_{\eta - \delta_x} [F(\xi, \eta(t))] = \sum_{x \in V} \mathbb{E}_\eta \left[\eta_x(t) \binom{\eta_x(t) - 1}{\xi_x} \prod_{\substack{y \in V \\ y \neq x}} \binom{\eta_y(t)}{\xi_y} \right] . \tag{10.25}$$

Using the elementary identity $n\binom{n-1}{k} = (n-k)\binom{n}{k}$, together with the conservation law $\sum_x \eta_x(t) = |\eta|$, the right-hand side of (10.25) is equal to

$$\sum_{x \in V} \mathbb{E}_\eta \left[(\eta_x(t) - \xi_x) \cdot \prod_{y \in V} \binom{\eta_y(t)}{\xi_y} \right] = (|\eta| - |\xi|) \cdot \mathbb{E}_\eta [F(\xi, \eta(t))] . \tag{10.26}$$

Combining together (10.25) and (10.26) we obtain (10.24). □

Via Lemma 10.4 we obtain that consistency implies recurrence relations for the factorial moment of order ξ of a system with n particles in terms of the factorial moments of the same order of systems with $n-1$ particles. This is true for all $\xi \in \Omega$ with the restriction $|\xi| \leq n - 1$.

In the next Theorem we obtain a relation for the weighted factorial moment of order ξ of a system with n particles in terms of the transition probabilities of a system with at most $n-1$ particles. To prove it, instead of directly iterating (10.24), we will use the relation (10.21).

From now on we will denote by Ω_n the set of configurations of n particles, i.e.

$$\Omega_n = \Big\{ \eta \in \Omega : |\eta| = n \Big\}, \qquad \text{with} \qquad |\eta| := \sum_{x \in V} \eta_x, \tag{10.27}$$

and we define the ordering between configurations:

$$\xi \leq \eta \qquad \text{if and only if} \qquad \xi_x \leq \eta_x \qquad \forall x \in V. \tag{10.28}$$

Theorem 10.5 (Recurrence Relation for the Weighted Factorial Moments) *A particle system $\{\eta(t),\ t \geq 0\}$ on a lattice V is consistent if and only if, for all $\eta, \xi \in \Omega$ such that $1 \leq |\xi| < |\eta|$, we have*

$$\mathbb{E}_\eta [F(\xi, \eta(t))] = \sum_{\substack{\varsigma \in \Omega_{|\xi|} \\ \varsigma \leq \eta}} F(\varsigma, \eta) \cdot \mathbb{P}_\varsigma (\varsigma(t) = \xi)\,. \tag{10.29}$$

Proof The process is consistent if and only if e^{S^-} is a symmetry of the semigroup, i.e. for all $f : \Omega \to \mathbb{R}$,

$$\left[\mathcal{S}(t)(e^{S^-} f)\right](\eta) = \left[e^{S^-} (\mathcal{S}(t) f)\right](\eta). \tag{10.30}$$

This is in turn true if and only if, for all $\xi \in \Omega$,

$$\left[\mathcal{S}(t)(e^{S^-} \delta_\xi)\right](\eta) = \left[e^{S^-} (\mathcal{S}(t) \delta_\xi)\right](\eta). \tag{10.31}$$

Using (10.21), we have that the l.h.s. of (10.31) is equal to

$$\mathbb{E}_\eta[F(\xi, \eta(t))] = [\mathcal{S}(t) F(\xi, \cdot)](\eta) = \left[\mathcal{S}(t)(e^{S^-} \delta_\xi)\right](\eta). \tag{10.32}$$

On the other hand, defining the functions

$$g_\xi(\eta) = \mathbb{P}_\eta(\eta(t) = \xi), \qquad \xi \in \Omega, \tag{10.33}$$

and using (10.22), we have that the r.h.s. of (10.31) is given by

$$\left[e^{S^-} (\mathcal{S}(t)\delta_\xi)\right](\eta) = [\mathcal{S}(t) g_\xi](\eta) = \sum_{\eta' \leq \eta} F(\eta', \eta) \cdot P_{\eta-\eta'}(\eta(t) = \xi) \tag{10.34}$$

where $\eta - \eta'$ is the configuration with occupancies $\eta_x - \eta'_x$ for all $x \in V$. Then (10.29) follows by taking the change of variable $\varsigma = \eta - \eta'$. □

Theorem 10.5 says that, for a consistent configuration process $\{\eta(t),\ t \geq 0\}$ it is possible to gain information about $\{\eta(t),\ t \geq 0\}$ in terms of the process starting from a configuration with fewer particles. More precisely, for the n-particles system, the factorial moment of order ξ, $|\xi| < n$, is given as a combination of the transition probabilities of the processes with $|\xi|$ particles.

Notice that, in the particular case $\xi = \delta_x$, the relation (10.24) gives the expectations of the occupancies in terms of the transition probabilities of a single random walker. To make this rigorous we first define the random walk associated to the process.

Definition 10.6 For an interacting particle system $\{\eta(t),\ t \geq 0\}$ we define the associated random walk $\{X^{\rm rw}(t),\ t \geq 0\}$ on V which is identified by the relation $\eta(t) = \delta_{X^{\rm rw}(t)}$ and the initial condition $X^{\rm rw}(0) = u$ with $\eta(0) = \delta_u$.

Then we have that, for a consistent process, it is possible to write, exactly as for self-dual processes, the occupation number expectations in terms of the single random walk transition probabilities.

Corollary 10.7 *Let $\{\eta(t),\ t \geq 0\}$ be a consistent process on a finite lattice V and let $\eta = \sum_{\kappa=1}^{n} \delta_{x_\kappa}$, then, for all $y \in V$ we have*

$$\mathbb{E}_\eta\left[\eta_y(t)\right] = \sum_{\kappa=1}^{n} \mathbf{P}_{x_\kappa}(X^{rw}(t) = y), \tag{10.35}$$

where $\mathbf{P}_x$, is the path space measure of the random walk $\{X^{rw}(t),\ t \geq 0\}$ on V starting from $x \in V$ associated to the process $\{\eta(t),\ t \geq 0\}$.

More generally the choice $\xi = \sum_{k=1}^{m} \delta_{x_k}$, with $x_1, x_2, \ldots, x_m$ m mutually distinct vertices in V in Theorem 10.5 will provide information about the m-point correlation functions $\mathbb{E}_\eta[\eta_{x_1}(t) \cdot \ldots \cdot \eta_{x_m}(t)]$. Whereas, taking $\xi = m\delta_x$, $m = 1, \ldots, n-1$ yields the moments of the x-th occupancy of order less than n, i.e. $\mathbb{E}[\eta_x^m(t)]$, for $m = 1, \ldots, n-1$.

Unfortunately the information provided by (10.29) is not complete. Indeed it is not possible to recover the full probability distribution of the process at time t from the collection of weighted factorial moments $\{\mathbb{E}_\eta[F(\xi, \eta(t))],\ 1 \leq \xi \leq |\eta| - 1\}$. This is due to the fact that information about moments of order $|\eta|$ is still missing, because, for the case $|\xi| = |\eta|$, (10.29) reduces to a tautological statement.

10.3.3 Consistency and Self-duality

Theorem 10.5 gives a relation of the type *from many to few*, as it relates a system with many particles to systems with fewer particles. It is in that sense very similar to the information contained in duality relations, especially for the "triangular" duality functions, where the dual configuration is smaller (in the sense of point-wise order) than the original configuration. However, we will see below that the consistency assumption is, weaker than duality, and indeed (10.29) holds for a larger class of processes than the self-dual ones considered so far. In the next theorem we will see the precise relation between the two properties. In particular we will prove that the additional condition to impose in order to guarantee duality is reversibility. This result is in the spirit of Theorem 2.15.

Theorem 10.8 (Consistency, Self-duality and Reversibility) *Let* $\{\eta(t),\ t \geq 0\}$ *be a particle system with state space* Ω*, let* ν *be a strictly-positive measure on* Ω *and define the function:*

$$D(\xi, \eta) = \frac{F(\xi, \eta)}{\nu(\xi)}, \tag{10.36}$$

with F *as in* (10.20)*. If two of the following three statements hold, then also the third one holds:*

(a) the process is consistent,
(b) the process is reversible with respect to ν*,*
(c) the process is self-dual with self-duality function D*.*

Proof Assume (a) and (b), then, using Theorem 10.5 we have

$$\begin{aligned}
\mathbb{E}_\eta\left[D(\xi, \eta(t))\right] &= \frac{1}{\nu(\xi)}\ \mathbb{E}_\eta\left[F(\xi, \eta(t))\right] \\
&= \sum_{\substack{\varsigma \in \Omega_{|\xi|} \\ \varsigma \leq \eta}} F(\varsigma, \eta) \cdot \frac{1}{\nu(\xi)}\ \mathbb{P}_\varsigma\left(\varsigma(t) = \xi\right) \\
&= \sum_{\substack{\varsigma \in \Omega_{|\xi|} \\ \varsigma \leq \eta}} F(\varsigma, \eta) \cdot \frac{1}{\nu(\varsigma)}\ \mathbb{P}_\xi\left(\xi(t) = \varsigma\right) \\
&= \sum_{\substack{\varsigma \in \Omega_{|\xi|} \\ \varsigma \leq \eta}} D(\varsigma, \eta) \cdot \mathbb{P}_\xi\left(\xi(t) = \varsigma\right) = \mathbb{E}_\xi[D(\xi(t), \eta)],
\end{aligned} \tag{10.37}$$

which is item (c) and where in the third equality we used the assumed reversibility. The other two implications can be proven in a similar way. □

10.4 Probabilistic Interpretation of Consistency

In this section we will study the coordinate description (i.e., going from unlabeled to labeled particle systems) of the dynamics introduced in Sect. 10.1, with the aim of giving a more probabilistic interpretation of the notion of consistency.

In this chapter we will call a *configuration process* on the lattice V, denoted by $\{\eta(t) = \{\eta_x(t),\ x \in V\},\ t \geq 0\}$, a Markov process taking values in $\Omega \subseteq \mathbb{N}^V$. The variable $\eta_x(t)$ represents the number of particles at site $x \in V$ at time t.

We switch now to the description via particle positions, where, if the system contains n particles, the coordinates are n-tuples $\mathbf{x} = (x_1, \ldots, x_n) \in V^n$ where x_i is the position of the i-th particle. We shall call a *coordinate process* on V, with n particles, denoted by $\{X^{(n)}(t),\ t \geq 0\}$, a Markov process taking values in V^n

that describes the positions of particles in the course of time. More precisely, for $i = 1, \ldots, n$, the random variable $X_i^{(n)}(t)$ denotes the position of the ith particle at time $t \geq 0$. We denote by $\{X(t) : t \geq 0\}$ a family of coordinate-processes $\{\{X^{(n)}(t),\ t \geq 0\}, n \in \mathbb{N}\}$.

In order to establish a link between the configuration and coordinate descriptions of the process, we define a function φ mapping the n-tuples, with arbitrary n, to configurations in $\mathbb{N}^V$, i.e., $\varphi : \cup_{n=1}^{\infty} V^n \to \mathbb{N}^V$ defined as follows. For $\mathbf{x} = (x_1, \ldots, x_n) \in V^n$ the associated configuration is denoted by

$$\varphi(\mathbf{x}) := \sum_{i=1}^{n} \delta_{x_i}. \tag{10.38}$$

Nevertheless the state-space Ω of a *configuration process* $\{\eta(t) : t \geq 0\}$ is not always equal to $\mathbb{N}^V$, but more generally of the type $\Omega = \times_{x \in V} \Upsilon_x$, with $\Upsilon_x \subseteq \mathbb{N}$. Thus, if $\Upsilon_x \neq \mathbb{N}$ for some $x \in V$, then not all the elements of V^n give rise to allowed configurations of Ω. For this reason we have to restrict the domain of the map φ by defining the set V_n of n-tuples $\mathbf{x} \in V^n$ such that the associated configuration $\varphi(\mathbf{x})$ is an element of $\Omega_n := \{\eta \in \Omega : |\eta| = n\}$.

10.4.1 Compatibility Between Labeled and Unlabeled Particle Systems

The configuration notation is, in general, the standard description for particle systems and is also the one that we so far we used throughout this book. A coordinate process naturally induces, under suitable conditions, a configuration process via the map φ defined in (10.38). This configuration process necessarily conserves the total number of particles. The map is not one-to-one, so there can be several coordinate processes whose image under the map φ yields the same configuration process. This leads us to the following definition.

Definition 10.9 (Compatibility) A family of coordinate processes $\{X(t),\ t \geq 0\}$ and a configuration process $\{\eta(t),\ t \geq 0\}$ are compatible if, for all $n \in \mathbb{N}$, the following holds: whenever $\varphi(X_1(0), \ldots, X_n(0)) = \eta(0)$ then

$$\{\varphi(X^{(n)}(t)),\ t \geq 0\} = \{\eta(t),\ t \geq 0\} \quad \text{in distribution.} \tag{10.39}$$

Notice that, in the coordinate process, particles are labeled, and thus distinguishable from each other throughout the dynamics. This information is lost when passing to the configuration process through the map φ. In the latter particles are indistinguishable, and then φ can be viewed as a projection from V_n to Ω_n. Of course it is not guaranteed that, starting from a coordinate process, that is Markov by definition, applying the map φ defined in (10.38) results in a Markov process on configuration space. To assure this we have to impose an additional requirement of permutation invariance.

On the other hand, if a coordinate process admits a compatible configuration process, then the latter is unique. From now on we will denote by $\mathscr{E}_n$ the set of functions $f : \Omega_n \to \mathbb{R}$ and by $\mathscr{C}_n$ the set of functions $g : V_n \to \mathbb{R}$. Moreover we denote by Σ_n the set of permutations of n elements.

Definition 10.10 (Permutation Invariance)

(a) A family of coordinate processes $\{X(t),\ t \geq 0\}$ is said to be *permutation-invariant* if, for all $n \in \mathbb{N}$, $\sigma \in \Sigma_n$,

$$\{(X_1^{(n)}(t), \ldots, X_n^{(n)}(t)),\ t \geq 0\} = \{(X_{\sigma(1)}^{(n)}(t), \ldots, X_{\sigma(n)}^{(n)}(t)),\ t \geq 0\}, \tag{10.40}$$

where equality hold in distribution.

(b) A family of probability measures $\boldsymbol{\mu} = \{\mu_n,\ n \in \mathbb{N}\}$, ($\mu_n$ probability measure on V_n) is called *permutation-invariant* if, for all $n \in \mathbb{N}$, $\sigma \in \Sigma_n$,

$$\mu_n(x_1, \ldots, x_n) = \mu_n(x_{\sigma(1)}, \ldots, x_{\sigma(n)}) \qquad \text{for all} \quad (x_1, \ldots, x_n) \in V_n. \tag{10.41}$$

(c) A function $g \in \mathscr{C}_n$ is said to be *permutation-invariant* if

$$g(x_1, \ldots, x_n) = g(x_{\sigma(1)}, \ldots, x_{\sigma(n)}) \qquad \text{for all} \quad (x_1, \ldots, x_n) \in V_n.$$

Equivalently, a function $g \in \mathscr{C}_n$ is permutation-invariant if there exists $f \in \mathscr{E}_n$ such that $g = f \circ \varphi$, where φ is given in (10.38).

Notice that, if a family of coordinate processes $\{X(t),\ t \geq 0\}$ is permutation invariant, then (10.40) holds true in particular at time 0. This means that, if $\boldsymbol{\mu} := \{\mu_n,\ n \in \mathbb{N}\}$ is the family of initial probability distributions of the process, μ_n being the probability distribution of $X^{(n)}(0)$, then also $\boldsymbol{\mu}$ is permutation-invariant.

We denote by L_n the infinitesimal generator of the n-particle coordinate process $\{X^{(n)}(t),\ t \geq 0\}$, and by $S_n(t)$ the related semigroup, i.e., for $g \in \mathscr{C}_n$

$$S_n(t)g(\mathbf{x}) := \mathbb{E}_{\mathbf{x}}[g(X^{(n)}(t))], \tag{10.42}$$

where $\mathbb{E}_{\mathbf{x}}$ denotes expectation when the process is started from $\mathbf{x} \in V_n$. The following proposition shows that, to any family of permutation-invariant coordinate-processes one can naturally associate a compatible configuration process.

Proposition 10.11 (Permutation Invariant Coordinate Process Is Equivalent to Configuration Process) *Let $\{X(t),\ t \geq 0\}$ be a family of permutation-invariant coordinate processes and let L_n be the generators of $\{X^{(n)}(t),\ t \geq 0\}$. Then there exists a unique configuration process compatible with $\{X(t),\ t \geq 0\}$ and its* *generator is the operator* L *whose action on functions* $f : \Omega \to \mathbb{R}$ *is given by the following relation. For all* $f \in \mathcal{E}_n$*,*

$$(Lf) \circ \varphi = L_n g, \qquad \textit{with} \quad g = f \circ \varphi. \tag{10.43}$$

Proof For a permutation $\sigma \in \Sigma_n$ and a function $g \in \mathcal{C}_n$ we define the operator

$$T_\sigma g(x_1, \ldots, x_n) = g(x_{\sigma(1)}, \ldots, x_{\sigma(n)}). \tag{10.44}$$

From the permutation invariance of the family of coordinate processes $\{X(t),\ t \geq 0\}$ it follows that

$$[L_n, T_\sigma] = 0 \qquad \text{for all} \quad n \in \mathbb{N} \quad \text{and} \quad \sigma \in \Sigma_n, \tag{10.45}$$

where $[\cdot, \cdot]$ denotes the commutator. Let $f : \Omega \to \mathbb{R}$ then, by definition, $g := f \circ \varphi$ is a permutation-invariant function. Hence, from (10.45) it follows that

$$T_\sigma L_n g = L_n T_\sigma g = L_n g. \tag{10.46}$$

This means that $L_n g$ is permutation-invariant, hence there exists a function $\tilde{f} : \Omega_n \to \mathbb{R}$ such that

$$L_n g(\mathbf{x}) = \tilde{f}(\varphi(\mathbf{x})) \qquad \forall \mathbf{x} \in V_n, \tag{10.47}$$

i.e., $L_n g = \tilde{f} \circ \varphi$. Then it is possible to define the operator L acting on functions $f : \Omega \to \mathbb{R}$ such that $Lf = \tilde{f}$, and then (10.43) is satisfied. From (10.46) we have that

$$T_\sigma S_n(t) g = S_n(t) g \tag{10.48}$$

and then also $S_n(t)g$ is a permutation-invariant function at all times. Now, if we denote by $\mathcal{S}(t)$ the semigroup associated to L, it follows that

$$\mathcal{S}(t) f(\eta) = S_n(t) g(\mathbf{x}) \qquad \text{for all} \quad \eta \in \Omega_n \text{ and } \mathbf{x} \in V_n \ : \ \varphi(\mathbf{x}) = \eta, \tag{10.49}$$

i.e.,

$$(\mathcal{S}(t) f) \circ \varphi = S_n(t) g \qquad \text{for all} \quad f \in \mathcal{E}_n. \tag{10.50}$$

From this it follows that L is the generator of a Markov process that is then the unique configuration process compatible with $\{X(t),\ t \geq 0\}$. □

10.4.2 Consistency for Coordinate Processes

At the level of configuration processes, one has consistency whenever the generator L commutes with the particle removal operator S^- (see Definition 10.1). In order to transport the notion of consistency to the coordinate level, we introduce a coordinate version of the particle removal operator in (10.2). We use here the notation $[n] = \{1, \ldots, n\}$ for the set of the first n natural numbers.

Definition 10.12 (Particle Removal Operators) For $n \in \mathbb{N}$, $i \in [n]$ we denote by $\pi_i^{(n)} : \mathscr{C}_{n-1} \to \mathscr{C}_n$ the removal operator of the ith labeled particle, acting on functions $g \in \mathscr{C}_{n-1}$ as follows:

$$(\pi_i^{(n)} g)(x_1, \ldots, x_n) = g(x_1, \ldots, x_{i-1}, x_{i+1}, \ldots, x_n) \qquad \text{for all} \quad x_i \in V, \tag{10.51}$$

and we denote by $\Pi^{(n)} : \mathscr{C}_{n-1} \to \mathscr{C}_n$ the operator acting on $g \in \mathscr{C}_{n-1}$ via

$$\Pi^{(n)} g = \sum_{i=1}^{n} \pi_i^{(n)} g. \tag{10.52}$$

In the next lemma we show that the commutation relation of the generator L with the operator S^- corresponds, in this new description, to an intertwining relation between the coordinate process with n particles and the coordinate process with $n-1$ particles.

Lemma 10.13 (Intertwining an n-particle System with an $(n-1)$-particle System) *Let L_n, $n \in \mathbb{N}$ be the generators of a family of permutation-invariant coordinate processes and let L be the Markov generator defined by the relation* (10.43), *then the following statements are equivalent:*

(a) for all $n \in \mathbb{N}$, L_n and L_{n-1} are intertwined via $\Pi^{(n)}$, i.e.

$$L_n \left(\Pi^{(n)}(g)\right) = \Pi^{(n)}\left(L_{n-1}(g)\right) \qquad \forall\, g \in \mathscr{C}_{n-1} \textit{ permutation-invariant}; \tag{10.53}$$

This statement is illustrated in the commutative diagram below

$$\begin{array}{ccc} \mathscr{C}_{n-1} & \xrightarrow{L_{n-1}} & \mathscr{C}_{n-1} \\ {\scriptstyle \Pi_n}\big\downarrow & & \big\downarrow{\scriptstyle \Pi_n} \\ \mathscr{C}_n & \xrightarrow{L_n} & \mathscr{C}_n. \end{array}$$

(b) $[L, S^-] = 0.$

Proof Let $\mathbf{x} = (x_1, \ldots, x_n)$ and $\eta := \varphi(\mathbf{x}) = \sum_{i=1}^n \delta_{x_i}$, then we have

$$\varphi(x_1, \ldots, x_{l-1}, x_{l+1}, \ldots, x_n) = \left(\sum_{i=1}^n \delta_{x_i}\right) - \delta_{x_l} = \eta - \delta_{x_l}. \tag{10.54}$$

As a consequence:

$$\begin{aligned}(\Pi^{(n)}(f \circ \varphi))(\mathbf{x}) &= \sum_{l=1}^n f(\eta - \delta_{x_l}) \\ &= \sum_{x \in V} \eta_x f(\eta - \delta_x), \end{aligned} \tag{10.55}$$

where the last step follows because every $x \in V$ is counted exactly η_x times in the sum $\sum_{l=1}^n f(\eta - \delta_{x_l})$. This proves that, for all $\eta \in \Omega_n$ and $\mathbf{x} \in V_n$ such that $\varphi(\mathbf{x}) = \eta$,

$$(\Pi^{(n)}(f \circ \varphi))(\mathbf{x}) = S^- f(\eta) . \tag{10.56}$$

Suppose now that $[L, S^-] = 0$, then on $\mathscr{E}_n$ we have that, for $g = f \circ \varphi$,

$$\begin{aligned} L_n \left(\Pi^{(n)} g\right)(\mathbf{x}) &= L_n \left[(S^- f) \left(\sum_{i=1}^n \delta_{x_i}\right)\right] \\ &= (LS^- f) \left(\sum_{i=1}^n \delta_{x_i}\right) \\ &= (S^- Lf) \left(\sum_{i=1}^n \delta_{x_i}\right) \\ &= \Pi^{(n)} Lf \left(\sum_{i=1}^n \delta_{x_i}\right) = \Pi^{(n)} \left(L_{n-1} g\right)(\mathbf{x}) \end{aligned} \tag{10.57}$$

where the equalities follow from (10.56), (10.43) and the commutation relation. Then (10.53) follows since, for all $g \in \mathscr{C}_{n-1}$ permutation-invariant, $g = f \circ \varphi$ for some $f \in \mathscr{E}_{n-1}$. The reverse implication is proved analogously. □

The meaning of the previous result is the following. On the left-hand side of (10.53) we remove a particle chosen at random at time 0, then evolve the process for a time $t \geq 0$, then evaluate a permutation-invariant function and take expectation. On the right-hand side, instead, we first evolve the process for a time t, then remove a randomly chosen particle, evaluate the same permutation-invariant function and take expectation. Then the intertwining relation (10.53) says that these two sequences of actions are equivalent. In other words, the operations "removing a randomly chosen particle" and "time evolution in the process followed by expectation" commute as long as we restrict to permutation-invariant functions. This suggests to give the following definition of consistent coordinate processes.

Definition 10.14 (Consistency for Cooordinate Process) A family of coordinate processes $\{X(t),\ t \geq 0\}$ is said to be *consistent* if, for all $n \in \mathbb{N}$, $n \geq 2$, and $i \in [n]$,

$$\begin{aligned} &\{(X_1^{(n)}(t), \ldots, X_i^{(n)}(t), X_{i+1}^{(n)}(t), \ldots, X_n^{(n)}(t)),\ t \geq 0\} \\ &\qquad = \{(X_1^{(n-1)}(t), \ldots, X_{n-1}^{(n-1)}(t)),\ t \geq 0\} \quad \text{in distribution.} \end{aligned} \tag{10.58}$$

Remark that, if $\{X(t),\ t \geq 0\}$ is a consistent coordinate process and $\boldsymbol{\mu} = \{\mu_n, n \in \mathbb{N}\}$ is the family of its initial distributions, then $\boldsymbol{\mu}$ is necessarily a consistent family of probability measures in the sense that, for all $n \in \mathbb{N}$, $n \geq 2$, $i \in [n]$ and $\mathbf{x} = (x_1, \ldots, x_n) \in V_n$,

$$\sum_x \mu_n(x_1, \ldots, x_i, x, x_{i+1}, \ldots, x_n) = \mu_{n-1}(x_1, \ldots, x_i, x_{i+1}, \ldots, x_n). \tag{10.59}$$

If $\{X(t),\ t \geq 0\}$ is also permutation-invariant, then also $\boldsymbol{\mu}$ is permutation invariant. In that case one has that *every m-dimensional marginal* of μ_n, $m \leq n$, coincides with μ_m, and, more generally, any m-dimensional marginal of $\{X^{(n)}(t),\ t \geq 0\}$, $m \leq n$, is equal in distribution to $\{X^{(m)}(t),\ t \geq 0\}$, i.e., for all $1 \leq i_1 < \ldots < i_m \leq n$,

$$\{X_{i_1}^{(n)}(t), \ldots, X_{i_m}^{(n)}(t),\ t \geq 0\} = \{X_1^{(m)}(t), \ldots, X_m^{(m)}(t),\ t \geq 0\} \quad \text{in distribution.} \tag{10.60}$$

Notice that one can simply find an example of such consistent permutation-invariant family $\boldsymbol{\mu} = \{\mu_n, n \in \mathbb{N}\}$ by taking

$$\mu_n = \frac{1}{n!} \sum_{\sigma \in \Sigma_n} \delta_{(x_{\sigma(1)}, \ldots, x_{\sigma(n)})} \qquad \text{for some } \mathbf{x} \in V_n. \tag{10.61}$$

In the following theorem we show the relation between the two notions of consistency, respectively in the configuration and in the coordinate variables.

Theorem 10.15 (Consistency of the Configuration Process Implies Consistency of the Coordinate Process) *Let $\{X(t),\ t \geq 0\}$ be a family of permutation-invariant coordinate processes initially distributed according to a consistent family of measures $\boldsymbol{\mu}$. If the unique compatible configuration process $\{\eta(t),\ t \geq 0\}$ is consistent, then also $\{X(t),\ t \geq 0\}$ is consistent.*

Proof Since $\{X(t),\ t \geq 0\}$ is permutation-invariant we have that also $\boldsymbol{\mu} = \{\mu_n, n \in \mathbb{N}\}$ is permutation-invariant. Assume that $\{\eta(t),\ t \geq 0\}$ is consistent, then $[L, S^-] = 0$, hence, from Lemma 10.13 we know that the intertwining relation

(10.53) holds true. Since $\{X(t),\ t \geq 0\}$ is permutation-invariant, in order to show that it is also consistent, it is sufficient to prove that for all $n \in \mathbb{N}$, $i \in [n]$ and $g \in \mathscr{C}_{n-1}$ permutation-invariant,

$$\begin{aligned}&\mathbb{E}_{\mu_n}^{(n)}\Big[g(X_1^{(n)}(t), \ldots, X_{i-1}^{(n)}(t), X_{i+1}^{(n)}(t), \ldots, X_n^{(n)}(t))\Big]\\&= \mathbb{E}_{\mu_{n-1}}^{(n-1)}\Big[g(X_1^{(n-1)}(t), \ldots, X_{n-1}^{(n-1)}(t))\Big]\end{aligned} \tag{10.62}$$

where $\mathbb{E}_{\mu_n}^{(n)}$ denotes expectation with respect to the coordinate process $\{X^{(n)}(t), t \geq 0\}$, started with distribution μ_n. Again from the hypothesis of permutation-invariance of the process it is sufficient to prove (10.62) for $i = n$. Fix $g \in \mathscr{C}_{n-1}$ permutation-invariant, then we have $\pi_l^{(n)} g = \pi_k^{(n)} g$ for all $k, l \in \{1, \ldots, n\}$. Hence, by consistency and permutation invariance of $\boldsymbol{\mu}$ one has

$$\int \pi_l g(x_1, \ldots, x_n)\mu_n(dx_1 \ldots dx_n) = \int g(x_1, \ldots, x_{n-1})\mu_{n-1}(dx_1 \ldots dx_{n-1}), \tag{10.63}$$

for all $n \in \mathbb{N}$ and all $l \in \{1, \ldots, n\}$. Therefore we have that

$$\begin{aligned}\mathbb{E}_{\mu_n}^{(n)}\Big[g(X_1^{(n)}(t), \ldots, X_{n-1}^{(n)}(t))\Big] &= \mathbb{E}_{\mu_n}^{(n)}\Big[(\pi_n^{(n)} g)(X_1^{(n)}(t), \ldots, X_{n-1}^{(n)}(t), X_n^{(n)}(t))\Big]\\&= \int [S_n(t)(\pi_n^{(n)} g)](x_1, \ldots, x_n)\mu_n(dx_1 \ldots dx_n)\\&= \frac{1}{n}\int S_n(t)(\Pi^{(n)} g)(x_1, \ldots, x_n)\mu_n(dx_1 \ldots dx_n).\end{aligned} \tag{10.64}$$

Now, from (10.53), we have that $S_n(t)(\pi_n^{(n)} g) = S_n(t)(\Pi^{(n)} g)$ and, as a consequence, we have that (10.64) is equal to

$$\begin{aligned}&\frac{1}{n}\int (\Pi^{(n)} S_{n-1}(t)g)(x_1, \ldots, x_n)\mu_n(dx_1 \ldots dx_n)\\&= \int (S_{n-1}(t)g)(x_1, \ldots, x_{n-1})\mu_{n-1}(dx_1 \ldots dx_n)\\&= \mathbb{E}_{\mu_{n-1}}^{(n-1)}\Big[g(X_1^{(n-1)}(t), \ldots, X_{n-1}^{(n-1)}(t))\Big].\end{aligned} \tag{10.65}$$

This proves that also $\{X(t),\ t \geq 0\}$ is consistent. □

Remark 10.16 One can wonder whether the inverse implication in Theorem 10.15 holds true, i.e., whether the consistency of a family of permutation-invariant coordinate processes $\{X(t),\ t \geq 0\}$ implies the consistency of the compatible configuration process $\{\eta(t),\ t \geq 0\}$. This is not exactly the case. Indeed, assuming that (10.62) holds true for all $g \in \mathscr{C}_{n-1}$ permutation-invariant, one can repeat the reasoning used in the previous proof to deduce that

$$\int S_n(t)(\Pi^{(n)}g)(x_1,\ldots,x_n)\mu_n(dx_1\ldots dx_n)$$
$$= \int (\Pi^{(n)}S_{n-1}(t)g)(x_1,\ldots,x_n)\mu_n(dx_1\ldots dx_n).$$

Then, differentiating with respect to t and evaluating at time 0 one gets

$$\int L_n(\Pi^{(n)}g)(x_1,\ldots,x_n)\mu_n(dx_1\ldots dx_n)$$
$$= \int (\Pi^{(n)}L_{n-1}g)(x_1,\ldots,x_n)\mu_n(dx_1\ldots dx_n),$$

for all $g \in \mathscr{C}_{n-1}$ permutation-invariant. This is however not sufficient to get (10.53) that is the condition needed to guarantee consistency of $\{\eta(t),\ t \geq 0\}$.

10.4.3 Combinatorial Interpretation of Consistency

In this section we will analyze the combinatorial meaning of the weighted factorial moments defined in Sect. 10.3. With this aim in mind, it is useful to rewrite the function $F(\xi,\eta)$ defined in (10.20) in coordinate notation. Let $m, n \in \mathbb{N}$ with $m \leq n$ we define the set $C_{m,n}$ of combinations of m elements chosen in $[n] := \{1,\ldots,n\}$,

$$C_{m,n} := \{(i_1,\ldots,i_m) : i_j \in [n],\ \forall\, j \in [m] \text{ s.t. } i_1 < i_2 < \ldots < i_m\} \subset [n]^m. \tag{10.66}$$

Let now $I := (i_1,\ldots,i_m)$ be an element of $[n]^m$, and let $\mathbf{x} \in V_n$ the vector of positions of n particles in the lattice V, then we denote by $\mathbf{x}_I$ the m-tuple:

$$\mathbf{x}_I := (x_{i_1},\ldots,x_{i_m}) \in V_m, \tag{10.67}$$

i.e. the vector of the positions of the subset of particles whose label is in I.Then the value $F(\xi,\eta)$ can be interpreted as the number of ways to choose, for each site $x \in V$, ξ_x particles out of η_x. Then, for every fixed particles labeling of the configuration η, i.e., for every $\mathbf{x} \in V_{|\eta|}$ such that $\varphi(\mathbf{x}) = \eta$, we can write

$$F(\xi,\eta) = \left|\{I \in C_{|\xi|,|\eta|} : \varphi(\mathbf{x}_I) = \xi\}\right| \mathbb{1}_{\{\xi\leq\eta\}}. \tag{10.68}$$

In other words, for any fixed labeling $\mathbf{x}$ of particles in the configuration η, $F(\xi, \eta)$ is the number of ways to select $|\xi|$ particles out of $|\eta|$ in such a way that the corresponding configuration in Ω is ξ.

Alternatively we can define the following ordering between elements of the coordinate state spaces. For $\mathbf{y} \in V_m$, and $\mathbf{x} \in V_n$ we say that

$$\mathbf{y} \leq \mathbf{x} \qquad \text{if and only if} \qquad m \leq n \quad \text{and} \quad \exists I \in C_{m,n} \quad \text{s.t.} \quad \mathbf{y} = \mathbf{x}_I. \tag{10.69}$$

In view of this, we can rewrite (10.68) as follows:

$$F(\xi, \eta) = \left|\{\mathbf{y} \in V_{|\xi|} : \mathbf{y} \leq \mathbf{x},\ \varphi(\mathbf{y}) = \xi\}\right| \mathbb{1}_{\{\xi \leq \eta\}}, \qquad \forall \mathbf{x} : \ \varphi(\mathbf{x}) = \eta. \tag{10.70}$$

This suggests that, summations of the type:

$$\sum_{\xi \in \Omega_m} F(\xi, \eta) f(\xi), \tag{10.71}$$

simplifies when switching from configurations to coordinate variables. Indeed, for every $\mathbf{x} \in V_{|\eta|}$ such that $\varphi(\mathbf{x}) = \eta$ we have

$$\begin{aligned} \sum_{\xi \in \Omega_m} F(\xi, \varphi(\mathbf{x})) f(\xi) &= \sum_{\substack{\xi \in \Omega_m \\ \xi \leq \varphi(\mathbf{x})}} \left|\{I \in C_{m,n} : \ \varphi(\mathbf{x}_I) = \xi\}\right| \cdot f(\xi) \\ = \sum_{\substack{\xi \in \Omega_m \\ \xi \leq \varphi(\mathbf{x})}} \sum_{\substack{I \in C_{m,n}: \\ \varphi(\mathbf{x}_I) = \xi}} f(\varphi(\mathbf{x}_I)) &= \sum_{I \in C_{m,n}} f(\varphi(\mathbf{x}_I)) = \sum_{\substack{\mathbf{y} \in V_m \\ \mathbf{y} \leq \mathbf{x}}} f(\varphi(\mathbf{y})). \end{aligned} \tag{10.72}$$

With the help of the coordinate notation it is then possible, using (10.72), to further simplify the consistency relation proven in Theorem 10.5. Indeed by applying (10.72) with $f(\varsigma) = \mathbb{P}_\varsigma(\varsigma(t) = \xi)$ we obtain an expression for the weighted factorial moments of a consistent particle system $\{\eta(t),\ t \geq 0\}$ on a lattice V. For all $\eta, \xi \in \Omega$ such that $1 \leq |\xi| < |\eta|$, we have

$$\mathbb{E}_\eta\left[F(\xi, \eta(t))\right] = \sum_{\substack{\mathbf{y} \in V_{|\xi|} \\ \mathbf{y} \leq \mathbf{x}}} \mathbb{P}_{\varphi(\mathbf{y})}(\xi(t) = \xi) \qquad \text{for all} \quad \mathbf{x} : \ \varphi(\mathbf{x}) = \eta. \tag{10.73}$$

In other words we have that the ξ-th order factorial moment of a process with $|\eta|$ particles at time t, is equal to a sum of transition probabilities for a system with $|\xi|$ particles, with $|\xi| < |\eta|$. Notice that, choosing $\xi = \delta_y$ we obtain again the result for expectations of occupancies obtained in Proposition 10.7.

10.5 The Reference Process

In this section we introduce a unifying notation which summarizes in one definition the class of processes studied in Chaps. 3, 4, 5 and 7. These processes share the consistency property and are all reversible with respect to homogeneous product measures. Then, using Theorem 10.8 a self-duality property can be deduced as a consequence of these two facts. We use the parameter $\theta \in \{-1, 0, 1\}$ to identify the type of interaction: $\theta = 0$ corresponds to IRW, $\theta = -1$ to SEP and $\theta = -1$ to SIP. The other parameters of the processes are fixed by the vector $\boldsymbol{\alpha} = \{\alpha_x,\ x \in V\}$. While for $\theta \in \{0, 1\}$, (IRW, SIP), we can let α_x vary in the set of positive real numbers $(0, +\infty)$, for $\theta = -1$ (SEP), the parameter α_x has also the meaning of maximal occupancy at site x, as a consequence it must be $\mathbb{N}$-valued (in such a way to have positive jump rates). So we define

$$\mathfrak{A}_\theta = \begin{cases} (0, \infty) & \text{for } \theta \in \{0, 1\} \\ \mathbb{N} & \text{for } \theta = -1. \end{cases} \tag{10.74}$$

that is the set where the parameters α_x take values in the different cases.

Definition 10.17 (The Reference Process) Let V be a finite set of vertices, $\theta \in \{-1, 0, 1\}$ and $\boldsymbol{\alpha} = \{\alpha_x,\ x \in V\}$ with $\alpha_x \in \mathfrak{A}_\theta$. Moreover we fix a symmetric irreducible transition function $p : V \times V \to [0, \infty)$. We say that an interacting particle system $\{\eta(t) : t \geq 0\}$ on the finite set V is a *reference process* with parameters $(\theta, \boldsymbol{\alpha}, p)$ if it has generator

$$L^{(\theta,\boldsymbol{\alpha},p)} = \frac{1}{2} \sum_{x,y \in V} p(x, y)\, L^{(\theta,\boldsymbol{\alpha})}_{x,y}, \tag{10.75}$$

with

$$L^{(\theta,\boldsymbol{\alpha})}_{x,y} f(\eta) = \eta_x(\theta\eta_y + \alpha_y)[f(\eta^{x,y}) - f(\eta)] + \eta_y(\theta\eta_x + \alpha_x)[f(\eta^{y,x}) - f(\eta)], \tag{10.76}$$

and state space

$$\Omega_{\theta,\boldsymbol{\alpha}} = \otimes_{x \in V} \Upsilon^{(\theta)}_x, \tag{10.77}$$

where $\Upsilon^{(\theta)}_x$ is the space of occupation numbers of the x-th site, i.e.

$$\Upsilon^{(\theta)}_x = \begin{cases} \mathbb{N} & \text{for } \theta \in \{0, +1\} \\ \{1, 2, \ldots, \alpha_x\} & \text{for } \theta = -1. \end{cases} \tag{10.78}$$

The reference processes are nothing else than the self-dual processes studied in Chaps. 3, 4, 5 and 7, namely:

$$L^{(\theta,\boldsymbol{\alpha},p)} = \begin{cases} L^{\text{IRW}(\boldsymbol{\alpha},p)} & \text{for } \theta = 0 \\ L^{\text{SIP}(\boldsymbol{\alpha},p)} & \text{for } \theta = +1 \\ L^{\text{SEP}(\boldsymbol{\alpha},p)} & \text{for } \theta = -1\,. \end{cases} \tag{10.79}$$

10.5.1 *Consistency of the Reference Process*

All the processes belonging to the class of reference processes defined above are consistent. This is due to the fact that, in all cases, the infinitesimal generator $L^{(\theta,\boldsymbol{\alpha},p)}$ commutes with the removal operator $S^- = \sum_{x\in V} a_x$, i.e.

$$[L^{(\theta,\boldsymbol{\alpha},p)}, S^-] = 0\,. \tag{10.80}$$

10.5.2 *Reversibility of the Reference Process*

The reference process $\{\eta(t) \,:\, t \geq 0\}$ with parameters $(\theta, \boldsymbol{\alpha}, p)$ as introduced in Definition 10.17 is reversible w.r.t. the one-parameter family of product probability measures

$$\{\nu_{\rho,\theta,\boldsymbol{\alpha}},\ \rho \in \mathcal{R}_\theta\} \qquad \text{with} \qquad \mathcal{R}_\theta := \begin{cases} [0, +\infty) & \text{for } \theta \in \{0, 1\} \\ [0, 1] & \text{for } \theta = -1, \end{cases} \tag{10.81}$$

where $\nu_{\rho,\theta,\boldsymbol{\alpha}} = \otimes_{x\in V}\, \nu_{\rho,\theta,\alpha_x}$ with marginals given by:

$$\nu_{\rho,\theta,\alpha}(n) = \begin{cases} \frac{(\rho\alpha)^n}{n!} \cdot e^{-\rho\alpha} & \text{for } \theta = 0, \quad \text{Pois}(\rho\alpha) \\[2ex] \frac{1}{(1+\rho)^\alpha} \cdot \left(\frac{\rho}{1+\rho}\right)^n \cdot \frac{\Gamma(\alpha+n)}{n!\Gamma(\alpha)} & \text{for } \theta = +1, \quad \text{DGamma}\left(\alpha,\, \frac{\rho}{1+\rho}\right) \\[2ex] (1-\rho)^\alpha \cdot \left(\frac{\rho}{1-\rho}\right)^n \cdot \frac{\Gamma(\alpha+1)}{n!\Gamma(\alpha+1-n)} & \text{for } \theta = -1, \quad \text{Bin}\,(\alpha, \rho)\,. \end{cases} \tag{10.82}$$

The parameter $\rho \in \mathcal{R}_\theta$ labeling the reversible measures has the meaning of a *weighted density*, indeed, for all sites $x \in V$, it is equal to the expected number of particles in that site divided by the corresponding intensity parameter α_x:

$$\rho = \mathbb{E}_{\nu_{\rho,\theta,\boldsymbol{\alpha}}}\left[\frac{\eta_x}{\alpha_x}\right] \qquad \text{for all} \quad \theta \in \{-1, 0, +1\}. \tag{10.83}$$

10.5.3 Self-duality of the Reference Process

As a consequence of Theorem 10.8 it follows that the references processes are self-dual with self-duality functions of the form $D(\xi,\eta) = F(\xi,\eta)/\nu(\xi)$ with F as in (10.20) and ν_ρ is one of its reversible measures defined in (10.82). In this way, we obtain that the duality function of the reference process with parameters $(\theta, \boldsymbol{\alpha}, p)$ are (modulo a factor depending only on the total number of dual particles $|\xi|$) given by:

$$D_{\theta,\boldsymbol{\alpha}}(\xi,\eta) = \prod_{i\in V} d_{\theta,\alpha_i}(\xi_i,\eta_i), \tag{10.84}$$

with

$$d_{\theta,\alpha}(k,n) = \frac{n!}{(n-k)!}\,\mathbb{1}_{\{k\le n\}}\cdot\begin{cases}\frac{1}{\alpha^k} & \text{for } \theta=0\\ \frac{\Gamma(\alpha)}{\Gamma(\alpha+k)} & \text{for } \theta=+1\\ \frac{\Gamma(\alpha+1-k)}{\Gamma(\alpha+1)} & \text{for } \theta=-1.\end{cases} \tag{10.85}$$

Consistently with the scheme (10.79), the function (10.84)–(10.85) reproduces, depending on θ, the triangular duality functions introduced in the previous chapters for the IRW ($\theta = 0$), for the inhomogeneous exclusion process ($\theta = -1$) and for the inhomogeneous inclusion process ($\theta = +1$).

10.5.4 Coordinate Reference Processes

In this section we provide a coordinate description of the reference process by labeling particles such that we have permutation invariance. Then Theorem 10.15 ensures the consistency of the latter also at the level of the coordinate process. For the reference process of parameters $(\theta, \boldsymbol{\alpha}, p)$ we define the set $\Omega_{n,\theta,\boldsymbol{\alpha}} = \{\eta \in \Omega_{\theta,\boldsymbol{\alpha}} : |\eta| = n\}$, i.e. the set of configurations of the state space consisting of n particles. We denote by $V_{n,\theta,\boldsymbol{\alpha}}$ the subset of n-tuples of V^n such that $\varphi(V_{n,\theta,\boldsymbol{\alpha}}) = \Omega_{n,\theta,\boldsymbol{\alpha}}$. For $\theta = 1$ or $\theta = 0$, $V_{n,\theta,\boldsymbol{\alpha}} = V^n$ whereas for $\theta = -1$, $V_{n,\theta,\boldsymbol{\alpha}}$ is the set of n-tuples such that at every site $x \in V$ there are at most α_x particles.

Definition 10.18 We define the coordinate reference process with parameters $(\theta, \boldsymbol{\alpha}, p)$ and $n \in \mathbb{N}$ particles the process $\{X^{(n)}(t),\ t \ge 0\}$ on $V_{n,\theta,\boldsymbol{\alpha}}$ whose generator is given by:

$$L_n^{(\theta,\boldsymbol{\alpha},p)} g(\mathbf{x}) = \sum_{y\in V}\sum_{\ell=1}^{n} p(\{x_\ell, y\})\left(\alpha_j + \theta\sum_{m=1}^{n} \mathbf{1}_{x_m=y}\right)\left(g(\mathbf{x}^{\ell\to y}) - g(\mathbf{x})\right), \tag{10.86}$$

where $\mathbf{x} = (x_1, \ldots, x_n) \in V_{n,\theta,\boldsymbol{\alpha}}$ and $\mathbf{x}^{\ell \to y}$ is obtained from $\mathbf{x}$ by moving the ℓ-th particle from its position x_ℓ to the site y. Moreover we call the collection $(\{X^{(n)}(t),\ t \geq 0\}, n \in \mathbb{N}) := \{X(t) : t \geq 0\}$ the family of coordinate reference processes with parameters $(\theta, \boldsymbol{\alpha}, p)$.

In the next proposition we will see that the family of coordinate reference processes is compatible with the corresponding reference process. Indeed projecting the particles jump rates appearing in the generator (10.86) to the configuration space, one clearly obtains the reference process jump rates given in (10.76). In the dynamics generated by (10.76) particles are indistinguishable. When there is a jump from a site x to a site y, the particle that makes the jump is uniformly chosen among all the particles hosted at site x. Thanks to this symmetry of the dynamics we can prove permutation-invariance. Even though (10.86) is not the only possible coordinate process compatible with the reference process, thanks to permutation-invariance, it seems to be the most natural choice for guaranteeing consistency.

Proposition 10.19 *The family of coordinate reference processes* $\{X(t),\ t \geq 0\}$ *with parameters* $(\theta, \boldsymbol{\alpha}, p)$ *is compatible with the reference process with the same parameters. Moreover* $\{X(t),\ t \geq 0\}$ *is a family of permutation-invariant consistent processes provided that the family of its initial distributions is consistent and permutation-invariant.*

Proof The compatibility statement immediately follows from the fact that the relation (10.43) holds true. In order to prove permutation-invariance, it is sufficient to show that

$$\mathbb{E}_{\mu_n}\Big[g(X_1^{(n)}(t), \ldots, X_n^{(n)}(t))\Big] = \mathbb{E}_{\mu_n}\Big[g(X_{\sigma(1)}^{(n)}(t), \ldots, X_{\sigma(n)}^{(n)}(t))\Big], \tag{10.87}$$

for all $\sigma \in \Sigma_n$ and $g \in \mathscr{C}_n$. This amounts to prove that

$$\begin{aligned} &\int [S_n(t)g](x_1, \ldots, x_n)\mu_n(dx_1 \ldots dx_n) \\ &= \int [S_n(t)\boldsymbol{\sigma} g](x_1, \ldots, x_n)\mu_n(dx_1 \ldots dx_n), \end{aligned} \tag{10.88}$$

where $\boldsymbol{\sigma}$ is the map $\boldsymbol{\sigma} : \mathscr{C}_n \to \mathscr{C}_n$ acting as follows:

$$\boldsymbol{\sigma} g(x_1, \ldots, x_n) = g(x_{\sigma(1)}, \ldots, x_{\sigma(n)}). \tag{10.89}$$

The relation (10.88) is true at time 0 by hypothesis, while, for positive times it immediately follows from the permutation-invariance of the generator:

$$\boldsymbol{\sigma}(L_n g) = L_n(\boldsymbol{\sigma} g), \tag{10.90}$$

which can be verified using the definition in (10.86). Indeed the jump rates when starting at $\mathbf{x}$ depend only on $\varphi(\mathbf{x})$. This concludes the proof of permutation invariance. Then, using Theorem 10.15, we have that the consistency of $\{X(t),\ t \geq 0\}$ follows as a consequence of permutation-invariance, compatibility with the reference process and the consistency of the latter. □

10.6 Consistency for Systems with Absorbing Sites

In this section we will see that the consistency property can be preserved when adding absorbing sites in the system, if the absorption rates are properly chosen. We will see how, for systems with absorbing sites, the recurrence relations (10.73) can be used to obtain information about the absorption probabilities. Notice that, in the presence of absorbing sites, the process does not admit a strictly-positive reversible measure. Therefore, the hypothesis of Theorem 10.8 are, thus, not met, and, as a consequence, we lack a strategy to produce a self-duality function in this context. Therefore the consistency property is fundamental here to obtain recurrence relations without having to use duality.

The interest in processes with absorbing sites arises primarily from their use in the setting of non-equilibrium systems. In Chap. 11 we will see that systems with absorbing boundaries arise as duals of systems with boundary reservoirs. As a consequence, in this setting, factorial moment relations of the type (10.29) will provide a lot of information about multivariate moments and correlation functions in non-equilibrium stationary state of boundary driven systems. Before analyzing the factorial moments, we will give an alternative description of consistent systems with absorbing sites that passes through the particle addition and removal operators a and $a^\dagger$ introduced in Chap. 3.

Fix a finite set of vertices V and an operator L^{bulk} that is the generator of a configuration process on V, with state space Ω. We then consider an extended set of vertices $V^* := V \cup V^{\text{abs}}$, with $V \cap V^{\text{abs}} = \emptyset$, where V^{abs} is a set of absorbing sites. We want to define a configuration process $\{\eta(t) : t \geq 0\}$ on V^* for which

$$\lim_{t\to\infty} \mathbb{P}_\eta(\eta(t) \in V^{\text{abs}}) = 1. \tag{10.91}$$

To do that we define the generator L as follows:

$$L = L^{\text{bulk}} + L^{\text{abs}}, \tag{10.92}$$

acting on functions $f : \Omega^* \to \mathbb{R}$, where

$$\Omega^* := \Omega \times \Omega^{\text{abs}} \qquad \text{with} \qquad \Omega^{\text{abs}} := \mathbb{N}^{V^{\text{abs}}}. \tag{10.93}$$

The bulk part of the generator L^{bulk} only acts on the variables $\{\eta_x, x \in V\}$, whereas L^{abs}, the absorption part of the generator, works on the whole Ω^* as follows

$$L^{\text{abs}} f(\eta) = \sum_{\substack{x \in V \\ y \in V^{\text{abs}}}} p(x, y)\, \eta_x \left[f(\eta^{x,y}) - f(\eta) \right], \tag{10.94}$$

where we extended the definition of the function p to the domain $V \times V^*$. More precisely the transition function p is a function $p : V \times V^* \to [0, \infty)$, whose restriction to the domain $V \times V$ is symmetric and irreducible.

The dynamics for the process $\{\eta(t), t \geq 0\}$ with generator L is then the following: inside V particles move according to the generator L^{bulk}, and additionally a particle at site $x \in V$ jumps at rate $p(x, y)$ to an absorbing site $y \in V^{\text{abs}}$, independently from the other particles. Once a particle reaches the set V^{abs} it is absorbed and does not move anymore. We thus obtain what we call the process with absorbing sites V^{abs} and absorption rates $p(x, y)$. We call such process an absorbing extension of the process in the bulk.

Definition 10.20 A configuration process on $V^* = V \cup V^{\text{abs}}$ with generator $L = L^{\text{bulk}} + L^{\text{abs}}$ (L^{abs} as in (10.92)) is called an absorbing extension of the configuration process on V with generator L.

We remark that it is possible to rewrite the absorption operator L^{abs} in terms of the creation and annihilation operators $a^\dagger$ and a of the Heisenberg algebra introduced in Chap. 3. We have the following expression:

$$L^{\text{abs}} = \sum_{\substack{x \in V \\ y \in V^{\text{abs}}}} p(x, y)[a_x a_y^\dagger - a_x a_x^\dagger], \tag{10.95}$$

where $a_x^\dagger$ and a_x are, respectively, the particle addition and particle removal operators at site x defined in (3.10)–(3.9). This has an important consequence in the context of consistent processes:

Lemma 10.21 *Let $\{\eta(t), t \geq 0\}$ be a consistent configuration process on a finite set of vertices V, then every absorbing extension to a set of vertices $V^* \supset V$ is a consistent process.*

Proof Let S_{bulk}^- be the particle removal operator in the bulk and S_{abs}^- the particle removal operator in V^{abs}, i.e.

$$S_{\text{bulk}}^- := \sum_{x \in V} a_x \qquad \text{and} \qquad S_{\text{abs}}^- := \sum_{x \in V^{\text{abs}}} a_x. \tag{10.96}$$

We need to prove that

$$[L, S_{\text{bulk}}^- + S_{\text{abs}}^-] = 0. \tag{10.97}$$

By assumption we already know that $[L^{\text{bulk}}, S^{-}_{\text{bulk}}] = 0$. We also know by construction that $[L^{\text{bulk}}, S^{-}_{\text{abs}}] = 0$ then it only remains to prove that

$$[L^{\text{abs}}, S^{-}_{\text{abs}} + S^{-}_{\text{bulk}}] = 0.$$

Using (10.95) and the fact that operators working on variables at different sites commute, it is sufficient to show that for all $x, y \in V^*$

$$\begin{aligned}[a_x a_y^{\dagger} - a_x a_x^{\dagger}, a_x + a_y] &= a_x[a_y^{\dagger}, a_y] - a_x[a_x^{\dagger}, a_x] \\ &= a_x - a_x = 0.\end{aligned}$$

Here we used the commutation relations $[a_x, a_y] = 0,\ [a_y^{\dagger}, a_x] = \delta_{x,y}$. □

10.6.1 Factorial Moments in Absorbing Systems

In what follows we denote by $\mathbb{P}_{\eta}(\eta(\infty) = \varsigma)$ the probability that eventually $\eta(t)$ settles in the absorbing configuration $\varsigma \in \Omega^{\text{abs}}$ starting from the initial configuration $\eta \in \Omega^*$, i.e.,

$$\mathbb{P}_{\eta}(\eta(\infty) = \varsigma) := \lim_{t\to\infty} \mathbb{P}_{\eta}(\eta(t) = \varsigma). \tag{10.98}$$

Similarly,

$$\mathbb{E}_{\eta}[f(\eta(\infty))] = \lim_{t\to\infty} \mathbb{E}_{\eta}[f(\eta(t))]. \tag{10.99}$$

As an immediate consequence of Lemma 10.21 and Theorem 10.5 we have the following formula for the factorial moments as $t \to \infty$.

Theorem 10.22 *Let $\{\eta(t),\ t \geq 0\}$ be a consistent configuration process on a finite set V^* with generator $L = L^{\text{bulk}} + L^{\text{abs}}$. Let $\eta \in \Omega^*$ and $\xi \in \Omega^{\text{abs}}$ with $1 \leq |\xi| \leq |\eta| - 1$, then*

$$\mathbb{E}_{\eta}\left[F(\xi, \eta(\infty))\right] = \sum_{\substack{\varsigma\in\Omega_{|\xi|} \\ \varsigma\leq\eta}} F(\varsigma, \eta) \cdot \mathbb{P}_{\varsigma}\left(\varsigma(\infty) = \xi\right). \tag{10.100}$$

Here the ξ-th order factorial moment of absorption probabilities for a system with $|\eta|$ particles is a sum of absorption probabilities for systems with less than $|\xi|$ particles, if $|\xi| < |\eta|$. Although these equations are not sufficient to determine the absorption probabilities in closed form, they are still considerably simplifying the problem of computing them, as they imply severe restrictions.

Notice that, choosing $\xi = \delta_y,\ y \in V^{\text{abs}}$ in (10.100) gives a statement for the expectations of occupation numbers:

$$\mathbb{E}_{\varphi(\mathbf{x})}\left[\eta_y(\infty)\right] = \sum_{k=1}^{n} \mathbf{P}_{x_k}(X^{\text{rw}}(\infty) = y), \tag{10.101}$$

where $\mathbf{P}_x$ is the path-space measure of the random walk $\{X^{\text{rw}}(t),\ t \geq 0\}$ on V^* starting from $x \in V^*$ associated to the "absorbed" configuration process $\{\eta(t),\ t \geq 0\}$ as in Definition 10.6.

10.6.2 Two Absorbing Sites

In case the system is in contact with only two absorbing sites, say ℓ and r, the limiting distribution $t \to \infty$ is completely determined by a single random variable, namely, the occupancy at one of the two absorbing sites, say $\xi_r(\infty)$. For this reason we can say that, to determine the limiting distribution, the system has, in a sense, one degree of freedom. Hence, in this case the problem of finding absorption probability drastically simplifies and Eq. (10.100) provide us a lot of information. We consider a consistent configuration process $\{\eta(t),\ t \geq 0\}$ on the extended vertex set $V^* = V \cup V^{\text{abs}}$ with V finite and absorbing set $V^{\text{abs}} = \{\ell, r\}$. The generator of the process L is of the form (10.92) and the state space is Ω^* as in (10.93).

In order to give some physical meaning to this setting we can think the bulk lattice as a one-dimensional chain $V = \{1, \ldots, N\}$ where particles can jump only to nearest-neighbouring sites, with interactions of the type "reference process" of Definition 10.17. We now extend the lattice with two extra sites ℓ, at the left of site 1 and r at the right of site N. These extra sites are absorbing and can be reached only from boundary sites, i.e. ℓ can be reached only from site 1 and r can be reached only from site N. This construction defines an absorbing extension of the reference process with absorbing set $V^* = \ell, r$. The generator of this process is given by:

$$\mathscr{L} = L^{\text{bulk}} + L^{\text{abs}}, \tag{10.102}$$

with

$$(L^{\text{bulk}} f)(\eta) = \sum_{x=1}^{N-1} \Big\{ \eta_x(\alpha_{x+1} + \theta\eta_{x+1})[f(\eta^{x,x+1}) - f(\eta)] + \eta_{x+1}(\alpha_x + \theta\eta_x)[f(\eta^{x+1,x}) - f(\eta)] \Big\}, \tag{10.103}$$

and

$$[L^{\text{abs}} f](\eta) = \alpha_\ell \, \eta_1 \left[f(\eta^{1,\ell}) - f(\eta) \right] + \alpha_r \, \eta_N \left[f(\eta^{N,r}) - f(\eta) \right]. \tag{10.104}$$

For the process above we can apply Theorem 10.22 with the choice $\xi = m\delta_r$, to obtain a formula for the mth factorial moment of the r-th occupancy $\eta_r(\infty)$. Let $\eta \in \Omega_n^*$ and $\mathbf{x} \in V_n^*$ be such that $\varphi(\mathbf{x}) = \eta$ then

$$\begin{aligned} \mathbb{E}_\eta \left[\binom{\eta_r(\infty)}{m} \right] &= \sum_{\substack{\varsigma \in \Omega_m \\ \varsigma \leq \eta}} F(\varsigma, \eta) \cdot \mathbb{P}_\varsigma \left(\varsigma_\ell(\infty) = 0 \right) \qquad (10.105) \\ &= \sum_{x \in C_{m,n}} \mathbb{P}_{\varphi(\mathbf{x}_I)} \left(\eta_\ell(\infty) = 0 \right) \quad \text{for} \quad m \in \{1, \ldots, |\eta| - 1\}. \end{aligned}$$

This can be viewed as a linear system of $n - 1$ independent equations in the $n + 1$ variables $\{\mathbb{P}_\eta \left(\eta_\ell(\infty) = m \right), \ m = 0, \ldots, n\}$. Complementing these with the normalization condition $\sum_{m=0}^{n} \mathbb{P}_\eta \left(\eta_\ell(\infty) = m \right) = 1$ we obtain n independent equations. This is still not sufficient to get a closed-form expression for the absorption probabilities (which are $n + 1$ unknowns), since, at each level, one independent equation is still missing.

Notice that the result above does not depend on the specific choice of the generator, but holds true for all absorbing extensions of consistent processes with only two absorbing sites.

10.7 A Consistent Non-reversible Particle System

In this final section we provide and example of a consistent process beyond the processes with generator of the form (10.75)–(10.76). Because this system is not reversible (its invariant measure is unknown, not of product form and probably complicated), the consistency does not lead to a simple self-duality relation. We consider $V = \{1, \ldots N\}$ and define a process with nearest-neighbor jumps whose generator is given by:

$$L = \sum_{x=1}^{N-1} L_{x,x+1}, \tag{10.106}$$

where

$$\begin{aligned} L_{x,x+1} f(\eta) &= \eta_x (\eta_{x+1} + \alpha) [f(\eta^{x,x+1}) - f(\eta)] \\ &\quad + \eta_{x+1} (\eta_x + \beta) [f(\eta^{x+1,x}) - f(\eta)], \end{aligned} \tag{10.107}$$

for some $\alpha, \beta > 0$. For $\alpha = \beta$ this coincides with the symmetric inclusion process on V with nearest-neighbour jumps. For $\alpha \neq \beta$ this defines an asymmetric version of the inclusion process. The asymmetry parameters are not introduced in the standard way, i.e. as factors multiplying respectively the right and the left jump rates. On the contrary they affect only the diffusive term of the rates and not the inclusion part of the interaction. We have that this is the correct choice for the asymmetry in order to guarantee a consistency property.

In order to prove consistency, it is convenient to rewrite the generator in terms of the operators K^+, K^- and K^0 that have been defined in (5.5). We recall here their action, making sure, this time, to keep track of the dependence on the parameter α. For all $\alpha > 0$ we denote by $K^{+,\alpha}$, K^- and $K^{0,\alpha}$ the operators working on functions $f : \mathbb{N} \to \mathbb{R}$ as follows:

$$\begin{aligned} K^{+,\alpha} f(n) &= (\alpha + n) f(\eta + 1), \\ K^- f(n) &= n f(n-1), \\ K^{0,\alpha} f(n) &= \left(\tfrac{\alpha}{2} + n\right) f(n). \end{aligned} \tag{10.108}$$

Notice that, differently from $K^{+,\alpha}$ and $K^{0,\alpha}$, the operator K^- does not depend on the parameter α. To prove consistency, we have to show that the generator L commutes with the removal operator $S^- = \sum_{x \in V} K_x^-$.

Analogously to what we have seen in Chap. 5, we write the single-edge generator $L_{x,x+1}$ in its abstract form as follows:

$$L_{x,x+1} = K_x^{+,\alpha} K_{x+1}^- + K_x^- K_{x+1}^{+,\beta} - 2K_x^{0,\alpha} K_{x+1}^{0,\beta} + \frac{\alpha\beta}{2}. \tag{10.109}$$

We recall the commutation relations:

$$\begin{aligned} [K_y^{0,\alpha}, K_x^-] &= K_x^- \delta_{x,y}, \\ [K_x^{+,\alpha}, K_y^-] &= 2K_x^{0,\alpha} \delta_{x,y}. \end{aligned} \tag{10.110}$$

In order to prove the consistency, i.e., $[L, S^-]=0$, it is sufficient to prove that, for all $x \in \{1, \ldots, N-1\}$,

$$[L_{x,x+1}, K_x^- + K_{x+1}^-] = 0. \tag{10.111}$$

This is obtained via the following computation

$$\begin{aligned} &\left[K_x^{+,\alpha} K_{x+1}^- + K_x^- K_{x+1}^{+,\beta} - 2K_x^{0,\alpha} K_{x+1}^{0,\beta} + \frac{\alpha\beta}{2}, K_x^- + K_{x+1}^-\right] \\ &= [K_x^{+,\alpha}, K_x^-] K_{x+1}^- - 2[K_x^{0,\alpha}, K_x^-] K_{x+1}^{0,\beta} \end{aligned}$$

$$+ K_x^- [K_{x+1}^{+,\beta}, K_{x+1}^-] - 2K_x^{0,\alpha}[K_{x+1}^{0,\beta}, K_{x+1}^-]$$
$$= 2K_x^{0,\alpha} K_{x+1}^- - 2K_x^- K_{x+1}^{0,\beta} + 2K_x^- K_{x+1}^{0,\beta} - 2K_x^{0,\alpha} K_{x+1}^- = 0. \quad (10.112)$$

It is possible to prove that the process with generator (10.106)–(10.107) does not admit a strictly positive product reversible measure. As a consequence, it does not fit into the scheme of Theorem 10.8 and therefore it is not possible to use consistency to deduce a self-duality property with duality functions factorizing over the sites.

10.8 Additional Notes

The original idea of consistency comes from an isomorphism property between the m-marginals transition probabilities of a system with n particles ($n > m$) and the transition probabilities of the system with m particles. This property is equivalent to consistency and was shown in Proposition 3.1 of [152] through combinatorial arguments. It is proven for the dual of the Kipnis-Marchioro-Presutti model that is an instantaneous-particle-redistribution-model with absorbing boundaries. The property is used there to show local equilibrium for the KMP. The dual process of the KMP is what we refer to as Th-SIP(1) with absorbing boundaries in Appendix C. The model fits in the $\mathfrak{su}(1, 1)$ Lie algebra scheme and it is nothing else than the instantaneous-thermalization limit of the SIP(1) with absorbing boundaries. As a consequence its consistency property immediately follows, in our setting, from the consistency of the inclusion process.

In [47] the authors characterize the interacting particle systems exhibiting consistency, find recurrence relations for factorial moments and investigate the link with duality. The work has a top-down perspective to the study of duality aiming to detect the whole class of processes satisfying a given property. Then self-dual processes with factorized triangular duality function emerge as a subclass of the class of consistent processes. The method recalls the approach used in [198] to characterize the class of self-dual interacting particle systems with factorized duality function.

An analogous notion of consistency has been developed in [137] for diffusion processes, and more precisely for a stochastic flow of kernels [6, 168] consisting of a family of sticky Brownian motions. Here the authors characterize the process via martingale problems. A consistency property for the same class of processes was proven in [167, 168] via a Dirichlet forms approach. The emergence of sticky Brownian motions in the scaling limit, in the condensation regime, of two SIP particles [6], raises the question of the relation between the two notions of consistency, i.e. consistency for families of sticky Brownian motions and consistency for interacting particle systems. The latter represents an open problem, as well as the understanding of consistent particle systems in terms of Dirichlet forms.

Chapter 11
Duality for Non-equilibrium Systems

In this chapter we extend the duality results obtained so far to the setting of *boundary-driven non-equilibrium systems*, i.e., systems driven out of equilibrium via the action of multiple reservoirs where particles can enter and leave the system. We will see that in this context, the processes are no longer self-dual, but dual to processes where the reservoirs are replaced with absorbing sites. In the first section we start with the simplest setting of independent random walkers moving on a one-dimensional chain, in contact with two external reservoirs placed at left and right ends of the system. In Sect. 11.3 we will consider the whole class of interacting particle systems studied in the previous chapters where we add the action of reservoirs. This class includes independent random walkers, symmetric inclusion process and symmetric exclusion process, in their inhomogeneous version, on general graphs and with a general set of reservoirs. For all these models we will prove duality between the process with reservoirs and the process with absorbing boundaries, both with "triangular" and with "orthogonal" duality functions. As applications of these duality results, we will prove existence and uniqueness of the so-called *non-equilibrium steady state*. We will use the triangular duality to prove properties of the "n-point correlation function", and we will use the orthogonal duality to obtain information about the multivariate centered moments. In Sect. 11.4 we will see how to add reservoirs for processes with continuous variables. We will focus on two models: the Brownian energy process with reservoirs, where we prove duality with the symmetric inclusion process with absorbing sites, and the deterministic energy process that we will show to be dual to independent random walkers with absorbing sites. Finally we will briefly discuss a *continuous-continuous* duality property, between continuous models with reservoirs and continuous models with absorbing sites.

C. Giardinà, F. Redig, *Duality for Markov Processes*, Grundlehren der mathematischen Wissenschaften 365, https://doi.org/10.1007/978-3-032-04099-2_11

11.1 Introduction

11.1.1 Non-equilibrium and Duality

One of the important applications of duality is the study of non-equilibrium interacting particle systems. This was the approach of the foundational paper [152] in which the Fourier law was proven for the KMP model (KMP standing for Kipnis, Marichioro, Presutti), using duality with a discrete KMP process with absorbing sites. Also in the context of the exclusion process, properties of a non-equilibrium system were proved using duality in [219]. In this book we restrict to non-equilibrium systems that are driven out-of-equilibrium by the addition of boundary reservoirs. When the "parameters" of these reservoirs are identical, the particle system driven by the reservoirs will be in equilibrium, i.e., reversible, with a reversible measure determined by the reservoir parameters. So the effect of the reservoirs is to select the "correct" reversible measure out of the one-parameter family of reversible measures. When the reservoir parameters are not identical, the system is "driven out-of-equilibrium", i.e., it is no longer reversible, and currents appear. The stationary measure which is still unique (because of the presence of the reservoirs) is then called a "non-equilibrium steady state" (NESS), where non-equilibrium refers to the absence of reversibility, or equivalently to the absence of detailed balance.

In a general setting it is very difficult to obtain precise information about a NESS. It is expected, and in some models also proven, that NESS behave in a very different way when compared to their equivalents in equilibrium which are Gibbs measures (see [113] for a thorough account on properties of Gibbs measures). As an example, a generic property of NESS is the presence of long-range correlations, whereas for Gibbs measures with short-range interaction, such long-range correlations are only present at critical points. Another property of NESS is the non-locality of large deviation functions (of e.g. the density profile) [27, 78]. These large deviation functions are very different when compared to large deviation functions in the context of Gibbs measures, which are differences of free energies and therefore local.

Our aim here is to define a class of boundary-driven non-equilibrium models for which duality still works. This requires a special choice of the action of reservoirs, which can be understood by direct computation, or by the algebraic structure of the model. Once one has duality where in the dual process the reservoirs are replaced by absorbing sites, computing moments of order n in the NESS can be reduced to computing absorption probabilities for n dual particles. For $n = 1$ this is a problem of absorption probabilities of a single random walk, which in some settings such as the chain can be computed explicitly. For $n \geq 2$, one can usually not obtain absorption probabilities in a closed form, but one can still derive many useful properties of the NESS from the dual with absorbing sites. Only for systems with extra structure such as integrability (see Chap. 12) one can obtain closed form formulas for all absorption probabilities. Such systems are, however usually restricted to chain geometry with nearest neighbor jumps. Non-equilibrium systems with duality form a much larger class, where results can be obtained on general graphs and reservoir geometry (i.e., multiple reservoirs coupled to different sites).

The place which our duality results have in the setting of non-equilibrium statistical mechanics is therefore in between "exactly solvable models" (such as the symmetric exclusion process [77], and the harmonic processes described in Chap. 12) and general (stochastic) non-equilibrium systems.

As already mentioned at the beginning, we restrict in this book to boundary driving as a source of non-equilibrium. There are various other ways in which systems can be driven out of equilibrium. These include bulk driving (e.g. asymmetric simple exclusion process) where a global drift is imposed [77], active particles [216], and chemical reactions where there is no longer conservation of the number of particles [39, 81]. In all these settings, non-equilibrium refers to non-reversiblity, and non-zero entropy production. We believe that in many of these settings special models with dualities can be identified but we leave this for future research.

In this chapter we focus on non-equilibrium steady states, i.e., we do not consider transient non-equilibrium (relaxation towards equilibrium) or time-dependent driving of systems such as thermodynamic transformations see e.g. [189]. In Chap. 12, we will discuss an exactly solvable model, where all moments in the NESS can be obtained in closed form. This model can be thought of as the equivalent of the SEP in the class of models with $SU(1,1)$ symmetry.

11.1.2 General Structure of the Models

In this chapter we consider particles moving on a finite set and we select a subset of sites where particles interact with the external reservoirs. The infinitesimal generator of the Markov process associated with models of this type can be generically expressed as the sum of two terms

$$L = L^{\text{bulk}} + L^{\text{res}} , \tag{11.1}$$

where L^{bulk} is the bulk dynamics part and L^{res} models the action of the reservoirs.

Our focus is on finding non-equilibrium models which have duality properties. These duality properties will then be exploited to gain insight in the structural properties of the non-equilibrium system. Having this in mind, it is convenient to start with a bulk dynamics that is self-dual (such as those studied in the previous chapters, i.e. SIP, SEP, IRW). Then we investigate the conditions on the entrance and exit rates of the particles from the reservoirs under which we still have the existence of a simpler dual process.

This simpler dual process will turn out to be absorbing, i.e., the reservoirs are replaced by absorbing sites. The dual configurations are defined on a vertex set consisting of the original vertex set V together with an additional vertex set V^{res} consisting of the absorbing sites. The duality functions are then of the form

$$D(\xi, \eta) = \prod_{x \in V} d(\xi_x, \eta_x) \prod_{x \in V^{\text{res}}} \rho_x^{\xi_x}. \tag{11.2}$$

Here ρ_x is the reservoir parameter associated to the reservoir at $x \in V^{\text{res}}$, ξ is a dual configuration, and η a configuration of the original process. That is, in the duality between the absorbing system and the system driven by reservoirs, the "standard" duality functions $\prod_{x \in V} d(\xi_x, \eta_x)$ are extended with an extra factor $\prod_{x \in V^{\text{res}}} \rho_x^{\xi_x}$ associated with the reservoirs.

For the sake of clarity, as a warm up, we will start in the next section by analyzing the case of independent particles, i.e., particles performing nearest-neighbour independent random walks. We will also assume that particles move on a one-dimensional chain that is in contact with only two reservoirs, coupled to the boundary sites, one at the left and the other at the right end of the chain. This example serves as an illustration of the method of associating to a boundary-driven system an absorbing system in the simplest possible setting, where the invariant measure is explicit. In Sect. 11.3 we will consider the setting of interacting particles moving on a graph that is in contact with a set of reservoirs. We will study in detail the case of the symmetric inclusion process (SIP), symmetric exclusion process (SEP) and independent random walkers, including their inhomogeneous versions. In Sect. 11.4 we consider interacting processes with continuous variables, and focus on the Brownian energy process with reservoirs.

11.2 Independent Random Walkers with Reservoirs

Here we start with the easy but illustrative example of independent random walkers on a chain with reservoirs at left and right end, and prove duality with a system of independent random walkers with absorbing sites. Using this duality, we then obtain that the unique stationary measure (NESS) is a inhomogeneous product of Poisson distributions. This provides the simplest context in which moments of the stationary distribution can be obtained via absorption probabilities of the dual process. The fact that the NESS is a product measure is a general property of zero range processes (see e.g. [27]), which in our context of processes with duality occurs only for independent random walkers. As we will see later, in all interacting cases (SIP, SEP, BEP) the NESS in non-product and exhibits long-range correlations.

We thus consider a linear chain $V = \{1, \ldots, N\}$ put in contact, at its left and right boundaries, with two particle reservoirs, say reservoir ℓ coupled to site 1 and reservoir r coupled to site N. Particles can thus leave or enter the system only through the boundary sites 1 and N. The interesting case is when the reservoirs impose on the boundary sites two different fixed particle densities, say $\rho_r > \rho_\ell \geq 0$. This generates, throughout the chain, a current of particles flowing from right to left.

In this situation we say that the system is driven out of equilibrium by the boundary reservoirs which act as external driving forces.

From Sect. 4.7 we know that, for independent particles, it is possible to allow for an asymmetry in the jump rates, and still have duality. We fix then two jump rates $p, q \geq 0$ and define a Markov generator L consisting, as in (11.1), of two terms:

$$L^{\text{bulk}} = L^{p,q} \qquad \text{and} \qquad L^{\text{res}} = pL_\ell + qL_r, \tag{11.3}$$

where $L^{p,q}$ describes independent particles in $\{1, \dots, N\}$ hopping at rate p to the right neighbour and at rate q to the left neighbour, i.e.

$$[L^{p,q} f](\eta) = \sum_{x=1}^{N-1} \left\{ p\eta_x (f(\eta^{x,x+1}) - f(\eta))) + q\eta_{x+1}(f(\eta^{x+1,x}) - f(\eta)) \right\}, \tag{11.4}$$

while L_ℓ and L_r describe the action of the reservoirs ℓ and r on the boundaries, where particles can leave and enter the system:

$$\begin{aligned} L_\ell &= \eta_1(f(\eta - \delta_1) - f(\eta)) + \rho_\ell(f(\eta + \delta_1) - f(\eta)) \\ L_r &= \eta_N(f(\eta - \delta_N) - f(\eta)) + \rho_r(f(\eta + \delta_N) - f(\eta)). \end{aligned} \tag{11.5}$$

Both L^{bulk} and L^{res} work on functions $f : \mathbb{N}^N \to \mathbb{R}$, and thus the state space of the process $\{\eta(t) : t \geq 0\}$ is $\Omega = \mathbb{N}^N$. The reservoir generators act in such a way as to "impose" to the boundary sites two fixed particle densities, ρ_ℓ at the left and ρ_r at the right. The stationary distribution of L_ℓ (resp. L_r) is the Poisson distribution with parameter ρ_ℓ (resp. ρ_r), as can be seen from the detailed balance relation:

$$\frac{\rho_\ell^n}{n!} e^{-\rho_\ell} n = \frac{\rho_\ell^{n-1}}{(n-1)!} e^{-\rho_\ell} \rho_\ell \qquad \text{for all} \quad n \in \mathbb{N}, n \geq 1.$$

As a consequence we can say that, throughout the evolution, each of the two reservoirs imposes on the corresponding boundary site, a number of particle that is distributed as a Poisson random variable with parameter ρ_ℓ at the left and ρ_r at the right.

From Sect. 4.7 we know that the bulk generator $L^{p,q}$ can be written in abstract form in terms of the addition and removal operators

$$\begin{aligned} a_x^\dagger f(\eta) &= f(\eta + \delta_x) \\ a_x f(\eta) &= \eta_x f(\eta - \delta_x). \end{aligned} \tag{11.6}$$

In terms of these operators, the bulk generator reads as follows:

$$L^{p,q} = \sum_{x=1}^{N-1} \left\{ p(a_x a_{x+1}^\dagger - a_x a_x^\dagger) + q(a_{x+1} a_x^\dagger - a_{x+1} a_{x+1}^\dagger) \right\}. \tag{11.7}$$

A similar rewriting can be done for the reservoir generators, which gives

$$\begin{aligned} L_\ell &= (a_1 - a_1 a_1^\dagger) + \rho_\ell (a_1^\dagger - I), \\ L_r &= (a_N - a_N a_N^\dagger) + \rho_r (a_N^\dagger - I), \end{aligned} \tag{11.8}$$

where I denotes the identity, i.e., $If = f$.

Our aim is then to find the analogue of the duality result found in Sect. 4.7. From Theorem 4.23 we know that the bulk term of the generator $L^{\text{bulk}} = L^{p,q}$ is dual to the generator $L^{q,p}$ of the process with exchanged jump rates, and with the duality function

$$D(\xi, \eta) = \prod_{x=1}^{N} d(\xi_x, \eta_x), \qquad \text{with} \qquad d(k, n) = \frac{n!}{(n-k)!} \mathbb{1}_{\{k \le n\}}. \tag{11.9}$$

We will see that in the presence of reservoirs, i.e., upon adding the L_ℓ, L_r to the bulk generator, there is still a duality property. However the process with reservoirs is no longer self-dual, instead, it has a dual process with absorbing sites. In other words the duality relation turns reservoirs into sinks. More precisely we will have an absorbing site corresponding to each reservoir, so $V^{\text{res}} = \{\ell, r\}$, where now ℓ and r have to be considered as "real" sites to be added to the set V. The dual particle system will then be a Markov process where particles move on the extended set $V^* = V \cup V^{\text{res}}$ with space state $\Omega^* = \mathbb{N}^{V^*}$. We denote by $\{\xi(t) : t \ge 0\}$ the dual process and we will show that it has generator L^{dual} of the form

$$L^{\text{dual}} = L^{q,p} + L^{\text{abs}} \qquad \text{with} \qquad L^{\text{abs}} = p\widehat{L}_\ell + q\widehat{L}_r, \tag{11.10}$$

working on functions $f : \Omega^* \to \mathbb{R}$, where the bulk dynamics is governed by $L^{q,p}$ that is obtained from (11.4) by switching the roles of p and q. Notice that $L^{q,p}$ works only on the bulk coordinates $\xi_1, \ldots, \xi_N$ of a configuration ξ. The generator L^{abs} works instead on the boundary sites and the extra sites corresponding the reservoirs, i.e. $\widehat{L}_\ell$ works on (ξ_ℓ, ξ_1) and $\widehat{L}_r$ works on (ξ_N, ξ_r) in the following way

$$[\widehat{L}_\ell f](\xi) = \xi_1 \left(f(\xi^{1,\ell}) - f(\xi) \right) \quad \text{and} \quad [\widehat{L}_r f](\xi) = \xi_N \left(f(\xi^{N,r}) - f(\xi) \right) \tag{11.11}$$

In terms of addition and removal operators, $\widehat{L}_\ell$ and $\widehat{L}_r$ read as follows:

$$\widehat{L}_\ell = a_1 a_\ell^\dagger - a_1 a_1^\dagger \qquad \text{and} \qquad \widehat{L}_r = a_N a_r^\dagger - a_N a_N^\dagger. \tag{11.12}$$

The generator L^{abs} is exactly of the form considered in (10.95): a particle jumps at rate p from 1 to ℓ and at rate q from N to r, different particles move independently. When a particle hits $V^{\mathrm{abs}} = V^{\mathrm{res}}$, it remains there forever.

In the following lemma we give a duality result for the boundary term L_ℓ of the generator. An analogous property holds for L_r.

Lemma 11.1 (Duality at the Boundaries) *Define the function*

$$D((\xi_\ell, \xi_1), \eta_1) = \rho_\ell^{\xi_\ell} \cdot \frac{\eta_1!}{(\eta_1 - \xi_1)!} \mathbb{1}_{\{\eta_1 \geq \xi_1\}}. \tag{11.13}$$

Then we have

$$[L_\ell D((\xi_\ell, \xi_1), \cdot)](\eta_1) = [\widehat{L}_\ell D((\cdot, \cdot), \eta_1)](\xi_\ell, \xi_1). \tag{11.14}$$

Proof We have

$$\begin{aligned}
&[L_\ell D((m,k),\cdot)](n)\\
&= \rho_\ell^m \left\{ \left(\frac{n!}{(n-k-1)!} - \frac{(n+1)!}{(n-k)!} \right) + \left(\frac{(n+1)!}{(n+1-k)!} - \frac{n!}{(n-k)!} \right) \rho_\ell \right\}\\
&= \frac{n!}{(n-k)!} \cdot \rho_\ell^m \cdot \left\{ ((n-k) - n) + \left(\frac{n+1}{n+1-k} - 1 \right) \rho_\ell \right\}\\
&= \frac{n!}{(n-k)!} \cdot \rho_\ell^m \cdot \left\{ -k + \frac{k}{n+1-k} \rho_\ell \right\}\\
&= k \cdot \left\{ \frac{n!}{(n-k+1)!} \cdot \rho_\ell^{m+1} - \frac{n!}{(n-k)!} \cdot \rho_\ell^m \right\}\\
&= [\widehat{L}_\ell D((\cdot,\cdot), n)](m,k),
\end{aligned} \tag{11.15}$$

from which we obtain (11.14). □

We are now ready to give the duality relation between the system with reservoirs and the system with absorbing boundaries. This is a consequence of the self-duality property of the system without reservoirs combined with the result of "boundary duality" proven in the previous lemma. The result is given in terms of generator duality, then semigroup duality follows from Theorem 2.5.

Theorem 11.2 (Duality for a Chain with Reservoirs) *The generators L and L^{dual} defined in* (11.3) *and* (11.10) *are dual with duality function*

$$D(\xi,\eta) = \rho_\ell^{\xi_\ell}\rho_r^{\xi_r}\prod_{x=1}^{N}\frac{\eta_x!}{(\eta_x-\xi_x)!}. \tag{11.16}$$

Proof The duality of the bulk generator follows from Theorem 4.23, and the duality of the boundary generator follows from Lemma 11.1. □

Remark 11.3 (Structure of the Duality Function) The duality function (11.16) can be written as follows

$$D(\xi,\eta) = \left(\int d(\xi_\ell,\eta_\ell)\nu_{\rho_\ell}(d\eta_\ell)\right)\left(\prod_{x=1}^{N} d(\xi_x,\eta_x)\right)\left(\int d(\xi_r,\eta_r)\nu_{\rho_r}(d\eta_r)\right), \tag{11.17}$$

where $d(k,n) = \frac{n!}{(n-k)!}$ is the single-site (triangular) duality function for the independent random walkers. I.e., the duality function can be obtained by integrating the usual duality function of independent random walkers over the "extra variables" η_ℓ, η_r w.r.t. the Poissonian equilibrium measure with the corresponding reservoir parameter (ρ_r for η_r, ρ_ℓ for η_ℓ). As we will see in later chapters, this structure of the duality function is general.

We can then apply the duality result to characterize the stationary measure of the process with reservoirs.

Proposition 11.4 (Stationary Measure) *The process of independent random walkers with reservoirs with generator given in* (11.1)–(11.3) *admits a unique stationary measure which is the inhomogeneous product of Poisson measures:*

$$\mu^{\text{st}}_{\rho_\ell,\rho_r} = \otimes_{x\in V}\text{Pois}(\rho_x). \tag{11.18}$$

Here the density profile $\{\rho_x, x\in V\}$ is given by

$$\rho_x = \rho_\ell + (\rho_r-\rho_\ell)\frac{x}{N+1}, \qquad x\in V \tag{11.19}$$

for the symmetric case $p=q$, and by

$$\rho_x = \rho_r - (\rho_r-\rho_\ell)\left(\frac{\left(\frac{p}{q}\right)^{N+1} - \left(\frac{p}{q}\right)^{x}}{\left(\frac{p}{q}\right)^{N+1}-1}\right), \qquad x\in V \tag{11.20}$$

for the asymmetric case $p \neq q$. If $\rho_l = \rho_r = \rho$ then the density profile is constant ($\rho_x = \rho$ for all $x \in V$) and in the case $p = q$ the corresponding homogeneous Poisson measure is also reversible.

Proof We prove the statement for the symmetric case $p = q$, since the proof of the asymmetric case is analogous. Since the particle number is not conserved, and all sites of the set are connected, the process is irreducible. As a consequence, if there exists a stationary probability measure, it is unique, and we denote it by $\mu^{\text{st}} = \mu^{\text{st}}_{\rho_\ell,\rho_r}$. Moreover from the ergodicity we have that

$$\lim_{t\to\infty} \mathbb{E}_\eta[D(\xi, \eta(t))] = \int D(\xi, \eta)\mu^{\text{st}}(d\eta). \tag{11.21}$$

On the other hand, using the duality relation (11.16), and since the dual particles are all eventually absorbed at sites ℓ and r, we have that

$$\begin{aligned}\lim_{t\to\infty} \mathbb{E}_\eta[D(\xi, \eta(t))] &= \lim_{t\to\infty} \mathbb{E}_\xi[D(\xi(t), \eta)] \\ &= \lim_{t\to\infty} \mathbb{E}_\xi[\rho_\ell^{\xi_\ell(t)} \rho_r^{\xi_r(t)}] := \mathbb{E}_\xi[\rho_\ell^{\xi_\ell(\infty)} \rho_r^{\xi_r(\infty)}]. \end{aligned}\tag{11.22}$$

It remains to compute the r.h.s. of (11.22). Dual particles move, in the bulk, as symmetric independent random walkers until they reach the absorbing sites ℓ and r. The probability that a dual particle starting from x is absorbed at ℓ is then equal to $1 - \frac{x}{N+1}$. Therefore, for any initial dual configuration ξ, using the independence of all particles, we have

$$\begin{aligned}\mathbb{E}_\xi[\rho_\ell^{\xi_\ell(\infty)} \rho_r^{\xi_r(\infty)}] &= \sum_{n=1}^{|\xi|} \mathbb{P}_\xi\,(\xi_\ell(\infty) = n)\ \rho_\ell^n\, \rho_r^{|\xi|-n} \\ &= \prod_{x\in V^*} \sum_{n_x=0}^{\xi_x} \binom{\xi_x}{n_x} \left[\left(1 - \tfrac{x}{N+1}\right)\rho_\ell\right]^{n_x} \left[\tfrac{x}{N+1}\,\rho_r\right]^{\xi_x - n_x} \\ &= \prod_{x\in V^*} \left(\rho_\ell + (\rho_r - \rho_\ell)\,\tfrac{x}{N+1}\right)^{\xi_x}. \end{aligned}\tag{11.23}$$

Defining the single-site densities as in (11.19) and using (11.21), we deduce that

$$\int D(\xi, \eta)\mu^{\text{st}}(d\eta) = \prod_{x\in V^*} \rho_x^{\xi_x}. \tag{11.24}$$

As we have seen in the previous chapters, (11.24) characterizes products of Poisson measures. Therefore, we conclude that the unique stationary measure $\mu^{\text{st}} = \mu^{\text{st}}_{\rho_\ell,\rho_r}$ is the inhomogeneous product of Poisson measures whose site-x marginal has density ρ_x. This concludes the proof of (11.18) for $q = p$. □

Remark 11.5 (Linear Density Profile) The linear density profile ρ_x is the solution of the boundary-value problem associated to the discrete Dirichlet Laplacian in $V = \{\ell, 1, \ldots, N, r\}$ associated to the dual process, with boundary conditions ρ_ℓ and ρ_r, i.e.

$$\begin{aligned} &q(\rho_{x+1} - \rho_x) + p(\rho_{x-1} - \rho_x) = 0, \qquad \forall\, 2 \le x \le N-1 \\ &q(\rho_2 - \rho_1) + p(\rho_\ell - \rho_1) = 0, \\ &q(\rho_r - \rho_N) + p(\rho_{N-1} - \rho_N) = 0. \end{aligned}$$

The linear density profile (11.19) obtained for the symmetric case will appear quite often in non-equilibrium steady states of symmetric systems with duality because a single dual particle behaves as a random walk also for interacting systems such as SEP or SIP.

11.3 Interacting Particle Systems with Reservoirs

In this section we generalize the duality result obtained for the IRW on a chain, by extending the analysis in several directions. We will restrict ourselves to considering systems with symmetric interaction. As we have seen in the previous chapters, the duality properties of symmetric processes do not depend on the geometry of the system. This is because they are derived via duality of the "single edge" generator, which is then copied along the edges of a general graph. Thus, as a first generalization, we will consider now particles moving on an arbitrary graph with vertex set V. The system is put in contact with a set of reservoirs to which we will associate an additional vertex set V^{res}. Each reservoir is in contact with a subset of bulk sites, through which particles can leave or enter the system.

Furthermore, we include an interaction component in the dynamics in such a way that for the bulk part of the generator we have (self)-duality. Therefore it is natural to assume that particles in the bulk jump with symmetric exclusion/inclusion rates. This guarantees the self-duality property, at least for the bulk part of the generator.

11.3.1 Set-up: The Reference Process with Reservoirs

We will use the setting of the "reference process" introduced in Definition 10.17, and add terms to the generator describing the interaction with reservoirs. We fix a finite set V and, as in Definition 10.17, a set of parameters $(\theta, \boldsymbol{\alpha}, p)$. For the readers' convenience, we briefly recall the meaning of the various parameters of the reference process.

1. The interaction parameter $\theta \in \{-1, 0, +1\}$ determines the nature of the interaction (SEP, IRW, SIP)
2. The parameter $\boldsymbol{\alpha} = \{\alpha_x,\ x \in V\}$ is a vector taking values in $\mathbb{N}^V$ when $\theta = -1$ and in that case represents the maximal number of particles allowed at each site x. It is a vector taking values $\mathbb{R}_+^V$ when $\theta \in \{0, +1\}$ and then determines the diffusion rate, i.e., a single particle moves at rate $p(x, y)\alpha_y$ from x to y, where $p(x, y)$ is the symmetric, irreducible transition function explained below.
3. We fix a symmetric, irreducible transition function $p : V \times V \to [0, +\infty)$ (as defined at the beginning of Sect. 3.1). This represents the "edge weights", and thus the geometry of the system.

We consider a particle system $\{\eta(t) : t \geq 0\}$ where particles move on the set V, with generator $L = L^{(\theta,\boldsymbol{\alpha},p)}$ of the type

$$L := L^{\text{bulk}} + L^{\text{res}}\ , \tag{11.25}$$

where L^{bulk}, resp. L^{res}, denote the bulk and the reservoir terms of the generator. We recall that the state space is

$$\Omega_{\theta,\boldsymbol{\alpha}} = \otimes_{x \in V} \Upsilon_x^{(\theta)}, \tag{11.26}$$

where, for $x \in V$, $\Upsilon_x^{(\theta)}$ is the space of occupation numbers of site x, i.e.

$$\Upsilon_x^{(\theta)} = \begin{cases} \mathbb{N} & \text{for } \theta \in \{0, +1\} \\ \{1, 2, \ldots, \alpha_x\} & \text{for } \theta = -1\ . \end{cases} \tag{11.27}$$

Bulk System

We will assume that the bulk term L^{bulk} is the generator of the *reference process* with parameters $(\theta, \boldsymbol{\alpha}, p)$ (see Definition 10.17), i.e.

$$L^{\text{bulk}} := \frac{1}{2} \sum_{x,y \in V} p(x, y)\, L_{x,y}^{\text{bulk}} \qquad \text{with} \tag{11.28}$$

$$L_{x,y}^{\text{bulk}} f(\eta) := \eta_x(\alpha_y + \theta\eta_y)[f(\eta^{x,y}) - f(\eta)] + \eta_y(\alpha_x + \theta\eta_x)[f(\eta^{y,x}) - f(\eta)]$$

acting on functions $f : \Omega_{\theta,\boldsymbol{\alpha}} \to \mathbb{R}$. More precisely L^{bulk} is, depending on whether θ is equal to -1, 0 or 1, respectively the generator of the partial exclusion process, independent random walkers or the inclusion process as described in the scheme (10.79).

Reservoirs

Regarding the action of reservoirs, we also want to consider the general setting where there can be multiple reservoirs which can be connected to multiple sites inside the bulk.

The second term of the generator L^{res}, models the action of these reservoirs placed at "external sites" contained in a finite set V^{res} disjoint from V, i.e., $V \cap V^{\text{res}} = \emptyset$. We also define the extended set $V^* = V \cup V^{\text{res}}$. The generator L^{res} reads

$$L^{\text{res}} = \sum_{\substack{x \in V \\ y \in V^{\text{res}}}} p(x, y)\, \alpha_y\, L_{x,y}, \tag{11.29}$$

where, for $(x, y) \in V \times V^{\text{res}}$,

$$L_{x,y} f(\eta) = \eta_x(1 + \theta\rho_y)[f(\eta - \delta_x) - f(\eta)] + \rho_y(\alpha_x + \theta\eta_x)[f(\eta + \delta_x) - f(\eta)], \tag{11.30}$$

acting on functions $f : \Omega_{\theta,\boldsymbol{\alpha}} \to \mathbb{R}$.

Notice the analogy between (11.30) and the jump rates in the bulk (11.28). Comparing them we see that the parameter ρ_x can be interpreted as the particle (weighted) density imposed by the reservoir $x \in V^{\text{res}}$. This is kept fixed throughout the dynamics and, in order to have reversibility, it has to be the same for all reservoirs, as we will see below. We collect the "reservoir (weighted) densities" in a vector that we will denote by $\boldsymbol{\rho}^{\text{res}} = \{\rho_x,\ x \in V^{\text{res}}\}$. The reason for which we use the expression weighted densities and not simply densities will be clarified later on.

Here we have also extended to the set V^* the vector of $\boldsymbol{\alpha} = \{\alpha_x,\ x \in V^*\}$. Similarly, the transition function $p : V \times V^* \to \mathbb{R}$ appearing in (11.29) is an extended version of the one appearing in the bulk generator. In order to assure the irreducibility of the Markov process, we assume that p is an "allowed transition function" as precised in the definition below.

Definition 11.6 (Allowed Edge Weight Function) We say that a function $p : V \times V^* \to [0, \infty)$ is an *allowed transition function* if its restriction to the domain $V \times V$ is symmetric and irreducible, and for all $y \in V^{\text{res}}$ there exists at least one $x \in V$ such that $p(x, y) > 0$. Moreover, let $x \in V$ be a bulk site and $y \in V^*$ a site of the extended set, we say that they are connected, $x \sim y$, if and only if $p(x, y) > 0$.

Notice that the relation $\sim$ is not an equivalence relation. Indeed the transition function $p(x, y)$ is not defined on a couple of reservoir sites, i.e. for $(x, y) \in V^{\text{res}} \times V^{\text{res}}$. This choice has the meaning that different reservoirs are not connected with each other. This is reasonable because reservoirs will have the role of sites whose densities are kept constant during the dynamics, so they are assumed not to interact with each other. Nevertheless the relation $\sim$ can be viewed as an equivalence relation if restricted to the bulk set V.

We shall call "boundary sites" those sites in the bulk that are in contact with at least one of the reservoirs, i.e. the sites $x \in V$ so that $x \sim y$ for some $y \in V^{\text{res}}$. The jump rates between a bulk site $x \in V$ and a reservoir site $y \in V^{\text{res}}$ only depend on the number of particles in x.

The Reference Process with Reservoirs

We conclude here by providing the explicit definition of the reference process with reservoirs. We remind the definitions of the sets

$$\mathcal{R}_\theta := \begin{cases} [0, +\infty) & \text{for } \theta \in \{0, 1\} \\ [0, 1] & \text{for } \theta = -1 \end{cases} \qquad \text{and} \qquad \mathfrak{A}_\theta = \begin{cases} (0, \infty) & \text{for } \theta \in \{0, 1\} \\ \mathbb{N} & \text{for } \theta = -1. \end{cases} \tag{11.31}$$

The meaning of these sets is resp. the possible values for the weighted densities, and the possible values of the parameters α_x. Here we remind that the weighted density is the expectation of $\frac{\eta_x}{\alpha_x}$.

Definition 11.7 (The Reference Process with Reservoirs) Let V, V^{res} be two disjoint sets, with V finite, fix $\theta \in \{-1, 0, 1\}$, $\boldsymbol{\alpha} = \{\alpha_x,\ x \in V^*\}$, $\alpha_x \in \mathfrak{A}_\theta$, and $\boldsymbol{\rho}^{\text{res}} = \{\rho_y,\ y \in V^{\text{res}}\}$, $\rho_y \in \mathcal{R}_\theta$. Fix moreover an allowed transition function $p : V \times V^* \to [0, \infty)$.

We say that a process $\{\eta(t) : t \geq 0\}$ on the set V is a *reference process in contact with reservoirs* V^{res} with (weighted) density reservoir profile $\boldsymbol{\rho}^{\text{res}}$ if it has state space $\Omega_{\theta,\boldsymbol{\alpha}}$ as in (10.77) and generator

$$L = L^{\text{bulk}} + L^{\text{res}}, \tag{11.32}$$

with

$$L^{\text{bulk}} = \frac{1}{2} \sum_{x,y \in V} p(x, y)\, L_{x,y}, \qquad L^{\text{res}} = \sum_{\substack{x \in V \\ y \in V^{\text{res}}}} p(x, y)\, \alpha_y\, L^{\text{res}}_{x,y},$$

where, for $x, y \in V$,

$$L_{x,y} f(\eta) = \eta_x(\alpha_y + \theta\eta_y)[f(\eta^{x,y}) - f(\eta)] + \eta_y(\alpha_x + \theta\eta_x)[f(\eta^{y,x}) - f(\eta)], \tag{11.33}$$

and, for $(x, y) \in V \times V^{\mathrm{res}}$,

$$L^{\mathrm{res}}_{x,y} f(\eta) = \eta_x(1+\theta\rho_y)[f(\eta-\delta_x) - f(\eta)] + \rho_y(\alpha_x + \theta\eta_x)[f(\eta+\delta_x) - f(\eta)], \tag{11.34}$$

acting on functions $f : \Omega_{\theta,\boldsymbol{\alpha}} \to \mathbb{R}$.

In what follows, in order to alleviate notation, we will often omit the dependence on the parameters $(\theta, \boldsymbol{\alpha}, \boldsymbol{\rho}^{\mathrm{res}}, p)$ in the notation for generator, duality function, and space state.

Equilibrium Setting

The system without reservoirs, i.e. the reference process with generator L^{bulk}, obtained by choosing $p(x, y) = 0$ for all $(x, y) \in V \times V^{\mathrm{res}}$, admits a one-parameter family of reversible product measures

$$\{\nu_{\rho,\theta,\boldsymbol{\alpha}},\ \rho \in \mathcal{R}_\theta\}, \tag{11.35}$$

where $\nu_{\rho,\theta,\boldsymbol{\alpha}} = \bigotimes_{x\in V} \nu_{\rho,\theta,\alpha_x}$ with marginals given by:

$$\nu_{\rho,\theta,\alpha}(n) = \begin{cases} \frac{(\rho\alpha)^n}{n!} \cdot e^{-\rho\alpha} & \text{for } \theta = 0, \quad \mathrm{Pois}(\rho\alpha) \\ \frac{1}{(1+\rho)^\alpha} \cdot \left(\frac{\rho}{1+\rho}\right)^n \cdot \frac{\Gamma(\alpha+n)}{n!\Gamma(\alpha)} & \text{for } \theta = +1, \quad \mathrm{DGamma}\left(\alpha, \frac{\rho}{1+\rho}\right) \\ (1-\rho)^\alpha \cdot \left(\frac{\rho}{1-\rho}\right)^n \cdot \frac{\Gamma(\alpha+1)}{n!\Gamma(\alpha+1-n)} & \text{for } \theta = -1, \quad \mathrm{Bin}(\alpha, \rho)\,. \end{cases} \tag{11.36}$$

In the next proposition we show that, when adding reservoirs with equal reservoir parameters, i.e. $\rho_x = \rho$ for all $x \in V^{\mathrm{res}}$ the system will have, as its unique stationary measure, the reversible product measure $\nu_{\rho,\theta,\boldsymbol{\alpha}}$ i.e. the reversible product measure of the bulk generator having parameter that is equal to the reservoir density ρ. Moreover the system will satisfy detailed balance and is therefore referred to as *equilibrium system.*

Proposition 11.8 (Equilibrium) *Consider the reference process in Definition 11.7. If the (weighted) reservoir density profile is constant, i.e., if* $\boldsymbol{\rho}^{\mathrm{res}} = \{\rho_x,\ x \in V^{\mathrm{res}}\}$, *with* $\rho_x = \rho$ *for all* $x \in V^{\mathrm{res}}$, *then the probability measure* $\nu_{\rho,\theta,\boldsymbol{\alpha}}$ *is the unique invariant measure and is also reversible .*

Proof We prove the statement for the case $\theta = 1$. The proof is similar for $\theta \in \{-1, 0\}$. We know that $\nu_{\rho,1,\boldsymbol{\alpha}}$ satisfies the detailed balance condition for L^{bulk}. Then, in order for it to be a reversible measure for the entire process, we have to verify the detailed balance condition for the reservoir term of the generator:

$$\alpha_y \rho(\alpha_x + \theta n) \nu_{\rho,1,\alpha_x}(n) = \alpha_y (n+1)\,(1+\rho)\,\nu_{\rho,1,\alpha_x}(n+1), \qquad \forall\, n \in \Upsilon_x\,. \tag{11.37}$$

This immediately follows from the definition of $\nu_{\rho,1,\boldsymbol{\alpha}}$. Indeed,

$$\frac{\nu_{\rho,1,\alpha_x}(n+1)}{\nu_{\rho,1,\alpha_x}(n)} = \frac{\rho}{n+1} \cdot \frac{1}{1+\rho} \cdot \frac{\Gamma(\alpha_x + n + 1)}{\Gamma(\alpha_x + n)} = \frac{\rho}{n+1} \cdot \frac{\alpha_x + n}{1+\rho} \tag{11.38}$$

for all $x \in V$, $y \in V^{\text{res}}$. This concludes the proof. □

We recall that the parameter ρ of the reversible measure $\nu_{\rho,\theta,\boldsymbol{\alpha}}$ has the meaning

$$\rho = \mathbb{E}_{\nu_{\rho,\theta,\boldsymbol{\alpha}}}\left[\frac{\eta_x}{\alpha_x}\right]. \tag{11.39}$$

This is the reason why we call ρ the *weighted density* imposed by the reservoirs.

Non-equilibrium Setting

We say that the process $\{\eta(t),\ t \geq 0\}$ with generator (11.32), is a non-equilibrium system, if the reservoir densities are not all equal, i.e., if there exist at least two different reservoirs $x, y \in V^{\text{res}}$ having different weighted densities $\rho_x \neq \rho_y$. Typically in this setting reversibility is lost and the system carries currents of particles flowing between reservoirs. By irreducibility, the system will, from any initial distribution, converge to a unique stationary measure that is then called *the non-equilibrium stationary state (abbreviated NESS)* and denoted by μ^{st}. Assuming, for the moment, the existence and uniqueness of μ^{st} (which will be proven below), we define the (weighted) stationary density profile as follows:

$$\boldsymbol{\rho}^{\text{st}} = \{\rho_x,\ x \in V\} \qquad \text{with} \qquad \rho_x^{\text{st}} := \mathbb{E}_{\mu^{\text{st}}}\left[\frac{\eta_x}{\alpha_x}\right]. \tag{11.40}$$

We have seen that, if all the reservoirs impose the same weighted density ρ, then the system is reversible with reversible measure ν_ρ, and then also the bulk density profile is flat, i.e. $\rho_x = \rho$ for all $x \in V$. Thus, in order to produce a non-homogeneous $\boldsymbol{\rho}$ we need to impose a non-homogeneous reservoir density profile

$$\boldsymbol{\rho}^{\text{res}} := \{\rho_y,\ y \in V^{\text{res}}\} \qquad \text{with} \quad \rho_y \in \mathcal{R}_\theta. \tag{11.41}$$

In this way each reservoir $y \in V^{\text{res}}$ has its own reversible measure $\nu_{\rho_y,\theta,\boldsymbol{\alpha}}$ and then it tries to induce, to every bulk site to which it is connected, a weighted density

ρ_y. In Sect. 11.3.2 below we will see that for any choice of the reservoir rates $\boldsymbol{\rho}^{\mathrm{res}}$ we have an absorbing dual process. Then, in Sect. 11.3.3 we will use duality to prove the existence and uniqueness of the stationary measure μ^{st}.

Geometrical Setting

The simplest geometrical setting of non-equilibrium is the one that we have covered in Sect. 11.2 for independent random walkers. More precisely, in that setting, a one-dimensional finite chain $\{1, \ldots, N\}$ is put in contact with a left reservoir, only interacting with site 1, and with a right reservoir, only interacting with site N, kept at fixed different densities ρ_ℓ and ρ_r. Of course this is not the only way to induce a current in a system, but still the easiest to treat. In the last few years an increasing interest has emerged for non-equilibrium models considering a system interacting with many (eventually infinite) reservoirs, non locally interacting with the bulk sites [18–21]. The duality results which we cover in this chapter do not depend on the number of reservoirs and on the way they are coupled to the bulk sites, for this reason we keep the setting as general as possible. The case of the chain in contact with two local reservoirs is then simply a particular case of this more general setting. We will treat the one-dimensional set-up with two reservoirs in more detail in Sect. 11.3.5.

11.3.2 Triangular Duality

Bulk Self-duality

We remind the reader that the system without reservoirs is self-dual with the duality function

$$D^{\mathrm{bulk}}_{\theta,\boldsymbol{\alpha}}(\xi, \eta) = \prod_{x \in V} d_{\theta,\alpha_x}(\xi_x, \eta_x), \tag{11.42}$$

with

$$d_{\theta,\alpha}(k, n) = \frac{n!}{(n-k)!} \mathbb{1}_{\{k \le n\}} \cdot \begin{cases} \frac{1}{\alpha^k} & \text{for } \theta = 0 \\ \frac{\Gamma(\alpha)}{\Gamma(\alpha+k)} & \text{for } \theta = +1 \\ \frac{\Gamma(\alpha+1-k)}{\Gamma(\alpha+1)} & \text{for } \theta = -1. \end{cases} \tag{11.43}$$

In other words D^{bulk} is a self-duality for the bulk-term of the generator L^{bulk}. In this chapter we will denote this function by $D^{\mathrm{bulk}}(\cdot, \cdot)$ in order to distinguish it from the duality function of the entire generator L that we are going to introduce below.

Duality with a System with Absorbing Sites

With the addition of reservoirs the reference process is no longer self-dual. Instead, we will see that it is dual to a process with absorbing sites of the type of those studied in Sect. 10.6. We start by defining the dual process.

We consider a Markov process $\{\xi(t) : t \geq 0\}$ modelling particles moving on the extended set $V^* = V \cup V^{\mathrm{res}}$ where now the sites in the set V^{res} are absorbing sites. In other words, V^{res} has now the role of what we denoted by V^{abs} in Sect. 10.6. Particles can reach the absorbing sites only passing through the bulk sites connected to them, i.e. only through the bonds $(x, y) \in V \times V^{\mathrm{res}}$, with $x \sim y$, where the relation $\sim$ is induced by the allowed transition function $p : V \times V^* \to [0, \infty)$ that is the same as the one appearing in Definition 11.7. Particles hitting V^{res} are absorbed and do not move anymore. The dual process depends on the parameters $(\theta, \boldsymbol{\alpha}, p)$ as in the previous section and it has generator given by:

$$L^{\mathrm{dual}} := L^{\mathrm{bulk}} + L^{\mathrm{abs}}, \tag{11.44}$$

with L^{bulk} as in (11.28) and

$$L^{\mathrm{abs}} := \sum_{\substack{x \in V \\ y \in V^{\mathrm{res}}}} p(x, y)\, \alpha_y\, L^{\mathrm{abs}}_{x,y}, \qquad L^{\mathrm{abs}}_{x,y} f(\xi) := \xi_x \left[f(\xi^{x,y}) - f(\xi) \right]. \tag{11.45}$$

Notice that, different from the reference process with reservoirs, the dual process has a generator that also acts on the variables of the absorbing sites ξ_y, $y, \in V^{\mathrm{res}}$. As a consequence, the process $\{\xi(t) : t \geq 0\}$ has as a state space, the extended space $\Omega^*_{\theta,\boldsymbol{\alpha}} = \Omega_{\theta,\boldsymbol{\alpha}} \times \Omega^{\mathrm{res}}$ with $\Omega^{\mathrm{res}} = \mathbb{N}^{V^{\mathrm{res}}}$ (see also (10.93)), that is different from the state space $\Omega_{\theta,\boldsymbol{\alpha}}$ of the reference process with reservoirs.

In what follows we will often omit, for the sake of notational simplicity, the dependence on the parameters $(\theta, \boldsymbol{\alpha}, p)$ in the notation of the generators, state spaces and functions depending on them. Nevertheless we will keep stressing the dependence of some of the parameters when we think it is necessary.

We are now ready to state and prove the duality result that we mentioned at the beginning of this section.

Theorem 11.9 (Triangular Duality with Absorbing Dual for Non-equilibrium Systems) *The reference process on the set V, in contact with reservoirs V^{res} and with reservoir density profile $\boldsymbol{\rho}^{\mathrm{res}}$ is dual to the process with absorbing sites V^{res} defined above with generator*(11.44)–(11.45) *with respect to the duality function $D_{\boldsymbol{\rho}^{\mathrm{res}},\theta,\boldsymbol{\alpha}} : \Omega^*_{\theta,\boldsymbol{\alpha}} \times \Omega_{\theta,\boldsymbol{\alpha}} \to \mathbb{R}$ given by*

$$D_{\boldsymbol{\rho}^{\mathrm{res}},\theta,\boldsymbol{\alpha}}(\xi, \eta) = D^{\mathrm{bulk}}_{\theta,\boldsymbol{\alpha}}(\xi, \eta) \prod_{y \in V^{\mathrm{res}}} \rho_y^{\xi_y}, \tag{11.46}$$

where $D^{\mathrm{bulk}}_{\theta,\boldsymbol{\alpha}}$ is the function defined in (11.42)–(11.43).

Proof We know that the process generated by the operator L^{bulk} is self-dual with duality function D^{bulk}. This means that the action of L^{bulk} on $D^{\text{bulk}}(\cdot, \eta)$ and on $D^{\text{bulk}}(\xi, \cdot)$ is the same. Thus, since L^{bulk} does not act on the V^{res}-components of ξ, we have that

$$[L^{\text{bulk}} D^{\text{bulk}}(\cdot, \eta)](\xi) = [L^{\text{bulk}} D^{\text{bulk}}(\xi, \cdot)](\eta). \tag{11.47}$$

It remains to verify that the actions of the boundary components of L and L^{dual} on the duality function are identical. More precisely, we have to verify that

$$L^{\text{res}}_{x,y} D(\xi, \cdot)(\eta) = L^{\text{abs}}_{x,y} D(\cdot, \eta)(\xi) \qquad \text{for all} \quad x \in V, \; y \in V^{\text{res}}. \tag{11.48}$$

For a fixed $(x, y) \in V \times V^{\text{res}}$, we have

$$\begin{aligned}
&[L^{\text{res}}_{x,y} D(\xi, \cdot)](\eta) \\
&= \rho_y(\alpha_x + \theta\eta_x)[D(\xi, \eta + \delta_x) - D(\xi, \eta)] + \eta_x(1 + \theta\rho_y) \\
&\quad \times [D(\xi, \eta - \delta_x) - D(\xi, \eta)] \\
&= D(\xi, \eta) \, \frac{(\eta_x - \xi_x)!}{\eta_x!} \cdot \Bigg\{ \rho_y(\alpha_x + \theta\eta_x) \left[\frac{(\eta_x + 1)!}{(\eta_x + 1 - \xi_x)!} - \frac{\eta_x!}{(\eta_x - \xi_x)!} \right] \\
&\qquad\qquad + \eta_x(1 + \theta\rho_y) \left[\frac{(\eta_x - 1)!}{(\eta_x - 1 - \xi_x)!} - \frac{\eta_x!}{(\eta_x - \xi_x)!} \right] \Bigg\} \\
&= D(\xi, \eta) \, \frac{\xi_x}{(\eta_x + 1 - \xi_x)} \cdot \left\{ \rho_y(\alpha_x + \theta\eta_x) - (1 + \theta\rho_y)(\eta_x + 1 - \xi_x) \right\} \\
&= D(\xi, \eta) \, \frac{\xi_x}{(\eta_x + 1 - \xi_x)} \cdot \left\{ \rho_y(\alpha_x + \theta\xi_x - \theta) - (\eta_x + 1 - \xi_x) \right\} \\
&= \xi_x \left[\rho_y \frac{(\alpha_x + \theta\xi_x - \theta)}{(\eta_x + 1 - \xi_x)} D(\xi, \eta) - D(\xi, \eta) \right] \\
&= \xi_x [D(\xi^{x,y}, \eta) - D(\xi, \eta)] = [L^{\text{abs}}_{x,y} D(\cdot, \eta)](\xi)
\end{aligned} \tag{11.49}$$

This concludes the proof. □

Remark 11.10 (The Dual Process Is a Consistent Process) We notice that the dual process with generator (11.45) belongs to the class of consistent processes with absorbing sites studied in Sect. 10.6. The bulk-term generator L^{bulk} is indeed the generator of a reference process that we know to be consistent, and $\{\xi(t) : t \geq 0\}$ is an absorbing extension of this to the set V^*. The process $\{\xi(t) : t \geq 0\}$ is therefore consistent as a consequence of Lemma 10.21.

Algebraic Description of the Dualities with Absorbing Sites

In this section we provide the Lie algebraic description of the reservoir generator L^{res}. This will then give the duality with the generator L^{abs} of the system with absorbing sites via a change of representation of a Lie algebra.

For the sake of simplicity we restrict to the case of the symmetric inclusion process ($\theta = 1$) so that the process is described via the $\mathfrak{su}(1,1)$ Lie algebra. We consider an edge between a boundary site $x \in V$ and a site $y \in V^{\text{res}}$ so that

$$L^{\text{res}}_{x,y} f(\eta) = \eta_x(1+\rho_y)[f(\eta-\delta_x) - f(\eta)] + \rho_y(\alpha_x+\eta_x)[f(\eta+\delta_x) - f(\eta)], \tag{11.50}$$

and

$$L^{\text{abs}}_{x,y} f(\xi) := \xi_x \left[f(\xi^{x,y}) - f(\xi)\right]. \tag{11.51}$$

To describe $L^{\text{res}}_{x,y}$ algebraically we introduce, for the boundary site $x \in V$, the operators K_x^+, K_x^- and K_x^0 working on functions $f : \mathbb{N}^V \to \mathbb{R}$ as:

$$\begin{aligned} K_x^+ f(\eta) &= (\alpha_x + \eta_x) f(\eta + \delta_x), \\ K_x^- f(\eta) &= \eta_x f(\eta - \delta_x), \\ K_x^0 f(\eta) &= \left(\tfrac{\alpha_x}{2} + \eta_x\right) f(\eta), \end{aligned} \tag{11.52}$$

which provide a representation (labeled by α_x) of the conjugate $\mathfrak{su}(1,1)$ Lie algebra. Furthermore, to describe the reservoir, we introduce for $y \in V^{\text{res}}$ the following additional operators: K_y^+, K_y^- and K_y^0 working on functions $f : \mathbb{N}^{V^{\text{res}}} \to \mathbb{R}$

$$\begin{aligned} K_y^+ f(\xi) &= (1 + \xi_y) f(\xi + \delta_y), \\ K_y^- f(\xi) &= \xi_y f(\xi - \delta_y), \\ K_y^0 f(\xi) &= \left(\tfrac{1}{2} + \xi_y\right) f(\xi), \end{aligned} \tag{11.53}$$

and $\mathcal{K}_y^+$, $\mathcal{K}_y^-$ and $\mathcal{K}_y^0$ working on functions $f : \mathbb{R}_+^{V^{\text{res}}} \to \mathbb{R}$

$$\begin{aligned} \mathcal{K}_y^+ f(\rho) &= \rho_y \left(\rho_y \frac{\partial}{\partial \rho_y} + 1\right) f(\rho), \\ \mathcal{K}_y^- f(\rho) &= \frac{\partial}{\partial \rho_y} f(\rho), \\ \mathcal{K}_y^0 f(\rho) &= \left(\rho_y \frac{\partial}{\partial \rho_y} + \frac{1}{2}\right) f(\rho). \end{aligned} \tag{11.54}$$

The operators in (11.53) provide a representation (with label $\alpha_y = 1$) of the conjugate $\mathfrak{su}(1,1)$ Lie algebra in terms of infinite dimensional matrices, while the operators in (11.54) provide a representation (also of label $\alpha_y = 1$) of the $\mathfrak{su}(1,1)$ Lie algebra in terms of differential operators.

We notice, in particular, that

$$\mathcal{K}_y^+\Big(\mathcal{K}_y^0 + \frac{1}{2}\Big)^{-1} f(\rho) = \rho_y f(\rho). \tag{11.55}$$

As a consequence, we may write

$$\begin{aligned} L_{x,y}^{\text{res}} &= \Big[1 + \mathcal{K}_y^+\Big(\mathcal{K}_y^0 + \frac{1}{2}\Big)^{-1}\Big]\Big[K_x^- - K_x^0 + \frac{\alpha_x}{2}\Big] \\ &+ \Big[\mathcal{K}_y^+\Big(\mathcal{K}_y^0 + \frac{1}{2}\Big)^{-1}\Big]\Big[K_x^+ - K_x^0 - \frac{\alpha_x}{2}\Big]. \end{aligned} \tag{11.56}$$

Similarly, observing that

$$\Big(K_y^0 + \frac{1}{2}\Big)^{-1} K_y^+ f(\xi) = f(\xi + \delta_y), \tag{11.57}$$

we may write

$$L_{x,y}^{\text{abs}} = \Big[\Big(K_y^0 + \frac{1}{2}\Big)^{-1} K_y^+\Big]K_x^- - \Big[K_x^0 - \frac{\alpha_x}{2}\Big]. \tag{11.58}$$

Now, for the boundary site $x \in V$ we recall from Proposition 5.14 that

$$\begin{aligned} K_x^+ &\xrightarrow{d_x} K_x^-, \\ K_x^- &\xrightarrow{d_x} K_x^+ - 2K_x^0 + K_x^-, \\ K_x^0 &\xrightarrow{d_x} K_x^0 - K_x^-, \end{aligned} \tag{11.59}$$

where

$$d_x(\xi_x, \eta_x) = \frac{\eta_x!}{(\eta_x - \xi_x)!}\frac{\Gamma(\alpha_x)}{\Gamma(\alpha_x + \xi_x)} \mathbb{1}_{\{\xi_x \le \eta_x\}}.$$

For the reservoir site $y \in V^{\text{res}}$ the following duality can easily be checked:

$$\begin{aligned} K_y^+ &\xrightarrow{d_y} \mathcal{K}_y^+, \\ K_y^- &\xrightarrow{d_y} \mathcal{K}_y^-, \\ K_y^0 &\xrightarrow{d_y} \mathcal{K}_y^0, \end{aligned} \tag{11.60}$$

where

$$d_y(\xi_y, \rho_y) = \rho_y^{\xi_y}.$$

Therefore, using the algebra dualities (11.59) and (11.60) and employing the general principle of inverting the order of operators when composing dualities, we obtain the duality relation for the Markov generators:

$$\begin{aligned} L_{x,y}^{\text{res}} &= \Big[1 + \mathcal{K}_y^+\Big(\mathcal{K}_y^0 + \frac{1}{2}\Big)^{-1}\Big]\Big[K_x^- - K_x^0 + \frac{\alpha_x}{2}\Big] \\ &\quad + \Big[\mathcal{K}_y^+\Big(\mathcal{K}_y^0 + \frac{1}{2}\Big)^{-1}\Big]\Big[K_x^+ - K_x^0 - \frac{\alpha_x}{2}\Big] \\ &\xrightarrow{D_{x,y}} \Big[1 + \Big(K_y^0 + \frac{1}{2}\Big)^{-1} K_y^+\Big]\Big(- K_x^0 + \frac{\alpha_x}{2}\Big) \\ &\quad + \Big(K_y^0 + \frac{1}{2}\Big)^{-1} K_y^+\Big(K_x^- + K_x^0 - \frac{\alpha_x}{2}\Big) \\ &= \Big(K_y^0 + \frac{1}{2}\Big)^{-1} K_y^+ K_x^- - \Big(K_x^0 - \frac{\alpha_x}{2}\Big) \\ &= L_{x,y}^{\text{abs}}, \end{aligned} \tag{11.61}$$

where

$$D_{x,y}((\xi_y, \xi_x), (\rho_y, \eta_x)) = \rho_y^{\xi_y} \frac{\eta_x!}{(\eta_x - \xi_x)!} \frac{\Gamma(\alpha_x)}{\Gamma(\alpha_x + \xi_x)} \mathbb{1}_{\{\xi_x \le \eta_x\}}. \tag{11.62}$$

11.3.3 Non-equilibrium Steady State

In this section, we use duality to prove existence and uniqueness of the stationary measure μ^{st} for all values of the parameter θ and to infer some properties of μ^{st}. In particular we will analyze the stationary density profile and we will study the two-point correlations proving that these are nonzero whenever $\theta \neq 0$.

Existence and Uniqueness of NESS

We start with the following simple result.

Proposition 11.11 (Irreducibility) *Let $\{\eta(t),\ t \ge 0\}$ be the reference process with reservoirs on a finite set V defined in Definition 11.7. Then $\{\eta(t),\ t \ge 0\}$ is an irreducible process.*

Proof The irreducibility of the reference process with reservoirs follows from the irreducibility of the transition function p restricted to the bulk sites. □

The proof of existence and uniqueness of μ^{st} is then immediate for the partial exclusion process ($\theta = -1$), which is an irreducible Markov chain on the finite state space $\Omega_{-1,\boldsymbol{\alpha}} = \prod_{x\in V}\{0, 1, \ldots, \alpha_x\}$. It remains to study the cases of IRW ($\theta = 0$) and SIP ($\theta = 1$), which are still irreducible Markov chains, but on the countable infinite state space $\mathbb{N}^V$. This will be proved below by using the duality property shown in the previous section.

In what follows we will show that for the boundary-driven systems under study there is a unique stationary measure and moreover, from any initial condition there is convergence in the course of time to that unique stationary measure. I.e., for all bounded functions $f : \Omega \to \mathbb{R}$ and for every configuration $\eta \in \Omega$ we have

$$\lim_{t\to\infty} \mathbb{E}_\eta \left[f(\eta(t))\right] = \mathbb{E}_{\mu^{\mathrm{st}}} \left[f(\eta)\right]. \tag{11.63}$$

We will prove (11.63), for our reference process with reservoirs, when choosing f equal to the duality polynomials: $f(\cdot) = D(\xi, \cdot)$, for all $\xi \in \Omega^*$ (notice that these are of course not bounded but sufficient to prove the convergence (11.63) because convergence in distributions is implied by convergence of moments). The duality moments $\{\mathbb{E}_{\mu^{\mathrm{st}}}[D(\xi, \eta)],\ \xi \in \Omega^*\}$ are then sufficient to characterize the unique stationary distribution μ^{st}.

In order to prove our existence and uniqueness result we use the following lemma, which ensures that under appropriate growth condition the expectations of the duality function are measure determining.

Lemma 11.12 (Growth Condition) *Let μ be a probability measure on Ω. If there exists $\rho \in (0, \infty)$ such that*

$$\mathbb{E}_\mu \left[D(\xi, \eta)\right] \le \rho^{|\xi|}, \tag{11.64}$$

for all $\xi \in \Omega^*$, then μ is uniquely determined by the expectations $\{\mathbb{E}_\mu[D(\xi, \eta)],\ \xi \in \Omega^*\}$.

Proof We start by expressing the moments of the occupancies η_x, $x \in V$, in terms of single-site triangular duality functions defined in (11.46)–(11.42)–(11.43). Then we will prove that the estimate (11.64) implies that the measure μ is uniquely determined by its multivariate moments, which boils down to an estimate on the growth of these moments.

Defining

$$w_x(m) = \begin{cases} \alpha_x^m & \text{for } \theta = 0 \\ \frac{\Gamma(\alpha_x+m)}{\Gamma(\alpha_x)} & \text{for } \theta = 1, \end{cases} \tag{11.65}$$

we have that, for all $x \in V$, and for all $k, n \in \mathbb{N}_0$,

$$n^k = \sum_{m=0}^{k} \left\{ \begin{matrix} k \\ m \end{matrix} \right\} \frac{n!}{(n-m)!} = \sum_{m=0}^{k} \left\{ \begin{matrix} k \\ m \end{matrix} \right\} d_{\alpha_x}(m,n)\, w_x(m), \tag{11.66}$$

where $\left\{ \begin{matrix} k \\ m \end{matrix} \right\}$ denotes the Stirling number of the second kind given by

$$\left\{ \begin{matrix} k \\ m \end{matrix} \right\} = \frac{1}{m!} \sum_{j=0}^{m} (-1)^{m-j} \binom{m}{j} j^k. \tag{11.67}$$

In view of (11.64), we obtain

$$\begin{aligned} \mathbb{E}_\mu\left[\eta_x^k\right] &= \sum_{m=0}^{k} \left\{ \begin{matrix} k \\ m \end{matrix} \right\} \mathbb{E}_\mu\left[D(m\delta_x,\eta)\right] w_x(m) \\ &\leq \sum_{m=0}^{k} \frac{w_x(m)}{m!} \mathbb{E}_\mu\left[D(m\delta_x,\eta)\right] \sum_{j=0}^{m} \binom{m}{j} j^k \\ &\leq k^k \sum_{m=0}^{k} \frac{(2\rho)^m}{m!} w_x(m). \end{aligned}$$

In both cases with $\theta = 0$ and $\theta = 1$, we get

$$\mathbb{E}_\mu\left[\eta_x^k\right] \leq (c_x k)^k, \tag{11.68}$$

for all $k \in \mathbb{N}$, with $c_x = (1 + 2\rho\alpha_x)$ for $\theta = 0$ and $c_x = \lfloor \alpha_x \rfloor !(1+2\rho)^{\lfloor \alpha_x \rfloor + 1}$ for $\theta = 1$. Therefore (11.68) yields

$$\sum_{k=1}^{\infty} \left(\mathbb{E}_\mu\left[\eta_x^{2k}\right]\right)^{-\frac{1}{2k}} \geq \frac{1}{c_x} \sum_{k=1}^{\infty} \frac{1}{2k} = \infty\,. \tag{11.69}$$

Because the above condition holds for all $x \in V$, the extended Carleman condition (see e.g. Theorem 14.19 in [208]) applies. Hence, μ is completely characterized by the moments $\{\mathbb{E}_\mu\left[\eta_x^k\right] : x \in V, k \in \mathbb{N}\}$. □

Theorem 11.13 (Existence and Uniqueness of the Non-equilibrium Steady State) *Let $\{\eta(t) : t \geq 0\}$ be a reference process with reservoirs, on the extended set $V^* = V \cup V^{\text{res}}$. Then the process admits a unique stationary measure μ^{st}.*

Proof Using duality with the absorbing dual, we obtain that, for all $\eta \in \Omega$ and $\xi \in \Omega^*$ with $|\xi| = k$,

$$\lim_{t\to\infty} \mathbb{E}_\eta \left[D(\xi, \eta(t))\right] = \lim_{t\to\infty} \mathbb{E}_\xi \left[D(\xi(t), \eta)\right] = \mathbb{E}_\xi\left[\prod_{y\in V^{\mathrm{res}}} \rho_y^{\xi_y(\infty)}\right]. \tag{11.70}$$

We note that the expression $\mathbb{E}_\xi\left[\prod_{y\in V^{\mathrm{res}}} \rho_y^{\xi_y(\infty)}\right]$ does not depend on $\eta \in \Omega$. Moreover we have the estimate

$$\lim_{t\to\infty} \mathbb{E}_\eta \left[D(\xi, \eta(t))\right] \le \left(\max_{y\in V^{\mathrm{res}}} \rho_y\right)^{|\xi|} \le \rho_*^{|\xi|}, \tag{11.71}$$

for all $\xi \in \Omega^*$ and for some $\rho_* \in \mathcal{R}_\theta$. Therefore, by Lemma 11.12, there exists a unique probability measure μ^{st} on Ω such that

$$\mathbb{E}_{\mu^{\mathrm{st}}} \left[D(\xi, \eta)\right] = \mathbb{E}_\xi\left[\prod_{y\in V^{\mathrm{res}}} \rho_y^{\xi_y(\infty)}\right], \tag{11.72}$$

then, from (11.70) it follows that μ^{st} is the unique stationary probability measure. □

If the reservoir density profile $\boldsymbol{\rho}^{\mathrm{res}}$ is flat, i.e. if $\rho_y = \rho$ for all $y \in V^{\mathrm{res}}$, then the stationary measure $\mu^{\mathrm{st}} = \nu_{\rho,\theta,\boldsymbol{\alpha}}$ is also a reversible measure (see Proposition 11.8). If there exist at least two reservoirs $y, y' \in V^{\mathrm{res}}$ with different densities $\rho_y \neq \rho_{y'}$, none of the measures in the set $\{\nu_{\rho,\theta,\boldsymbol{\alpha}},\ \rho \in \mathcal{R}_\theta\}$ is stationary. Moreover, as we will see in the course of this section, μ^{st} is not a product measure, except for the case of IRW ($\theta = 0$).

The Stationary Weighted Density Profile: One Dual Particle

As a first application of duality with an absorbing dual, we can compute the weighted density profile of the unique stationary measure μ^{st}.

The external reservoirs induce, in the long-time limit, a weighted density profile $\boldsymbol{\rho}^{\mathrm{st}} = \{\rho_x^{\mathrm{st}},\ x \in V\}$ in the bulk. We show below that this density profile solves a discrete Dirichlet problem, which corresponds to the absorption probabilities of a single dual particle. In the simplest setting of the chain with nearest neighbor jumps, this is a linear profile. On a general graph with multiple reservoirs, one can not expect to have a closed-form formula for the weighted density profile. We show below that it can even occur that in the bulk the stationary weighted density profile is constant, and the corresponding non-equilibrium steady state is not an equilibrium product measure.

From the duality relation we can extract relevant information about the stationary weighted density profile. To obtain this, we compute the duality moments plugging in dual configurations with a single dual particle, i.e. $\xi = \delta_x$, for some $x \in V$. Using (11.42)–(11.43), we have

$$D_\theta(\delta_x, \eta) = \frac{\eta_x}{\alpha_x} \qquad \text{for all} \quad \theta \in \{-1, 0, +1\}. \tag{11.73}$$

As a consequence, from the duality relation we have

$$\begin{aligned} \mathbb{E}_{\mu_\theta^{\text{st}}}\left[\frac{\eta_x}{\alpha_x}\right] &= \int D_\theta(\delta_x, \eta)\mu_\theta^{\text{st}}(d\eta) \\ &= \sum_{y \in V^{\text{res}}} \rho_y \cdot \mathbf{P}_x(X^{\text{rw}}(\infty) = y) = \mathbf{E}_x\left[\rho_{X^{\text{rw}}(\infty)}\right], \end{aligned} \tag{11.74}$$

where $\{X^{\text{rw}}(t),\, t \geq 0\}$ is the random walker on V^* associated to the configuration process $\{\xi(t) : t \geq 0\}$ corresponding to a single dual particle which is eventually absorbed in the set V^{res}. Then, recalling the definition of local stationary density ρ_x^{st}, $x \in V$, given in (11.40), from (11.74) we deduce that

$$\rho_x^{\text{st}} = \mathbf{E}_x\left[\rho_{X^{\text{rw}}(\infty)}\right], \qquad \text{for all} \quad x \in V^*. \tag{11.75}$$

We remark that the random walker $\{X^{\text{rw}}(t),\, t \geq 0\}$ has jump rates that do not depend on θ, hence it can be thought of as one of the random walkers of the system of independent random walkers corresponding to the case $\theta = 0$. As a consequence we have that the bulk density profile $\boldsymbol{\rho}^{\text{st}} = \{\rho_x^{\text{st}},\ x \in V\}$ is the same for the three cases $\theta \in \{-1, 0, 1\}$, and it only depends on the geometry of the system (determined by the parameters p and $\boldsymbol{\alpha}$) and on the reservoirs density profile $\boldsymbol{\rho}^{\text{res}}$.

Starting from (11.75) and using that the dual random walk moves form $x \in V$ to $y \in V^*$ at rate $\alpha_y p(x, y)$ we obtain that $\boldsymbol{\rho}^{\text{st}}$ satisfies the equation

$$\sum_{y \in V^*} p(x, y)\, \alpha_y\, (\rho_y^{\text{st}} - \rho_x^{\text{st}}) = 0, \qquad x \in V. \tag{11.76}$$

This is the discrete analogue of the boundary-value problem associated to the Dirichlet Laplacian. The analogue of the discrete Laplacian which appears in the nearest neighbour case is here the generator associated to a single dual walker. Notice that the bulk densities $\boldsymbol{\rho}^{\text{st}} = \{\rho_x^{\text{st}},\ x \in V\}$ are the unknown of the system, whereas the reservoir densities $\boldsymbol{\rho}^{\text{res}} = \{\rho_x,\ x \in V^{\text{res}}\}$ can be thought of as boundary conditions.

In the proposition below, we derive sufficient conditions on the parameters of the process and the reservoir parameters such that in the bulk the stationary weighted density profile is constant. Notice that this is of course the case when the reservoir densities are constant, but as the proposition shows it can happen that also in a non-equilibrium setting, i.e., for non-constant reservoir densities the profile is constant, but then the measure μ^{st} is not the equilibrium product measure.

Proposition 11.14 (Constant Density Profile) *The stationary weighted density profile $\boldsymbol{\rho}^{st}$ is homogeneous if and only the reservoir density profile $\boldsymbol{\rho}^{res} = \{\rho_y,\ y \in V^{res}\}$ satisfies*

$$\sum_{y \in V^{res}} \frac{\alpha_y\, p(x,y)}{\sum_{v \in V^{res}} \alpha_v\, p(x,v)}\, \rho_y = \rho \qquad \text{for all} \quad x \in V, \tag{11.77}$$

for some $\rho \in \mathcal{R}_\theta$.

Proof This follows directly from (11.76). □

The term in the r.h.s. of (11.77) is the average density imposed on the bulk site $x \in V$ by the whole set of reservoirs to which it is connected. According to (11.77), this average density imposed from outside has to be the same for all bulk sites $x \in V$. Notice that, in general, the condition (11.77) is weaker than the condition for reversibility with respect to a strictly positive product reversible measure that requires a flat reservoir density profile $\rho_y = \rho$ for all $y \in V^{res}$. As a consequence, we have that an non-constant reservoir density profile $\boldsymbol{\rho}^{res}$ can still induce a constant bulk density profile $\boldsymbol{\rho}^{st}$, even if it cannot induce reversibility. Only in the case of two reservoirs, which will be dealt with in Sect. 11.3.5, a homogeneous (=constant) bulk density profile can be produced if and only if the two reservoir densities are equal.

Non-product Nature of the Stationary Measure

In this subsection we prove that the non-equilibrium steady state μ^{st} is a product measure only in the case $\theta = 0$ (independent random walkers) or in the case that the reservoir density profile is constant, in which case we have a reversible equilibrium measure. I.e., in all interacting cases $\theta \in \{-1, 1\}$ and non-constant reservoir density, the non-equilibrium steady state is not a product measure.

Definition 11.15 (Product Measures Associated to the Reference Process) The *product measure associated to the reference process* is defined as

$$\nu_{\boldsymbol{\rho}^{st}} = \nu_{\boldsymbol{\rho}^{st},\theta,\boldsymbol{\alpha}} := \otimes_{x \in V}\, \nu_{\rho_x^{st},\theta,\alpha_x}\,, \tag{11.78}$$

where $\boldsymbol{\rho}^{st} := \{\rho_x^{st} : x \in V\}$ is the stationary weighted density profile of the process and $\nu_{\rho,\theta,\alpha_x}$ is the measure defined in (11.36).

Theorem 11.16 (Non-product Nature of the Stationary Measure) *Let $\{\eta(t) : t \geq 0\}$ be a reference process with reservoir density profile $\boldsymbol{\rho}^{res} = \{\rho_y,\ y \in V^{res}\}$. Suppose moreover that the conditions of Theorem 11.13 are satisfied and let μ^{st} be the unique stationary measure of the process. Then*

(a) for $\theta = 0$,

$$\mu^{\text{st}} = \nu_{\boldsymbol{\rho}^{\text{st}}} , \tag{11.79}$$

(b) for $\theta \in \{-1, +1\}$, μ^{st} *is a product measure if and only if* $\boldsymbol{\rho}^{\text{res}}$ *is constant, i.e.* $\rho_y^{\text{res}} = \rho$ *for all* $y \in V^{\text{res}}$*; in this case,* $\mu^{\text{st}} = \nu_{\boldsymbol{\rho}^{\text{res}}}$ *which is then also a reversible measure.*

Proof We first prove item (a). Fix $\theta = 0$, we have to prove that, for all $\xi \in \Omega^*$,

$$\int_\Omega [Lf](\eta)\, \nu_{\boldsymbol{\rho}^{\text{st}}}(\mathrm{d}\eta) = 0, \tag{11.80}$$

for all measurable f. For simplicity we will denote by $d_x(\cdot,\cdot)$ the one-site duality function defined in (11.43) and by ν_{x,ρ_x} the marginal $\nu_{\rho_x,0,\alpha_x}$ defined in (11.36) that is a Poisson measure of parameter $\alpha_x \rho_x$. Then the following relation holds

$$\sum_{n\in\Upsilon_x} d_x(k,n)\, \nu_{x,\rho_x}(n) = \rho_x^k , \qquad \text{for all} \quad k \in \Upsilon_x. \tag{11.81}$$

Then, using duality we obtain, for all $\xi \in \Omega^*$,

$$\begin{aligned}
&\int_\Omega [LD(\xi,\cdot)]\, \nu_{\boldsymbol{\rho}^{\text{st}}}(\mathrm{d}\eta) \\
&= \int_\Omega [L^{\text{dual}} D(\cdot,\eta)](\xi)\, \nu_{\boldsymbol{\rho}^{\text{st}}}(\mathrm{d}\eta) \\
&= \frac{1}{2}\sum_{x,y\in V} p(x,y) \int_\Omega \left\{\xi_x(\alpha_y + \theta\xi_y)[D(\xi^{x,y},\eta) - D(\xi,\eta)]\right\}\, \nu_{\boldsymbol{\rho}^{\text{st}}}(\mathrm{d}\eta) \\
&+ \frac{1}{2}\sum_{x,y\in V} p(x,y) \int_\Omega \left\{\xi_y(\alpha_x + \theta\xi_x)[D(\xi^{y,x},\eta) - D(\xi,\eta)]\right\}\, \nu_{\boldsymbol{\rho}^{\text{st}}}(\mathrm{d}\eta) \\
&+ \sum_{\substack{x\in V \\ y\in V^{\text{res}}}} p(x,y) \int_\Omega \xi_x \left[D(\xi^{x,y},\eta) - D(\xi,\eta)\right]\, \nu_{\boldsymbol{\rho}^{\text{st}}}(\mathrm{d}\eta).
\end{aligned} \tag{11.82}$$

Hence, using the fact that

$$\begin{aligned}
\int_\Omega D(\xi^{x,y},\eta)\, \nu_{\boldsymbol{\rho}^{\text{st}}}(\mathrm{d}\eta) &= \rho_y \int_\Omega D(\xi - \delta_x,\eta)\, \nu_{\boldsymbol{\rho}^{\text{st}}}(\mathrm{d}\eta) \\
\int_\Omega D(\xi,\eta)\, \nu_{\boldsymbol{\rho}^{\text{st}}}(\mathrm{d}\eta) &= \rho_x \int_\Omega D(\xi - \delta_x,\eta)\, \nu_{\boldsymbol{\rho}^{\text{st}}}(\mathrm{d}\eta)
\end{aligned} \tag{11.83}$$

we obtain

$$\begin{aligned}
&\int_\Omega [LD(\xi,\cdot)]\, \nu_{\rho^{\text{st}}}(\mathrm{d}\eta) \\
&= \sum_{x\in V}\left(\int_\Omega D(\xi-\delta_x,\eta)\,\nu_{\rho^{\text{st}}}(\mathrm{d}\eta)\right)\Bigg\{\sum_{y\in V^{\text{res}}} p(x,y)\,\xi_x(\rho_y-\rho_x) \\
&+\tfrac{1}{2}\sum_{y\in V} p(x,y)\,\big[(\rho_y-\rho_x)\xi_x(\alpha_y+\theta\xi_y)+(\rho_x-\rho_y)\xi_y(\alpha_x+\theta\xi_x)\big]\Bigg\} \\
&= \sum_{x\in V}\left(\int_\Omega D(\xi-\delta_x,\eta)\,\nu_{\rho^{\text{st}}}(\mathrm{d}\eta)\right)\xi_x\cdot\sum_{y\in V^*} p(x,y)\,\alpha_y\,(\rho_y-\rho_x)=0,
\end{aligned} \tag{11.84}$$

where in the last identity we used the Eq. (11.76). Then (11.80) follows because the products of Poisson distributions are completely characterized by their factorial moments

$$\int_\Omega D(\xi,\eta)\,\nu_{\rho^{\text{st}}}(\mathrm{d}\eta), \qquad \xi\in\Omega^*. \tag{11.85}$$

We prove now item (b). We will prove that if $\theta\neq 0$ and if there are at least two different reservoir densities, then the covariance between η_x and η_y is not zero for at least one pair x, y. This implies that the unique stationary measure μ^{st} cannot be product.

For $\theta\neq 0$, from Proposition 11.8 we know that if $\boldsymbol{\rho}^{\text{res}}$ is a flat profile equal to ρ everywhere, then ν_ρ is a the unique stationary measure that is product and also reversible. Suppose now that μ^{st} is a product measure, then, in particular, all two-point correlations are zero, i.e. for all $x, y\in V$ with $x\neq y$,

$$\mathbb{E}_{\mu^{\text{st}}}\left[\frac{\eta_x}{\alpha_x}\cdot\frac{\eta_y}{\alpha_y}\right]=\mathbb{E}_{\mu^{\text{st}}}\big[D(\delta_x+\delta_y,\eta)\big]=\rho_x\rho_y. \tag{11.86}$$

Let us abbreviate

$$\rho_x'' := \mathbb{E}_{\mu^{\text{st}}}\left[D(2\delta_x,\eta)\right], \tag{11.87}$$

by stationarity, duality and (11.86), we obtain, for all $x\in V$,

$$\begin{aligned}
\int_\Omega L[D(2\delta_x,\cdot)](\eta)\,\mu^{\text{st}}(\mathrm{d}\eta) &= \int_\Omega [L^{\text{dual}}D(\cdot,\eta)](2\delta_x)\,\mu^{\text{st}}(\mathrm{d}\eta) \\
&= 2\sum_{y\in V^*} p(x,y)\,\alpha_y\,(\rho_y\,\rho_x-\rho_x'')=0.
\end{aligned} \tag{11.88}$$

By adding and subtracting $2\sum_{y\in V^*} p(x, y)\alpha_y\rho_x^2$ to the identity above and using Eq. (11.76), we get

$$2\left(\rho_x^2 - \rho_x''\right)\sum_{y\in V^*} p(x, y)\alpha_y = 0 \quad \text{for all} \quad x \in V. \tag{11.89}$$

Then, since $\alpha_y > 0$ for all $y \in V^*$, from the fact that $p(\cdot, \cdot)$ is an allowed transition function, we deduce that

$$\rho_x'' = \rho_x^2\,, \quad \text{for all} \quad x \in V\,. \tag{11.90}$$

Fix now $x, y \in V, x \neq y$. Using (11.86), (11.90), stationarity of μ^{st} and duality, we get

$$\begin{aligned}
0 &= \int_\Omega [LD(\delta_x + \delta_y, \cdot)](\eta)\,\mu^{\mathrm{st}}(\mathrm{d}\eta) = \int_\Omega [L^{\mathrm{dual}} D(\cdot, \eta)](\delta_x + \delta_y)\,\mu^{\mathrm{st}}(\mathrm{d}\eta) \\
&= \rho_y \sum_{u\in V^*} p(x, u)\,\alpha_u\,(\rho_u - \rho_x) + \rho_x \sum_{u\in V^*} p(y, u)\,\alpha_u\,(\rho_u - \rho_y) \\
&\quad + \theta\, p(x, y)\,(\rho_x - \rho_y)^2 \\
&= \theta\, p(x, y)\,(\rho_x - \rho_y)^2 \qquad \text{for all} \quad x, y \in V,
\end{aligned} \tag{11.91}$$

where in the last identity we used (11.76). Since $\theta \neq 0$, the identity above implies that for all fixed $x \in V$,

$$\sum_{y\in V} p(x, y)\,(\rho_x - \rho_y)^2 = 0. \tag{11.92}$$

From the fact that $p(\cdot, \cdot)$ is an allowed transition function, we have that (11.92) can be true if and only if

$$\rho_x = \rho_y\,, \quad \text{for all} \quad x, y \in V. \tag{11.93}$$

This concludes the proof. □

Higher Order Moments: *n* Dual Particles

When $\theta \neq 0$ and the reservoirs impose non-homogeneous densities, the stationary measure μ^{st} is not of a product form. This makes it difficult to compute it explicitly. Nevertheless, thanks to duality, it is possible to find an expression for the stationary moments in terms of the absorption probabilities of the dual process. More precisely, using duality and (11.63) we obtain

$$\int D_\theta(\xi,\eta)\mu_\theta^{\mathrm{st}}(d\eta) = \lim_{t\to\infty} \mathbb{E}_\mu\left[D_\theta(\xi,\eta(t))\right] = \mathbb{E}_\xi\Big[\prod_{y\in V^{\mathrm{res}}} \rho_y^{\xi_y(\infty)}\Big]. \tag{11.94}$$

Here we stress only the dependence on the parameter θ, in order to underline the different properties of the cases $\theta = 0$ (IRW) and $\theta \neq 0$ (SEP and SIP). For $\theta \neq 0$ and $\boldsymbol{\rho}^{\mathrm{res}}$ non-homogeneous, we know that μ^{st} is no longer a product measure and then the correlation functions do not factorize. Nevertheless for $\xi = \sum_{i=1}^n \delta_{x_i}$ with x_i mutually different vertices from (11.94) we have that

$$\mathbb{E}_{\mu_\theta^{\mathrm{st}}}\left[\prod_{x=1}^n \frac{\eta_{x_i}}{\alpha_{x_i}}\right] = \mathbf{E}_\xi\Big[\prod_{y\in V^{\mathrm{res}}} \rho_y^{\xi_y(\infty)}\Big]. \tag{11.95}$$

In Sect. 11.3.5 below we will see that, when there are only two reservoirs, thanks to the consistency of the dual process, it is possible to write these multivariate moments in terms of the probability that all the dual particles are absorbed in one of the two sinks. In Chap. 13 we will consider a new class of non-equilibrium systems where the non-equilibrium steady state μ^{st} can be obtained explicitly in the form of a mixture of product measures.

11.3.4 Orthogonal Duality

Duality between systems driven by reservoirs and systems with absorbing boundaries with triangular duality functions allows to express multivariate (factorial) moments in the non-equilibrium steady state in terms of absorption probabilities of dual particles. Besides moments, we are also interested in correlation functions and cumulants, which measure the deviation between the non-equilibrium steady state and a product measure, and whose scaling behavior is relevant for studying the deviation from local equilibrium. In order to understand the behavior of cumulants, it is natural to turn to orthogonal duality functions, and to search for the analogue of (11.46) where in the bulk instead of triangular duality functions we have orthogonal ones.

We shall prove below that this is indeed possible. The dual process associated with the orthogonal duality function will be the same as the dual process associated to the triangular duality function. Furthermore, the orthogonal duality functions of the reference system with reservoirs are of the form

$$D(\xi,\eta) = \left(\prod_{x\in V} d_\rho^{\mathrm{or}}(\xi_x,\eta_x)\right)\left(\prod_{x\in V^{\mathrm{res}}} (\rho_x - \rho)^{\xi_x}\right), \tag{11.96}$$

where $(\rho_x)_{x\in V^{\mathrm{res}}}$ are the reservoir parameters, and $\rho > 0$ is an additional free parameter. This additional parameter labels the family of reversible product measure of the system without reservoirs.

The duality functions (11.96) do not satisfy an orthogonality relation in the $L^2(\mu^{\mathrm{st}})$ space corresponding to the stationary measure. Instead, the orthogonal duality functions in (11.96) are orthogonal in the $L^2(\mu_\rho)$ space corresponding to the reversible product measures with parameter ρ (which also explains the presence of this additional parameter ρ in (11.96)). It turns out that choosing appropriate values of ρ will yield useful information about the behavior of centered moments, and cumulants in the stationary state μ^{st} (see e.g. Theorem 11.30 below).

In Chap. 9 we have seen that the triangular duality functions (11.42)–(11.43) are not the only self-dualities of the reference processes without reservoirs. Indeed, the reference process also admits a class of orthogonal self-duality functions. These are the functions $\mathcal{D}_\rho$, $\rho \in \mathcal{R}_\theta$, identified in Theorem 9.5. In this section we investigate the possibility to preserve a duality relation of the same type when adding reservoirs to a reference process. We want the dual process to be again the process with absorbing sites (11.44)–(11.45). Here we will use the notation $\mathcal{D}^{\mathrm{bulk}}_\rho = \mathcal{D}^{\mathrm{bulk}}_{\rho,\theta,\boldsymbol{\alpha}}$ for the bulk duality functions of Theorem 9.5, where, when necessary, we will stress also the dependence on the parameters $\theta \in \{-1, 0, +1\}$ determining the process and $\boldsymbol{\alpha} = \{\alpha_x,\ x \in V\}$. We recall the explicit expression of the function $\mathcal{D}^{\mathrm{bulk}}_{\rho,\theta,\boldsymbol{\alpha}} : \Omega \times \Omega \to \mathbb{R}$:

$$\mathcal{D}^{\mathrm{bulk}}_{\rho,\theta,\boldsymbol{\alpha}}(\xi,\eta) = \prod_{x\in V} d^{\mathrm{or}}_{\rho,\theta,\alpha_x}(\xi_x,\eta_x) \tag{11.97}$$

$$\text{with} \qquad d^{\mathrm{or}}_{\rho,\theta,\alpha}(k,n) = (-\rho)^k \cdot \begin{cases} {}_2F_0\left[\begin{matrix}-k,-n\\ -\end{matrix};-\frac{1}{\rho\alpha}\right] & \theta = 0,\\ {}_2F_1\left[\begin{matrix}-k,-n\\ \theta\alpha\end{matrix};-\frac{\theta}{\rho}\right] & \theta\in\{-1,1\},\end{cases} \tag{11.98}$$

that is thus a self-duality function for the reference process (without reservoirs) with parameters $(\theta, \boldsymbol{\alpha}, p)$. We recall, moreover, that the set of functions $\{\mathcal{D}^{\mathrm{bulk}}_{\rho,\theta,\boldsymbol{\alpha}}(\xi,\cdot),\ \xi \in \Omega\}$ is an orthogonal basis of $L^2(\Omega, \nu_{\rho,\theta,\boldsymbol{\alpha}})$, where $\nu_{\rho,\theta,\boldsymbol{\alpha}}$ is the reversible product measure defined in (11.36).

We want to find now orthogonal polynomial duality functions between the reference process with reservoirs and the reference process with absorbing boundaries. As we did for triangular duality functions in Sect. 11.3.2 we look in the class of functions obtained by multiplying the bulk orthogonal dualities $\mathcal{D}^{\mathrm{bulk}}_{\rho,\theta,\boldsymbol{\alpha}}$ by a term depending only on absorbing sites occupancies, ξ_y, $y \in V^{\mathrm{res}}$. In other words we look for functions $\mathcal{D}_{\rho,\theta,\boldsymbol{\alpha}} : \Omega^* \times \Omega \to \mathbb{R}$ of the type

$$\mathcal{D}_{\rho,\theta,\boldsymbol{\alpha}}(\xi,\eta) = \mathcal{D}^{\mathrm{bulk}}_{\rho,\theta,\boldsymbol{\alpha}}(\xi,\eta)\cdot \prod_{y\in V^{\mathrm{res}}} d^{\mathrm{res}}_y(\xi_y), \tag{11.99}$$

for some $d^{\mathrm{res}}_y : \mathbb{N} \to \mathbb{R}$, $y \in V^{\mathrm{res}}$.

The idea is to derive it from the triangular duality $D_{\theta,\boldsymbol{\alpha}}$ given in (11.46), by acting with a suitable symmetry of the generator, in the spirit of item 2) of Theorem 2.13. We start by considering the bulk terms of the duality functions D^{bulk} and $\mathfrak{D}^{\text{bulk}}_{\rho}$ and looking for a symmetry of the bulk generator, L^{bulk}, intertwining between them. The hint is given by the following equation relating the single-site orthogonal duality function and the triangular ones: for all $\rho \in \mathcal{R}_\theta$ one has

$$d^{\text{or}}_{\rho,\theta,\alpha}(k,n) = [e^{-\rho a} d_{\theta,\alpha}(\cdot,n)](k) = \sum_{m=0}^{k} \binom{k}{m} d_{\theta,\alpha}(m,n)\,(-\rho)^{k-m}, \tag{11.100}$$

where a is the single site removal operator (see also (10.22)). As a consequence we have that

$$\mathfrak{D}^{\text{bulk}}_{\rho,\theta,\boldsymbol{\alpha}}(\xi,\eta) = [e^{-\rho S^-} D^{\text{bulk}}_{\theta,\boldsymbol{\alpha}}(\cdot,\eta)](\xi), \tag{11.101}$$

where $S^- := \sum_{x\in V} a_x$ is the removal operator on V that, from consistency, is a symmetry of L^{bulk}. From item 2 of Theorem 2.9 and (11.101) we can deduce that $\mathfrak{D}^{\text{bulk}}$ is a self-duality for L^{bulk} from the fact that D^{bulk} and $e^{-\rho S^-}$ are, respectively, a self-duality and a symmetry of L^{bulk}.

The fact that absorbing extensions preserve the commutation with the removal operator, as proved in Lemma 10.21, allows to apply the argument above to the duality between the system with reservoirs and the system with absorbing sites, as shown in the next theorem.

Theorem 11.17 (Orthogonal Duality with Absorbing Dual for Non-equilibrium Particle Systems) *The reference process with parameters $(\theta,\boldsymbol{\alpha},p)$ and reservoir density profile $\boldsymbol{\rho}^{\text{res}} = \{\rho_y,\ y\in V^{\text{res}}\}$ is dual to the reference process with absorbing sites V^{res} and same parameters, with respect to the duality function $\mathfrak{D}_{\rho,\boldsymbol{\rho}^{\text{res}},\theta,\boldsymbol{\alpha}} : \Omega^*_{\theta,\boldsymbol{\alpha}} \times \Omega_{\theta,\boldsymbol{\alpha}} \to \mathbb{R}$ given by*

$$\mathfrak{D}_{\rho,\boldsymbol{\rho}^{\text{res}},\theta,\boldsymbol{\alpha}}(\xi,\eta) = \mathfrak{D}^{\text{bulk}}_{\rho,\theta,\boldsymbol{\alpha}}(\xi,\eta)\cdot \prod_{y\in V^{\text{res}}} (\rho_y-\rho)^{\xi_y}, \tag{11.102}$$

for all $\rho\in\mathcal{R}_\theta$.

Proof From Lemma 10.21 we know that the removal operator $S^- := \sum_{x\in V^*} a_x$ is a symmetry of L^{dual}. Therefore, for any fixed $\rho\in\mathcal{R}_\theta$, also $e^{-\rho S^-}$ is a symmetry of L^{dual}. Moreover from Theorem 11.9 we know that D is a duality function between L and L^{dual}, then, from item 2) of Theorem 2.13 it follows that also the function $\mathfrak{D}_\rho$ defined by

$$\mathfrak{D}_\rho(\xi,\eta) := [e^{-\rho S^-} D(\cdot,\eta)](\xi) \tag{11.103}$$

is a duality between L and L^{dual}. It remains to compute the r.h.s. of (11.103) and show that it coincides with the expression in the r.h.s. of (11.102). To this aim we split the removal operator in its bulk and absorbing parts:

$$S^- = S^-_{\mathrm{bulk}} + S^-_{\mathrm{abs}}, \qquad \text{with} \qquad S^-_{\mathrm{bulk}} := \sum_{x \in V} a_x, \qquad S^-_{\mathrm{abs}} := \sum_{x \in V^{\mathrm{res}}} a_x. \tag{11.104}$$

Then, recalling that also D factorizes in a bulk and in a reservoir term (see (11.46)), we have that the r.h.s. of (11.103) is equal to

$$[e^{-\rho S^-_{\mathrm{bulk}}} D^{\mathrm{bulk}}(\cdot, \eta)](\xi) \cdot e^{-\rho S^-_{\mathrm{abs}}} \prod_{y \in V^{\mathrm{res}}} \rho_y^{\xi_y}. \tag{11.105}$$

From (11.101) we know that the first factor in (11.105) is equal to $\mathcal{D}^{\mathrm{bulk}}(\xi, \eta)$. In order to compute the second factor, it is sufficient to use the factorized form of the symmetry $e^{-\rho S^-_{\mathrm{abs}}}$ together with the following computation

$$e^{-\rho a_y} \rho_y^k = \sum_{m=0}^{k} \binom{k}{m} \rho_y^m (-\rho)^{k-m} = (\rho_y - \rho)^k \qquad \text{for} \quad y \in V^{\mathrm{res}}. \tag{11.106}$$

Combination of (11.105) and (11.106) then concludes the proof. □

Remark 11.18 Notice that, in case there exists a reservoir $y \in V^{\mathrm{res}}$ with density ρ, i.e. with $\rho_y = \rho$, then the duality function (11.102) is zero for all dual configurations admitting particles in y. More generally, if $V^{\mathrm{res}}_\rho \subseteq V^{\mathrm{res}}$ is the set of reservoirs with density ρ, $V^{\mathrm{res}}_\rho := \{y \in V^{\mathrm{res}} : \rho_y = \rho\}$, then

$$\mathcal{D}_{\rho, \boldsymbol{\rho}^{\mathrm{res}}}(\xi, \eta) = \mathcal{D}^{\mathrm{bulk}}_\rho(\xi, \eta) \cdot \prod_{y \in V^{\mathrm{res}} \setminus V^{\mathrm{res}}_\rho} (\rho_y - \rho)^{\xi_y} \cdot \mathbb{1}_{\{\xi|_{V^{\mathrm{res}}_\rho} = 0\}}. \tag{11.107}$$

We recall the fact that, the duality functions in (11.102) are *orthogonal* with respect to the product measure ν_ρ defined in (11.36) with the corresponding parameter ρ (and not w.r.t. μ^{st} unless all reservoir densities are equal to ρ). More precisely, for each fixed $\rho \in \mathcal{R}_\theta$, the set of duality functions

$$\{\mathcal{D}_\rho(\xi, \cdot) : \xi \in \widehat{\Omega}\} \qquad \text{with} \qquad \widehat{\Omega} = \Omega \times \{0\}^{V^{\mathrm{res}}}, \tag{11.108}$$

is an orthogonal basis of $L^2(\Omega, \nu_\rho)$. This will be fully exploited in Theorem 11.25 below.

The measure ν_ρ is, in general, not stationary, except for the case in which the reservoirs density profile $\boldsymbol{\rho}^{\mathrm{res}}$ is constant with $\rho_x = \rho$ for all $x \in V^{\mathrm{res}}$. Only in this case we can say that the duality functions are orthogonal with respect to the stationary measure. Nevertheless, due to the uniqueness of the stationary measure

$\mu^{\rm st}$, it is possible to obtain useful information about it from the orthogonality (11.108) by initializing the system with the measure ν_ρ and letting time go to infinity. It is possible to further generalize this idea thanks to the following equation:

$$\sum_{n\in\Upsilon_{x,\theta,\alpha}} d^{\rm or}_{\rho,\theta,\alpha}(k,n)\,\nu_{\bar\rho,\theta,\alpha}(n) = (\bar\rho-\rho)^k \qquad \text{for all} \qquad x\in V,$$
$$\text{and} \quad \rho,\bar\rho\in\mathcal{R}_\theta, \tag{11.109}$$

relating the single-site orthogonal duality function with parameter ρ to the marginal reversible measure with a different parameter, $\nu_{\bar\rho}$, $\bar\rho\neq\rho$ (see e.g. Sect. 4 of [198] for the proof). This suggests that it is still possible to obtain an expression for orthogonal duality moments when the system is initialized with an inhomogeneous product measure of the form:

$$\nu_{\bar{\boldsymbol\rho},\theta,\boldsymbol\alpha} := \otimes_{x\in V}\,\nu_{\bar\rho_x,\theta,\alpha_x}, \qquad \text{for some} \qquad \bar{\boldsymbol\rho}=\{\bar\rho_x,\ x\in V\}, \tag{11.110}$$

where the densities $\bar\rho_x$ can take any values in $\mathcal{R}_\theta$ and the marginals $\nu_{\rho,\theta,\alpha}$ are the marginal measures defined in (11.36). Notice that, if $\bar\rho_x=\rho$ for all $x\in V$, then $\nu_{\bar{\boldsymbol\rho}}$ coincides with the measure ν_ρ, that is reversible for the system at equilibrium. Whereas, choosing $\bar{\boldsymbol\rho}=\boldsymbol\rho^{\rm st}=\{\rho^{\rm st}_x,\ x\in V\}$ the stationary density profile, we have that $\nu_{\bar{\boldsymbol\rho}}$ recovers the product measure $\nu_{\boldsymbol\rho^{\rm st}}$ defined in Definition 11.15. The duality moments can be written in terms of suitable generating functions of the dual process, as we will see in the next proposition.

Proposition 11.19 (Expectation of Orthogonal Duality Functions) *Let $\{\eta(t):t\geq 0\}$ be a reference process with reservoir density profile $\boldsymbol\rho^{\rm res}=\{\rho_y,\ y\in V^{\rm res}\}$ satisfying the hypothesis of Theorem 11.13 and let $\{\xi(t):\ t\geq 0\}$ be its dual. Then, for all $\rho\in\mathcal{R}_\theta$, $\bar{\boldsymbol\rho}=\{\bar\rho_x,\ x\in V\}$, $\bar\rho_x\in\mathcal{R}_\theta$, $\xi\in\Omega^*$, we have for all $t\geq 0$*

$$\mathbb{E}_{\nu_{\bar{\boldsymbol\rho}}}\left[\mathcal{D}_\rho(\xi,\eta(t))\right] = \mathbf{E}_\xi\left[\prod_{y\in V^{\rm res}}(\rho_y-\rho)^{\xi_y(t)}\cdot\prod_{x\in V}(\bar\rho_x-\rho)^{\xi_x(t)}\right], \tag{11.111}$$

and

$$\mathbb{E}_{\mu^{\rm st}}\left[\mathcal{D}_\rho(\xi,\eta)\right] = \mathbf{E}_\xi\left[\prod_{y\in V^{\rm res}}(\rho_y-\rho)^{\xi_y(\infty)}\right]. \tag{11.112}$$

Proof Let $\xi\in\Omega^*$. Then, by duality we have

$$\mathbb{E}_{\nu_{\bar{\boldsymbol\rho}}}\left[\mathcal{D}_\rho(\xi,\eta(t))\right] = \sum_{\xi'\in\Omega^*} p_t(\xi,\xi')\,\mathbb{E}_{\nu_{\bar{\boldsymbol\rho}}}\left[\mathcal{D}_\rho(\xi',\eta)\right]. \tag{11.113}$$

with

$$\mathbb{E}_{\nu_{\bar{\rho}}}\left[\mathcal{D}_\rho(\xi',\eta)\right] = \prod_{y\in V^{\text{res}}} (\rho_y-\rho)^{\xi'_y} \cdot \prod_{x\in V} \sum_{\eta_x \in \Upsilon_{x,\theta,\alpha}} d^{\text{or}}_{\rho,\alpha_x}(\xi_x,\eta_x)\, \nu_{\bar{\rho}_x,\alpha_x}(\eta_x)$$

$$= \prod_{y\in V^{\text{res}}} (\rho_y-\rho)^{\xi'_y} \cdot \prod_{x\in V} (\bar{\rho}_x-\rho)^{\xi_x}, \tag{11.114}$$

where this last identity is a consequence of (11.109). Then (11.111) follows immediately and, taking the limit as $t \to \infty$, we obtain (11.112) as a consequence of the uniqueness of the stationary measure μ^{st} (that is a consequence of the fact that we are under the hypothesis of Theorem 11.13). □

Remark 11.20 Considering the case $\bar{\rho}_x = \rho$ for all $x \in V$ in (11.111), one obtains the duality moments for the system started with the homogeneous measure ν_ρ:

$$\mathbb{E}_{\nu_\rho}\left[\mathcal{D}_\rho(\xi,\eta(t))\right] = \mathbf{E}_\xi\left[\prod_{y\in V^{\text{res}}} (\rho_y-\rho)^{\xi_y(t)} \cdot \mathbb{1}_{\{\xi(t)|_V=0\}}\right], \tag{11.115}$$

while, taking $\bar{\boldsymbol{\rho}} = \boldsymbol{\rho}^{\text{st}} = \{\rho^{\text{st}}_x,\ x\in V\}$ one gets

$$\mathbb{E}_{\nu_{\rho^{\text{st}}}}\left[\mathcal{D}_\rho(\xi,\eta(t))\right] = \mathbf{E}_\xi\left[\prod_{x\in V^*} (\rho^{\text{st}}_x-\rho)^{\xi_x(t)}\right]. \tag{11.116}$$

11.3.5 Systems with Two Reservoirs

The simplest geometrical setting is the one introduced in Sect. 11.2 for independent particles, i.e. particles moving on a one-dimensional chain $V = \{1,\dots,N\}$ in contact at its right and left end with two reservoirs $V^{\text{res}} = \{\ell, r\}$ kept at different particle densities $\rho_\ell \neq \rho_r$. Here ℓ is the left reservoir, exchanging particles only with the leftmost site 1 and r is the right reservoir in contact with the rightmost bulk site N. In Sect. 11.2 we assumed moreover that particles could jump only to nearest-neighbouring sites.

The symmetric case of the example covered in Sect. 11.2 is a special case of our reference process with reservoirs, with the following choices: V and V^{res} chosen as above and the transition function p is given by:

$$p(x,y) = \mathbb{1}_{\{|x-y|=1\}}, \qquad \text{for} \quad x,y\in V=\{1,\dots,N\},$$

$$p(x,\ell) = \mathbb{1}_{\{x=1\}} \qquad \text{and} \qquad p(x,r) = \mathbb{1}_{\{x=N\}}. \tag{11.117}$$

Notice that this is an allowed transition function (see Definition 11.6). Further, choosing parameters $\theta = 0$ and $\alpha_x = 1, \forall x \in V^*$, we recover the symmetric version of the system of independent random walkers on a chain studied in Sect. 11.2. For general θ and $\boldsymbol{\alpha}$ we simply get a system of *reference processes on a chain* in contact with two reservoirs. Processes with this geometry have been studied extensively as models of non-equilibrium. A paradigmatic case is the standard symmetric exclusion process (obtained for $\theta = -1$ and $\alpha_x = 1$ for all $x \in V$) for which a full characterization of the stationary measure μ^{st} has been found in terms of the matrix ansatz solution (see e.g. [77] and [173, Part III. Section 3]).

In view of this, we continue our analysis specializing to the case $|V^{\mathrm{res}}| = 2$, i.e., two reservoirs but keeping the geometry as general as possible. Systems on a chain will then be a special case, for which we can do more explicit computations. The main result will be Theorem 11.140, where we show that centered moments of order n in the non-equilibrium steady state behave as $(\rho_r - \rho_l)^n$ multiplied with a function that does not depend on the reservoir parameters but only on the geometry of the system and the nature of the interaction.

Therefore, we consider a reference process $\{\eta(t) : t \geq 0\}$ where particles move on a set V in contact with two reservoirs $V^{\mathrm{res}} = \{\ell, r\}$. If $(\theta, \boldsymbol{\alpha}, p)$ are the parameters of the process and $\boldsymbol{\rho}^{\mathrm{res}} = \{\rho_\ell, \rho_r\}$, $\rho_\ell, \rho_r \in \mathcal{R}_\theta$, $\boldsymbol{\alpha}^{\mathrm{res}} = \{\alpha_r, \alpha_\ell\}$, $\alpha_r, \alpha_\ell > 0$ are the parameters of the reservoirs, the generator is given by:

$$\mathcal{L} = L^{\mathrm{bulk}} + L^{\mathrm{res}}, \tag{11.118}$$

with

$$L^{\mathrm{bulk}} = \frac{1}{2} \sum_{x,y \in V} p(x,y) L^{\mathrm{bulk}}_{x,y}, \qquad L^{\mathrm{res}} = \sum_{\substack{x \in V \\ y \in \{\ell, r\}}} p(x,y)\, \alpha_y\, L^{\mathrm{res}}_{x,y},$$

$$L^{\mathrm{bulk}}_{x,y} f(\eta) = \eta_x(\alpha_y + \theta\eta_y)[f(\eta^{x,y}) - f(\eta)] + \eta_y(\alpha_x + \theta\eta_x)[f(\eta^{y,x}) - f(\eta)]$$

$$L^{\mathrm{res}}_{x,y} f(\eta) = \eta_x(1 + \theta\rho_y)[f(\eta - \delta_x) - f(\eta)] + \rho_y(\alpha_x + \theta\eta_x)[f(\eta + \delta_x) - f(\eta)]. \tag{11.119}$$

Here for simplicity we keep calling ℓ and r the two reservoirs even if this notation does not have a geometric significance any longer. A generalization of the chain system (11.117) preserving the geometric interpretation of left and right reservoirs, but allowing long range jumps, can be obtained, for instance, by taking $V = \{1, \ldots, N\}$, $\ell = 0$, $r = N + 1$ and $p(x, y) = \mathfrak{p}(|x - y|)$ for some decreasing function $\mathfrak{p}$ of the sites distance.

We denote by $\{\xi(t), t \geq 0\}$, the dual process with absorbing sites $V^{\mathrm{res}} = \{\ell, r\}$ whose generator is given by:

$$L^{\mathrm{abs}} = L^{\mathrm{bulk}} + L^{\mathrm{abs}}, \qquad L^{\mathrm{abs}} = \sum_{\substack{x \in V \\ y \in \{\ell, r\}}} p(x, y)\, \alpha_y\, L^{\mathrm{abs}}_{x,y}$$

$$L^{\mathrm{abs}}_{x,y} f(\eta) = \xi_x \left[f(\xi^{x,y}) - f(\xi) \right]. \tag{11.120}$$

These two processes are dual with respect to the triangular and orthogonal duality functions:

$$\begin{aligned} D(\xi, \eta) &= \rho_\ell^{\xi_\ell} \cdot D^{\mathrm{bulk}}(\xi, \eta) \cdot \rho_r^{\xi_r}, \\ \mathfrak{D}_\rho(\xi, \eta) &= (\rho_\ell - \rho)^{\xi_\ell} \cdot \mathfrak{D}^{\mathrm{bulk}}_\rho(\xi, \eta) \cdot (\rho_r - \rho)^{\xi_r}. \end{aligned} \tag{11.121}$$

Weighted Density Profile in the Stationary Measure

Hence, in this case, using (11.75), we have that, for all $x \in V^*$,

$$\begin{aligned} \mathbb{E}_{\mu^{\mathrm{st}}} \left[\frac{\eta_x}{\alpha_x} \right] &= \rho_x^{\mathrm{st}} \\ &= \mathbf{E}_x \left[\rho_{X^{\mathrm{rw}}(\infty)} \right] \\ &= \rho_r + (\rho_\ell - \rho_r) \cdot \mathbf{P}_x \left(X^{\mathrm{rw}}(\infty) = \ell \right), \end{aligned} \tag{11.122}$$

where $\{X^{\mathrm{rw}}(t) : t \geq 0\}$ is the random walk absorbed in $\{\ell, r\}$, associated to the dual process. Moreover from (11.76) we know that the stationary weighted density profile $\boldsymbol{\rho}^{\mathrm{st}} = \{\rho_x^{\mathrm{st}},\ x \in V\}$ is the solution of the system of equations:

$$p(x, \ell)\, \alpha_\ell (\rho_\ell - \rho_x^{\mathrm{st}}) + \left(\sum_{y \in V} p(x, y)\, \alpha_y\, (\rho_y^{\mathrm{st}} - \rho_x^{\mathrm{st}}) \right) + p(x, r)\, \alpha_r (\rho_r - \rho_x^{\mathrm{st}}) = 0, \tag{11.123}$$

that is the analogue of the boundary-value problem associated to the Dirichlet Laplacian. This system, in general, is not trivial to solve, except for the special case of a chain (11.117) which we discuss below.

Non-equilibrium Systems on a Chain

Density Profile For a chain, i.e. choosing $V = \{1, \ldots, N\}$ and the transition function p as in (11.117) the solution of (11.123) can be explicitly computed and is given by the following expression. Put

$$\psi(x) = \sum_{y=x}^{N} \frac{1}{p(\{y, y+1\})\alpha_y \alpha_{y+1}}. \tag{11.124}$$

Then we have

$$\rho_x^{\text{st}} = \rho_r + \frac{\psi(x)}{\psi(0)} (\rho_\ell - \rho_r), \tag{11.125}$$

where we identified the site ℓ with 0 and the site r with $N+1$. If we further specialize to the case where $\alpha_x = \alpha$, for all $x \in V^*$ and $p(x, x+1) = p(1, \ell) = p(N, r) = 1$ for all $x \in \{1, \dots, N-1\}$, then $\psi(x) = N + 1 - x$, and the profile $\{\rho_x^{\text{st}}, x \in V\}$ is linear:

$$\rho_x^{\text{st}} = \rho_r + \left(1 - \frac{x}{N+1}\right) (\rho_\ell - \rho_r). \tag{11.126}$$

This is consistent with what we found in Sect. 11.2 for the case of IRW (see (11.19)). Notice that, using (11.122), we can deduce that the factor between brackets multiplying $(\rho_\ell - \rho_r)$ is, in both formulas above, exactly equal to the dual-process absorption probability $\mathbf{P}_x\,(X^{\text{rw}}(\infty) = \ell)$.

Two-Point Covariance The two-point correlations are known, instead, only for the homogeneous case, i.e. for $\alpha_x = \alpha$ for all $x \in V^*$. These have been found in [235], via duality, by computing the absorption probabilities of two dual particles. The non-equilibrium covariance is given by:

$$\text{cov}_{\mu_\theta^{\text{st}}}(\eta_x, \eta_y) = (\rho_r - \rho_\ell)^2 \frac{\theta\, x\,(N + 1 - y)}{(N+1)^2(\alpha(N+1) + \theta)}, \qquad \text{for } y > x. \tag{11.127}$$

In the next two subsections we will show some applications of the duality relations obtained so far. We will look at certain observables and, first using the triangular dualities and then using the orthogonal ones, we will obtain expressions for suitable moments in terms of the dual absorption probabilities. The analysis will give, in particular, the exact nature of the dependence of these moments on the reservoir densities ρ_ℓ and ρ_r.

Triangular Duality and Correlations

As we have seen in the previous analysis, if p is an allowed transition function, then a stationary measure $\mu_\theta^{\text{st}} = \mu_{\rho_\ell,\rho_r,\theta}^{\text{st}}$ exists and is unique and by Theorem 11.16, it is not a product measure if $\rho_\ell \neq \rho_r$ and $\theta \neq 0$.

Here we want to obtain more information about the n-point correlations in the stationary measure μ^{st}, or, more generally, about the stationary expectation of the triangular duality functions that are a slight modification of the factorial moments. In this section we will often stress the dependence on θ of the stationary measure and the duality function, in order to underline the special role of the case $\theta = 0$, where μ^{st} is a product of Poisson measures.

We start by observing that, for all $\theta \in \{-1, 0, 1\}$, by the fact that the absorbing dual preserves the total number of dual particles, we have

$$\int D_\theta(\xi, \eta)\mu_\theta^{\mathrm{st}}(d\eta) = \rho_\ell^{|\xi|} \cdot \mathbb{E}_\xi\left[\left(\frac{\rho_r}{\rho_\ell}\right)^{\xi_r(\infty)}\right] = \rho_\ell^{|\xi|} \cdot \mathbb{E}_\xi\left[\left(\tfrac{\rho_r}{\rho_\ell}\right)^{\xi_r(\infty)}\right]. \tag{11.128}$$

In the next theorem we give a formula for the difference between the ξ-th order stationary expectation of the triangular duality function of the process with parameter $\theta \in \{-1, +1\}$ (interacting case) and the one of the IRW ($\theta = 0$, non-interacting case). We will show that this is, as a function of the reservoir densities ρ_ℓ and ρ_r, equal to $\rho_\ell^{|\xi|}$ times a polynomial in the variable $(\frac{\rho_r}{\rho_\ell} - 1)$. For this polynomial the coefficients of the terms of degree 0 and 1 are zero. The coefficients of the polynomial are given in terms of the factorial moments of the dual process.

Theorem 11.21 (Comparison to Independent Walkers) *Let* $\xi \in \Omega^*$, *then*

$$\int D_\theta(\xi, \eta)\, \mu_\theta^{\mathrm{st}}(d\eta) - \int D_0(\xi, \eta)\, \mu_0^{\mathrm{st}}(d\eta) \tag{11.129}$$

$$= \sum_{\kappa=2}^{|\xi|}\left(\mathbb{E}_\xi\left[\binom{\xi_r(\infty)}{\kappa}\right] - \mathbb{E}_\xi^{\mathrm{irw}}\left[\binom{\xi_r(\infty)}{\kappa}\right]\right) \cdot (\rho_r - \rho_\ell)^\kappa\, \rho_\ell^{|\xi|-\kappa},$$

for all $\kappa \in \{2, \dots, |\xi|\}$.

Proof Putting $z := \frac{\rho_r}{\rho_\ell}$, we have that the l.h.s. of (11.129) is equal to

$$\rho_\ell^{|\xi|}\left(\mathbb{E}_\xi^{(\theta)}[z^{\xi_r(\infty)}] - \mathbb{E}_\xi^{(0)}[z^{\xi_r(\infty)}]\right). \tag{11.130}$$

To prove the result we write

$$\begin{aligned}
\mathbb{E}_\xi^{(\theta)}\left[z^{\xi_r(\infty)}\right] &= \mathbb{E}_\xi^{(\theta)}\left[(z-1+1)^{\xi_r(\infty)}\right] \\
&= \mathbb{E}_\xi^{(\theta)}\left[\sum_{\kappa=0}^{\xi_r(\infty)}(z-1)^\kappa \cdot \binom{\xi_r(\infty)}{\kappa}\right] \\
&= \sum_{\kappa=0}^{|\xi|}(z-1)^\kappa \cdot \mathbb{E}_\xi^{(\theta)}\left[\binom{\xi_r(\infty)}{\kappa}\right],
\end{aligned} \tag{11.131}$$

where we denoted by $\mathbb{E}_\xi^{(\theta)}$ the expectation with respect to the process with parameter θ started from ξ. Then in particular $\mathbb{E}_\xi^{(0)} = \mathbb{E}_\xi^{\text{irw}}$. The terms $\kappa = 0$ and $\kappa = 1$ of the summation in (11.131) are the same for all values of θ, indeed,

$$\mathbb{E}_\xi^{(\theta)}\left[\binom{\xi_r(\infty)}{0}\right] = 1 \qquad \text{and}$$
$$\mathbb{E}_\xi^{(\theta)}\left[\binom{\xi_r(\infty)}{1}\right] = \mathbb{E}_\xi^{(\theta)}\left[\xi_r(\infty)\right] = \mathbb{E}_\xi^{(0)}\left[\xi_r(\infty)\right], \tag{11.132}$$

for all θ, where the second identity follows from (10.101) i.e. the fact that, from the consistency of the dual process, the expectation of $\xi_r(\infty)$ is a sum of absorption probabilities of the random walk $\{X^{\text{rw}}(t),\ t \geq 0\}$ associated to $\{\xi(t),\ t \geq 0\}$ (as in Definition 10.6), that is the same for all values of θ. As a consequence, when taking the difference of the generating functions one can start the summation from $\kappa = 2$, i.e.

$$\begin{aligned}&\mathbb{E}_\xi^{(\theta)}\left[z^{\xi_r(\infty)}\right] - \mathbb{E}_\xi^{(0)}\left[z^{\xi_r(\infty)}\right]\\ &= \sum_{\kappa=2}^{|\xi|}(z-1)^\kappa \cdot \left(\mathbb{E}_\xi^{(\theta)}\left[\binom{\xi_r(\infty)}{\kappa}\right] - \mathbb{E}_\xi^{(0)}\left[\binom{\xi_r(\infty)}{\kappa}\right]\right)\end{aligned} \tag{11.133}$$

from which the statement of the theorem follows. □

Remark 11.22 Notice that, thanks to the consistency of $\{\xi(t) : t \geq 0\}$, we can also write:

$$\begin{aligned}&\mathbb{E}_\xi\left[\binom{\xi_r(\infty)}{\kappa}\right] - \mathbb{E}_\xi^{\text{irw}}\left[\binom{\xi_r(\infty)}{\kappa}\right]\\ &= \sum_{\substack{\varsigma\in\Omega_\kappa\\ \varsigma\leq\xi}} F(\varsigma,\xi)\cdot\left(\mathbb{P}_\varsigma\left(\varsigma_\ell(\infty)=0\right) - \mathbb{P}_\varsigma^{\text{irw}}\left(\varsigma_\ell(\infty)=0\right)\right),\end{aligned} \tag{11.134}$$

for all $\kappa = 2, \ldots, |\xi|$, where we remind the reader that F appearing in the r.h.s. of (11.134) is defined in (10.20). This further simplifies the formula as, in order to compute the r.h.s. of (11.129) one only needs to know the deviation (from the case $\theta = 0$) of the probability that all the dual particles are absorbed in the left reservoirs, and not the whole probability distribution of $\xi_r(\infty)$.

Specializing Theorem 11.21 to the case of dual configurations in which particles occupy different vertices one obtains a result for the stationary correlation functions.

Corollary 11.23 (Multivariate Centered Moments) *For every* $\mathbf{x} = (x_1, \ldots, x_n) \in V_n$ *such that* $x_i \neq x_j$ *for all* $i, j \in \{1, \ldots, n\}, i \neq j$, *we have*

$$\mathbb{E}_{\mu_\theta^{\mathrm{st}}}\left[\prod_{x=1}^{n} \frac{\eta_{x_i}}{\alpha_{x_i}} - \prod_{i=1}^{n} \rho_{x_i}\right]$$
$$= \sum_{\kappa=2}^{n} \left(\mathbb{E}_\xi\left[\binom{\xi_r(\infty)}{\kappa}\right] - \mathbb{E}_\xi^{\mathrm{irw}}\left[\binom{\xi_r(\infty)}{\kappa}\right]\right) \cdot (\rho_r - \rho_\ell)^\kappa \, \rho_\ell^{n-\kappa}. \tag{11.135}$$

Here $\xi = \varphi(\mathbf{x})$, *where we remind the reader that* $\varphi(\mathbf{x})$ *denotes the configuration* $\sum_{i=1}^n \delta_{x_i}$ *and* $\boldsymbol{\rho} = \{\rho_x,\ x \in V\}$ *is the stationary density profile.*

Proof For all θ, we have that

$$\mathbb{E}_{\mu_\theta^{\mathrm{st}}}\left[\prod_{x=1}^{n} \frac{\eta_{x_i}}{\alpha_{x_i}}\right] = \int D_\theta(\varphi(\mathbf{x}), \eta)\mu_\theta^{\mathrm{st}}(d\eta). \tag{11.136}$$

In particular, for $\theta = 0$, from the product-nature of $\mu_0^{\mathrm{st}} = \nu_{\boldsymbol{\rho}^{\mathrm{st}}}$, we have

$$\int D_0(\varphi(\mathbf{x}), \eta)\mu_0^{\mathrm{st}}(d\eta) = \prod_{x=1}^{n} \mathbb{E}_{\mu_0^{\mathrm{st}}}\left[\frac{\eta_{x_i}}{\alpha_{x_i}}\right] = \prod_{i=1}^{n} \rho_{x_i}^{\mathrm{st}}. \tag{11.137}$$

The result then follows from Theorem 11.21. □

Remark 11.24 Using the corollary to $n = 2$, and using (11.134), we obtain information about the stationary covariance. The dependence on the boundary densities is, in this case, exactly a quadratic function of their difference: for $x \neq y$,

$$\mathrm{cov}_{\mu_\theta^{\mathrm{st}}}\left(\frac{\eta_x}{\alpha_x}, \frac{\eta_y}{\alpha_y}\right) = (\rho_r - \rho_\ell)^2 \cdot \left(\mathbb{P}_{\varphi(x,y)}(\xi_\ell(\infty) = 0) - \mathbb{P}_{\varphi(x,y)}^{\mathrm{irw}}(\xi_\ell(\infty) = 0)\right). \tag{11.138}$$

The multiplying factor here is the difference of the two absorption probabilities (between the interacting and the independent case) in the left reservoir. This does not depend on ρ_ℓ and ρ_r, but only on the underlying geometry of the graph and the reservoirs. It is non-positive for exclusion particles, $\theta = -1$ (by Liggett's inequality [172], Chapter 8) and non-negative for inclusion particles, $\theta = +1$ (by the analogue of Ligget's inequality from [117]).

Orthogonal Duality and Centered Moments

In (11.138) we have seen that the two-point correlations are proportional to $(\rho_r - \rho_\ell)^2$, i.e., they are equal to $(\rho_r - \rho_\ell)^2$ multiplied with a function that depends only on the geometry of the system and the interaction parameter, but not on the reservoir parameters. This is not true for higher order correlations, i.e. for $n \geq 3$, the n-point correlations are not proportional to $(\rho_r - \rho_\ell)^n$ (see Corollary 11.23). We are therefore interested in finding the correct observables which have the property that their expectation under μ^{st} is $(\rho_r - \rho_\ell)^n$ multiplied with a function that does not depend on the reservoir parameters.

An hint to answer this question is provided by the orthogonal duality relation. Indeed, using (11.112), and specializing it to the case of two reservoirs, we know that for any fixed $\rho \in \mathcal{R}$, the duality moment with parameter ρ can be written as $\mathbb{E}_\xi[(\rho_\ell - \rho)^{\xi_\ell(\infty)}(\rho_r - \rho)^{\xi_r(\infty)}]$. Then, choosing $\rho = \frac{\rho_\ell + \rho_r}{2}$, one obtains the desired property.

From (11.98) we know that

$$d^{\text{or}}_{\rho,\alpha}(1,n) = \frac{n}{\alpha} - \rho, \tag{11.139}$$

and then, choosing $\xi = \varphi(\mathbf{x})$, $\mathbf{x} = (x_1, \ldots, x_n)$, with $x_i \neq x_j$ for $i \neq j$, in (11.112), one obtains an expression for the moments:

$$\mathbb{E}_{\mu^{\text{st}}}\left[\prod_{x=1}^{m}\left(\frac{\eta_{x_i}}{\alpha_{x_i}} - \rho\right)\right], \qquad \text{with} \quad x_i \neq x_j \tag{11.140}$$

where ρ is a fixed parameter. It is natural then to ask whether it is possible to "center" these moments and get a formula for:

$$\mathbb{E}_{\mu^{\text{st}}}\left[\prod_{x=1}^{m}\left(\frac{\eta_{x_i}}{\alpha_{x_i}} - \rho^{\text{st}}_{x_i}\right)\right], \qquad \text{with} \quad x_i \neq x_j. \tag{11.141}$$

The idea is then to "modulate" the single-site duality function with the local stationary density ρ^{st}_x.

We are now ready to state the following main result of the section that provides expression for these centered moments. The proof of this result will be given at the end of this section, where it is shown to follow from a more general result (Theorem 11.30 below).

Theorem 11.25 (Multivariate Centered Moments) *For the reference process we have*

$$\mathbb{E}_{\nu_{\rho^{\text{st}}}}\left[\prod_{i=1}^{n}\left(\frac{\eta_{x_i}(t)}{\alpha_{x_i}} - \rho^{\text{st}}_{x_i}\right)\right] = (\rho_\ell - \rho_r)^n\; \psi_t(\varphi(\mathbf{x}))\,, \tag{11.142}$$

with

$$\psi_t(\varphi(\mathbf{x})) = \mathbb{P}^{\text{irw}}_{\varphi(\mathbf{x})}(\xi_\ell(\infty)=0)\cdot\sum_{\varsigma\le\varphi(\mathbf{x})}\mathbb{E}_\varsigma\left[(-1)^{|\varsigma|-\varsigma_r(t)}\cdot\frac{\mathbb{P}^{\text{irw}}_{\varsigma(t)}(\varsigma_\ell(\infty)=0)}{\mathbb{P}^{\text{irw}}_{\varsigma}(\varsigma_\ell(\infty)=0)}\right]. \tag{11.143}$$

Moreover, in the stationary state we have

$$\mathbb{E}_{\mu^{\text{st}}}\left[\prod_{i=1}^n\left(\frac{\eta(x_i)}{\alpha_{x_i}}-\rho^{\text{st}}_{x_i}\right)\right] = (\rho_\ell-\rho_r)^n\,\psi(\varphi(\mathbf{x}))\,, \tag{11.144}$$

where

$$\psi(\varphi(\mathbf{x})) = \mathbb{P}^{\text{irw}}_{\varphi(\mathbf{x})}(\xi_\ell(\infty)=0)\cdot\sum_{\varsigma\le\varphi(\mathbf{x})}\frac{\mathbb{P}_\varsigma\,(\varsigma_\ell(\infty)=0)}{\mathbb{P}^{\text{irw}}_{\varsigma}(\varsigma_\ell(\infty)=0)}. \tag{11.145}$$

Remark 11.26 Notice that the centered stationary moments (11.144) are, in general, different from the correlation functions obtained in (11.135), while the two coincide for the case $|\xi|=2$ (see (11.138)).

The proof of Theorem 11.25 is given at the end of this section. The main idea in the proof is the introduction of the functions $\mathcal{H}_{\bar{\rho}} = \mathcal{H}_{\bar{\rho},\theta,\boldsymbol{\alpha}}$

$$\mathcal{H}_{\bar{\rho}}(\xi,\eta) := \prod_{x\in V} d^{\text{or}}_{\bar{\rho}_x,\alpha_x}(\xi_x,\eta_x)\cdot\mathbb{1}_{\{\xi\in\widehat{\Omega}\}}, \tag{11.146}$$

where $\widehat{\Omega}$ is the set of dual configurations with no particles in the reservoirs defined in (11.108).

Specializing these functions to the case $\xi=\varphi(\mathbf{x})$, where $\mathbf{x}=(x_1,\dots,x_n)\in V_n$, with $x_i\neq x_j$ for $i\neq j$, for which one has the centered moments that we want:

$$\mathcal{H}_{\rho}(\varphi(\mathbf{x}),\eta) = \prod_{i=1}^n\left(\frac{\eta_{x_i}}{\alpha_{x_i}}-\rho_{x_i}\right). \tag{11.147}$$

The functions (11.146) have some advantages compared to $\mathcal{D}_\rho$. Choosing $\bar{\rho}=\boldsymbol{\rho}^{\text{st}}$, the expectation with respect to μ^{st} of $\mathcal{H}_{\boldsymbol{\rho}}(\varphi(\mathbf{x}),\eta)$, with $\mathbf{x}=(x_1,\dots,x_m)$, $x_x\neq x_j$, are exactly the centered moments (11.141). The other advantage lies in the fact that they are orthogonal with respect to the non-homogeneous product measure $\nu_{\bar{\rho}}$, i.e. $\{\mathcal{H}_{\bar{\rho}}(\xi,\eta),\ \xi\in\widehat{\Omega}\}$ is an orthogonal basis of $L^2(\Omega,\nu_{\bar{\rho}})$. Choosing $\bar{\rho}=\boldsymbol{\rho}^{\text{st}}$ this means orthogonality with respect to the local equilibrium measure $\nu_{\boldsymbol{\rho}^{\text{st}}}$. On the other hand, a disadvantage is the fact that (11.146) is no longer a duality function. Nevertheless, as we will see in the next lemma, it is possible to expand $\mathcal{H}_{\bar{\rho}}(\xi,\cdot)$ in terms of $\{\mathcal{D}_\rho(\varsigma,\cdot),\ \varsigma\le\xi\}$, where the ordering is pointwise, in the sense of (10.28).

Lemma 11.27 *For all* $\rho \in \mathcal{R}$, $\bar{\boldsymbol{\rho}} = \{\bar{\rho}_x,\ x \in V\}$, $\bar{\rho}_x \in \mathcal{R}$, *we have, for all configurations* $\eta \in \Omega$ *and* $\xi \in \widehat{\Omega}$,

$$\mathcal{H}_{\bar{\boldsymbol{\rho}}}(\xi, \eta) = \sum_{\varsigma \le \xi} \mathcal{D}_\rho(\varsigma, \eta) \cdot \prod_{x \in V} \binom{\xi_x}{\varsigma_x} (\rho - \bar{\rho}_x)^{\xi_x - \varsigma_x}. \tag{11.148}$$

Proof For $\xi \in \widehat{\Omega}$ we have that $\xi_y = 0$ for all $y \in V^{\text{res}}$ and then, by (11.103), we have

$$\mathcal{D}_\rho(\xi, \eta) = \mathcal{D}_\rho^{\text{bulk}}(\xi, \eta) = [e^{-\sum_{x \in V} \rho a_x} D^{\text{bulk}}(\cdot, \eta)](\xi), \tag{11.149}$$

where a_x is the one-particle removal operator at site x. On the other hand, using (11.100), we have that

$$d^{\text{or}}_{\bar{\rho}_x, \alpha_x}(k, n) = [e^{-\bar{\rho}_x a_x} d_{\alpha_x}(\cdot, n)](k), \qquad \text{for all} \quad x \in V, \tag{11.150}$$

and, as a consequence, we can write

$$\begin{aligned}
\mathcal{H}_{\bar{\boldsymbol{\rho}}}(\xi, \eta) &= [e^{-\sum_{x \in V} \bar{\rho}_x a_x} D^{\text{bulk}}(\cdot, \eta)](\xi) \\
&= e^{-\sum_{x \in V} (\bar{\rho}_x - \rho) a_x} [e^{-\rho S^-} D^{\text{bulk}}(\cdot, \eta)](\xi) \\
&= [e^{-\sum_{x \in V} (\bar{\rho}_x - \rho) a_x} \mathcal{D}_\rho^{\text{bulk}}(\cdot, \eta)](\xi).
\end{aligned} \tag{11.151}$$

Now, using that

$$[e^{-(\bar{\rho}_x - \rho) a_x} d^{\text{or}}_{\rho, \alpha_x}(\cdot, n)](k) = \sum_{\ell=0}^{k} \binom{k}{\ell} d^{\text{or}}_{\rho, \alpha_x}(\ell, n) \, (\rho - \bar{\rho}_x)^{k-\ell}, \tag{11.152}$$

we can write

$$\mathcal{H}_{\bar{\boldsymbol{\rho}}}(\xi, \eta) = \sum_{\varsigma \in \Omega} \mathcal{D}_\rho^{\text{bulk}}(\varsigma, \eta) \cdot \prod_{x \in V} \binom{\xi_x}{\varsigma_x} (\rho - \bar{\rho}_x)^{\xi_x - \varsigma_x}. \tag{11.153}$$

Therefore (11.148) follows because, for $\varsigma \in \widehat{\Omega}$, $\mathcal{D}_\rho^{\text{bulk}}(\varsigma, \cdot) = \mathcal{D}_\rho(\varsigma, \cdot)$. □

The idea is now that combining Proposition 11.19 and Lemma 11.27, one can get an expression for the centered moments (11.141). We will do this in full details, restricting for simplicity, to the case of systems in contact with only two reservoirs.

In this section we will go beyond the stationary result and use the orthogonal duality to compute time-dependent $\mathcal{H}_{\bar{\boldsymbol{\rho}}}$-moments in terms of suitable observables of the dual process. Here $\mathcal{H}_{\bar{\boldsymbol{\rho}}}$ is the function defined in (11.146), with $\bar{\boldsymbol{\rho}}$ a general density profile. Choosing $\bar{\boldsymbol{\rho}} = \boldsymbol{\rho}^{\text{st}}$ i.e. the stationary density profile, we will obtain the centered moments (11.141) that we will see to be equal to powers of $(\rho_\ell - \rho_r)$ times a function of the other parameters.

In view of this aim, we combine Proposition 11.19 and Lemma 11.27 where we choose the parameter $\rho \in \mathcal{R}$ of the orthogonal duality function $\mathcal{D}_\rho$ to be of the form

$$\rho = \rho(\beta) := \rho_r + \beta(\rho_\ell - \rho_r), \tag{11.154}$$

for some fixed $\beta \in (0, 1)$, independent on the reservoir densities ρ_ℓ and ρ_r. In a similar way we fix a vector

$$\boldsymbol{\beta} := \{\beta_x,\ x \in V\} \quad \text{for some} \quad \beta_x \in (0,1),\ x \in V \quad \text{and} \quad \beta_\ell = 1,\ \beta_r = 0, \tag{11.155}$$

and choose

$$\bar{\rho} = \bar{\rho}(\boldsymbol{\beta}) = \{\rho(\beta_x),\ x \in V\}, \qquad \rho(\beta_x) := \rho_r + \beta_x(\rho_\ell - \rho_r) \in \mathcal{R}_\theta, \tag{11.156}$$

as the vector labeling the measure $\nu_{\bar{\rho}}$ in Proposition 11.19 and the function $\mathcal{H}_{\bar{\rho}}$ in Lemma 11.27. Here the values β_x, $x \in V$ have to be thought of as independent of ρ_ℓ and ρ_r and the boundary values β_ℓ and β_r have been chosen in such a way that $\rho(\beta_\ell) = \rho_\ell$ and $\rho(\beta_r) = \rho_r$, so that $\rho(\beta_x)$ can be interpreted as an interpolation between the two reservoir densities.

Notice that, independently from the geometry of the system, the stationary density vector $\boldsymbol{\rho}^{\text{st}} = \{\rho_x^{\text{st}},\ x \in V\}$ is a particular case of (11.156), indeed, from (11.122), we know that, choosing

$$\beta_x = \mathbf{P}_x(X^{\text{rw}}(\infty) = \ell), \tag{11.157}$$

with $\{X^{\text{rw}}(t) : t \geq 0\}$ the random walker associated to the dual process, one has $\rho(\beta_x) = \rho_x^{\text{st}}$.

With these choices we can prove that the expectation of orthogonal duality functions for the system initialized with $\nu_{\bar{\rho}}$, as functions of the reservoir densities, are equal to powers of the difference between the reservoir densities, multiplied with suitable functions.

Proposition 11.28 *Let $\{\eta(t) : t \geq 0\}$ be a reference process with reservoirs $\{\ell, r\}$ and densities $\{\rho_\ell, \rho_r\}$ and let $\{\xi(t) : t \geq 0\}$ be its dual process. Let $\boldsymbol{\beta} \in (0,1)^V$ and $\bar{\rho} = \bar{\rho}(\boldsymbol{\beta})$ as in* (11.156)*, then, for all $\xi \in \Omega^*$, we have*

$$\mathbb{E}_{\nu_{\bar{\rho}}}\left[\mathcal{D}_\rho(\xi, \eta(t))\right] = (\rho_\ell - \rho_r)^{|\xi|} \cdot \mathbb{E}_\xi\left[\prod_{x \in V^*} (\beta_x - \beta)^{\xi_x(t)}\right], \qquad \forall \quad t \geq 0 \tag{11.158}$$

and

$$\mathbb{E}_{\mu^{\mathrm{st}}}\left[\mathcal{D}_{\rho,\theta}(\xi,\eta)\right] = [(\rho_\ell - \rho_r)(1-\beta)]^{|\xi|} \cdot \mathbb{E}_\xi\left[\left(\frac{\beta}{\beta-1}\right)^{\xi_r(\infty)}\right] \tag{11.159}$$

for all $\beta \in (0,1)$.

The proposition is an immediate consequence of Proposition 11.19. It claims that the ξ-th order orthogonal duality moment, under the measure $\nu_{\bar{\rho}}$, is, at any time, equal to $(\rho_\ell - \rho_r)^{|\xi|}$ times a function of β, $\boldsymbol{\beta}$, θ and the underlying geometry of the system.

Remark 11.29 For the choice $\beta_x = \beta = \frac{1}{2}$ for all $x \in V$ and, thus, $\rho_{\beta_x} = \rho_{1/2} = \frac{\rho_\ell+\rho_r}{2}$, (11.158) and (11.159) further simplify as

$$\mathbb{E}_{\nu_\rho}\left[\mathcal{D}_{\rho,\theta}(\xi,\eta(t))\right] = \left(\frac{\rho_r-\rho_\ell}{2}\right)^{|\xi|} \mathbb{E}_\xi\left[(-1)^{\xi_r(t)}\, \mathbb{1}_{\{\xi_\ell(t)+\xi_r(t)=|\xi|\}}\right] \tag{11.160}$$

and

$$\mathbb{E}_{\mu^{\mathrm{st}}}\left[\mathcal{D}_{\rho,\theta}(\xi,\eta)\right] = \left(\frac{\rho_r-\rho_\ell}{2}\right)^{|\xi|} \mathbb{E}_\xi\left[(-1)^{\xi_r(t)}\right]. \tag{11.161}$$

We want to use now the previous result to compute the $\mathcal{H}_{\bar{\rho}}$-moments and, as a consequence, obtain an expression for the m-points centered moments.

Let us fix $\beta \in (0,1)$, then combining (11.158) and (11.148) we obtain

$$\begin{aligned}
&\mathbb{E}_{\nu_{\bar{\rho}}}\left[\mathcal{H}_{\bar{\rho}}(\xi,\eta(t))\right] \\
&= \sum_{\varsigma\in\widehat{\Omega}} \mathbb{E}_{\nu_{\bar{\rho}}}\left[\mathcal{D}_\rho(\varsigma,\eta(t))\right] \cdot \prod_{x\in V}\binom{\xi_x}{\varsigma_x}(\rho-\bar{\rho}_x)^{\xi_x-\varsigma_x} \\
&= \sum_{\varsigma\in\widehat{\Omega}} (\rho_\ell-\rho_r)^{|\varsigma|}\cdot \mathbb{E}_\varsigma\left[\prod_{y\in V^*}(\beta_y-\beta)^{\varsigma_y(t)}\right] \\
&\quad \cdot \prod_{x\in V}\binom{\xi_x}{\varsigma_x}\Big((\beta-\beta_x)(\rho_\ell-\rho_r)\Big)^{\xi_x-\varsigma_x} \\
&= (\rho_\ell-\rho_r)^{|\xi|}\cdot \sum_{\varsigma\in\widehat{\Omega}} \mathbb{E}_\varsigma\left[\prod_{y\in V^*}(\beta_y-\beta)^{\varsigma_y(t)}\right]\cdot \prod_{x\in V}\binom{\xi_x}{\varsigma_x}(\beta-\beta_x)^{\xi_x-\varsigma_x} \\
&:= \psi_{\beta,\boldsymbol{\beta},t}(\xi).
\end{aligned} \tag{11.162}$$

Since the l.h.s. of (11.162) does not depend on β, also the function $\psi_{\beta,\boldsymbol{\beta},t}(\xi)$ must be independent on this parameter, which implies

$$\frac{d}{d\beta}\psi_{\beta,\boldsymbol{\beta},t}(\xi) = 0 \qquad \text{for all} \quad \boldsymbol{\beta} \in \mathbb{R}^V,\ \xi \in \Omega^*,\ t \geq 0. \tag{11.163}$$

This equation gives information about the absorption probabilities of the dual process, comparable with the recurrence relations which we obtained in Sect. 10.6 for factorial moments of consistent systems with absorbing boundaries. The following result follows immediately from the computation above.

Theorem 11.30 *Let $\{\eta(t) : t \geq 0\}$ be a reference process with reservoirs $\{\ell, r\}$ and densities $\{\rho_\ell, \rho_r\}$ and let $\{\xi(t) : t \geq 0\}$ be its dual process. Let $\boldsymbol{\beta} \in (0,1)^V$ and $\bar{\boldsymbol{\rho}} = \bar{\boldsymbol{\rho}}(\boldsymbol{\beta})$ as in (11.156), then, for any $\xi \in \widehat{\Omega}$, $t \geq 0$, we have*

$$\mathbb{E}_{\nu_{\bar{\rho}}}\left[\mathcal{H}_{\bar{\boldsymbol{\rho}},\theta}(\xi, \eta_t)\right] = (\rho_\ell - \rho_r)^{|\xi|}\, \psi_{\boldsymbol{\beta},t}(\xi)\,, \tag{11.164}$$

where

$$\psi_{\boldsymbol{\beta},t}(\xi) = \sum_{\varsigma \in \widehat{\Omega}} \mathbb{E}_\varsigma\left[\mathbb{1}_{\varsigma_\ell(t)=0} \cdot \prod_{y \in V}(\beta_y - 1)^{\varsigma_y(t)}\right] \cdot \prod_{x \in V}\binom{\xi_x}{\varsigma_x}(1-\beta_x)^{\xi_x - \varsigma_x}. \tag{11.165}$$

Moreover

$$\mathbb{E}_{\mu^{\mathrm{st}}}\left[\mathcal{H}_{\bar{\boldsymbol{\rho}},\theta}(\xi, \eta_t)\right] = (\rho_\ell - \rho_r)^{|\xi|}\, \psi_{\boldsymbol{\beta}}(\xi)\,, \tag{11.166}$$

with

$$\psi_{\boldsymbol{\beta}}(\xi) = \lim_{t\to\infty}\psi_{\boldsymbol{\beta},t}(\xi) = \sum_{\varsigma \in \widehat{\Omega}} \mathbf{P}_\varsigma\left(\varsigma_\ell(\infty) = 0\right) \cdot \prod_{x \in V}\binom{\xi_x}{\varsigma_x}(1-\beta_x)^{\xi_x - \varsigma_x}. \tag{11.167}$$

Proof The statement follows from (11.164) by choosing $\beta = 1$ and defining $\psi_{\boldsymbol{\beta},t} = \psi_{1,\boldsymbol{\beta},t}$. □

Notice that the functions $\psi_{\boldsymbol{\beta},t}(\xi) \in \mathbb{R}$ and $\psi_{\boldsymbol{\beta}}(\xi) \in \mathbb{R}$ depend neither on ρ_ℓ nor on ρ_r, but only on $\theta \in \{-1, 0, 1\}$ and the underlying geometry of the system (i.e. on the parameters p, $\boldsymbol{\alpha}$ and $\boldsymbol{\alpha}^{\mathrm{res}}$).

We want to obtain now, suitably specializing Theorem 11.30, an expression for the centered moments (11.141). This is achieved by choosing β_x equal to the absorption probabilities in (11.157) for which we have $\bar{\boldsymbol{\rho}}(\boldsymbol{\beta}) = \boldsymbol{\rho}^{\mathrm{st}} = \{\rho_x^{\mathrm{st}},\ x \in V\}$ the stationary density profile. Under this choice the measure $\nu_{\bar{\rho}}$ coincides with the local equilibrium measure $\nu_{\rho^{\rm st}}$. The result is given in the following corollary. We recall the notation (10.20) from the previous chapter: $F(\varsigma,\xi)=\prod_{x\in V}\binom{\xi_x}{\varsigma_x}$.

Corollary 11.31 *Under the hypothesis of Theorem 11.30, let* $\xi\in\widehat{\Omega}$, *then*

$$\mathbb{E}_{\nu_{\rho^{\rm st}}}\left[\mathcal{H}_{\boldsymbol{\rho}}(\xi,\eta(t))\right]=(\rho_\ell-\rho_r)^{|\xi|}\,\psi_t(\xi)\,, \tag{11.168}$$

where

$$\psi_t(\xi)=\mathbb{P}^{\rm irw}_{\xi}(\xi_\ell(\infty)=0)\cdot\sum_{\varsigma\in\widehat{\Omega}}F(\varsigma,\xi)\cdot\mathbb{E}_\varsigma\left[(-1)^{|\varsigma|-\varsigma_r(t)}\cdot\frac{\mathbb{P}^{\rm irw}_{\varsigma(t)}(\varsigma_\ell(\infty)=0)}{\mathbb{P}^{\rm irw}_{\varsigma}(\varsigma_\ell(\infty)=0)}\right]. \tag{11.169}$$

Then, in particular,

$$\mathbb{E}_{\mu^{\rm st}}\left[\prod_{i=1}^{n}\left(\frac{\eta_{x_i}}{\alpha_{x_i}}-\rho^{\rm st}_{x_i}\right)\right]=(\rho_\ell-\rho_r)^{|\xi|}\,\psi(\xi)\,, \tag{11.170}$$

where

$$\psi(\xi)=\lim_{t\to\infty}\psi_t(\xi)=\mathbb{P}^{\rm irw}_{\xi}(\xi_\ell(\infty)=0)\cdot\sum_{\varsigma\in\widehat{\Omega}}F(\varsigma,\xi)\cdot\frac{\mathbb{P}_\varsigma\,(\varsigma_\ell(\infty)=0)}{\mathbb{P}^{\rm irw}_{\varsigma}(\varsigma_\ell(\infty)=0)}. \tag{11.171}$$

Proof The function ψ_t above is equal to the function $\psi_{\boldsymbol{\beta},t}$ defined in (11.165) with $\boldsymbol{\beta}=\{\beta_x,\ x\in V\}$, $\beta_x=\mathbf{P}_x(X^{\rm rw}(\infty)=\ell)$ where $\{X^{\rm rw}(t)\,:\,t\geq 0\}$ is the random walker associated to the dual process. Then

$$\begin{aligned}
&\psi_t(\xi)=\\
&=\sum_{\varsigma\in\widehat{\Omega}}\mathbb{E}_\varsigma\left[(-1)^{|\varsigma|-\varsigma_r(t)}\mathbf{1}_{\varsigma_\ell(t)=0}\cdot\prod_{y\in V}\mathbf{P}_y(X^{\rm rw}(\infty)=r)^{\varsigma_y(t)}\right]\\
&\cdot\prod_{x\in V}\binom{\xi_x}{\varsigma_x}\mathbf{P}_x(X^{\rm rw}(\infty)=r)^{\xi_x-\varsigma_x}\\
&=\sum_{\varsigma\in\widehat{\Omega}}\prod_{x\in V}\binom{\xi_x}{\varsigma_x}\cdot\mathbb{E}_\varsigma\left[(-1)^{|\varsigma|-\varsigma_r(t)}\mathbf{1}_{\varsigma_\ell(t)=0}\cdot\mathbb{P}^{\rm irw}_{\varsigma(t)}(\varsigma_\ell(\infty)=0)\right]\\
&\cdot\frac{\mathbb{P}^{\rm irw}_{\xi}(\xi_\ell(\infty)=0)}{\mathbb{P}^{\rm irw}_{\varsigma}(\varsigma_\ell(\infty)=0)},
\end{aligned} \tag{11.172}$$

from which we obtain (11.169). Then (11.171) follows from the fact that

$$\mathbb{E}_\varsigma \left[(-1)^{|\varsigma|-\varsigma_r(t)} \cdot \frac{\mathbb{P}^{\text{irw}}_{\varsigma(t)}(\varsigma_\ell(\infty)=0)}{\mathbb{P}^{\text{irw}}_{\varsigma}(\varsigma_\ell(\infty)=0)} \right]$$
$$= \frac{1}{\mathbb{P}^{\text{irw}}_{\varsigma}(\varsigma_\ell(\infty)=0)} \mathbb{E}_\varsigma \left[(-1)^{\varsigma_\ell(t)} \cdot \mathbb{P}^{\text{irw}}_{\varsigma(t)}(\varsigma_\ell(\infty)=0) \right] \tag{11.173}$$

and taking the limit as $t \to \infty$,

$$\lim_{t\to\infty} \mathbb{E}_\varsigma \left[(-1)^{\varsigma_\ell(t)} \cdot \mathbb{P}^{\text{irw}}_{\varsigma(t)}(\varsigma_\ell(\infty)=0) \right]$$
$$= \lim_{t\to\infty} \mathbb{E}_\varsigma \left[(-1)^{\varsigma_\ell(t)} \mathbb{1}_{\varsigma_\ell(t)=0} \cdot \mathbb{P}^{\text{irw}}_{\varsigma(t)}(\varsigma_\ell(\infty)=0) \right]$$
$$= \mathbb{E}_\varsigma \left[(-1)^{\varsigma_\ell(\infty)} \mathbb{1}_{\varsigma_\ell(\infty)=0} \right] = \mathbb{P}_\varsigma(\varsigma_\ell(\infty)=0)\,. \tag{11.174}$$

This concludes the proof. □

Proof of Theorem 11.25 It immediately follows by specializing Corollary 11.31 to the case $\xi = \varphi(\mathbf{x})$, where $\mathbf{x} = (x_1, \ldots, x_n) \in V_n$, with $x_i \neq x_j$ for $i \neq j$, for which one has

$$\mathcal{H}_\rho(\varphi(\mathbf{x}), \eta) = \prod_{i=1}^{n} \left(\frac{\eta_{x_i}}{\alpha_{x_i}} - \rho_x \right). \tag{11.175}$$

□

11.4 Brownian Energy Process with Reservoirs

The previous sections were devoted to particle systems driven out of equilibrium. The action of reservoirs was modeled with input and output rates appropriately chosen in order to have an absorbing dual. If one wants to model heat conduction, then it is natural to have models with continuous variables (energies, momenta). These include interacting diffusion models such as BEP, BMP, and continuous mass transport models such as KMP. Also in such models we want to add the modeling of reservoir action, i.e., input and output of mass at sites connected to the reservoir. The parameters associated to reservoirs will now be called "temperatures" (rather than densities) and will be denoted by T_y where the subindex y refers to the location of the reservoir.

In this last section we consider the BEP to which we add reservoir action, which we show to be dual to an absorbing SIP (continuous-discrete duality). We also discuss some consequences of this duality on the correlation functions in the NESS.

Finally, in Sect. 11.4.2 we show a continuous-continuous duality between boundary driven BEP and a process which we call BEP with absorbing sites, which are sites in which mass can disappear according to a deterministic dynamics.

In order to be consistent with the notation used in the previous section, we rewrite the generator of the BEP($\boldsymbol{\alpha}$) including the parameter θ introduced in the previous chapter (see e.g. the scheme in (10.79)) for particle systems. This allows to study, in a unified way, both the BEP and the system of linear ODE that has been introduced in Sect. 4.1 as the scaling limit and dual of IRW. The latter is, in this way recovered choosing $\theta = 0$, while the BEP is obtained for $\theta = 1$.

In the spirit of the previous sections, we fix a finite set $V =$ and a set of reservoirs V^{res} disjoint from it. We put $V^* = V \cup V^{\text{res}}$ and fix an allowed transition function $p : V \times V^* \to [0, \infty)$ as in Definition 11.6. Moreover we set $(\boldsymbol{\alpha}, \boldsymbol{\alpha}^{\text{res}}) = \{\alpha_x,\ x \in V^*\}$, with $\alpha_x \in (0, \infty)$ and $\theta \in \{0, +1\}$, we define the diffusion process $\{\zeta(t),\ t \ge 0\}$ on the set V with reservoirs V^{res}, as the process with state space $\Omega := [0, \infty)^V$ and generator $\mathscr{L} = \mathscr{L}^{(\theta, \boldsymbol{\alpha}, p)}$ given by

$$\mathscr{L} := \mathscr{L}^{\text{bulk}} + \mathscr{L}^{\text{res}}, \tag{11.176}$$

acting on functions $f : \Omega \to \mathbb{R}$ as follows

$$\mathscr{L}^{\text{bulk}} := \frac{1}{2} \sum_{x,y \in V} p(x, y) \mathscr{L}^{\text{bulk}}_{x,y}, \qquad \text{with} \tag{11.177}$$

$$\mathscr{L}^{\text{bulk}}_{x,y} = \theta \zeta_x \zeta_y \left(\frac{\partial}{\partial \zeta_y} - \frac{\partial}{\partial \zeta_x} \right)^2 + (\alpha_y \zeta_x - \alpha_x \zeta_y) \left(\frac{\partial}{\partial \zeta_y} - \frac{\partial}{\partial \zeta_x} \right), \tag{11.178}$$

and

$$\mathscr{L}^{\text{res}} := \sum_{\substack{x \in V \\ y \in V^{\text{res}}}} p(x, y)\, \alpha_y\, \mathscr{L}^{\text{res}}_{x,y} \qquad \text{with} \tag{11.179}$$

$$\mathscr{L}^{\text{res}}_{x,y} := (T_y \alpha_x - \zeta_x) \frac{\partial}{\partial \zeta_x} + \theta\, T_y \zeta_x \frac{\partial^2}{\partial \zeta_x^2}, \tag{11.180}$$

where $T_y \in [0, \infty)$ is the temperature imposed by the reservoir $y \in V^{\text{res}}$ to the bulk sites connected to it. We will use the notation $\boldsymbol{T}^{\text{res}} = \{T_y,\ x \in V^{\text{res}}\}$ for the reservoir temperature profile. As done in the previous sections for particle systems, the reservoir action $\mathscr{L}^{\text{res}}$ has been chosen in such a way to preserve a duality property and, only in the equilibrium set-up (i.e. when the external temperatures are all the same: $T_y = T$ for all $y \in V^{\text{res}}$), also the reversibility.

11.4.1 Continuous-discrete Duality

The following theorem gives a duality result which extends the duality relation already known for the system without reservoirs. The bulk duality function $\mathfrak{D}^{\text{bulk}}_{\theta,\boldsymbol{\alpha}}$ is the duality between the processes without reservoirs. More precisely, for $\theta = 1$, $\mathfrak{D}^{\text{bulk}}_{\theta,\alpha}$ is the duality function between the BEP($\boldsymbol{\alpha}$) and the SIP($\boldsymbol{\alpha}$) on V introduced in Sect. 6.4, while, for $\theta = 0$, $\mathfrak{D}^{\text{bulk}}_{\theta,\boldsymbol{\alpha}}$ is the duality function between IRW($\boldsymbol{\alpha}$) and the deterministic system introduced in Sect. 4.1. To this bulk duality function is added the part of the duality function coming from the reservoirs.

Theorem 11.32 (Duality Between Energy Transport Processes with Reservoirs and Discrete Processes with Absorbing Sites) *For all values of the parameters* $(\theta, \boldsymbol{\alpha}, p)$, *the generator* $\mathcal{L} = \mathcal{L}^{(\theta,\boldsymbol{\alpha},p)}$ *defined in* (11.176)–(11.177)–(11.179) *is dual to the generator* $L^{\text{dual}} = L^{\text{dual},(\theta,\boldsymbol{\alpha},p)}$ *(defined in* (11.44)–(11.45)*) of the reference process on* V *with absorbing sites* V^{res} *and parameters* $(\theta, \boldsymbol{\alpha}, p)$. *The duality function* $\mathfrak{D} = \mathfrak{D}_{\theta,\boldsymbol{\alpha}}$ *is given by:*

$$\mathfrak{D}_{\theta,\boldsymbol{\alpha}}(\xi,\zeta) = \mathfrak{D}^{\text{bulk}}_{\theta,\boldsymbol{\alpha}}(\xi,\zeta)\cdot \prod_{y\in V^{\text{res}}} T_y^{\xi_y}, \qquad \mathfrak{D}^{\text{bulk}}_{\theta,\boldsymbol{\alpha}}(\xi,\zeta) = \prod_{x\in V} \mathfrak{d}_{\theta,\alpha_x}(\xi_x,\zeta_x),$$

$$\textit{with} \qquad \mathfrak{d}_{\theta,\alpha}(k,n) = z^k \cdot \begin{cases} \frac{1}{\alpha^k} & \textit{for } \theta = 0 \\ \\ \frac{\Gamma(\alpha)}{\Gamma(\alpha+k)} & \textit{for } \theta = +1 \end{cases} \tag{11.181}$$

Proof The duality property follows if we prove

$$[\mathcal{L}\mathfrak{D}^{\text{bulk}}(\xi,\cdot)](\zeta) = [L^{\text{dual}}\mathfrak{D}^{\text{bulk}}(\cdot,\zeta)](\xi), \qquad \forall \zeta \in \Omega,\ \xi \in \Omega^*, \tag{11.182}$$

where Ω^* is the extended state space $\Omega^* = \Omega \times \Omega^{\text{abs}}$ defined in (10.93).

The BEP($\boldsymbol{\alpha}$) with reservoirs V^{res}, say $\{\zeta(t),\ t \geq 0\}$, is dual to the SIP($\boldsymbol{\alpha}$) on the extended set $V^* = V \cup V^{\text{res}}$ with absorbing sites V^{res}. In other words, the set of absorbing sites in the dual process coincides with the set of reservoirs in the BEP($\boldsymbol{\alpha}$). Analogously, the result for $\theta = 0$, tells us that the IRW($\boldsymbol{\alpha}$) on the extended set $V^* = V \cup V^{\text{res}}$, with absorbing set V^{res}, is dual to the deterministic system with reservoirs set $V^{\text{res}} = V^{\text{res}}$ whose evolution is governed by a transport equation with transport operator $\mathcal{L}^{(0,\boldsymbol{\alpha},p)}$.

From the duality between the processes without reservoirs we know that

$$[\mathcal{L}^{\text{bulk}}\mathfrak{D}^{\text{bulk}}(\xi,\cdot)](\zeta) = [L^{\text{bulk}}\mathfrak{D}^{\text{bulk}}(\cdot,\zeta)](\xi). \tag{11.183}$$

Then, in order to prove the statement, we only need to verify that

$$\mathcal{L}^{\text{res}}_{x,y}\mathfrak{D}(\xi,\cdot)(\zeta) = L^{\text{abs}}_{x,y}\mathfrak{D}(\cdot,\zeta)(\xi) \qquad \text{for all} \quad (x,y) \in V \times V^{\text{res}}. \tag{11.184}$$

We have

$$
\begin{aligned}
&\mathcal{L}^{\text{res}}_{x,y}\mathfrak{D}(\xi,\cdot)(\zeta) = \\
&= \mathfrak{D}(\xi,\zeta)\cdot\xi_x\left\{T_y\cdot\frac{\alpha_x+\theta(\xi_x-1)}{\zeta_x}-1\right\} \\
&= \xi_x\left\{\mathfrak{D}(\xi^{x,y},\zeta)-\mathfrak{D}(\xi,\zeta)\right\} \\
&= L^{\text{abs}}_{x,y}\mathfrak{D}(\cdot,\zeta)(\xi)
\end{aligned}
\tag{11.185}
$$

□

Notice that for $\theta = 0$, $\mathcal{L}$ is the generator of a deterministic dynamics. As we will see below, the study of this degenerate and simpler case will be instrumental to understand the more interesting case $\theta = +1$. As we will see in the next paragraphs, indeed, it is possible to extract information about the stationary correlation functions of the BEP($\boldsymbol{\alpha}$) with reservoirs by treating the case $\theta = 0$ as a term of comparison.

Reversible Measure at Equilibrium

The process with generator $\mathcal{L}^{\text{bulk}}$ has a one-parameter family of reversible measures $\nu_T = \nu_{T,\theta,\boldsymbol{\alpha}}$, $\boldsymbol{\alpha} = \{\alpha_x,\ x\in V\}$, $T>0$ that are, for $\theta = 0$, products of Dirac-delta measures and, for $\theta = 1$, products of Gamma distributions:

$$
\nu_{T,1,\boldsymbol{\alpha}} \sim \otimes_{x\in V}\ \mathtt{Gamma}\left(\alpha_x,\frac{1}{T}\right) \qquad \text{and} \qquad \nu_{T,0,\boldsymbol{\alpha}} \sim \otimes_{x\in V}\ \delta_{T\alpha_x}, \tag{11.186}
$$

i.e. the measures

$$
\nu_{T,1,\boldsymbol{\alpha}}(d\zeta) = \prod_{x\in V}\frac{1}{T^{\alpha_x}}\cdot\frac{\zeta_x^{\alpha_x-1}}{\Gamma(\alpha_x)}\, e^{-\zeta_x/T}\, d\zeta_x \qquad \text{and} \quad \nu_{T,0,\boldsymbol{\alpha}}(d\zeta) = \prod_{x\in V}\delta_{T\alpha_x}(d\zeta_x). \tag{11.187}
$$

Notice that in both cases the expectation of the energy at site x is given by

$$
\int\frac{\zeta_x}{\alpha_x}\cdot\nu_{T,\theta,\boldsymbol{\alpha}}(d\zeta) = T, \qquad \theta\in\{0,1\} \tag{11.188}
$$

this enables us to interpret the parameter T as a weighted single-site temperature.

The reservoir terms of the generator are chosen in such a way that they still satisfy the detailed-balance condition. Indeed it is possible to verify that, for both $\theta\in\{0,1\}$, and for all fixed $(x,y)\in V\times V^{\text{res}}$, the single-edge generator $\mathcal{L}^{\text{res}}_{x,y}$ is self-adjoint with respect to the measure $\nu_{T_y,\theta,\alpha_x}$. Equation (11.188) provides an explanation for the interpretation of T_y as the (weighted) temperature of the y-th reservoir.

Analogously to what we had for particle systems, if all the reservoirs are kept at the same temperature, i.e. if $T_y = T$ for all $y \in V^{\text{res}}$, the system eventually reaches the stationary measure $\nu_{T,\theta,\alpha}$ that is reversible, and, due to irreducibility (if $p(\cdot,\cdot)$ is an allowed transition function), also the unique stationary measure of the process. Out of equilibrium, i.e. when the reservoir temperature profile $\boldsymbol{T}^{\text{res}}$ is not homogeneous, the process still admits a unique stationary measure $\mu^{\text{st}} = \mu^{\text{st}}_{\boldsymbol{T}^{\text{res}},\theta,\boldsymbol{\alpha},\boldsymbol{\alpha}^{\text{res}}}$. While for $\theta = 0$ this is still a product measure, for $\theta = 1$ this is no longer the case.

Correlation Functions in the Non-equilibrium Stationary State

Suppose from now on that $p : V \times V^* \to [0,\infty)$ is an allowed transition function. Suppose moreover that $\boldsymbol{T}^{\text{res}}$ is not homogeneous. Then, analogously to what we have seen for the reference process in Theorem 11.16, there exists a unique stationary measure μ^{st}, that is in product form if and only if $\theta = 0$. More precisely, we have that

$$\mu^{\text{st}}_{\boldsymbol{T}^{\text{res}},0,\boldsymbol{\alpha}} = \otimes_{x \in V}\, \delta_{T_x \alpha_x}, \tag{11.189}$$

where $\boldsymbol{T} = \{T_x,\ x \in V\}$ is the bulk stationary temperature profile induced by $\boldsymbol{T}^{\text{res}}$. The local stationary temperatures T_x, $x \in V$ can be computed using duality as we will see below.

For both cases $\theta \in \{0,1\}$, we can write the duality moments under μ^{st}_θ in terms of the absorption probabilities of the dual process:

$$\int \mathfrak{D}_\theta(\xi,\zeta)\mu^{\text{st}}_\theta(d\zeta) = \mathbb{E}_\xi\Big[\prod_{y \in V^{\text{res}}} T_y^{\xi_y(\infty)}\Big], \qquad \text{for} \quad \theta \in \{0,1\}. \tag{11.190}$$

Choosing the dual configuration to be $\xi = \delta_x$, $x \in V$ we get

$$\mathfrak{D}_\theta(\delta_x,\zeta) = \frac{\zeta_x}{\alpha_x}, \qquad \text{for} \quad \theta \in \{0,1\}, \tag{11.191}$$

and then we have that the local stationary temperatures are given by:

$$\begin{aligned} T_x = \mathbb{E}_{\mu^{\text{st}}_\theta}\left[\frac{\zeta_x}{\alpha_x}\right] &= \int \mathfrak{D}_\theta(\delta_x,\zeta)\mu^{\text{st}}(d\zeta) \\ &= \sum_{y \in V^{\text{res}}} T_y \cdot \mathbf{P}_x(X^{\text{rw}}(\infty) = y) = \mathbf{E}_x\left[T(X^{\text{rw}}(\infty))\right], \end{aligned} \tag{11.192}$$

where $\{X^{\text{rw}}(t),\ t \geq 0\}$ is the single random walker associated to the dual process in the sense of Definition 10.6. This performs a random walk on $V^* = V \cup V^{\text{res}}$ with rates that are compatible with the ones of L^{dual} and is eventually absorbed in one of the reservoirs in V^{res}. We remark that the generator of the random walker appearing in (11.192) does not depend on the value of θ, and, as a consequence, the bulk temperature profile only depends on the geometry of the system and on the external reservoirs:

$$\boldsymbol{T} = \boldsymbol{T}(\boldsymbol{T}^{\text{res}}, \boldsymbol{\alpha}, \boldsymbol{\alpha}^{\text{res}}, p). \tag{11.193}$$

Let us take now the dual configuration $\xi = \varphi(\mathbf{x})$ where $\mathbf{x} = (x_1, \ldots, x_n)$ and all x_i are mutually different sites of V, then we have

$$\mathfrak{D}_\theta(\xi, \zeta) = \prod_{i=1}^{n} \frac{\zeta_{x_i}}{\alpha_{x_i}}, \qquad \text{for all} \quad \theta \in \{0, 1\}. \tag{11.194}$$

Thus, from the product nature of μ_0^{st} we have

$$\mathbb{E}_{\mu_0^{\text{st}}}\left[\prod_{i=1}^{n} \frac{\zeta_{x_i}}{\alpha_{x_i}}\right] = \int \mathfrak{D}_0\Big(\sum_{i=1}^{n} \delta_{x_i}, \zeta\Big)\mu_0^{\text{st}}(d\zeta) = \prod_{i=1}^{n} \mathbb{E}_{\mu_0^{\text{st}}}\left[\frac{\zeta_{x_i}}{\alpha_{x_i}}\right] = \prod_{i=1}^{n} T_{x_i}, \tag{11.195}$$

while, for $\theta > 0$,

$$\begin{aligned}\mathbb{E}_{\mu_\theta^{\text{st}}}\left[\prod_{i=1}^{n} \frac{\zeta_{x_i}}{\alpha_{x_i}}\right] &= \int \mathfrak{D}_\theta\Big(\varphi(\mathbf{x}), \zeta\Big)\mu_\theta^{\text{st}}(d\zeta) \\ &= \sum_{\varsigma \in \Omega^{\text{abs}}} \prod_{y \in V^{\text{res}}} T_y^{\varsigma_y} \cdot \mathbf{P}_{\varphi(\mathbf{x})}(\xi(\infty) = \varsigma) \\ &= \mathbf{E}_{\varphi(\mathbf{x})}\left[\prod_{y \in V^{\text{res}}} T_y^{\xi_y(\infty)}\right].\end{aligned} \tag{11.196}$$

In view of these considerations, we can obtain, as we did for particle systems, substantial information. In particular we consider the case of systems with only two reservoirs.

Two Reservoirs

Here we assume that the set V is in contact with only two reservoirs that we denote by ℓ and r, $V^{\text{res}} = \{\ell, r\}$ and we call $T_\ell, T_r > 0$ their temperatures. In this case the duality function with the reference process with two absorbing sites is

$$\mathfrak{D}_\theta(\xi, \zeta) = T_\ell^{\xi_\ell} T_r^{\xi_r} \cdot \mathfrak{D}_\theta^{\text{bulk}}(\zeta, \xi) \tag{11.197}$$

thus

$$
\begin{aligned}
E_{\mu_\theta^{\rm st}}\left[\frac{\zeta_x}{\alpha_x}\right] &= T_x \\
&= \mathbf{E}_x\left[\rho(X^{\rm rw}(\infty))\right] \\
&= T_r + (T_\ell - T_r)\cdot \mathbf{P}_x\left(X^{\rm rw}(\infty) = \ell\right), \qquad (11.198)
\end{aligned}
$$

where $\{X^{\rm rw}(t) : t \geq 0\}$ is the single random walker associated to the dual process, and (11.190) becomes:

$$
\int \mathfrak{D}_\theta(\xi,\zeta)\mu_\theta^{\rm st}(d\zeta) = T_r^{|\xi|}\cdot \mathbb{E}_\xi\left[\left(\frac{T_\ell}{T_r}\right)^{\xi_\ell(\infty)}\right]. \qquad (11.199)
$$

We then have the following result whose proof is similar to the proof of Theorem 11.21.

Theorem 11.33 (Comparison to Independent Walkers) *Let $\xi \in \Omega$, then*

$$
\begin{aligned}
&\int \mathfrak{D}_\theta(\xi,\zeta)\ \mu_\theta^{\rm st}(d\eta) - \int \mathfrak{D}_0(\xi,\zeta)\ \mu_0^{\rm st}(d\eta) \\
&= \sum_{\kappa=2}^{|\xi|}\left(\mathbb{E}_\xi\left[\binom{\xi_r(\infty)}{\kappa}\right] - \mathbb{E}_\xi^{\rm irw}\left[\binom{\xi_r(\infty)}{\kappa}\right]\right)\cdot (T_r - T_\ell)^\kappa\ T_\ell^{|\xi|-\kappa}.
\end{aligned}
$$

(11.200)

As a consequence of this theorem we deduce that the correlation functions satisfy the relation below:

Corollary 11.34 (Multivariate Moments) *For every $\mathbf{x} = (x_1,\ldots,x_n) \in V_n$ such that $x_i \neq x_j$ for all $i, j \in \{1,\ldots,n\}, i \neq j$, we have*

$$
\begin{aligned}
&\mathbb{E}_{\mu_\theta^{st}}\left[\prod_{i=1}^n \frac{\zeta_{x_i}}{\alpha_{x_i}} - \prod_{i=1}^n T_{x_i}\right] \\
&= \sum_{\kappa=2}^{|\xi|}\left(\mathbb{E}_\xi\left[\binom{\xi_r(\infty)}{\kappa}\right] - \mathbb{E}_\xi^{\rm irw}\left[\binom{\xi_r(\infty)}{\kappa}\right]\right)\cdot (T_r - T_\ell)^\kappa\ T_\ell^{n-\kappa} \qquad (11.201)
\end{aligned}
$$

with $\xi = \varphi(\mathbf{x})$.

Specializing the formula to the case $n = 2$ we obtain, for $x \neq y$,

$$
\mathrm{cov}_{\mu_\theta^{\rm st}}\left(\frac{\zeta_x}{\alpha_x}, \frac{\zeta_y}{\alpha_y}\right) = (T_r - T_\ell)^2\cdot\left(\mathbb{P}_{\delta_x+\delta_y}(\xi_\ell(\infty) = 0) - \mathbb{P}^{\rm irw}_{\delta_x+\delta_y}(\xi_\ell(\infty) = 0)\right).
$$

(11.202)

11.4.2 Continuous-continuous Duality

In item 3 of Theorem 6.7 we proved that the Brownian energy process without reservoirs, besides being dual to the symmetric inclusion process is also self-dual. Here we wonder whether, in analogy with the discrete case, it is possible to construct suitable continuous versions of "absorbing sites" V^{res}, to be added to the bulk-term of the generator of the BEP. In this way we would like to prove a duality result between the BEP with reservoirs and the BEP with absorbing sites. We will see that the "absorbing" term of the generator that one has to add has the following form:

$$\mathcal{L}^{\text{abs}} = \sum_{\substack{x \in V \\ y \in V^{\text{res}}}} p(x, y)\, \alpha_y\, \mathcal{L}^{\text{abs}}_{x,y} \qquad \text{with} \qquad \mathcal{L}^{\text{abs}}_{x,y} f(v) = v_x \left(\frac{\partial}{\partial v_y} - \frac{\partial}{\partial v_x} \right) f(v). \tag{11.203}$$

The single-edge generator $\mathcal{L}^{\text{abs}}_{x,y}$ can be interpreted as a continuous version of the "discrete absorbing generator" defined in (11.120). It is indeed the deterministic flow associated to the transport equation

$$\frac{d}{dt} f(v) = v_x \left(\frac{\partial}{\partial v_y} - \frac{\partial}{\partial v_x} \right) f(v), \qquad x \in V,\ y \in V^{\text{res}} \tag{11.204}$$

is given by the solution of the following system of ODEs:

$$\begin{cases} \frac{d}{dt}\, v_x(t) = -v_x(t) \\ \frac{d}{dt}\, v_y(t) = v_x(t), \end{cases} \tag{11.205}$$

whose solutions are

$$\begin{cases} v_x(t) = v_x(0)\, e^{-t} \\ v_y(t) = (v_x(0) + v_y(0)) - v_x(0)\, e^{-t}. \end{cases} \tag{11.206}$$

From (11.206) we can see that the total mass $v_x(0) + v_y(0)$ initially present in the bulk-reservoir edge (x, y) flows exponentially fast from the bulk site x to the reservoir y. And then in the long run all the mass is "absorbed" in the site $y \in V^{\text{res}}$. Also in this setting, in the dual dynamics, the sites in the set V^{res} can be interpreted as absorbing sites. In the following theorem we show that the Brownian energy process with reservoirs is dual to the Brownian energy process with absorbing boundaries.

Theorem 11.35 (Duality Between Energy Transport Processes with Reservoirs and Continuous Processes with Absorbing Sites) *The process with generator* $\mathcal{L} = \mathcal{L}^{\text{bulk}} + \mathcal{L}^{\text{res}}$ *given in* (11.176)–(11.177)–(11.179) *is dual to the process with absorbing sites whose generator is*

$$\mathcal{L}^{\text{dual}} = \mathcal{L}^{\text{bulk}} + \mathcal{L}^{\text{abs}}, \tag{11.207}$$

with $\mathscr{L}^{\text{abs}}$ *given by* (11.203). *The duality function is*

$$\mathscr{D}_{\boldsymbol{\alpha},T^{\text{res}}}(v,\zeta) = \mathscr{D}_{\boldsymbol{\alpha}}^{\text{bulk}}(v,\zeta) \cdot \prod_{y\in V^{\text{res}}} e^{(T_y-1)v_y}, \tag{11.208}$$

and $\mathscr{D}_{\boldsymbol{\alpha}}^{\text{bulk}}(\cdot,\cdot)$ *given by*

$$\mathscr{D}_{\boldsymbol{\alpha}}^{\text{bulk}}(v,\zeta) = \prod_{x\in V^{\text{res}}} d_{\alpha_x}(v_x,\zeta_x) \qquad \textit{with} \qquad d_\alpha(v,z) = e^{-v}\,{}_0F_1\left[\begin{matrix} - \\ \theta\alpha \end{matrix}; \theta z v\right]. \tag{11.209}$$

Proof From the self-duality of the process without reservoirs we know that

$$[\mathscr{L}^{\text{bulk}}\mathscr{D}^{\text{bulk}}(v,\cdot)](\zeta) = [\mathscr{L}^{\text{bulk}}\mathscr{D}^{\text{bulk}}(\cdot,\zeta)](v). \tag{11.210}$$

Hence, in order to prove the statement we only need to verify that

$$\mathscr{L}_{x,y}^{\text{res}}\mathscr{D}(v,\cdot)(\zeta) = \mathscr{L}_{x,y}^{\text{abs}}\mathscr{D}(\cdot,\zeta)(v) \qquad \text{for all} \quad (x,y)\in V\times V^{\text{res}}. \tag{11.211}$$

We have

$$\begin{aligned} &\mathscr{L}_{x,y}^{\text{res}}\mathscr{D}(v,\cdot)(\zeta) = \\ &= \left\{(T_y\alpha-\zeta_x)\frac{\partial}{\partial\zeta_x} + \theta\zeta_x\frac{\partial^2}{\partial\zeta_x^2}\right\}\mathscr{D}(v,\zeta) \\ &= v_x\left(\frac{\partial}{\partial v_y} - \frac{\partial}{\partial v_x}\right)\mathscr{D}(v,\zeta) \\ &= \mathscr{L}_{x,y}^{\text{abs}}\mathscr{D}(\cdot,\zeta)(v) \end{aligned} \tag{11.212}$$

that concludes the proof. □

An analogous duality result between the process with reservoirs and the process with absorbing sites holds true for an orthogonal duality function. See Appendix C for more details.

11.5 Additional Notes

Interacting particle systems and, more generally, stochastic models of transport, have been widely used as prototype models for the understanding non-equilibrium phenomena. Pioneer works of this field are [152] and [219] treating, respectively, the KMP model and the symmetric exclusion process. In the course of the last forty years models of this type have been used to prove typical phenomena of non-equilibrium such as the emergence of long-range correlations [77, 79, 119, 219] and the non-locality of large deviation free energies [23–28, 79]. An important role in the study of non-equilibrium systems is played by exactly solvable models, i.e. systems for which there is a full knowledge of the non-equilibrium steady state. These includes to the symmetric exclusion process with reservoirs [77] and some multispecies generalizations [4], for which the non-equilibrium stationary correlations are given in a closed form in terms of a Matrix Product Ansatz, allowing to compute several quantities in great details, such as fluctuations and large deviations of density and current [75, 76, 78]. Other works treating the problem from an exact integrability point of view are [65, 68, 119, 122, 163–165, 232, 237]. As explained in this chapter, the class of interacting particle systems with reservoirs showing a duality processes includes the SEP model but contains several processes that are not exactly solvable via Matrix Product Ansatz.

A duality property was shown in the context of non-equilibrium in [152] where the KMP is proven to be dual to the absorbing discrete KMP process. Various duality relations for boundary driven systems were obtained in [115, 116]. A comprehensive review on duality for boundary-driven processes is provided in [40] where duality properties and applications are shown for a whole class of models of non-equilibrium. In [94] the authors study the same class of processes and generalize the results by finding also orthogonal polynomial duality functions, thus completing the picture given in [40]. In [188] the author generalizes the duality result for ASEP by allowing a more general set of boundary conditions. In [170, 187] the authors find a duality property and study local equilibrium for a generalized version of KMP in which the components have different degrees of freedom and the rate of interaction depends on the spatial location.

Duality techniques for the study of scaling limits of the exclusion process with reservoirs have been broadly used (see e.g. [9, 85, 120, 121, 160]). In the last few years duality results have been used for the study of a broader class of processes. In [190, 229] local equilibrium properties of the SIP with reservoirs are studied via duality. In [100] the authors study the hydrodynamic limit of the inclusion process in contact with slow reservoirs. In the last few years the interest for systems in contact with multiples reservoirs with long range interactions has emerged [18–21].

For boundary-driven interacting particle systems that are dual to systems with absorbing sites, it has been identified an *isospectrality* property between non-equilibrium systems (reservoirs with different parameters) and equilibrium systems (reservoirs with the same parameters) [107, 108, 227]. This implies the *existence* of a mapping between the non-equilibrium and equilibrium processes for all models with an absorbing duality and for all spatial structures, i.e. on all graphs. This mapping, which turns out to be non-local, could be explicitly described for *integrable* one-dimensional systems with two reservoirs [106, 107]. We shall show in Chap. 13 how the mapping allowed the computations of all moments with respects to the stationary measure. From the explicit knowledge of all moments, a mixture structure of the non-equilibrium steady state has been inferred.

Chapter 12
Duality and Macroscopic Fields

In this chapter we study consequences of duality in the study of hydrodynamic limits, and more generally macroscopic fluctuation fields. The approach is in the spirit of [71] where the macroscopic limits are studied via duality functions, but we focus on simple applications, mostly emphasizing what extra information one can obtain via duality, such as the deviation from local equilibrium. First we show that the scaling limit of a single dual particle determines the macrosopic equation for the particle density. Then, starting with independent walkers we study the time dependent variance of the density field via two dual particles and compare with the solution of the limiting Ornstein Uhlenbeck process. We show how the propagation of local equilibrium is related to the scaling behavior of an arbitrary number of dual particles. We then turn to the interacting case, where we essentially show the same results, using coupling with independent particles. Finally we consider higher order macrosopic fields, and provide a new application of orthogonal polynomial duality, namely a quantitative version of the Boltzmann-Gibbs principle.

12.1 Introduction

In this chapter, we will use duality to obtain precise information on macroscopic fields, in particular the density field. We emphasize that for the systems under study, the hydrodynamic limit, as well as the fluctuations of the density field have been understood and belong to the standard "repertoire" of hydrodynamic limits, see e.g. [71, 151].

The aim of this chapter is therefore not so much to add to the general knowledge of hydrodynamic limits, but more to highlight what one can do with duality in this context. More precisely, we discuss the connection between the behavior of macroscopic fields and the scaling properties of dual particles. Generally speaking, the scaling limit of a single dual particle determine the hydrodynamic equation,

C. Giardinà, F. Redig, *Duality for Markov Processes*, Grundlehren der mathematischen Wissenschaften 365, https://doi.org/10.1007/978-3-032-04099-2_12

in the sense that the expectation of the density field converges to the solution of the hydrodynamic equation, which in our context is the linear heat equation. The control of the variance of the density field is related to the behavior of two dual particles. From the scaling properties of their joint dynamics, one can understand both the stationary and nonstationary behavior of the variance of the density fluctuation field. In particular, quantities such as the effect of deviation from local equilibrium (on the variance) become accessible. Therefore, we provide a detailed study of the time dependent (co)variance of the density field and study it in different regimes (equilibrium, local equilibrium, non-equilibrium). After the study of the variance, the natural next step is "propagation of chaos" or "propagation of local equilibrium". Local equilibrium is the property that around each macroscopic point, the distribution is close to an equilibrium product measures, with a parameter depending on the macroscopic point. In the approach to the hydrodynamic limit of [71], one of the aims is to show that local equilibrium propagates on the macroscopic time scale. As we will see, we can understand this property of propagation of local equilibrium via the study of the scaling properties of n dual particles, more precisely via the comparison of n dual particles with n independent particles. Under a general coupling condition, we then obtain propagation of local equilibrium. Next, we consider higher order fields, which correspond to pair and n-tuple empirical distributions of particles, and show that they converge to solutions of the n-dimensional heat equation. Once more, this follows as a consequence of the scaling limit of n dual particles.

Finally, we use orthogonal polynomial duality to prove a quantitative version of the Boltzmann-Gibbs principle. The Boltzmann-Gibbs principle is an approximation result which states that in an appropriate sense, the fluctuation field of a general local function can be replaced by its "projection" on the density field. This corresponds to the physical intuition that the density field is the field associated to the conserved quantity, and therefore the "slowest fluctuating field". As a consequence, on macroscopic space-time scales, other fields become a function of the density field. In the context where one can use duality with orthogonal polynomials, this projection result then can be restated by saying that the fluctuation fields of all orthogonal duality polynomials of order ≥ 2 are negligible. We show that this property can be understood from the behavior of dual particles: more precisely from the fact that on large time scales, the proportion of time that dual particles spend close to each other is negligible.

This chapter is organized as follows. In the next section we start with the case of independent random walkers, which is very simple but already shows interesting features for the variance and propagation of local equilibrium. We then turn to the interacting case (Sect. 12.3) and show how with coupling assumptions essentially the same results as in the independent random walkers case can be recovered. In the final sections we discuss higher order fields (Sect. 12.4) and the Boltzmann-Gibbs principle (Sect. 12.5).

12.2 The Case of Independent Random Walkers

12.2.1 Preliminaries

In this section, we consider independent walkers and restrict to the vertex set $V = \mathbb{Z}^d$, and consider nearest neighbor random walk. I.e., for the single particle transition rate we choose $p(x, y) = 1\!\!1_{\{x\sim y\}}$, where $x \sim y$ denotes that x and y are nearest neighbors.

Remark 12.1 This can be generalized without effort to the case where the single particle transition rate $p(x, y)$ is symmetric and translation invariant, i.e.,

$$p(x, y) = \pi(y - x) = \pi(x - y), \tag{12.1}$$

where $\sum_{x\in\mathbb{Z}^d} \pi(x) = 1$ and furthermore

$$\sum_{x\in\mathbb{Z}^d} |x|^2 \pi(x) < \infty. \tag{12.2}$$

From the symmetry of $p(x, y)$, we have that $\sum_x x\pi(x) = 0$. More precisely, by (12.2) the sum $\sum_x |x|\pi(x) < \infty$ and then, by symmetry, $\sum_x x\pi(x) = 0$. The diffusive scaling limit of the random walk jumping with rates $p(x, y)$ between x and y is the diffusion process with generator $\sum_{ij} D_{ij} \frac{\partial^2}{\partial_{x_i} \partial x_j}$ with diffusion matrix $D_{ij} = \sum_{z\in\mathbb{Z}^d} z_i z_j \pi(z)$, i.e., the process $\sqrt{D} W(t)$ where $W(t)$ is a standard d-dimensional Brownian motion and $\sqrt{D}$ denotes the square root of the positive matrix D.

In the rest of the section, we will continue with the nearest neighbor case, where $D = I$, where I denotes the identity matrix. Then, the associated continuous-time random walk $\{X(t), t \geq 0\}$ scales to standard d-dimensional Brownian motion, i.e.,

$$\epsilon X(\epsilon^{-2} t) \to W(t) \qquad \text{as} \qquad \epsilon \to 0, \tag{12.3}$$

where $W(t)$ is a d-dimensional standard Brownian motion, and where the convergence is weak convergence in path space. The d-dimensional standard Brownian motion has as a generator $\frac{1}{2}\Delta$, and is connected to the heat equation

$$\frac{\partial \rho(t, x)}{\partial t} = \frac{1}{2}\Delta\rho(t, x), \tag{12.4}$$

via $\rho(t, x) = \mathbb{E}(\rho(0, x + W(t))$.

In order to define macroscopic fields, we introduce a scaling parameter $\epsilon > 0$ that intuitively represents the ratio between the microscopic and the macroscopic length scale. The random fields which we consider act as continuous linear functionals on a class of test functions $\varphi : \mathbb{R}^d \to \mathbb{R}$ which we choose to be the Schwartz functions, i.e., C^∞ functions of which all derivatives converge to zero at infinity faster than any polynomial. We denote by $\mathcal{S}$ the set of these test functions and $\mathcal{S}'$ the dual, i.e., the set of Schwartz distributions. The fields that we consider are then random elements of $\mathcal{S}'$.

We can then define the notion of macroscopic fields associated to local functions.

Definition 12.2 (Macroscopic Fields)

1. For a local function $f : \Omega = \mathbb{N}^{\mathbb{Z}^d} \to \mathbb{R}$, a configuration $\eta \in \Omega$ and a scale $\epsilon > 0$ we define the field of f at scale ϵ as the distribution which maps $\varphi \in \mathcal{S}$ to

$$\mathcal{X}_\epsilon(\varphi, f, \eta) = \epsilon^d \sum_x \varphi(\epsilon x)\tau_x f(\eta), \tag{12.5}$$

where τ_x denotes shift over x, i.e., $\tau_x f(\eta) = f(\tau_x \eta)$ and $(\tau_x \eta)_y = \eta(y + x)$.

2. When $f(\eta) = \eta_0 = D(\delta_0, \eta)$ the field $\mathcal{X}_\epsilon$ is called the density field:

$$\mathcal{X}_\epsilon(\varphi, \eta) = \mathcal{X}_\epsilon(\varphi, D(\delta_0, \cdot), \eta) = \epsilon^d \sum_x \varphi(\epsilon x)\, \eta_x. \tag{12.6}$$

3. The density fluctuation field is defined as

$$\begin{aligned}\mathcal{Y}_\epsilon(\varphi, \eta) &= \epsilon^{d/2} \sum_x \varphi(\epsilon x)\, (\eta_x - \mathbb{E}(\eta_x)) \\ &= \epsilon^{-d/2} \left(\mathcal{X}_\epsilon(\varphi, \eta) - \mathbb{E}\left(\mathcal{X}_\epsilon(\varphi, \eta)\right)\right).\end{aligned} \tag{12.7}$$

In this definition, we implicitly assume that the configuration η is such that the sum in the rhs of (12.5) is absolutely convergent. Otherwise, we say that the field does not exist. For the existence of the density field, a sufficient condition for the convergence of the rhs of (12.6) is that the configurations η_x does not increase faster than a polynomial as $x \to \infty$, i.e., there exists $C > 0, k > 0$ such that $\eta_x \le C\|x\|^k$. Later on, as a consequence of precise assumptions on the distribution of η, this bound will hold with probability one.

The prefactor ϵ^d means that we consider the macroscopic fields on the scale of the law of large numbers. This is also called the hydrodynamic spatial scaling, i.e., the scaling corresponding to the "hydrodynamic limit" which is a deterministic partial differential equation. In the density fluctuation field we have the prefactor $\epsilon^{d/2}$ which corresponds to central limit scaling, and the limit will therefore be stochastic, and the solution of stochastic partial differential equation of Ornstein-Uhlenbeck type.

For the distribution of η we choose a scale-dependent probability measure on the state space $\mathbb{N}^{\mathbb{Z}^d}$, which we denote μ^ϵ. This distribution has to be chosen appropriately, such that the fields have well-defined limiting expectations, as we specify below.

Definition 12.3 (Assumptions on the Initial Distribution) The family of probability measures $\{\mu_\epsilon : \epsilon > 0\}$ is said to satisfy the uniform finite moments condition (abbreviation UFMC) if for all $k \in \mathbb{N}$,

$$\sup_{\epsilon>0} \sup_{x\in\mathbb{Z}^d} \int \eta_x^k d\mu_\epsilon(\eta) =: C_k < \infty, \tag{12.8}$$

where C_k satisfies the Carleman growth conditions, i.e., $\sum_k C_{2k}^{-1/2k} < \infty$.

Remark 12.4 Notice that (12.8) is satisfied if there exists $\theta > 0$ such that we have a uniform bound on the moment generating function

$$\sup_x \sup_{\epsilon>0} \mathbb{E}_{\mu_\epsilon}\left(e^{\theta\eta_x}\right) < \infty. \tag{12.9}$$

We will in this section always work with families $\{\mu_\epsilon : \epsilon > 0\}$ satisfying UFMC, which guarantees that for $f(\eta) = \eta_0^k$ the fields (12.5) are well-defined with probability one. We can then define the notion of compatibility between a family $\{\mu_\epsilon : \epsilon > 0\}$ and a density profile.

We remind the reader that the reversible ergodic measures for the system of independent walkers are homogeneous Poisson product measures ν_ρ with marginals $\nu_\rho(\eta_x = n) = \frac{\rho^n}{n!}e^{-\rho}$. We call these measures "the equilibrium product measures". We also remind the reader the classical self-duality polynomials for the system of independent random walkers: $D(\xi, \eta) = \prod_{x\in\mathbb{Z}^d} d(\xi_x, \eta_x)$ with $d(k, n) = n!/(n-k)!$, and where ξ is a finite configuration, i.e., such that $\sum_x \xi_x < \infty$.

For a finite configuration ξ, we recall the notation $\xi = \sum_{i=1}^n \delta_{x_i}$, where $x_1, \ldots, x_n \in \mathbb{Z}^d$, which allows us to view $D(\xi, \eta)$ as a function of the finite configurations ξ, or equivalently as a symmetric functions of n-tuples $(x_1, \ldots, x_n) \in \mathbb{Z}^{dn}$.

We say that a family of probability measures μ_ϵ D-converges to a limit μ if for all ξ finite configurations we have

$$\lim_{\epsilon\to 0} \int D(\xi, \eta) d\mu_\epsilon(\eta) = \int D(\xi, \eta) d\mu. \tag{12.10}$$

Because in this section we will only focus on this form of convergence, we will simply call it "convergence" (instead of D-convergence) and denote it $\mu_\epsilon \to \mu$. Notice that this form of convergence is stronger than convergence in distribution and it implies convergence of moments.

Definition 12.5 (Compatible Density Profiles and Local Equilibrium) Let $\rho : \mathbb{R}^d \to [0, \infty)$ be a bounded smooth function. We think of ρ as a macroscopic density profile.

1. We say that the family of probability measures μ_ϵ, $\epsilon > 0$ has expected density compatible with ρ if for all $\epsilon > 0$, and $x \in \mathbb{Z}^d$

$$\int \eta_x d\mu_\epsilon(\eta) = \rho(\epsilon x). \tag{12.11}$$

2. We say that $\{\mu_\epsilon : \epsilon > 0\}$ satisfies the local equilibrium property with density profile ρ if for all $x \in \mathbb{R}^d$

$$\tau_{\lfloor \epsilon^{-1} x \rfloor} \mu_\epsilon \to \nu_{\rho(x)}, \tag{12.12}$$

as $\epsilon \to 0$.

Remark 12.6 Notice that the equality for all ϵ in (12.11) is for convenience only, and can be replaced with some extra effort by the condition

$$\lim_{\epsilon \to 0} \left(\int \eta_x d\mu_\epsilon(\eta) - \rho(\epsilon x) \right) = 0. \tag{12.13}$$

The idea of local equilibrium is that around each macro-point $x \in \mathbb{R}^d$, corresponding to the micro-point $\lfloor \epsilon^{-1} x \rfloor \in \mathbb{Z}^d$, the distribution is Poisson with a parameter $\rho(x)$ depending on the choice of the macro-point. Of course, the equilibrium distribution ν_ρ is a special case of a local equilibrium with constant density profile.

A classical example of a local equilibrium with density profile ρ is the inhomogeneous product of Poisson distributions with slowly varying parameter $\rho(\epsilon x)$ at $x \in \mathbb{Z}^d$, i.e., the family

$$\mu_\epsilon = \otimes_{x \in \mathbb{Z}^d} \nu_{\rho(\epsilon x)}. \tag{12.14}$$

Let us show in the easy proposition below that (12.14) indeed satisfies the definition (12.12).

Proposition 12.7 *The family defined via* (12.14) *is a local equilibrium.*

Proof Recall that for a inhomogeneous product of Poisson measures

$$\nu^f := \otimes_{i \in \mathbb{Z}^d} \nu_{f(i)}, \tag{12.15}$$

where $f : \mathbb{Z}^d \to [0, \infty)$, we have the characterizing property

$$\int D\left(\sum_{i=1}^n \delta_{x_i}, \eta \right) \nu^f(d\eta) = \prod_{i=1}^n f(x_i). \tag{12.16}$$

Therefore, in order to show (12.12), we have to show that

$$\lim_{\epsilon \to 0} \int \tau_{\lfloor \epsilon^{-1} x \rfloor} D\left(\sum_{i=1}^n \delta_{y_i}, \eta \right) \mu_\epsilon(d\eta) = \rho(x)^n. \tag{12.17}$$

From this we see that for the measures (12.14) and for any choice $y_1, \ldots, y_n \in \mathbb{Z}^d$, we have, using the assumed smoothness of ρ:

$$\begin{aligned}\lim_{\epsilon\to 0}\int \tau_{\lfloor \epsilon^{-1}x\rfloor} D\left(\sum_{i=1}^{n}\delta_{y_i},\eta\right)\mu_\epsilon(d\eta) &= \lim_{\epsilon\to 0}\prod_{i=1}^{n}\rho(\epsilon(y_i+\lfloor \epsilon^{-1}x\rfloor))\\ &= \rho(x)^n = \int D\left(\sum_{i=1}^{n}\delta_{y_i},\eta\right)\nu_{\rho(x)}(d\eta),\end{aligned} \tag{12.18}$$

which shows (12.17). Finally, notice that clearly, (12.14) satisfies the condition UFMC by the assumed boundedness of ρ, as can be seen e.g. via (12.9). □

12.2.2 *Expected Density Field: From a Single Dual Particle to the Heat Equation*

In this section we show that the diffusive scaling of a single dual particle to Brownian motion implies the convergence of the expected density field at macroscopic times to the solution of the heat equation.

Theorem 12.8 (Expected Density Field) *Let $\rho : \mathbb{R}^d \to [0, \infty)$ be a bounded smooth function. Let $\{\mu_\epsilon, \epsilon > 0\}$ be a family of probability measures on $\mathbb{N}^{\mathbb{Z}^d}$ which has expected density compatible with ρ in the sense of* (12.11)*. Let η be distributed according to μ_ϵ. Then for the expected density field at a macroscopic time $t > 0$ we have*

$$\lim_{\epsilon\to 0}\mathbb{E}_{\mu_\epsilon}(\mathcal{X}_\epsilon(\varphi,\eta(\epsilon^{-2}t))) = \int \varphi(x)\rho(t,x)dx, \tag{12.19}$$

where $\rho(t, x)$ is the solution of the heat equation

$$\frac{\partial\rho(t,x)}{\partial t} = \frac{1}{2}\Delta\rho(t,x), \tag{12.20}$$

where Δ denotes the Laplacian in $\mathbb{R}^d$, and where the initial condition is given by $\rho(0, x) = \rho(x)$.

Proof We compute, using duality with one dual particle, and the fact that $D(\delta_x, \eta) = \eta_x$ (for D as in (3.45)–(3.44))

$$\begin{aligned}\mathbb{E}_{\mu_\epsilon}[\eta_x(\epsilon^{-2}t)] &= \mathbb{E}_{\mu^\epsilon} D(\delta_x, \eta(\epsilon^{-2}t)) \\ &= \int \mathbb{E}_x^{\mathrm{RW}} D(\delta_{X(\epsilon^{-2}t)}, \eta)\mu^\epsilon(d\eta) \\ &= \mathbb{E}_0^{\mathrm{RW}} \rho(\epsilon x + \epsilon X(\epsilon^{-2}t)), \end{aligned} \tag{12.21}$$

where $\mathbb{E}_x^{\mathrm{RW}}$ denotes the expectation w.r.t. the random walk $\{X(t) : t \geq 0\}$ starting from x at time $t = 0$, and where in the last step we used translation invariance of this random walk. Then the expected density field at macroscopic times is given by

$$\mathbb{E}_{\mu_\epsilon}\left[\mathcal{X}_\epsilon(\varphi, \eta(\epsilon^{-2}t))\right] = \epsilon^d \sum_x \varphi(\epsilon x)\, \mathbb{E}_0^{\mathrm{RW}} \rho(\epsilon x + \epsilon X(\epsilon^{-2}t)). \tag{12.22}$$

Define now, for $x \in \mathbb{R}^d, t > 0$

$$\rho(t, x) := \mathbb{E}^{\mathrm{BM}}\left[\rho(W(t) + x)\right] = \int_{\mathbb{R}^d} \frac{e^{-\frac{(x-y)^2}{2t}}}{(2\pi t)^{d/2}} \rho(y)dy, \tag{12.23}$$

where $\mathbb{E}^{\mathrm{BM}}$ refers to the expectation w.r.t. the standard Brownian motion $W(t)$, then, as we already mentioned before, $\rho(t, x)$ is the solution of the heat equation with initial condition $\rho(0, x) = \rho(x)$. In what follows we will denote by Q_t the semigroup of Brownian motion, i.e., $Q_t f(x) = \mathbb{E} f(W(t) + x)$. Using this notation we then have, using (12.23), that $\rho(t, x) = Q_t \rho(0, x)$.

Starting from (12.22), using the invariance principle and using dominated convergence, we obtain

$$\begin{aligned}\lim_{\epsilon\to 0} \mathbb{E}_{\mu_\epsilon}\left[\mathcal{X}_\epsilon(\varphi, \eta(\epsilon^{-2}t))\right] &= \lim_{\epsilon\to 0} \epsilon^d \sum_x \varphi(\epsilon x)\, \mathbb{E}_0^{\mathrm{RW}} \rho(\epsilon x + \epsilon X(\epsilon^{-2}t)) \\ &= \lim_{\epsilon\to 0} \epsilon^d \sum_x \varphi(\epsilon x)\, \mathbb{E}_0^{BM} \rho(\epsilon x + W(t))) \\ &= \lim_{\epsilon\to 0} \epsilon^d \sum_x \varphi(\epsilon x)\rho(t, \epsilon x) = \int \varphi(x)\rho(t, x)dx. \end{aligned} \tag{12.24}$$

□

Remark 12.9 In the proof we used translation invariance of the random walk, i.e., when denoting $X^x(t)$ the random walk starting from $x \in \mathbb{Z}^d$ at time $t > 0$, we used the fact that $X^x(t) = x + X^0(t)$. In the non-translation invariant case, we need the invariance principle from an arbitrary starting point, i.e., we need that if $x_\epsilon \in \mathbb{Z}^d$

such that $\epsilon x_\epsilon \to y \in \mathbb{R}^d$ as $\epsilon \to 0$, then $\epsilon X^{x_\epsilon}(\epsilon^{-2}t) \to y + W(t)$, which is slightly more than the invariance principle starting from the origin. This is used to prove the hydrodynamic limit in a context where one has duality, but no translation invariance, such as a (dynamic or static) random environment. See [96, 202] for examples for the partial exclusion process in random environment. The basic idea to pass from the one-particle invariance principle to the hydrodynamic limit is due to Nagy [186], see [92] for a more recent implementation of this idea for particle systems in random environment.

12.2.3 Variance of the Density Field: Two Dual Particles

We now compute the variance of the density field. We will consider various cases for the initial measure, and from the study of the variance which amounts to consider two dual particles, we can already access some interesting quantities such as the deviation from local equilibrium. This will become clear in the discussion after the computation of Lemma 12.11.

We start from a family $\{\mu_\epsilon, \epsilon > 0\}$ of probability measures which satisfies UFMC, and which is compatible with a macroscopic density profile $\rho : \mathbb{R}^d \to [0, \infty)$ which is bounded and smooth, i.e., $\mathbb{E}_{\mu_\epsilon}(\eta_x) = \rho(\epsilon x)$. In order to deal with the second moment of the density field we define the second order expectations of μ_ϵ via

$$\rho_2^\epsilon(\epsilon x, \epsilon y) = \int D(\delta_x + \delta_y, \eta)\mu_\epsilon(d\eta) - \rho(\epsilon x)\rho(\epsilon y). \tag{12.25}$$

If $x \neq y$, then $\mathrm{Cov}_{\mu_\epsilon}(\eta_x, \eta_y) = \rho_2^\epsilon(\epsilon x, \epsilon y)$. Remark that if μ_ϵ is a Poisson product measure, then ρ_2^ϵ is equal to zero. If μ_ϵ is a general product measure, then $\rho_2^\epsilon(\epsilon x, \epsilon y) = 0$ for $x \neq y$, and

$$\rho_2^\epsilon(\epsilon x, \epsilon x) = \mathrm{Var}_{\mu_\epsilon}(\eta_x) - \rho(\epsilon x). \tag{12.26}$$

Definition 12.10 (Decaying Covariances) Consider the family of probability measure $\{\mu_\epsilon, \epsilon > 0\}$.

1. **Decaying covariances.** We say that the family $\{\mu_\epsilon, \epsilon > 0\}$ has decaying covariances if there exists a continuous $\psi : [0, \infty) \to [0, \infty)$ such that $\lim_{r\to\infty} \psi(r) = 0$ and such that for all $\epsilon > 0, x, y \in \mathbb{Z}^d$

$$\mathrm{Cov}_{\mu_\epsilon}(\eta_x, \eta_y) \leq \psi(\|x - y\|), \tag{12.27}$$

where $\|x\|$ denotes the Euclidean norm in $\mathbb{R}^d$.

2. **Slowly varying covariances.** We say that the family $\{\mu_\epsilon, \epsilon > 0\}$ has slowly varying covariances if

$$\begin{aligned}\mathrm{Var}_{\mu_\epsilon}(\eta_x, \eta_x) &= V(\epsilon x)\\ \mathrm{Cov}_{\mu_\epsilon}(\eta_x, \eta_y) &= C(\epsilon x, \epsilon y)\epsilon^d \text{ for } x \neq y,\end{aligned} \tag{12.28}$$

for some smooth bounded functions $V : \mathbb{R} \to [0, \infty)$ and $C : \mathbb{R}^2 \to R$.

We further denote

$$\rho^\epsilon(t, \epsilon x) = \mathbb{E}_{\mu_\epsilon}(\eta_x(\epsilon^{-2}t)). \tag{12.29}$$

By Theorem 12.8, we have, for a test function $\varphi \in \mathcal{S}$

$$\epsilon^d \sum_x \varphi(\epsilon x)\rho^\epsilon(t, \epsilon x) = \int \varphi(x)\rho(t, x)dx + o(1), \tag{12.30}$$

where $\rho(t, x)$ is the solution of the heat Eq. (12.20), with initial condition $\rho(0, x) = \rho(x)$, and where $o(1) \to 0$ as $\epsilon \to 0$. In the following lemma we compute the time-dependent covariances of particle occupations.

Lemma 12.11 (Time-dependent Variance of the Density-field)

$$\begin{aligned}&\mathbb{E}_{\mu^\epsilon}\left[\eta_x(\epsilon^{-2}t)\eta_y(\epsilon^{-2}t)\right] - \rho^\epsilon(t, \epsilon x)\rho^\epsilon(t, \epsilon y)\\ &= \mathbb{E}^{\mathrm{IRW}}_{x,y}\left[\rho^\epsilon_2(\epsilon X_1(\epsilon^{-2}t), \epsilon X_2(\epsilon^{-2}t))\right] + 1\!\!1_{\{x=y\}}\, \rho^\epsilon(t, \epsilon x),\end{aligned} \tag{12.31}$$

where $\mathbb{E}^{\mathrm{IRW}}_{x,y}$ denotes expectation w.r.t. two independent random walkers starting at x, y at time $t = 0$. As a consequence,

$$\begin{aligned}&Var_{\mu_\epsilon}\left(\mathcal{X}_\epsilon(\varphi, \eta(\epsilon^{-2}t))\right)\\ &= \epsilon^{2d}\sum_{x,y}\varphi(\epsilon x)\varphi(\epsilon y)\, \mathbb{E}^{\mathrm{IRW}}_{x,y}\left[\rho^\epsilon_2(\epsilon X_{\epsilon^{-2}t}, \epsilon Y_{\epsilon^{-2}t})\right] + \epsilon^{2d}\sum_x \varphi(\epsilon x)^2\, \rho^\epsilon(t, \epsilon x).\end{aligned} \tag{12.32}$$

Proof Denote as before by $X^x(t)$ the position of the random walk starting at x at time t. First use that

$$\eta_x\eta_y = D(\delta_x + \delta_y, \eta) + 1\!\!1_{\{x=y\}}D(\delta_x, \eta). \tag{12.33}$$

Then, as a consequence, using self-duality and (12.25), (12.29) we have

$$
\begin{aligned}
&\mathbb{E}_{\mu_\epsilon}(\eta_x(\epsilon^{-2}t)\eta_y(\epsilon^{-2}t)) \\
&= \mathbb{E}_{\mu_\epsilon}\left(D(\delta_x+\delta_y,\eta(\epsilon^{-2}t)) + \mathbb{1}_{\{x=y\}}D(\delta_x,\eta(\epsilon^{-2}t))\right) \\
&= \mathbb{E}_{\mu_\epsilon}\mathbb{E}^{\mathrm{IRW}}\left(\left(D(\delta_{X_1^x(\epsilon^{-2}t)}+\delta_{X_2^y(\epsilon^{-2}t)},\eta) + \mathbb{1}_{\{x=y\}}D(\delta_{X_1^x(\epsilon^{-2}t)},\eta\right)\right) \\
&= \rho^\epsilon(t,\epsilon x)\rho^\epsilon(t,\epsilon y) + \mathbb{E}^{\mathrm{IRW}}\rho_2^\epsilon(\epsilon X_1^x(\epsilon^{-2}t),\epsilon X_2^y(\epsilon^{-2}t)) \\
&+ \mathbb{1}_{\{x=y\}}\rho^\epsilon(t,\epsilon x).
\end{aligned}
\tag{12.34}
$$

Here $\mathbb{E}^{\mathrm{IRW}}$ denotes expectations w.r.t. the two independent walkers $X_1^x(t)$, $X_2^y(t)$, starting from x, resp. y. Now (12.31) follows immediately. To prove the consequence, compute

$$
\begin{aligned}
&\mathrm{Var}_{\mu_\epsilon}\left(\mathcal{X}_\epsilon(\varphi,\eta(\epsilon^{-2}t))\right) \\
&= \epsilon^{2d}\sum_{x,y\in\mathbb{Z}^d}\varphi(\epsilon x)\varphi(\epsilon y)\left(\mathbb{E}_{\mu^\epsilon}\left[\eta_x(\epsilon^{-2}t)\eta_y(\epsilon^{-2}t)\right] - \rho^\epsilon(t,\epsilon x)\rho^\epsilon(t,\epsilon y)\right).
\end{aligned}
\tag{12.35}
$$

Then use (12.31). □

12.2.4 *Scaling Limits of the Variance of the Density Field*

In this subsection we provide detailed properties of the variance of the density fluctuation field, in various non-equilibrium settings determined by different choices of the initial measure μ_ϵ. In particular, on the level of the variance, one can understand the "deviations from local equilibrium". More precisely, when the initial measure is not a product of Poisson with parameter $\rho(\epsilon x)$, then the time-dependent variance of the density fluctuation field contains correction terms related to the deviation from local equilibrium.

In the next discussion we compute the variance of the density field $\mathcal{X}_\epsilon$ and we remark that the variance of the density fluctuation field is related to that via

$$
Var(\mathcal{Y}_\epsilon) = \epsilon^{-d}Var(\mathcal{X}_\epsilon). \tag{12.36}
$$

We distinguish five cases.

(a) **Local equilibrium: the initial measure μ_ϵ is a product of Poisson measures.** Then we have that $\rho_2^\epsilon = 0$ and the variance of the density field at time $\epsilon^{-2}t$ equals

$$\mathrm{Var}_{\mu_\epsilon}(\mathcal{X}_\epsilon\left(\varphi, \eta(\epsilon^{-2}t))\right) = \epsilon^{2d}\sum_x \varphi(\epsilon x)^2 \rho^\epsilon(t, \epsilon x) \approx \epsilon^{2d}\sum_x \varphi(\epsilon x)^2 \rho(t, \epsilon x). \tag{12.37}$$

This quantity is of order $O(\epsilon^d)$ and corresponds to the variance of

$$\epsilon^d \sum_x \varphi(\epsilon x) Z_x, \tag{12.38}$$

where Z_x are independent Poisson random variables with expectation $\rho(t, \epsilon x)$. This is because a product of Poisson measures is propagated into a product of Poisson measures (cf. Theorem 4.14). This exact propagation of local equilibrium is an exceptional situation which holds only for independent walkers. As we have seen before, it corresponds to the fact that the intertwining dynamics in the continuum is deterministic, or equivalently via Doob's theorem. As a consequence, the limiting variance of the density fluctuation field $\mathcal{Y}_\epsilon(\varphi, \eta(\epsilon^{-2}t))$ equals

$$\lim_{\epsilon\to 0} Var\left(\mathcal{Y}_\epsilon(\varphi, \eta(\epsilon^{-2}t))\right) = \int \varphi(x)^2 \rho(t, x) dx. \tag{12.39}$$

(b) **The initial measure is a general product measure.** Then $\rho_2^\epsilon(\epsilon x, \epsilon y) = 0$ for $x \neq y$ and using (12.26) we obtain for the variance of the density field at time $\epsilon^{-2}t$

$$\mathrm{Var}_{\mu_\epsilon}(\mathcal{X}_\epsilon\left(\varphi, \eta(\epsilon^{-2}t))\right) \tag{12.40}$$

$$= \epsilon^{2d}\sum_x \varphi(\epsilon x)^2 \rho^\epsilon(t, \epsilon x) + \epsilon^{2d}\sum_{x,y} \varphi(\epsilon x)\varphi(\epsilon y)$$

$$\cdot \mathbb{E}^{\mathrm{IRW}}_{x,y}\left(\mathbb{1}_{\{X_1(\epsilon^{-2}t) = X_2(\epsilon^{-2}t)\}}\left(\mathrm{Var}_{\mu^\epsilon}(\eta_{X(\epsilon^{-2}t)}) - \rho(\epsilon X(\epsilon^{-2}t))\right)\right).$$

The first term corresponds to the local equilibrium contribution and is of order $O(\epsilon^d)$, whereas the second term comes from the deviation from local equilibrium and is also of order $O(\epsilon^d)$, whenever the variance $\mathrm{Var}_{\mu^\epsilon}(\eta_x)$ is uniformly bounded in x and ϵ. This can be seen via

$$\mathbb{P}^{IRW}_{x,y}\left(X_1(\epsilon^{-2}t) = X_2(\epsilon^{-2}t)\right) = p_{2\epsilon^{-2}t}(x - y, 0) \approx \frac{e^{-\frac{(x-y)^2}{4\epsilon^{-2}t}}}{(4\pi\epsilon^{-2}t)^{d/2}} = O(\epsilon^d), \tag{12.41}$$

where $p_t(x, y)$ denotes the transition probability for the random walk to move from x to y in time t. Here the first equality follows from the fact that $X_1(t) - X_2(t)$ has the same law as $X(2t)$ by translation invariance, whereas the last step follows via the local limit theorem.
Now if we assume that the variance is also slowly varying in space, i.e., of the form

$$\mathrm{Var}_{\mu_\epsilon}(\eta_x) = V(\epsilon x), \tag{12.42}$$

with $V : \mathbb{R} \to [0, \infty)$ a bounded smooth function, then we can further work out the second term in term in (12.40) and find, using the local limit theorem, and abbreviating $q_t(x, y) = (2\pi t)^{-d/2} e^{-(x-y)^2/2t}$ the transition probability density of Brownian motion:

$$\begin{aligned}
&\lim_{\epsilon\to 0} \epsilon^{-d} \epsilon^{2d} \sum_{x,y} \varphi(\epsilon x)\varphi(\epsilon y) \\
&\cdot \mathbb{E}^{\mathrm{IRW}}_{x,y}\left(1\!\!1_{\{X_1(\epsilon^{-2}t)=X_2(\epsilon^{-2}t)\}} \left(\mathrm{Var}_{\mu^\epsilon}(\eta_{X(\epsilon^{-2}t)}) - \rho(\epsilon X(\epsilon^{-2}t))\right)\right) \\
&= \lim_{\epsilon\to 0} \epsilon^{d} \sum_{x,y,z\in\mathbb{Z}^d} \varphi(\epsilon x)\varphi(\epsilon y)(V(\epsilon z) - \rho(\epsilon z))p_{\epsilon^{-2}t}(x, z)p_{\epsilon^{-2}t}(y, z) \\
&= \int \varphi(x)\varphi(y)(V(z) - \rho(z))q_t(x, z)q_t(y, z)dxdydz.
\end{aligned} \tag{12.43}$$

So in this case, we also conclude that the variance of the density field is of order ϵ^d, and that the limiting variance of the density fluctuation field equals

$$\begin{aligned}
&\lim_{\epsilon\to 0} \mathrm{Var}\left(\mathcal{Y}_\epsilon(\varphi, \eta(\epsilon^{-2}t))\right) \\
&= \int \varphi(x)^2\rho(t, x)dx \\
&+ \int \varphi(x)\varphi(y)(V(z) - \rho(z))q_t(x, z)q_t(y, z)\, dxdydz.
\end{aligned} \tag{12.44}$$

The term

$$\int \varphi(x)\varphi(y)(V(z) - \rho(z))q_t(x, z)q_t(y, z)\, dxdydz, \tag{12.45}$$

corresponds to the correction coming from the deviation from local equilibrium. Indeed, notice that for the local equilibrium distribution, $V(z) - \rho(z) = 0$, so this term is not present.

We can rewrite the limiting variance using $Q_t f(x) = \int q_t(x, y) f(y) dy$ for the semigroup of Brownian motion and the fact that $\rho(t, \cdot) = Q_t(\rho(0, \cdot))$, combined with self-adjointness of Q_t in $L^2(dx)$:

$$\lim_{\epsilon\to 0} \text{Var}\left(\mathcal{Y}_\epsilon(\varphi, \eta(\epsilon^{-2}t))\right) = \int Q_t(\varphi^2)(x)\rho(x)dx$$
$$+ \int (Q_t(\varphi))^2(z)(V(z) - \rho(z))dz. \quad (12.46)$$

We will see below that this expression coincides with the variance that one computes from the limiting infinite dimensional Ornstein Uhlenbeck process which describes the evolution of the fluctuation field in the limit $\epsilon \to 0$.

(c) **The initial measure has decaying covariance.**
Then, as in the previous two cases, in this case the limiting variance of the density fluctuation field equals zero, as $\epsilon \to 0$. Indeed, using the defining property (12.27), the correction, corresponding to the deviation from local equilibrium, can now be estimated by

$$\epsilon^{2d} \sum_{x,y} \varphi(\epsilon x)\varphi(\epsilon y)\mathbb{E}^{\text{IRW}}_{x,y}\left[\psi(|X_{\epsilon^{-2}t} - Y_{\epsilon^{-2}t}|)\right], \quad (12.47)$$

which goes to zero as $\epsilon \to 0$. Indeed, with probability close to 1 as $\epsilon \to 0$, $|X_{\epsilon^{-2}t} - Y_{\epsilon^{-2}t}| \to +\infty$, and therefore, using the assumption $\psi(r) \to 0$ as $r \to \infty$, the factor $\mathbb{E}^{\text{IRW}}_{x,y}\left[\psi(|X_{\epsilon^{-2}t} - Y_{\epsilon^{-2}t}|)\right]$ tends to zero as $\epsilon \to 0$.

(d) **The initial measure has slowly varying covariances.** If we assume that the initial distribution μ_ϵ has a slowly varying variance, as well as a suitable scaling form for the covariance, as defined in (12.28), then we can repeat an analogous computation as we did in item b) for a product initial measure, and we find that the limiting variance of the density fluctuation field equals

$$\lim_{\epsilon\to 0} \text{Var}_{\mu_\epsilon}\left(\epsilon^{-d/2}\mathcal{X}_\epsilon(\varphi, \eta(\epsilon^{-2}t))\right)$$
$$= \int Q_t(\varphi^2)(x)\rho(x)dx + \int (Q_t(\varphi)(x))^2(V(x) - \rho(x))dx$$
$$+ \int (Q_t\varphi)(x)Q_t(\varphi)(y)C(x, y)dxdy, \quad (12.48)$$

where Q_t denotes the semigroup of Brownian motion.

(e) **The initial measure has macroscopic covariances.** This means that

$$\rho_2^\epsilon(\epsilon x, \epsilon y) \approx \rho_2(\epsilon x, \epsilon y) \quad (12.49)$$

does not converge to zero as $\epsilon \to 0$ for $x \neq y$ but has a limiting scaling form (where we remind the reader that $\rho_2^\epsilon(\epsilon x, \epsilon y)$ is defined in (12.25)). In words, (12.49) means that under the measure μ_ϵ the particle numbers η_x, η_y have "macroscopic" covariances (i.e., of order ϵ^{-1}). In that case the limiting variance equals

$$\lim_{\epsilon\to 0} \mathrm{Var}_{\mu_\epsilon}\left(\mathcal{X}_\epsilon\left(\varphi, \eta(\epsilon^{-2}t)\right)\right)$$

$$= \int \varphi(x)\varphi(y)\mathbb{E}^{\mathrm{BM}}\left[\rho_2(x + W_1(t), y + W_2(t))\right] dxdy, \qquad (12.50)$$

where $\mathbb{E}^{\mathrm{BM}}$ is now the expectation with respect to two independent standard d-dimensional Brownian motions, W_1 and W_2. Because the variance of the density field does not vanish in the limit $\epsilon \to 0$, in that setting the limiting density field is no longer deterministic.

12.2.5 Connection with the SPDE for the Limiting Density Fluctuation Field

Here we show that the limiting time-dependent variance of the density fluctuation field, as we computed in the previous section in various cases, corresponds (in all these cases) to the limiting linear Ornstein-Uhlenbeck stochastic partial differential equation (SPDE) for the fluctuation field, which is given by

$$dY(t,x) = \frac{1}{2}\Delta Y(t,x)dt - \nabla.\left(\sqrt{\rho(t,x)}d\mathcal{W}(t,x)\right). \qquad (12.51)$$

Here $\mathcal{W}(t,x)$ denotes space-time white noise. The term $\nabla.\left(\sqrt{\rho(t,x)}d\mathcal{W}(t,x)\right)$ has to be interpreted as follows: for a test function $f \in \mathcal{S}$, the quantity

$$\int_0^T \nabla.\left(\sqrt{\rho(t,x)}d\mathcal{W}(t,x)\right)[\varphi], \qquad (12.52)$$

is a martingale with quadratic variation $\int_0^T \int_{\mathbb{R}} \rho(s,x)(\nabla\varphi(x))^2 dxdt$.

The mild solution of (12.51) is given by

$$Y_t(\varphi) = Y_0(Q_t\varphi) + \int_0^t \nabla(Q_{t-s}\varphi)(x).\sqrt{\rho(s,x)}d\mathcal{W}(s,x), \qquad (12.53)$$

where we remind that $Q_t f(x) = \int q_t(x,y)f(y)dy$ denotes the semigroup of Brownian motion. Here we denoted $Y_t(\varphi)$ the pairing of the distribution $Y(t,\cdot)$ with the test function φ.

From (12.53) we can compute the variance of $Y_t(\varphi)$

$$\mathrm{Var}\,(Y_t(\varphi)) = \mathrm{Var}(Y_0(Q_t\varphi)) + \int_0^t \int (\nabla(Q_{t-s}\varphi))^2 \rho(x,s)dxds. \qquad (12.54)$$

Then, we have to match (12.54) to (12.46)

We show that the right hand side of (12.46) fits with the solution of the SPDE (12.51), where as initial condition we have for $Y(0,\cdot)$ white noise with variance $\mathrm{Var}\,(Y_0(\varphi)) = \int V(x)\varphi^2(x)dx$, which by the central limit theorem corresponds to the limit of $\mathcal{Y}_\epsilon(\varphi, \eta(\epsilon^{-2}t))$ at time $t = 0$, because $\eta(0)$ is by assumption distributed as a product measure.

We then have to show that

$$\int \varphi^2(x)\rho(t,x)dx - \int (Q_t\varphi)^2(x)\rho(x)dx = \int \int_0^t (\nabla(Q_{t-s}\varphi))^2 \rho(x,s)dsdx. \qquad (12.55)$$

This follows from the identity

$$\int \frac{d}{ds}\left((Q_{t-s}\varphi)^2 \rho(x,s)\right) dx = \int (\nabla(Q_{t-s}\varphi))^2 \rho(x,s)dx, \qquad (12.56)$$

which in turn can be derived from the fact $\frac{d}{ds}Q_s = \frac{1}{2}\Delta Q_s$ combined with partial integration, and the fact $\rho(x,s) = Q_s\rho(x,0)$.

Similarly, the variance computed in (12.48) fits with the solution of (12.51) starting from initial condition $Y(0,\cdot)$ Gaussian white noise with variance

$$\int \varphi(x)\varphi(y)C(x,y)dxdy + \int V(x)\varphi(x)^2. \qquad (12.57)$$

12.2.6 Summary

Combining the computations for the expected density field with the computations for the variance of the density field, we can summarize as follows.

1. **Hydrodynamic limit in L^2.**
   If the initial measure μ_ϵ has the following two properties:
   (i) well-defined (limiting) density profile

$$\mathbb{E}_{\mu^\epsilon}(\eta_x) = \rho(\epsilon x); \qquad (12.58)$$

(ii) decaying covariance as in item c) of the previous discussion

$$\mathrm{Cov}_{\mu^\epsilon}(\eta_x, \eta_y) \le \psi(\|x-y\|) \qquad \text{with} \quad \psi(r) \to 0 \quad \text{as} \quad r \to \infty, \tag{12.59}$$

or slowly varying variance and covariance as in item d) of the previous discussion. Then we conclude that

$$\lim_{\epsilon \to 0} \mathrm{Var}_{\mu_\epsilon}(\mathcal{X}_\epsilon\left(\varphi, \eta(\epsilon^{-2}t))\right) = 0. \tag{12.60}$$

Together with the result on the expected density field (12.24), this implies that, at every macroscopic time t, the density field converges in L^2 to the deterministic quantity

$$\int \varphi(x)\, \rho(t,x)\, dx, \tag{12.61}$$

where $\rho(t,x)$ is the solution of the heat Eq. (12.20), i.e.

$$\lim_{\epsilon \to 0} \mathbb{E}\left[\mathcal{X}_\epsilon\left(\varphi, \eta(\epsilon^{-2}t)\right) - \int \varphi(x)\, \rho(t,x)\, dx\right]^2 = 0. \tag{12.62}$$

2. **Limiting variance of the density fluctuation field.**
If the initial measure μ_ϵ has a scaling form for the variance and covariance in the sense (12.28), then the density fluctuation field $\mathcal{Y}_\epsilon(\varphi, \eta(\epsilon^{-2}t))$ has a limiting variance which fits with the mild solution of the stochastic partial differential Eq. (12.51) with initial condition $Y(0,\cdot)$ given by white noise with variance $\mathbb{E}(Y_0(\varphi)^2) = \int C(x,y)\varphi(x)\varphi(y)dxdy + \int V(x)\varphi(x)dx$.

12.2.7 Local Equilibrium: n Dual Particles

Local equilibrium as defined in (12.12) is expected to be a property that sets in at any positive macroscopic time, and is preserved in the course of (macroscopic) time. This should be the case if one starts from a reasonable family of measures $\{\mu_\epsilon, \epsilon > 0\}$, i.e., a family compatible with a density profile, and such that there is some decay of covariances. If there is duality, then the main reason why local equilibrium sets in is the fact that at macroscopic times dual particles will typically be at large distances from each other and therefore move independently.

Let us start with a sketchy computation which explains this intuition and contains the main idea of the proof of propagation of local equilibrium. Let us assume, for simplicity, that we start from a product measure μ_ϵ, compatible with a density profile ρ, and suppose that we want to compute the expectation at macroscopic time t (i.e., microscopic time $\epsilon^{-2}t$) of a classical duality polynomial around the macroscopic point $x \in \mathbb{R}^d$, i.e., around the microscopic point $\lfloor \epsilon^{-1} x \rfloor$:

$$\int \mathbb{E}_\eta D\left(\sum_{i=1}^{n} \delta_{\lfloor \epsilon^{-1} x \rfloor + y_i}, \eta(\epsilon^{-2} t)\right) \mu_\epsilon(d\eta), \tag{12.63}$$

where $y_1, \ldots, y_n \in \mathbb{Z}^d$. Using self-duality we find that this expectation can be rewritten as

$$\int \mathbb{E}^{\text{IRW}} D\left(\sum_{i=1} \delta_{X_i^{\lfloor \epsilon^{-1} x \rfloor + y_i}(\epsilon^{-2} t)}, \eta\right) \mu_\epsilon(d\eta), \tag{12.64}$$

where the expectation $\mathbb{E}^{\text{IRW}}$ is w.r.t. the independent random walkers initially at positions $\lfloor \epsilon^{-1} x \rfloor + y_i, i = 1, \ldots, n$. Because we let the walkers evolve for a time $\epsilon^{-2} t$, we have with probability close to one that all their locations at time $\epsilon^{-2} t$, i.e., $X_i^{\lfloor \epsilon^{-1} x \rfloor + y_i}(\epsilon^{-2} t)$ are different (typically they will be at distance ϵ^{-1} from each other). Therefore, using the fact that for $y_1, \ldots, y_n$ mutually different elements of $\mathbb{Z}^d$

$$D\left(\sum_{i=1}^{n} \delta_{y_i}, \eta\right) = \prod_{i=1}^{n} \eta_{y_i}, \tag{12.65}$$

we can approximate (12.64) by

$$\int \mathbb{E}^{\text{IRW}} \left(\prod_{i=1}^{n} \eta_{X_i^{\lfloor \epsilon^{-1} x \rfloor + y_i}(\epsilon^{-2} t)}\right) \mu_\epsilon(d\eta), \tag{12.66}$$

and because μ_ϵ is a product measure, also by

$$\mathbb{E}^{\text{IRW}} \left(\prod_{i=1}^{n} \int \eta_{X_i^{\lfloor \epsilon^{-1} x \rfloor + y_i}(\epsilon^{-2} t)} \mu_\epsilon(d\eta)\right), \tag{12.67}$$

and because μ_ϵ is compatible with profile ρ, by

$$\prod_{i=1}^{n} \rho(t, \epsilon(\lfloor \epsilon^{-1} x \rfloor + y_i)) \approx \rho(t, x)^n. \tag{12.68}$$

In other words, around the macroscopic point x (corresponding to the microscopic lattice point $\lfloor \epsilon^{-1} x \rfloor$, at macroscopic time $t > 0$ the distribution is approximately (when $\epsilon \to 0$) a product of Poisson with density $\rho(t, x)$, where $\rho(t, x)$ is the solution of the heat equation starting from the initial density profile.

The following theorem formalizes this computation. For simplicity we assume that the measures μ_ϵ consists of product measures. This can be replaced easily by decaying covariance (of general polynomials of the particle numbers η_x) assumptions.

Theorem 12.12 (Local Equilibrium) *Let the family* $\{\mu_\epsilon, \epsilon > 0\}$ *have expected density compatible with the density profile* $\rho : \mathbb{R}^d \to [0, \infty)$*. Assume moreover that* μ_ϵ *is a product measure for all* $\epsilon > 0$ *and satisfies UFMC* (12.8)*. Let us denote by* $\mu_\epsilon S(\epsilon^{-2}t)$ *the distribution at time* $\epsilon^{-2}t$ *when started from* μ_ϵ*. Then the family* $\{\mu_\epsilon S(\epsilon^{-2}t) : \epsilon > 0\}$ *is a local equilibrium with density profile* $\rho(t, x)$ *given by the solution of the heat Eq.* (12.20) *with initial condition* $\rho(0, x) = \rho(x)$.

We start with a simple lemma.

Lemma 12.13 *Let us denote by* A_n *the set of n-tuples of lattice points with all n entries different. Then we have the following estimate*

$$\sup_{x_1,\ldots,x_n \in \mathbb{Z}^d} \mathbb{P}^{\mathrm{IRW}}_{x_1,\ldots,x_n}((X^{x_1}(t), \ldots, X^{x_n}(t)) \in A_n^c) \leq n(n-1)p_{2t}(0,0). \tag{12.69}$$

Proof

$$\begin{aligned}\mathbb{P}^{\mathrm{IRW}}_{x_1,\ldots,x_n}((X^{x_1}(t), \ldots, X^{x_n}(t)) \in A_n^c) &\leq \sum_{i,j=1, i\neq j}^{n} \mathbb{P}^{\mathrm{IRW}}_{x_i,x_j}\left(X^{x_i}(t) = X^{x_j}(t)\right) \\ &\leq n(n-1)p_{2t}(0,0),\end{aligned} \tag{12.70}$$

where in the last step we used that $X_t^{x_i} - X_t^{x_j}$ has the same distribution as $X_{2t}^{x_i - x_j}$ which follows from independence of the walks, and we also used that $p_t(x, y) \leq p_t(0, 0)$. □

We can now prove Theorem 12.12.

Proof We have to prove that for all choices of $y_1, \ldots, y_n \in \mathbb{Z}^d$, and $x \in \mathbb{R}^d$

$$\lim_{\epsilon \to 0} \int \mathbb{E}_\eta D\left(\sum_{i=1}^{n} \delta_{\lfloor \epsilon^{-1}x \rfloor + y_i}, \eta(\epsilon^{-2}t)\right) \mu_\epsilon(d\eta) = \rho(t, x)^n. \tag{12.71}$$

By duality this is equivalent to showing that

$$\lim_{\epsilon \to 0} \int \mathbb{E}^{\mathrm{IRW}} D\left(\sum_{i=1} \delta_{X_i^{\lfloor \epsilon^{-1}x \rfloor + y_i}(\epsilon^{-2}t)}, \eta\right) \mu_\epsilon(d\eta) = \rho(t, x)^n. \tag{12.72}$$

Abbreviate

$$\mathbf{x}(\epsilon, t) = \left(X^{\lfloor \epsilon^{-1}x \rfloor + y_1}(\epsilon^{-2}t), \ldots X^{\lfloor \epsilon^{-1}x \rfloor + y_n}(\epsilon^{-2}t)\right). \tag{12.73}$$

Denoting A_n the set of n-tuples of lattice points with all n entries different, we may write

$$\begin{aligned}
&\mathbb{E}^{\mathrm{IRW}}\left(D\left(\sum_{i=1}\delta_{X_i^{\lfloor\epsilon^{-1}x\rfloor+y_i}(\epsilon^{-2}t)},\eta\right)\right)\\
&=\mathbb{E}^{\mathrm{IRW}}\left(\prod_{i=1}^{n}D\left(\delta_{X_i^{\lfloor\epsilon^{-1}x\rfloor+y_i}(\epsilon^{-2}t)},\eta\right)1\!\!1_{A_n}(\mathbf{x}(\epsilon,t))\right)\\
&+\mathbb{E}^{\mathrm{IRW}}\left(\prod_{i=1}^{n}D\left(\delta_{X_i^{\lfloor\epsilon^{-1}x\rfloor+y_i}(\epsilon^{-2}t)},\eta\right)1\!\!1_{A_n^c}(\mathbf{x}(\epsilon,t))\right).
\end{aligned}\tag{12.74}$$

Next estimate, using the inequality $0\le D(\sum_{i=1}^n\delta_{z_i},\eta)\le\prod_{i=1}^n\eta_{z_i}$ which holds for all choices $z_1,\ldots,z_n\in\mathbb{Z}^d$, the Cauchy-Schwarz inequality, and the condition UFMC

$$\begin{aligned}
&\mathbb{E}^{\mathrm{IRW}}\int\left(\prod_{i=1}^{n}D\left(\delta_{X_i^{\lfloor\epsilon^{-1}x\rfloor+y_i}(\epsilon^{-2}t)},\eta\right)1\!\!1_{A_n^c}(\mathbf{x}(\epsilon,t))\mu_\epsilon(d\eta)\right)\\
&\le(\mathbb{P}^{\mathrm{IRW}}(\mathbf{x}(\epsilon,t)\notin A_n))^{1/2}C_n,
\end{aligned}\tag{12.75}$$

where

$$C_n=\left(\sup_{z_1,\ldots,z_n\in\mathbb{Z}^d}\int(\prod_{i=1}^{n}\eta_{z_i})^2\mu_\epsilon(d\eta)\right)^{1/2}<\infty.\tag{12.76}$$

By Lemma 12.13 we conclude that

$$\lim_{\epsilon\to0}\mathbb{E}^{\mathrm{IRW}}\left(\int\left(\prod_{i=1}^{n}D\left(\delta_{X_i^{\lfloor\epsilon^{-1}x\rfloor+y_i}(\epsilon^{-2}t)},\eta\right)\right)1\!\!1_{A_n^c}(\mathbf{x}(\epsilon,t))\mu_\epsilon(d\eta)\right)=0.\tag{12.77}$$

Then, by the product character of the measure μ_ϵ, and the assumed density profile, i.e., $\mathbb{E}_{\mu_\epsilon}(\eta_x)=\rho(\epsilon x)$

$$\begin{aligned}
&\mathbb{E}^{\mathrm{IRW}}\int\left(\prod_{i=1}^{n}D\left(\delta_{X_i^{\lfloor\epsilon^{-1}x\rfloor+y_i}(\epsilon^{-2}t)},\eta\right)\mu_\epsilon(d\eta)1\!\!1_{A_n}(\mathbf{x}(\epsilon,t))\right)\\
&=\mathbb{E}^{\mathrm{IRW}}\left(1\!\!1_{A_n}(\mathbf{x}(\epsilon,t))\prod_{i=1}^{n}\rho(\epsilon X^{\lfloor\epsilon^{-1}x\rfloor+y_i}(\epsilon^{-2}t))\right).
\end{aligned}\tag{12.78}$$

By the assumed boundedness of the profile ρ, we have

$$\mathbb{E}^{\mathrm{IRW}}\left(1\!\!1_{A_n^c}(\mathbf{x}(\epsilon,t))\prod_{i=1}^{n}\rho(\epsilon X^{\lfloor\epsilon^{-1}x\rfloor+y_i}(\epsilon^{-2}t))\right)$$

$$\le \left(\mathbb{E}^{\mathrm{IRW}}1\!\!1_{A_n^c}(\mathbf{x}(\epsilon,t))\right)^{1/2}\mathbb{E}^{\mathrm{IRW}}\left(\prod_{i=1}^{n}\rho^2(\epsilon X^{\lfloor\epsilon^{-1}x\rfloor+y_i}(\epsilon^{-2}t))\right)^{1/2}$$

$$\le K_n\left(\mathbb{P}^{\mathrm{IRW}}(\mathbf{x}(\epsilon,t)\notin A_n)\right)^{1/2}, \tag{12.79}$$

where $K_n = \|\rho\|_\infty^n$. Using Lemma 12.13 once more, we conclude that

$$\lim_{\epsilon\to 0}\mathbb{E}^{\mathrm{IRW}}\left(1\!\!1_{A_n^c}(\mathbf{x}(\epsilon,t))\prod_{i=1}^{n}\rho(\epsilon X^{\lfloor\epsilon^{-1}x\rfloor+y_i}(\epsilon^{-2}t))\right)=0. \tag{12.80}$$

Therefore we can remove the indicator $1\!\!1_{A_n}$ in the rhs of (12.78), and arrive at

$$\mathbb{E}^{\mathrm{IRW}}\prod_{i=1}^{n}\rho(\epsilon X^{\lfloor\epsilon^{-1}x\rfloor+y_i}(\epsilon^{-2}t)) = \prod_{i=1}^{n}\mathbb{E}^{\mathrm{RW}}_{\lfloor\epsilon^{-1}x\rfloor+y_i}(\rho(\epsilon X(\epsilon^{-2}t))$$
$$\to \rho(t,x)^n, \qquad \text{when } \epsilon\to 0, \tag{12.81}$$

where the last step follows from the invariance principle, i.e., the random walk starting from $\lfloor\epsilon^{-1}x\rfloor + y_i$ at time $\epsilon^{-2}t$ converges to $x + W(t)$ for all $i \in \{1,\ldots,n\}$ as $\epsilon \to 0$. □

12.2.8 Stationary Fluctuation Field

In this section we consider the independent random walkers starting from a homogeneous Poisson product measure ν_ρ and look at the stationary density fluctuation field

$$\mathcal{Y}_\epsilon(\varphi,\eta(\epsilon^{-2}t)) = \epsilon^{d/2}\sum_x \varphi(\epsilon x)(\eta_x(\epsilon^{-2}t)-\rho). \tag{12.82}$$

As we have already seen in the computation of the variance of the density field, we expect that as $\epsilon \to 0$, this field converges to an infinite dimensional Ornstein Uhlenbeck process given by the SPDE

$$dY(t,x) = \frac{1}{2}\Delta Y(t,x)dt + \sqrt{\rho}\nabla d\mathcal{W}(t,x), \tag{12.83}$$

where $\mathcal{W}(t,x)$ is space-time white noise. The solution of the SPDE (12.83) is a stationary Gaussian distribution valued process with covariance

$$\mathrm{Cov}(Y_t(\varphi), Y_s(\varphi)) = \rho\langle \varphi, Q_{|t-s|}\varphi\rangle. \tag{12.84}$$

We will use duality with the deterministic system to obtain this result in the sense of convergence of finite dimensional distributions. The duality with the deterministic system can be used to compute the moment generating function. Notice that a stronger result, namely convergence in path space, is proved in [151], Chapter 11, using martingale methods. The martingale method is more suitable to obtain tightness in path space via the control of the quadratic variation.

Theorem 12.14 *For all $a, b \in \mathbb{R}$*

$$\lim_{\epsilon\to 0} \log \mathbb{E}_{\nu_\rho} e^{a\mathcal{Y}_\epsilon(\varphi,\eta(0))+b\mathcal{Y}_\epsilon(\varphi,\eta(\epsilon^{-2}t))} = \frac{\rho}{2}\left((a^2+b^2)\langle\varphi,\varphi\rangle + 2ab\langle\varphi, Q_t\varphi\rangle\right), \tag{12.85}$$

where $\langle,\rangle$ denotes inner product in $L^2(dx)$ and where Q_t is the semigroup of Brownian motion. As a consequence, for all choices of $t_1,\dots,t_n$, the limit in distribution of the vector $(\mathcal{Y}_\epsilon(\varphi,\eta(t_1)),\dots\mathcal{Y}_\epsilon(\varphi,\eta(t_n)))$ is a stationary Gaussian vector $(Y_{t_1},\dots,Y_{t_n})$ with covariances

$$cov(Y_{t_i}, Y_{t_j}) = \rho\langle\varphi, Q_{|t_i-t_j|}\varphi\rangle. \tag{12.86}$$

Proof We start by computing

$$\mathbb{E}_\eta\left(e^{b\mathcal{Y}_\epsilon(\varphi,\eta(\epsilon^{-2}t))}\right) = \mathbb{E}_\eta\left(\prod_x e^{b\varphi(\epsilon x)\eta_x(\epsilon^{-2}t)}\right)\prod_x e^{-b\rho\epsilon^{d/2}\varphi(\epsilon x)}. \tag{12.87}$$

Using duality with the deterministic system we obtain

$$\mathbb{E}_\eta\left(\prod_x e^{b\varphi(\epsilon x)\eta_x(\epsilon^{-2}t)}\right) = \prod_x\left(\sum_y p_{\epsilon^{-2}t}(x,y)e^{\epsilon^{d/2}b\varphi(\epsilon y)}\right)^{\eta_x}. \tag{12.88}$$

Indeed, putting $z_x = e^{b\varphi(\epsilon x)}$ the product in the lhs of (12.88) is of the form $\prod_x z_x^{\eta_x(\epsilon^{-2}t}$), and hence the $\mathbb{E}_\eta$ expectation is by duality with the deterministic system equal to $\prod_x z_x(t)^{\eta_x}$, with $z_x(t) = \sum_y p_{\epsilon^{-2}t}(x,y)z_y$. Using the invariance principle, the sum

$$\sum_y p_{\epsilon^{-2}t}(x,y)e^{\epsilon^{d/2}b\varphi(\epsilon y)} =: Q_t^\epsilon(e^{\epsilon^{d/2}b\varphi(\cdot)})(\epsilon x), \tag{12.89}$$

which is well-approximated by $Q_t(e^{\epsilon^{d/2} b\varphi(\cdot)})(\epsilon x)$ where Q_t is the semigroup of Brownian motion. Therefore,

$$\mathbb{E}_\eta e^{a\mathcal{Y}_\epsilon(\varphi,\eta(0))+b\mathcal{Y}_\epsilon(\varphi,\eta(\epsilon^{-2}t))}, \tag{12.90}$$

can be rewritten as

$$e^{\epsilon^{d/2} a\sum_x(\eta_x-\rho)\varphi(\epsilon x)} e^{-b\epsilon^{d/2}\rho\sum_x\varphi(\epsilon x)} \prod_x (Q_t^\epsilon(e^{\epsilon^{d/2} b\varphi(\cdot)})(\epsilon x))^{\eta_x}). \tag{12.91}$$

We now take the expectation of this expression w.r.t. the Poisson product measure, using $\mathbb{E}_{\nu_\rho} z^{\eta_x} = e^{\rho(z-1)}$, and obtain

$$\begin{aligned}
&\log \mathbb{E}_{\nu_\rho} e^{a\mathcal{Y}_\epsilon(\varphi,\eta(0))+b\mathcal{Y}_\epsilon(\varphi,\eta(\epsilon^{-2}))}\\
&= \sum_x \Big[\Big((-\epsilon^{d/2}(a+b)\rho\varphi(\epsilon x)\Big)\\
&+ \rho\Big(\exp\Big(\varphi(\epsilon x)\epsilon^{d/2}a + \log Q_t^\epsilon(e^{\epsilon^{d/2} b\varphi(\cdot)})(\epsilon x)\Big) - 1\Big)\Big].
\end{aligned} \tag{12.92}$$

We then remark

$$\log Q_t^\epsilon e^{\epsilon^{d/2} b\varphi(\cdot)}(\epsilon x) - 1 = \epsilon^{d/2} b Q_t^\epsilon \varphi(\epsilon_x) + \frac{1}{2}\epsilon^d b^2 Q_t^\epsilon(\varphi^2)(\epsilon x) + o(\epsilon^d). \tag{12.93}$$

Substituting this expression into (12.92), and expanding the exponentials, hereby neglecting all terms $o(\epsilon^d)$, and using that $Q_t^\epsilon \to Q_t$ we obtain (12.85). Using the Markov property, and iteratively using the duality with the deterministic system, the extension to multiple times is straightforward. We leave this to the reader. □

Remark 12.15 (Tightness) Theorem 12.14 implies that the trajectory of the random distribution $\mathcal{Y}_\epsilon(\cdot, \eta(\epsilon^{-2}t))$ converges in the sense of finite dimensional distributions to the stationary solution of (12.83). In order to establish convergence in path space (in $D([0,\infty), \mathcal{S}')$), one has to prove tightness. This can be done along the same lines by estimating the $\mathbb{E}_{\nu_\rho}(\mathcal{Y}_\epsilon(\varphi, \eta(\epsilon^{-2}t)) - \mathcal{Y}_\epsilon(\varphi, \eta(0)))^4$ and obtaining an upper bound of order t^2 when $t \to 0$. Then by Kolmogorov's tightness criterion, $\{\mathcal{Y}_\epsilon(\varphi, \eta(\epsilon^{-2}t)), t \geq 0\}$ is tight in $D([0,\infty), \mathbb{R})$, which implies tightness of $\{\mathcal{Y}_\epsilon(\cdot, \eta(\epsilon^{-2}t)), t \geq 0\}$ in $D([0,\infty), \mathcal{S}')$. We omit the details here, because the by now standard methodology based on martingales gives a much more elegant and robust approach based on Aldous' tightness criterion.

12.3 The Interacting Case

In this section we show that many of the results for independent random walkers can be generalized to interacting systems with duality.

Following the line of thought of the case of independent random walkers, we will discuss the following items.

1. If the scaling limit of a single dual particle is Brownian motion, then the expected density field converges to the solution of the heat equation. This is exactly the same for interacting systems with duality as it is for independent random walkers because the statement depends only on a distribution of the trajectory of a single dual particle, and therefore the interaction between dual particles does not play any role in this statement. Therefore Theorem 12.8 holds in the interacting case as well.
2. The variance of the density field and its scaling behavior, both stationary and non-stationary can be obtained from the study of two dual particles.
3. The macroscopic propagation of local equilibrium ("propagation of chaos") can be obtained from a suitable coupling between n dual particles and n independent particles. How local equilibrium sets in and propagates is then analogous to the case of independent random walkers, because the dual particles are with probability close to 1 at different locations at any macroscopic time.

For the sake of simplicity we will restrict to three models: independent random walkers (already covered), symmetric exclusion, and symmetric inclusion. Similar results can be derived for diffusion processes dual to SIP or thermalized models such as KMP.

For the three models under study the underlying single particle random walk $\{X(t) : t \geq 0\}$ has jump rates $\alpha p(x, y)$. We assume that $p(x, y)$ is of finite range, symmetric and translation invariant, and such that the associated random walk $\epsilon X(\epsilon^{-2}t)$ converges to a Brownian motion with variance αt, i.e., to $W(\alpha t)$, where $W(t)$ denotes standard Brownian motion.

We recall the generator of the models under consideration:

$$Lf(\eta) = \sum_{x,y\in\mathbb{Z}^d} p(x, y)\eta_x(\alpha + \sigma\eta_y)(f(\eta^{x,y}) - f(\eta)), \tag{12.94}$$

where $\sigma = 0, 1, -1$ for independent random walkers, symmetric inclusion, symmetric partial exclusion respectively.

We will start with the computation of the time-dependent covariance using two dual particles. This will help us in obtaining estimates for the variance of the density field.

12.3.1 Time Dependent Covariance of Particle Numbers: Two Dual Particles

In order to study the behavior of the variance of the density field in terms of two dual particles (next section), in this section we compute time dependent covariance of the particle numbers at different locations, as well as of the time dependent variance at a single location. We believe that these explicit computations are of independent use, as we discuss after having performed them.

We start with some notation and then proceed with the computation of variances and covariances. We put

$$\begin{aligned} D(\delta_x, \eta) &= c_1 \eta_x \\ D(2\delta_x, \eta) &= c_2 \eta_x(\eta_x - 1). \end{aligned} \tag{12.95}$$

We have

1. For SIP(α): $c_1 = \frac{1}{\alpha}$, $c_2 = \frac{1}{\alpha(\alpha+1)}$.
2. For SEP(α), with $\alpha \geq 2$, $c_1 = \frac{1}{\alpha}$ $c_2 = \frac{1}{\alpha(\alpha-1)}$.
3. For independent random walkers $c_1 = c_2 = 1$.

We start the process with the configuration at time zero $\eta(0)$ distributed according to a probability measure ν and further denote

$$\begin{aligned} \rho(x) &= \int \eta_x d\nu = \frac{1}{c_1} \int D(\delta_x, \eta) d\nu, \\ \rho_2(x) &= \int \eta_x(\eta_x - 1) = \frac{1}{c_2} \int D(2\delta_x, \eta) d\nu, \\ C(x, y) &= \int \eta_x \eta_y d\nu - \int \eta_x d\nu \int \eta_y d\nu. \end{aligned} \tag{12.96}$$

Notice that for the reversible homogeneous product measures $\nu = \nu_\rho$ (with marginal Poisson for $\sigma = 0$, Negative Binomial for $\sigma = 1$ and Binomial for $\sigma = -1$), we have $\rho(x) = \rho$ and, because as we saw earlier $\int D(2\delta_x, \eta)\nu_\rho(d\eta) = \left(\int D(\delta_x, \eta)\nu_\rho(d\eta)\right)^2$ we also have $c_2 \rho_2 = (c_1 \rho)^2$.

Furthermore, we denote as before by $p_t(x, y)$ the transition probability for a single dual particle to go from x to y in time t and by $p_t(x, y; u, v)$ the transition probability for two (labeled) dual particles to go from x, y to u, v in time t, and

$$\rho_t(x) = \sum_y p_t(x, y)\rho(y) = \int \mathbb{E}_\eta(\eta_x(t))\nu(d\eta). \tag{12.97}$$

Finally we denote the time dependent covariance of the particle numbers at x and y when initially started from ν as

$$\Xi(t, x, y; \nu) = \int \mathbb{E}_\eta(\eta_x(t)\eta_y(t))\nu(d\eta) - \rho_t(x)\rho_t(y). \tag{12.98}$$

Computation of the Time Dependent Covariance of Particle Occupation Numbers

We have the following useful computational lemma.

Lemma 12.16 *We have the following equalities:*

1. *First, for $x \neq y$:*

$$\mathbb{E}_\eta(\eta_x(t)\eta_y(t)) = \sum_{u\neq v} p_t(x, y; u, v)\eta_u\eta_v + \frac{c_2}{c_1^2} p_t(x, y; u, u)\eta_u(\eta_u - 1). \tag{12.99}$$

2. *Second:*

$$\mathbb{E}_\eta(\eta_x(t)^2) = \frac{c_1^2}{c_2} \sum_{u\neq v} p_t(x, x; u, v)\eta_u\eta_v + \sum_u p_t(x, x; u, u)\eta_u(\eta_u - 1)$$
$$+ \sum_u p_t(x, u)\eta_u. \tag{12.100}$$

3. *Third, for the time-dependent covariance starting from the initial measure ν, we get, for $x \neq y$:*

$$\begin{aligned}\Xi(t, x, y; \nu) &= \int \mathbb{E}_\eta(\eta_x(t)\eta_y(t))\nu(d\eta) - \rho_t(x)\rho_t(y)\\ &= \sum_{u,v}(p_t(x, y; u, v) - p_t(x, u)p_t(y, v))\rho(u)\rho(v)\\ &+ \sum_u p_t(x, y; u, u)\left(\frac{c_2}{c_1^2}\rho_2(u) - \rho(u)^2\right)\\ &+ \sum_{u\neq v} p_t(x, y; u, v)C(u, v).\end{aligned} \tag{12.101}$$

4. *Fourth, for the time-dependent variance starting from the initial measure* ν*, we get*

$$
\begin{aligned}
\Xi(t,x,x;\nu) &= \int \mathbb{E}_\eta(\eta_x(t)^2)\nu(d\eta) - \rho_t(x)^2 \\
&= \frac{c_1^2}{c_2}\sum_{u,v}(p_t(x,x;u,v) - p_t(x,u)p_t(x,v))\rho(u)\rho(v) \\
&+ \sum_u p_t(x,x;u,u)\left(\rho_2(u) - \frac{c_1^2}{c_2}\rho(u)^2\right) \\
&+ \left(\frac{c_1^2}{c_2} - 1\right)\rho_t(x)^2 \\
&+ \frac{c_1^2}{c_2}\sum_{u\neq v} p_t(x,x;u,v)C(u,v).
\end{aligned}
\tag{12.102}
$$

Proof We start with the first equality

$$
\begin{aligned}
\mathbb{E}_\eta(\eta_x(t)\eta_y(t)) &= \frac{1}{c_1^2}\mathbb{E}_\eta D(\delta_x + \delta_y, \eta(t)) \\
&= \frac{1}{c_1^2}\sum_{u,v} p_t(x,y;u,v)D(\delta_u + \delta_v, \eta) \\
&= \sum_{u\neq v} p_t(x,y;u,v)\eta_u\eta_v \\
&+ \frac{c_2}{c_1^2}\sum_u p_t(x,y;u,u)\eta_u(\eta_u - 1).
\end{aligned}
\tag{12.103}
$$

The third equality follows from the first equality, by integrating the η-variable over ν, subtracting $\rho_t(x)\rho_t(y)$ and using that $\rho_t(x) = \sum_u p_t(x,u)\rho(u)$.

For the second equality:

$$
\begin{aligned}
\mathbb{E}_\eta(\eta_x(t)^2) &= \mathbb{E}_\eta(\eta_x(t)(\eta_x(t) - 1)) + \mathbb{E}_\eta(\eta_x(t)) \\
&= \frac{1}{c_2}\mathbb{E}_\eta D(2\delta_x, \eta(t)) + \sum_u p_t(x,u)\eta_u \\
&= \frac{c_1^2}{c_2}\sum_{u\neq v} p_t(x,x;u,v)\eta_u\eta_v \\
&+ \sum_u p_t(x,x;u,u)\eta_u(\eta_u - 1) + \sum_u p_t(x,u)\eta_u.
\end{aligned}
\tag{12.104}
$$

The fourth equality follows by integrating the second equality over ν, subtracting $\rho_t(x)^2$ and using that $\rho_t(x) = \sum_u p_t(x, u)\rho(u)$. $\square$

Let us denote by $\mathbb{E}^C_{x,y;x,y}$ a coupling (upper-index C referring to coupling) of two dual particles with two independent walkers (where a single walker moves as a single dual particle) starting both from x, y, and by $\mathbb{E}_{x,y}$ expectation of two dual particles starting from x, y. In this coupling we denote by $X(t), Y(t)$, resp. $\widetilde{X}(t), \widetilde{Y}(t)$ the positions of the two dual particles, resp. the two independent particles. Then we can summarize our expression for $\Xi(t, x, y; \nu)$ as follows

$$\begin{aligned}\Xi(t, x, y, \nu) = & \left[1 + \frac{c_1^2 - c_2}{c_2}\delta_{x,y}\right]\Bigg[\mathbb{E}^C_{x,y;x,y}\left[\rho(X(t))\rho(Y(t)) - \rho(\widetilde{X}(t))\rho(\widetilde{Y}(t))\right] \\ & + \mathbb{E}_{x,y}\left(\mathbb{1}_{\{X(t)=Y(t)\}}\left(\frac{c_2}{c_1^2}\rho_2(X(t)) - \rho(X(t))^2\right)\right)\Bigg] \\ & + \delta_{x,y}\left(\frac{(c_1^2 - c_2)}{c_2}\rho_t(x)^2 + \rho_t(x)\right) \\ & + \mathbb{E}_{x,y}(\mathbb{1}_{\{X(t)\neq Y(t)\}}C(X(t), Y(t))). \end{aligned} \tag{12.105}$$

Discussion of the Time-dependent Covariance (12.105)

Let us start by some discussion of the various terms encountered in the computational lemma, and in (12.105)

1. First notice that if we take the homogeneous stationary product measure, for which $\int D(2\delta_x, \eta)d\nu = \left(\int D(\delta_x, \eta)d\nu\right)^2 = c_1^2\rho^2$ and hence
$$\rho_2 = \frac{c_1^2}{c_2}\rho^2, \tag{12.106}$$
we find $\Xi(t, x, y; \nu) = 0$ as it should.
2. If we consider the independent random walk case, and start from a inhomogeneous Poisson product measure, the term $\left(\frac{c_2}{c_1^2}\rho_2(u) - \rho(u)^2\right)$ vanishes because $c_1 = c_2 = 1$ and $\rho_2(u) = \rho(u)^2$. Therefore, in that case $\Xi(t, x, y; \nu)$ vanishes, as it should be, because at time t the measure is again product Poisson with parameter $\rho_t(x)$ at x.
3. Again for the independent random walkers: the term
$$\sum_u p_t(x, y; u, u)\left(\frac{c_2}{c_1^2}\rho_2(u) - \rho(u)^2\right), \tag{12.107}$$

measures the deviation from Poisson and has no definite sign, i.e., it can be positive or negative depending on ν. This shows that even for independent random walkers, when started from a non-Poissonian product measure, correlations at time t can be positive or negative.

4. In case we start from the homogeneous stationary product measure ν, we find

$$\Xi(t, x, x; \nu) = \rho - \rho^2 + \frac{c_1^2}{c_2}\rho^2, \qquad (12.108)$$

as it should, because the stationary variance of η_0 equals

$$\int \eta_0^2 \, d\nu - \rho^2 = \int \eta_0(\eta_0 - 1) d\nu + \rho - \rho^2 = c_2^{-1}(c_1\rho)^2 + \rho - \rho^2. \qquad (12.109)$$

5. In the independent random walk case, when starting from an inhomogeneous product of Poisson measures with density $\rho(x)$ at x, the first two terms vanish, and the third term equals $\rho_t(x)$, as it should because the measure at time t is inhomogeneous Poisson with density $\rho_t(x)$ at x, and the variance of $\eta_x(t)$ then indeed equals $\rho_t(x)$.
6. We see that the first term in (12.105) measures the difference between two dual particles and two independent particles, it vanishes if we consider the homogeneous case, i.e., when $\rho(x) = \rho_t(x) = \rho$ does not depend on x. It also vanishes for independent random walkers. Moreover, by Proposition 5.30, we have that the second term becomes larger if we replace X_t, Y_t by independent random walkers, provided $\left(\frac{c_2}{c_1^2}\rho_2(x) - \rho(x)^2\right) \geq 0$. For such initial ν we conclude that η_x and η_y are positively correlated at time t when we evolve according to SIP(α).
7. In the independent random walkers case, the second term measure the deviation from local equilibrium, i.e., starting from a non-Poissonian product measure, in the limit of large t the term

$$\left[\mathbb{E}_{x,y}\left(\mathbb{1}_{\{X(t)=Y(t)\}}\left(\frac{c_2}{c_1^2}\rho_2(X(t)) - \rho(X(t))^2\right)\right)\right] \qquad (12.110)$$

vanishes, hence in that limit the covariance of $\eta_x(t)$ and $\eta_y(t)$ for $x \neq y$ vanishes, and the variance becomes the Poissonian variance $\rho_t(x)$.
8. The expression

$$\left[\mathbb{E}^C_{x,y;x,y}\left[\rho(X(t))\rho(Y(t)) - \rho(\widetilde{X}(t))\rho(\widetilde{Y}(t))\right]\right] \qquad (12.111)$$

is non-negative for SIP, and non-positive for SEP. This follows from the correlation inequalities which we proved in Sect. 5.8. Indeed $(x, y) \mapsto \rho(x)\rho(y)$ is a symmetric and positive definite function, and therefore, expectation of this term under SIP (resp. SEP) is larger (resp. smaller) or equal than the independent random walk expectation.

12.3.2 Variance of the Density Fluctuation Field

We now consider the setting where the initial distribution of particle numbers, denoted μ_ϵ, satisfies

$$\begin{aligned}\rho(\epsilon x) &= \int \eta_x \mu_\epsilon(d\eta),\\ \rho_2(\epsilon x) &= \int \eta_x(\eta_x - 1)\mu_\epsilon(d\eta),\\ \epsilon^d C(\epsilon x, \epsilon y) &= \int \eta_x \eta_y d\mu_\epsilon - \int \eta_x d\mu_\epsilon \int \eta_y d\mu_\epsilon, \end{aligned} \tag{12.112}$$

where with a small abuse of notation (w.r.t. the previous section) now $\rho : \mathbb{R}^d \to \mathbb{R}^+, \rho_2 : \mathbb{R}^d \to \mathbb{R}^+, C : \mathbb{R}^{2d} \to \mathbb{R}$, and where we always implicitly assume that in the case SEP(α) the image of ρ, ρ_2 is $[0, \alpha], [0, \alpha(\alpha - 1)]$. We assume for simplicity that ρ, ρ_2 are smooth and bounded, with bounded derivatives up to order three, and C is bounded and continuous. An important special case is when μ_ϵ is the local equilibrium product measure corresponding to the density profile ρ. In that case $C = 0$ and $\rho_2(x) = \frac{c_1^2}{c_2}\rho(x)$, cf. (12.106).

We further denote

$$\rho^\epsilon(t, \epsilon x) = \sum_y p_{\epsilon^{-2}t}(x, y)\rho(\epsilon y) = \mathbb{E}_x(\rho(\epsilon X(\epsilon^{-2}t))). \tag{12.113}$$

Then we have, using the invariance principle for a single dual particle (which is a random walk), for $t > 0, x \in \mathbb{R}^d$ that $\rho^\epsilon(t, x) \approx \rho(t, x)$ where $\rho(t, x)$ solves the heat equation with diffusion constant α, i.e.

$$\frac{\partial \rho(t, x)}{\partial t} = \frac{\alpha}{2} \Delta \rho(t, x). \tag{12.114}$$

We further denote the density field at macroscopic time t by

$$\mathcal{X}_\epsilon(\varphi, \eta(\epsilon^{-2}t)) = \epsilon^d \sum_x \varphi(\epsilon x)\eta_x(\epsilon^{-2}t), \tag{12.115}$$

and the density fluctuation field by

$$\mathcal{Y}_\epsilon(\varphi, \eta(\epsilon^{-2}t)) = \epsilon^{d/2} \sum_x \varphi(\epsilon x) \left(\eta_x(\epsilon^{-2}t) - \rho_{\epsilon^{-2}t}(\epsilon x) \right). \tag{12.116}$$

Then for the expectation of the density field we obtain, using duality with one dual particle:

$$\mathbb{E}_{\mu_\epsilon} \mathcal{X}_\epsilon(\varphi, \eta(\epsilon^{-2}t)) = \epsilon^d \sum_x \varphi(\epsilon x) \sum_y p_{\epsilon^{-2}t}(x, y) \rho(\epsilon y), \tag{12.117}$$

which converges to $\int \varphi(x)\rho(t, x)$ where $\rho(t, x)$ solves the heat Eq. (12.114).

Then using the formula (12.105) we obtain the following expression for the variance of the density fluctuation field at macroscopic times.

$$\begin{aligned}
&\mathbb{E}_\nu \left(\left(\mathcal{Y}_\epsilon(\varphi, \eta(\epsilon^{-2}t)) \right)^2 \right) \\
&= \epsilon^d \sum_{x,y \in \mathbb{Z}^d} \varphi(\epsilon x)\varphi(\epsilon y) \Xi(t, x, y, \nu) \\
&= \epsilon^d \sum_{x,y \in \mathbb{Z}^d} \varphi(\epsilon x)\varphi(\epsilon y) \left[1 + \frac{c_1^2 - c_2}{c_2} \delta_{x,y} \right] \\
&\left[\mathbb{E}^C_{x,y;x,y} \left[\rho(\epsilon X(\epsilon^{-2}t))\rho(\epsilon Y((\epsilon^2 t)) - \rho(\epsilon \widetilde{X}(\epsilon^{-2}t))\rho(\epsilon \widetilde{Y}(\epsilon^{-2}t)) \right] \right. \\
&\left. + \mathbb{E}_{x,y} \left(\mathbb{1}_{\{X(\epsilon^{-2}t) = Y(\epsilon^{-2}t)\}} \left(\frac{c_2}{c_1^2} \rho_2(\epsilon X(\epsilon^2 t)) - \rho(\epsilon X(\epsilon^{-2}t))^2 \right) \right) \right] \\
&+ \epsilon^d \sum_{x \in \mathbb{Z}^d} \varphi(\epsilon x)^2 \left(\frac{(c_1^2 - c_2)}{c_2} \rho^\epsilon(t, \epsilon x)^2 + \rho^\epsilon(t, \epsilon x) \right) \\
&+ \epsilon^d \sum_{x,y \in \mathbb{Z}^d} \varphi(\epsilon x)\varphi(\epsilon y) \mathbb{E}_{x,y}(\mathbb{1}_{\{X(\epsilon^{-2}t) \neq Y(\epsilon^{-2}t)\}} \epsilon^d C(\epsilon X(\epsilon^{-2}t), \epsilon Y(\epsilon^{-2}t))).
\end{aligned} \tag{12.118}$$

Discussion of the Variance of the Density Fluctuation Field

Let us discuss the behavior of the various terms appearing in the rhs of (12.118).

We start with the last three terms of (12.118). Under the assumption that the couple $(\epsilon X(\epsilon^{-2}t), \epsilon Y(\epsilon^{-2}t)))$ converges to two independent Brownian motions

$(B_1(t), B_2(t)$, the last three terms in (12.118) converge to (as $\epsilon \to 0$)

$$\int\int\int \varphi(x)\varphi(y)q_t(x,u)q_t(y,u)\left(\frac{c_2}{c_1^2}\rho_2(u) - \rho(u)^2\right)dxdydu$$
$$+\int \varphi(x)^2\left(\frac{(c_1^2-c_2)}{c_2}\rho(t,x)^2 + \rho(t,x)\right)dx$$
$$+\int\int \varphi(x)\varphi(y)\mathbb{E}C(x+B_1(t), y+B_2(t))dxdy, \tag{12.119}$$

with $q_t(x, y)$ the transition probability density of Brownian motion. The first term measures the influence of deviation from local equilibrium, and vanishes when we start from the local equilibrium product measure see (12.106). The second term is the variance of the time-evolved local equilibrium. The third term measures the effect of covariance in the initial distribution. If we denote $Q_t f(x) = \mathbb{E}f(x+B(t))$ the semigroup of Brownian motion, and if we further denote

$$\chi(\rho) = \left(\frac{(c_1^2-c_2)}{c_2}\rho^2 + \rho\right), \tag{12.120}$$

which is the variance under the equilibrium product measure μ_ρ with $\mathbb{E}_{\mu_\rho}(\eta_x) = \rho$, then the three terms in (12.119) can be rewritten as follows

$$\int (Q_t\varphi(u))^2\left(\frac{c_2}{c_1^2}\rho_2(u) - \rho(u)^2\right)du$$
$$+\int \varphi(x)^2\chi(\rho(t,x))dx$$
$$+\int\int Q_t\varphi(x)Q_t\varphi(y)C(x,y)dxdy. \tag{12.121}$$

In order to deal with the first term of (12.118) we have to introduce some assumptions.

Definition 12.17 (Suitable Coupling, Two Particles) We say that there exists a suitable coupling of 2 dual particles and 2 independent particles if under this coupling, we have the following property: under the coupling path space measure $\mathbb{P}^C_{x,y;x,y}$ we have, for all $t > 0$ the following control on the discrepancies:

$$\epsilon\left((X(\epsilon^{-2}t), Y(\epsilon^{-2}t)) - (\widetilde{X}(\epsilon^{-2}t), \widetilde{Y}(\epsilon^{-2}t))\right) \to (0,0) \text{ in probability.} \tag{12.122}$$

Remark 12.18 The condition (12.122) is requiring that on the diffusive space-time rescaling, the discrepancies (i.e., the differences between X and $\widetilde{X}$ and between Y and $\widetilde{Y}$) are negligible. Because the interaction between the particles acts only when they are within the range of the kernel $p(x, y)$, this is also a very natural condition. Both conditions are met for the coupling between SIP and independent particles constructed in [190] as well as for the stirring coupling of exclusion particles constructed in [71].

We then have the following

Lemma 12.19 *If there exists a suitable coupling of 2 dual particles and 2 independent particles, then we have*

$$\lim_{\epsilon\to 0} \epsilon^{2d} \sum_{x,y\in\mathbb{Z}^d} \varphi(\epsilon x)\varphi(\epsilon y) \tag{12.123}$$

$$\left[\mathbb{E}^C_{x,y;x,y}\left[\rho(\epsilon X(\epsilon^{-2}t))\rho(\epsilon Y((\epsilon^{-2}t)) - \rho(\epsilon\widetilde{X}(\epsilon^{-2}t))\rho(\epsilon\widetilde{Y}(\epsilon^{-2}t))\right]\right] = 0.$$

As a consequence,

$$\lim_{\epsilon\to 0} Var(\mathcal{X}_\epsilon(\varphi, \eta(\epsilon^{-2}t))) = 0. \tag{12.124}$$

Proof From a simple Taylor expansion argument, using (12.122) we see that

$$\mathbb{E}^C_{x,y;x,y}\left[\rho(\epsilon X(\epsilon^{-2}t))\rho(\epsilon Y((\epsilon^{-2}t)) - \rho(\epsilon\widetilde{X}(\epsilon^{-2}t))\rho(\epsilon\widetilde{Y}(\epsilon^{-2}t))\right] = o(1), \tag{12.125}$$

where $o(1)$ is uniformly bounded and goes to zero as $\epsilon \to 0$. Therefore, using that φ is a smooth test function, we have that $\epsilon^{2d}\sum_{x,y\in\mathbb{Z}^d}\varphi(\epsilon x)\varphi(\epsilon y)$ is uniformly bounded in ϵ and converges to the integral $\int\int|\varphi(x)\varphi(y)|dxdy$. As a consequence,

$$\epsilon^{2d} \sum_{x,y\in\mathbb{Z}^d} \varphi(\epsilon x)\varphi(\epsilon y)$$

$$\left[\mathbb{E}^C_{x,y;x,y}\left[\rho(\epsilon X(\epsilon^{-2}t))\rho(\epsilon Y(\epsilon^{-2}t)) - \rho(\epsilon\widetilde{X}(\epsilon^{-2}t))\rho(\epsilon\widetilde{Y}(\epsilon^{-2}t))\right]\right]$$

$$\le o(1) \sum_{x,y\in\mathbb{Z}^d} |\varphi(\epsilon x)\varphi(\epsilon y)| \to 0, \tag{12.126}$$

which gives (12.123). The consequence follows from (12.118), combined with the discussion above providing the behavior of the last three terms in the rhs of that equation. □

Remark 12.20

1. Later on we will prove that

$$\limsup_{\epsilon\to 0} \epsilon^d \sum_{x,y\in\mathbb{Z}^d} \varphi(\epsilon x)\varphi(\epsilon y)$$
$$\cdot\left[\mathbb{E}^C_{x,y;x,y}\left[\rho(\epsilon X(\epsilon^{-2}t))\rho(\epsilon Y((\epsilon^{-2}t)) - \rho(\epsilon \widetilde{X}(\epsilon^{-2}t))\rho(\epsilon\widetilde{Y}(\epsilon^{-2}t))\right]\right] < \infty, \tag{12.127}$$

which implies that the variance of the density fluctuation field remains bounded (as a function of ϵ). This cannot be proved via a simple coupling argument.
This, combined with the discussion of the three last terms in (12.118) implies that the variance of the density field is in fact of order ϵ^d, i.e., behaves roughly as if the summands defining it were independent.

2. If we put $\chi(\rho) = \left(\frac{(c_1^2-c_2)}{c_2}\rho^2 + \rho\right)$, which is the variance under the equilibrium product measure μ_ρ with $\mathbb{E}_{\mu_\rho}(\eta_x) = \rho$, then the limiting fluctuation field is expected to satisfy the SPDE

$$dY(t,x) = \frac{1}{2}\Delta Y(t,x)dt - \nabla.\left(\sqrt{\chi(\rho(t,x))}\mathcal{W}(t,x)\right). \tag{12.128}$$

When we start from a homogeneous product measure, the first term in (12.118) vanishes, and in that case it is easy to verify that the obtained variance of the density fluctuation field matches with the mild solution of (12.128).

Let us summarize the result of our computations.

Theorem 12.21 (Hydrodynamic Limit) *Let* $\{\mu_\epsilon : \epsilon > 0\}$ *be a family of measures such that* (12.112) *holds. Assume additionally that there exists a suitable coupling in the sense of Definition 12.17. Then for all* $t > 0$ *the density field* $\mathcal{X}_\epsilon(\varphi, \eta(\epsilon^{-2}t))$ *converges in* L^2 *to* $\int \varphi(x)\rho(t,x)dx$ *where* $\rho(t,x)$ *is the solution of the heat Eq.* (12.114) *and with initial condition* $\rho(0,x) = \rho(x)$.

Proof For the expectation we have, by the invariance principle

$$\mathbb{E}_{\mu_\epsilon}(\mathcal{X}_\epsilon(\varphi, \eta(\epsilon^{-2}t))) = \epsilon^d \sum_x \varphi(\epsilon x)\mathbb{E}_x \rho(\epsilon X(\epsilon^{-2}t)) \to \int \varphi(x)\rho(t,x)dx. \tag{12.129}$$

For the variance we obtain, thanks to the computations in Sect. 12.3.2 that

$$\lim_{\epsilon\to 0} Var\left(\mathcal{X}_\epsilon(\varphi, \eta(\epsilon^{-2}t)\right) = 0. \tag{12.130}$$

□

To conclude this section, we prove (12.127) in the nearest neighbor case., which shows that the variance of the density fluctuation field stays bounded as $\epsilon \to 0$. The proof is inspired by a method from [196].

Proposition 12.22 *Let $p(x, y) = \frac{1}{2d} \mathbb{1}_{\{x \sim y\}}$ be nearest neighbor random walk (we denoted $x \sim y$ for x, y being neighbors). Then we have*

$$\limsup_{\epsilon \to 0} \epsilon^d \sum_{x,y \in \mathbb{Z}^d} \varphi(\epsilon x)\varphi(\epsilon y) \tag{12.131}$$

$$\cdot \left[\mathbb{E}^C_{x,y;x,y} \left[\rho(\epsilon X(\epsilon^{-2}t))\rho(\epsilon Y((\epsilon^{-2}t))) - \rho(\epsilon \widetilde{X}(\epsilon^{-2}t))\rho(\epsilon \widetilde{Y}(\epsilon^{-2}t)) \right] \right] < \infty.$$

Proof Before we start the proof, we briefly explain the idea. The sum in (12.131) has two summation indices, and therefore appears at first sight to be of order ϵ^{-d}. However, if we can show that "effectively" the sum over y is a "restricted" sum (e.g. by imposing that $x \sim y$), then this problem disappears. Because the coupling is between independent and interacting particles, effectively, the coupling only will create "discrepancies" whenever the particles are at neighboring positions, i.e., interact. This "effectively" creates a restriction on the double sum, which is enough to guarantee that the double sum actually behaves as a single sum, for which the normalization ϵ^d is enough.

To formalize this we need some notation. Denote

$$\psi(\epsilon, x, y, t) = \mathbb{E}^C_{x,y;x,y} \left[\rho(\epsilon X(\epsilon^{-2}t))\rho(\epsilon Y((\epsilon^{-2}t))) - \rho(\epsilon \widetilde{X}(\epsilon^{-2}t))\rho(\epsilon \widetilde{Y}(\epsilon^{-2}t)) \right]. \tag{12.132}$$

Further denote by L_2 the generator of 2 dual particles. Let us furthermore abbreviate

$$\begin{aligned} U(\epsilon, x, y, t) &= \mathbb{E}^{\mathrm{IRW}}_{x,x}(\rho(\epsilon X(\epsilon^{-2}t))\rho(\epsilon Y(\epsilon^{-2}t))) \\ &+ \mathbb{E}^{\mathrm{IRW}}_{y,y}(\rho(\epsilon X(\epsilon^{-2}t))\rho(\epsilon Y(\epsilon^{-2}t))) \\ &- 2\mathbb{E}^{\mathrm{IRW}}_{x,y}(\rho(\epsilon X(\epsilon^{-2}t))\rho(\epsilon Y(\epsilon^{-2}t))). \end{aligned} \tag{12.133}$$

Then we compute

$$\begin{aligned} \frac{d}{dt}\psi(\epsilon, x, y, t) &= \epsilon^{-2} L_2 \psi(\epsilon, x, y, t) \\ &\quad - \sigma\epsilon^{-2} \mathbb{1}_{\{x \sim y\}} U(\epsilon, x, y, t) \\ &= \epsilon^{-2} L_2 \psi(\epsilon, x, y, t) + K(\epsilon, x, y, t), \end{aligned} \tag{12.134}$$

where we denoted

$$
\begin{aligned}
K(\epsilon,x,y,t) &= \sigma\epsilon^{-2}\mathbb{1}_{\{x\sim y\}}U(\epsilon,x,y,t)\\
&= \sigma\epsilon^{-2}\mathbb{1}_{\{x\sim y\}}\left(\mathbb{E}_x^{\mathrm{RW}}(\rho(\epsilon X(\epsilon^{-2}t)) - \mathbb{E}_y^{RW}(\rho(\epsilon X(\epsilon^{-2}t))\right)^2.
\end{aligned}
\tag{12.135}
$$

Now notice that because of the smoothness and boundedness of ρ and the local limit theorem we can estimate

$$
|\mathbb{E}_x^{\mathrm{RW}}(\rho(\epsilon X(\epsilon^{-2}t)) - \mathbb{E}_y^{\mathrm{RW}}(\rho(\epsilon X(\epsilon^{-2}t))| \le C\epsilon|x-y|, \tag{12.136}
$$

which implies the pointwise upperbound

$$
K(\epsilon,x,y,t) \le C\mathbb{1}_{\{x\sim y\}}. \tag{12.137}
$$

Starting from (12.134) and using the classical variation of constants method, together with the fact that $\psi(\epsilon,x,y,0) = 0$ we obtain, also using the bound (12.137)

$$
\begin{aligned}
\psi(\epsilon,x,y,t) &= \int_0^t e^{\epsilon^{-2}(t-s)L_2}K(\epsilon,x,y,s)ds\\
&\le C\int_0^t e^{\epsilon^{-2}(t-s)L_2}\mathbb{1}_{\{x\sim y\}}\\
&= C\sum_{x',y'}\int_0^t p^{(2)}_{\epsilon^{-2}(t-s)}(x,y;x'y')\mathbb{1}_{\{x'\sim y'\}},
\end{aligned}
\tag{12.138}
$$

where in the last line we denoted by $p_r^{(2)}(x,y;x'y')$ the transition probability for two dual particles to move from x, y to x', y' in time r. Because the process of two dual particles has a reversible σ-finite measure $\pi_{\sigma,\alpha}(x,y)$ which only on the diagonal $x = y$ differs from the counting measure, and where the weight $\pi_{\sigma,\alpha}(x,x)$ does not depend on x, one easily obtains the bound

$$
p^{(2)}_{\epsilon^{-2}(t-s)}(x,y;x'y') \le Bp^{(2)}_{\epsilon^{-2}(t-s)}(x',y';x,y), \tag{12.139}
$$

where B is a constant (i.e., not depending on $x, y, x', y', \epsilon, t$). Using this together with (12.138), we arrive at

$$
\begin{aligned}
&\epsilon^d\sum_{x,y}|\varphi(\epsilon x)\varphi(\epsilon y)|\psi(\epsilon,x,y,t)\\
&\le \epsilon^d C\sum_{x,y,x'\sim y'}|\varphi(\epsilon x)\varphi(\epsilon y)|\int_0^t p^{(2)}_{\epsilon^{-2}(t-s)}(x,y;x'y')
\end{aligned}
$$

$$= B\epsilon^d C \sum_{x,y,x'\sim y'} |\varphi(\epsilon x)\varphi(\epsilon y)| \int_0^t p^{(2)}_{\epsilon^{-2}(t-s)}(x',y';x,y)$$

$$\leq \|\varphi\|_\infty \epsilon^d BC \sum_{x,x'\sim y'} |\varphi(\epsilon x) \int_0^t p_{\epsilon^{-2}(t-s)}(x',x), \tag{12.140}$$

where in the last step we used $\sum_y p^{(2)}_{\epsilon^{-2}(t-s)}(x',y';x,y) = p_{\epsilon^{-2}(t-s)}(x',x)$ because the first marginal of two particles is a random walk. Now if we denote

$$\sum_x |\varphi(\epsilon x)| \int_0^t p_{\epsilon^{-2}(t-s)}(x',x) := V(t,\epsilon,\epsilon x'), \tag{12.141}$$

by the invariance principle, V is bounded and smooth function of x'. As a consequence we obtain

$$\|\varphi\|_\infty \epsilon^d BC \sum_{x,x'\sim y'} |\varphi(\epsilon x)| \int_0^t p_{\epsilon^{-2}(t-s)}(x',x) \leq \|\varphi\|_\infty \epsilon^d BC \sum_{x'\sim y'} V(t,\epsilon,\epsilon x'), \tag{12.142}$$

and the rhs of (12.142) clearly remains bounded as $\epsilon \to 0$. □

12.3.3 Propagation of Local Equilibrium

We have seen in the independent random walk case that propagation of local equilibrium is a consequence of the fact that the dual particles spread out over the lattice, i.e., have negligible probabilities to be at the same location at macroscopic times, combined with the fact that n independent walkers scale to n independent Brownian motions under diffusive rescaling. In the interacting case, a similar reasoning holds, provided one has a suitable coupling between n dual particles and n independent particles. Under that setting, we can proceed as in Sect. 12.2.7.

First we define what we call a suitable coupling of n dual particles and n independent particles.

Definition 12.23 (Suitable Coupling, n Particles) We say that

$$\{(X_1(t), \dots X_n(t); Y_1(t), \dots Y_n(t)) : t \geq 0\} \tag{12.143}$$

is a suitable coupling of n particles with n independent particles if we have the following two properties.

1. Correct marginals: $\{(X_1(t), \dots, X_n(t)) : t \geq 0\}$ equals in distribution the n particle process started from $(X_1(0), \dots, X_n(0))$, and $\{(Y_1(t), \dots, Y_n(t)) : t \geq 0\}$ equals in distribution n independent particles starting from $(Y_1(0), \dots, Y_n(0))$.

2. Suitable behavior of discrepancies: as $\epsilon \to 0$

$$(\epsilon X_1(\epsilon^{-2}t), \ldots \epsilon X_n(\epsilon^{-2}t)) - (\epsilon Y_1(\epsilon^{-2}t), \ldots, \epsilon Y_n(\epsilon^{-2}t)) \to \quad 0 \text{ in probability.} \tag{12.144}$$

This is the natural generalization of what we defined for 2 particles before. Next we recall the definition a family of measure associated to density profile ρ. For simplicity we will restrict to product measures here. This can be generalized to measures with decaying covariances without too much effort.

Definition 12.24 (Measures Compatible with a Profile) Let $\rho : \mathbb{R}^d \to [0, \infty)$ denote a bounded and smooth density profile. We say that a family of probability measures $\{\mu_\epsilon, \epsilon > 0\}$ is compatible with the profile if

1. Tempered: uniform moments bound, that is μ_ϵ is UFMC (in the sense of (12.8)) for all $\epsilon > 0$.
2. Density profile:

$$\int \eta_x \mu_\epsilon(d\eta) = \rho(\epsilon x). \tag{12.145}$$

We recall the definition of convergence of duality polynomials:

Definition 12.25 (D-convergence) $\mu_\epsilon \to \mu$ in the sense of convergence of duality polynomials if for all ξ finite configurations $\int D(\xi, \eta)\mu_\epsilon(d\eta) \to \int D(\xi, \eta)\mu(d\eta)$.

Then we have the following result on propagation of local equilibrium.

Theorem 12.26 (Local Equilibrium) *Let $\rho : \mathbb{R}^d \to [0, \infty)$ denote a bounded and smooth density profile. Let $\{\mu_\epsilon : \epsilon > 0\}$ be a family of product measures compatible with the profile ρ. Then as $\epsilon \to 0$, for all $t > 0$, the time evolved measure $\mu_\epsilon S(\epsilon^{-2}t)$ is a local equilibrium distribution with density profile ρ_t, i.e., for all $x \in \mathbb{R}^d$, in the sense of convergence of duality polynomials*

$$\tau_{\lfloor \epsilon^{-1}x \rfloor}(\mu_\epsilon S(\epsilon^{-2}t)) \to \mu_{\rho(t,x)}. \tag{12.146}$$

Here ρ_t is the solution of the heat Eq. (12.114), *and initial condition ρ.*

Proof Let us denote for $(x_1, \ldots x_n)$ denote by $(X_1^{x_1}(t), \ldots, X_n^{x_n}(t))$ denote the positions of n (labeled) dual particles initially located at $x_1, \ldots, x_n$. We remind the reader that $D(\delta_x, \eta) = \eta_x/\alpha$. Therefore, in order to obtain (12.146) we have to prove that for all $x_1, \ldots, x_n \in \mathbb{Z}^d$, and $x \in \mathbb{R}^d$

$$\lim_{\epsilon \to 0} \int \mathbb{E}_\eta D\left(\delta_{\lfloor \epsilon^{-1}x \rfloor + y_1} + \ldots \delta_{\lfloor \epsilon^{-1}x \rfloor + y_n}, \eta(\epsilon^{-2}t)\right) \mu_\epsilon(d\eta) = \frac{\rho_t(x)^n}{\alpha^n}. \tag{12.147}$$

Indeed, the reversible product measure μ_ρ is characterized by

$$\int \eta_x d\mu_\rho = \rho,$$

$$\int D(\xi,\eta)d\mu_\rho(\eta) = \left(\int D(\delta_0,\eta)d\mu_\rho(\eta)\right)^{|\xi|}. \tag{12.148}$$

Before the limit $\epsilon \to 0$, the lhs of (12.147) equals, using duality

$$\int \mathbb{E}_{(\lfloor \epsilon^{-1}x\rfloor + y_1,\ldots,\lfloor \epsilon^{-1}x\rfloor + y_n)} D\left(\delta_{X^{\lfloor \epsilon^{-1}x\rfloor + y_1}}(\epsilon^{-2}t) + \ldots \delta_{X^{\lfloor \epsilon^{-1}x\rfloor + y_n}(\epsilon^{-2}t)}, \eta\right) \mu_\epsilon(d\eta). \tag{12.149}$$

As a first step, we have to show that with probability converging to 1 as $\epsilon \to 0$ we have that the n locations of the dual particles $(X^{\lfloor \epsilon^{-1}x\rfloor + y_1}(\epsilon^{-2}t), \ldots, X^{\lfloor \epsilon^{-1}x\rfloor + y_n}(\epsilon^{-2}t))$ are all different. To prove this, denote by $\xi = \delta_{\lfloor \epsilon^{-1}x\rfloor + y_1} + \ldots \delta_{\lfloor \epsilon^{-1}x\rfloor + y_n}$, then we have to prove that

$$\lim_{t\to\infty} \mathbb{P}_\xi(\exists x \in \mathbb{Z}^d : \xi_x(t)(\xi_x(t) - 1) \geq 1) = 0. \tag{12.150}$$

This is proved in Lemma 12.27 below for SIP (the proof for SEP being analogous). Then we can proceed as in the proof of Theorem 12.12 to arrive that the expression (12.149) equals

$$\mathbb{E}_{(\lfloor \epsilon^{-1}x\rfloor + y_1,\ldots,\lfloor \epsilon^{-1}x\rfloor + y_n)}\left(\prod_{i=1}^n \rho(\epsilon X^{\lfloor \epsilon^{-1}x\rfloor + y_1}(\epsilon^{-2}t)\right) + o(1), \tag{12.151}$$

where $o(1) \to 0$ for all $x, y_1, \ldots, y_n$. Arrived there, we use the assumed suitable coupling to conclude that the expression in (12.151) equals

$$\mathbb{E}^{\mathrm{IRW}}_{(\lfloor \epsilon^{-1}x\rfloor + y_1,\ldots,\lfloor \epsilon^{-1}x\rfloor + y_n)}\left(\prod_{i=1}^n \rho(\epsilon X^{\lfloor \epsilon^{-1}x\rfloor + y_1}(\epsilon^{-2}t)\right) + o(1), \tag{12.152}$$

where the expectation is now over independent random walkers. Then using the independence, and the invariance principle, we arrive at $\rho_t(x)^n + o(1)$. □

We finally state and prove Lemma 12.27.

Lemma 12.27 *Let ξ be a configuration with n particles. Then we have an upperbound*

$$\mathbb{P}_\xi(\exists i \in \mathbb{Z}^d : \xi_i(t)(\xi_i(t) - 1) \geq 1) \leq C_n(t), \tag{12.153}$$

where $C_n(t)$ only depends on ξ via the number of particles n, and tends to zero as $t \to \infty$.

Proof We remind that we assumed $p(x, y)$ to be of finite range, i.e., such that $p(x, y) = 0$ for $|x - y| > R$. Notice that the duality polynomial with two dual particles equals $d(2, n) = Cn(n - 1)$, with $C = 1/\alpha\alpha(+1)$. Therefore, using the Markov inequality there is $C_1 > 0$ such that

$$\mathbb{P}_\xi(\xi_i(t)(\xi(t) - 1) \geq 1) \leq C_1 \mathbb{E}_\xi D(2\delta_i, \xi(t)) = C_1 \mathbb{E}_{i,i} D(\delta_{X_1(t)} + \delta_{X_2(t)}, \xi), \tag{12.154}$$

where in the last step we used self-duality. Let us denote by $p_t(i, i; u, v)$ the transition probability for two SIP particles initially at (i, i) to move to (u, v) in time t. Because SIP with two particles admits a strictly positive reversible σ-finite measure $\lambda(x, y)$ which is uniformly bounded, and uniformly bounded from below, it follows that

$$\mathbb{P}_\xi(\xi_i(t)(\xi(t) - 1) \geq 1) \leq C_2 \sum_{x,y:\xi_x\xi_y \geq 1} p_t(x, y; i, i). \tag{12.155}$$

Summing over i gives then the upperbound

$$C_2 \sum_{x,y:\xi_x\xi_y \geq 1} \mathbb{P}_{x,y}(X_1^x(t) = X_2^y(t)) \leq C_2 n(n-1) \sup_{x,y} \mathbb{P}_{x,y}(X_1^x(t) - X_2^y(t) = 0), \tag{12.156}$$

where in the last inequality we used that there at most n particles to choose from in the sum $\sum_{x,y:\xi_x\xi_y \geq 1}$. Now because the difference process $X_1^x(t) - X_2^y(t)$ is Markov and jumps like a random walk, except when at distance R from the origin (where R is the assumed finite range of the random walk kernel $p(x, y)$), it follows that $\lim_{t \to \infty} \sup_{x,y} \mathbb{P}_{x,y}(X_1^x(t) - X_2^y(t) = 0) = 0$. □

Remark 12.28 In the spirit of propagation of local equilibrium one can also obtain a slightly stronger statement of "local ergodicity". If we start from a local equilibrium product measure μ_ϵ, corresponding to a density profile ρ then at later times we have that the macroscopic field of duality polynomials

$$\epsilon^d \sum_x \varphi(\epsilon x) D(\xi, \tau_x(\eta)(\epsilon^{-2} t)), \tag{12.157}$$

converges as $\epsilon \to 0$ to the deterministic quantity

$$\int dx \varphi(x) \rho(t, x)^n, \tag{12.158}$$

where $n = \sum_x \xi_x$ is the number of dual particles. For the expectation this follows immediately by the definition of local equilibrium. To show that the variance of (12.157) goes to zero when $\epsilon \to 0$ one can proceed as before by using duality and the fact that dual particles at macroscopic times are typically at different locations.

12.4 Higher Order Hydrodynamic Fields

The hydrodynamic limit, which is the law of large numbers for the density field can be stated equivalently as the convergence of the field

$$\epsilon^d \sum_x D(\delta_x, \eta)\varphi(\epsilon x), \tag{12.159}$$

associated to the first duality polynomial, because for the models under consideration, $D(\delta_x, \eta) = \eta_x/\alpha$. Similarly, the density fluctuation field in the stationary setting, i.e., started from the stationary product measure μ_ρ, can be rewritten (up to a multiplicative constant) as

$$\epsilon^{d/2} \sum_x D_\rho(\delta_x, \eta)\varphi(\epsilon x), \tag{12.160}$$

where $D_\rho(\delta_x, \eta)$ is the first order orthogonal polynomial, which for a single dual particle at x is proportional to $\eta_x - \rho$. This suggests that by passing to higher order duality polynomials, we can define higher order hydrodynamic fields and higher order fluctuation fields. In order to proceed, it is convenient to weight the sums defining higher order fields with the reversible sigma-finite measure on $(\mathbb{Z}^d)^n$, which we denote by $\pi_{\sigma,\alpha}(x_1, \ldots, x_n)$, where we remind the reader that σ, α are the parameters of the model (12.94). This measure is such that the semigroup of n particle motion is self-adjoint in $l_2((\mathbb{Z}^d)^n)$. This measure is unique up to a multiplicative constant, which we fix by requiring that whenever the n particles are at different locations, the weight $\pi(x_1, \ldots, x_n) = 1$. For SIP($\alpha$) this measure is given by

$$\pi_{\sigma,\alpha}(x_1, \ldots, x_n) = \frac{1}{\alpha^n} \prod_{x \in \mathbb{Z}^d} \frac{\Gamma(\alpha + \xi_x)}{\Gamma(\alpha)\xi_x!}, \tag{12.161}$$

where ξ is the configuration $\sum_i \delta_{x_i}$. Notice that the infinite product is well defined because ξ is a finite configuration, i.e., only finitely many terms of the product are different from 1. Also notice the factors $\xi_x!$ in the denominator which are needed because (12.161) is the measure on the level of the labeled particle positions. For

SEP(α) this sigma-finite measure is given by

$$\pi_{\sigma,\alpha}(x_1, \ldots, x_n) = \prod_x \binom{\alpha}{\xi_x}. \tag{12.162}$$

For independent random walkers, the measure is given by $\prod_{x \in \mathbb{Z}^d} \frac{1}{\xi_x!}$. Notice that the measure π depends only on the particle configuration "combinatorics", and not on the precise locations of the particles.

We then define the fields of order n as follows

$$\epsilon^{nd} \sum_{x_1,\ldots,x_n} \varphi(\epsilon x_1, \ldots, \epsilon x_n) D_{\sigma,\alpha}\left(\delta_{x_1} + \ldots \delta_{x_n}, \eta\right) \pi_{\sigma,\alpha}(\epsilon x_1, \ldots, \epsilon x_n), \tag{12.163}$$

where $\varphi : \mathbb{R}^{nd} \to \mathbb{R}$ is a Schwartz function and $D_{\sigma,\alpha}$ is the triangular self-duality function. Without loss of generality we can restrict to symmetric functions φ, and further to tensor products $\varphi(x_1) \ldots \varphi(x_n) =: \varphi^{\otimes n}(x_1, \ldots, x_n)$, because linear combinations of those span all symmetric functions. Next, we note that for the three models the duality polynomials equal to the same function divided by the reversible measure π. More precisely, as we saw before the duality polynomial for the model with parameters σ, α reads

$$D_{\sigma,\alpha}(\xi, \eta) = \frac{\mathcal{D}(\xi, \eta)}{\pi_{\sigma,\alpha}(\xi)} \tag{12.164}$$

where

$$\mathcal{D}(\xi, \eta) = \prod_x \binom{\eta_x}{\xi_x} \tag{12.165}$$

Our discussion above then motivates the following definition.

Definition 12.29 (n-th Order Hydrodynamic Field) The hydrodynamic field of order n at scale ϵ is defined as

$$\mathcal{X}(n, \epsilon, \varphi, \eta) = \epsilon^{nd} \sum_{x_1,\ldots,x_n} \varphi^{\otimes n}(\epsilon x_1, \ldots, \epsilon x_n) \mathcal{D}(x_1, \ldots, x_n; \eta), \tag{12.166}$$

where $\mathcal{D}$ is given in (12.165).

Definition 12.30 (Consistent Profile) Let $\{\mu_\epsilon, \epsilon > 0\}$ denote a family of probability measures on the configuration space, indexed by the scaling parameter $\epsilon > 0$. Let $\rho(n, \cdot) : \mathbb{R}^{nd} \to [0, \infty)$. We say that $\{\mu_\epsilon, \epsilon > 0\}$ is consistent with the n-th order profile if

$$\lim_{\epsilon \to 0} \int \mathcal{X}(n, \epsilon, \varphi, \eta) \mu_\epsilon(d\eta) = \int_{\mathbb{R}^{nd}} \rho(n, x_1, \ldots, x_n) \varphi^{\otimes n}(x_1, \ldots, x_n) dx_1 \ldots dx_n. \tag{12.167}$$

We can then define "n-th order hydrodynamics", which in words means that whenever at time zero a family $\{\mu_\epsilon, \epsilon > 0\}$ of probability measures on the configuration space is consistent with a n-th order profile, then at later macroscopic times we have consistency with a new n-th order profile which the solution of the nd-dimensional heat equation with diffusion constant α. We will state a theorem proving this in the case that $\{\mu_\epsilon, \epsilon > 0\}$ is a family of product measures. The generalization to families of measures with a "controllable" covariance is straightforward, and as in the $n = 1$ case, obtained via a study of the variance of the nth-order field.

12.4.1 Hydrodynamic Equation for the n-th Order Field

We then have the following result on the evolution of the n-th order profile on macroscopic time scales.

Theorem 12.31 (Hydrodynamic Limit of the n-th Order Field) *Assume that there exists a suitable coupling of n dual particles with n independent particles in the sense of Definition 12.23. Let $\{\mu_\epsilon, \epsilon > 0\}$ denote a family of probability measures on the configuration space consistent with a smooth and bounded profile $\rho(n, \cdot) : \mathbb{R}^{nd} \to \mathbb{R}$. Then, for all $t > 0$ the time evolved n-th order field*

$$\mathcal{X}(n, \epsilon, \varphi, \eta(\epsilon^{-2}t)) \to \int \rho(n, t, x_1, \ldots, x_n)\varphi^{\otimes n}(x_1, \ldots, x_n)dx_1 \ldots dx_n, \tag{12.168}$$

where the convergence is in L^2 sense and where $\rho(n, t, x_1, \ldots, x_n)$ is the solution of (12.114) *in $\mathbb{R}^{dn}$.*

Proof The proof proceeds in two steps:

1. Step 1: convergence of expectation of $\mathcal{X}(n, \epsilon, \varphi, \eta(\epsilon^{-2}t))$.
2. Step 2: control of the variance: $\text{Var}(\mathcal{X}(n, \epsilon, \varphi, \eta(\epsilon^{-2}t))) \to 0$ as $\epsilon \to 0$.

We prove the first step in the general case, and the second in the case of independent random walkers, the other models being dealt with analogously via the assumed existence of a suitable coupling, in the spirit of the case $n = 1$ proved earlier.

Step 1: Expectation
Denote $D_{\sigma,\alpha}(\xi, \eta) = D_{\sigma,\alpha}(x_1, \ldots, x_n; \eta)$ whenever $\xi = \sum_{i=1}^n \delta_{x_i}$. Let us further denote $Q_t^{n,\sigma,\alpha}$ the semigroup of n dual particles, and Q_t^n the semigroup of n independent Brownian motions with diffusion constant α, i.e., of the process $B_1(\alpha t), \ldots, B_n(\alpha t))$ where $B_1, \ldots, B_n$ are independent standard Brownian motions on $\mathbb{R}^d$. Let us also denote $Q_t^{n,\text{IRW}}$ the semigroup of n independent walkers with jumping rates $\alpha p(x, y)$ between x and y. Further, $o(1)$ denotes a quantity

converging to zero when $\epsilon \to 0$, and $\sum^{\neq}_{x_1,\ldots,x_n}$ denotes the sum over $x_1, \ldots, x_n$ where $x_i \neq x_j$ for $i \neq j$. Finally, we abbreviate $\varphi^{\otimes n}(x_1, \ldots, x_n) =: \Phi(x_1, \ldots, x_n)$

Then we have, using (12.164), duality and the reversibility of $\pi_{\alpha,\sigma}$:

$$
\begin{aligned}
&\mathbb{E}_{\mu_\epsilon}\left(\mathcal{X}(n, \epsilon, \varphi, \eta(\epsilon^{-2}t))\right) \\
&= \epsilon^{nd} \sum_{x_1,\ldots,x_n \in \mathbb{Z}^d} \mathbb{E}_{\mu_\epsilon}\left(\mathcal{D}(x_1, \ldots, x_n, \eta(\epsilon^{-2}t)\right) \Phi(\epsilon x_1, \ldots, \epsilon x_n) \\
&= \epsilon^{nd} \sum_{x_1,\ldots,x_n \in \mathbb{Z}^d} \mathbb{E}_{\mu_\epsilon}\left(D_{\sigma,\alpha}(x_1, \ldots, x_n, \eta(\epsilon^{-2}t)\right) \\
&\quad \Phi(\epsilon x_1, \ldots, \epsilon x_n) \pi_{\sigma,\alpha}(x_1, \ldots, x_n) \\
&= \epsilon^{nd} \sum_{x_1,\ldots,x_n \in \mathbb{Z}^d} Q^{n,\sigma,\alpha}_{\epsilon^{-2}t}\left(\int \left(D_{\sigma,\alpha}(x_1, \ldots, x_n, \eta)\right) \mu_\epsilon(d\eta)\right) \\
&\quad \Phi(\epsilon x_1, \ldots, \epsilon x_n) \pi_{\sigma,\alpha}(x_1, \ldots, x_n) \\
&= \epsilon^{nd} \sum_{x_1,\ldots,x_n \in \mathbb{Z}^d} \int \left(D_{\sigma,\alpha}(x_1, \ldots, x_n, \eta)\right) \mu_\epsilon(d\eta) \\
&\quad Q^{n,\sigma,\alpha}_{\epsilon^{-2}t} \Phi(\epsilon x_1, \ldots, \epsilon x_n) \pi_{\sigma,\alpha}(x_1, \ldots, x_n) \\
&= \epsilon^{nd} \sum_{x_1,\ldots,x_n \in \mathbb{Z}^d} \rho(n, \epsilon x_1, \ldots, \epsilon x_n) \\
&\quad Q^{n,\mathrm{IRW}}_{\epsilon^{-2}t} \Phi(\epsilon x_1, \ldots, \epsilon x_n) \pi_{\sigma,\alpha}(x_1, \ldots, x_n) + o(1).
\end{aligned}
\tag{12.169}
$$

Here in the third line, the semigroup $Q^{n,\sigma,\alpha}_{\epsilon^{-2}t}$ is acting on the variables $x_1, \ldots, x_n$, and in the last step we used the assumption that there exists a suitable coupling of n dual particles with n independent walkers. Then we can continue using the invariance principle which implies that we can approximate $Q^{n,\mathrm{IRW}}_{\epsilon^{-2}t}$ by Q^n_t:

$$
\begin{aligned}
&\mathbb{E}_{\mu_\epsilon}\left(\mathcal{X}(n, \epsilon, \varphi, \eta(\epsilon^{-2}t))\right) \\
&= \epsilon^{nd} \sum_{x_1,\ldots,x_n \in \mathbb{Z}^d} \rho(n, \epsilon x_1, \ldots, \epsilon x_n) Q^n_t \Phi(\epsilon x_1, \ldots, \epsilon x_n) \pi_{\sigma,\alpha}(x_1, \ldots, x_n) + o(1) \\
&= \epsilon^{nd} \sum^{\neq}_{x_1,\ldots,x_n \in \mathbb{Z}^d} \rho(n, \epsilon x_1, \ldots, \epsilon x_n) Q^n_t \Phi(\epsilon x_1, \ldots, \epsilon x_n) \pi_{\sigma,\alpha}(x_1, \ldots, x_n) + o(1) \\
&= \epsilon^{nd} \sum^{\neq}_{x_1,\ldots,x_n \in \mathbb{Z}^d} \rho(n, \epsilon x_1, \ldots, \epsilon x_n) Q^n_t \Phi(\epsilon x_1, \ldots, \epsilon x_n) + o(1)
\end{aligned}
$$

$$= \int \rho(n, x_1, \ldots, x_n) Q_t^n \Phi(x_1, \ldots, x_n) dx_1 \ldots dx_n + o(1)$$

$$= \int Q_t^n \rho(n, x_1, \ldots, x_n) \Phi(x_1, \ldots, x_n) dx_1 \ldots dx_n + o(1). \tag{12.170}$$

Here in the last steps we used the fact that $\pi = 1$ on configurations with n different locations for the n particles.

Step 2: Variance

To control the variance, as earlier announced, we restrict to the independent particle case. We have

$$\begin{aligned}
&\mathbb{E}_{\mu_\epsilon}\left(\mathcal{X}^2(n, \epsilon, \varphi, \eta(\epsilon^{-2}t))\right) \\
&= \sum_{x_1,\ldots,x_n} \sum_{y_1,\ldots,y_n} \Phi(\epsilon x_1, \ldots, \epsilon x_n)\Phi(\epsilon y_1, \ldots, \epsilon y_n) \\
&\quad \mathbb{E}_{\mu_\epsilon}\left(\mathcal{D}(x_1, \ldots, x_n; \eta(\epsilon^{-2}t))\mathcal{D}(y_1, \ldots, y_n; \eta(\epsilon^{-2}t))\right) \\
&= \sum_{x_1,\ldots,x_n}^{\neq} \sum_{y_1,\ldots,y_n:\{y_1,\ldots,y_n\}\cap\{x_1,\ldots,x_n\}=\emptyset}^{\neq} \Phi(\epsilon x_1, \ldots, \epsilon x_n)\Phi(\epsilon y_1, \ldots, \epsilon y_n) \\
&\quad \cdot \left[\mathbb{E}_{\mu_\epsilon}\left(\mathcal{D}(x_1, \ldots, x_n; \eta(\epsilon^{-2}t))\mathcal{D}(y_1, \ldots, y_n; \eta(\epsilon^{-2}t))\right)\right] + o(1) \\
&= \sum_{x_1,\ldots,x_{2n}}^{\neq} Q_{\epsilon^{-2}t}^{2n,\mathrm{IRW}} \int \mathcal{D}(x_1, \ldots, x_{2n}, \eta)\mu_\epsilon(d\eta)\Phi(\epsilon x_1, \ldots, \epsilon x_{2n}) + o(1) \\
&= \sum_{x_1,\ldots,x_{2n}}^{\neq} \int \mathcal{D}(x_1, \ldots, x_{2n}, \eta)\mu_\epsilon(d\eta) Q_{\epsilon^{-2}t}^{2n,\mathrm{IRW}} \Phi(\epsilon x_1, \ldots, \epsilon x_{2n}) + o(1).
\end{aligned} \tag{12.171}$$

Using now the invariance principle combined with the fact that for $x_1, \ldots, x_{2n}$ mutually different points of $\mathbb{Z}^d$

$$\mathcal{D}(x_1, \ldots, x_{2n}, \eta) = \mathcal{D}(x_1, \ldots, x_n, \eta)\mathcal{D}(x_{n+1}, \ldots, x_{2n}, \eta), \tag{12.172}$$

together with

$$\Phi(x_1, \ldots, x_{2n}) = \varphi^{\otimes 2n}(x_1, \ldots, x_{2n}), \tag{12.173}$$

we arrive at

$$
\begin{aligned}
&\mathbb{E}_{\mu_\epsilon}\left(\mathcal{X}^2(n,\epsilon,\varphi,\eta(\epsilon^{-2}t))\right)\\
&=\sum_{x_1,\ldots,x_{2n}}^{\neq}\int \mathcal{D}(x_1,\ldots,x_{2n},\eta)\mu_\epsilon(d\eta)Q_t^{2n}\Phi(\epsilon x_1,\ldots,\epsilon x_{2n})+o(1)\\
&=\left(\int \rho(n,x_1,\ldots,x_n)Q_t^n\Phi(\epsilon x_1,\ldots,\epsilon x_n)\right)^2+o(1)\\
&=\left(\mathbb{E}_{\mu_\epsilon}\left(\mathcal{X}(n,\epsilon,\varphi,\eta(\epsilon^{-2}t))\right)\right)^2+o(1).
\end{aligned}
\tag{12.174}
$$

This shows that $Var\left(\mathcal{X}(n,\epsilon,\varphi,\eta(\epsilon^{-2}t))\right)=o(1)$.

□

12.5 Boltzmann Gibbs Principle and Orthogonal Polynomial Duality

In the study of fluctuation fields, an important principle is the so-called Boltzmann Gibbs principle which states that the density fluctuation field is the slowest varying field, and other fluctuation fields can be replaced by their "projection" on the density fluctuation field. This general idea appeared first in a paper by Brox and Rost, [36], and was later generalized and refined, see [151], Chapter 11, for a general statement and proof. For systems with orthogonal polynomial duality, it is natural to study the fluctuation fields of orthogonal polynomials. These "span" the fluctuation fields of general functions. We will see that in a sense to be defined below, the fluctuation fields of all orthogonal polynomials of order at least two are negligible. This implies that the fluctuation field of a general function can be replaced by its projection on orthogonal polynomials of order 1, which is in turn the density fluctuation field.

We recall that for a general local function f and a reversible product measure ν_ρ, we define its stationary fluctuation field via

$$
\mathcal{Y}_\epsilon(f,\varphi;\eta)=\epsilon^{d/2}\sum_x\left(\tau_x f(\eta)-\int f\,d\nu_\rho\right)\varphi(\epsilon x),
\tag{12.175}
$$

and when $f=\eta_0$ we denote the field by $\mathcal{Y}_\epsilon(\varphi;\eta)$, and call it the (stationary) density fluctuation field. Strictly speaking, these fields also depend on ρ, but we suppress that dependence to lighten notation. The Boltzmann Gibbs principle states that for a general f, at macroscopic times, the field

$$
\mathcal{Y}_\epsilon(f,\varphi;\eta(\epsilon^{-2}t))
\tag{12.176}
$$

can be well approximated by a constant times the density fluctuation field, i.e., by

$$C(\rho, f)\mathcal{Y}_\epsilon(\varphi; \eta(\epsilon^{-2}t)), \tag{12.177}$$

where $C(\rho, f)$ is a constant depending on f and ρ:

$$C(\rho, f) = \frac{d}{d\rho'}\left(\int f d\mu_{\rho'}\right)\bigg|_{\rho'=\rho}. \tag{12.178}$$

If we are in a context where orthogonal polynomial duality is valid, we can first consider f which are orthogonal duality polynomials, and expand every other f in this "basis". The statement of the Boltzmann-Gibbs principle can then be reformulated as the statement that the fluctuation fields of orthogonal duality polynomials of order ≥ 2 are negligible, (in a sense to be described below, see Theorem 12.34) .

In order to fix ideas, we start with a simple example which illustrates in which sense fluctuations of orthogonal polynomials of order 2 are negligible. Let us denote by $\mathcal{D}_\rho(x_1, \ldots, x_n; \eta)$ the orthogonal (in $L^2(\nu_\rho)$) duality polynomial corresponding to the dual configuration $\xi = \sum_{i=1}^n \delta_{x_i}$.

Then we start with the following example which shows that the fluctuation field of $\mathcal{D}_\rho(0, 0; \eta)$ is negligible. The example contains already the main idea needed in the proof of the general case. In order to proceed, we need some notation. Let $X(t), Y(t)$ denote the positions of two dual particles, then by translation invariance of the interaction, the difference process $Z(t) = X(t) - Y(t)$ is a Markov process. In the independent random walk case, this is again a random walk moving at twice the speed, whereas in the interacting case with $p(x, y) = \pi(y - x)$ and finite range R, it is a Markov process with generator

$$Lf(z) = \sum_{r\in A} 2\pi(r)(\alpha + \sigma \mathbb{1}_{\{r=-z\}})(f(z + r) - f(z)), \tag{12.179}$$

where A denotes the finite set $[-R, R]^d \cap \mathbb{Z}^d$. This process is moving as a random walk as long as it is at distance larger than R from the origin, and has different rates within the set A of locations at distance at most R from the origin. We denote $\mathbb{E}_z^Z$ expectation in this process $\{Z(t) : t \geq 0\}$ starting from z. Let us denote by $l_t(z) = \int_0^t \mathbb{1}_{\{Z(s)=z\}}$ the local time of this process at $z \in \mathbb{Z}^d$. By the fact that the process behaves outside the ball of radius R as an ordinary finite range translation invariant random walk, we have for all $z, z' \in \mathbb{Z}^d$

$$\lim_{t\to\infty} \frac{1}{t}\mathbb{E}_z l_t(z') = 0. \tag{12.180}$$

Indeed, this is the case for ordinary random walk, and since one can decompose the difference process in "epochs" (excursions) where it behaves as an ordinary random walk, followed by periods in which it is in the ball $B(0, R)$, which is left after a time bounded above by an exponentially distributed random variable, one concludes (12.180) for the difference process as well.

We then prove the following.

Proposition 12.32 *For all $T > 0$, and for all Schwartz functions φ we have*

$$\lim_{\epsilon\to 0} \mathbb{E}_{\nu_\rho} \left(\epsilon^{d/2} \sum_x \int_0^T \mathcal{D}(x, x; \eta(\epsilon^{-2}t))\varphi(\epsilon x)dt \right)^2 = 0. \tag{12.181}$$

Proof First we write, abbreviating $C_\rho = \|\mathcal{D}_\rho(0, 0; \eta)\|^2_{L^2(\nu_\rho)} = \|\mathcal{D}_\rho(x, x; \eta)\|^2_{L^2(\nu_\rho)}$:

$$\begin{aligned}
&\mathbb{E}_{\nu_\rho} \left(\epsilon^{d/2} \sum_x \int_0^T \mathcal{D}_\rho(x, x; \eta(\epsilon^{-2}s))\varphi(\epsilon x) \right)^2 \\
&= 2\epsilon^d \int_0^T dt \int_0^t ds \sum_x \sum_y \varphi(\epsilon x)\varphi(\epsilon y) \\
&\quad \mathbb{E}_{\nu_\rho} \left(\mathcal{D}_\rho(x, x; \eta(\epsilon^{-2}(t-s))\mathcal{D}_\rho(y, y; \eta(0)) \right) \\
&= 2\epsilon^d C_\rho \int_0^T dt \int_0^t ds \sum_x \sum_y \varphi(\epsilon x)\varphi(\epsilon y) p_{\epsilon^{-2}(t-s)}(x, x; y, y) \\
&\le 2\epsilon^d C_\rho \int_0^T dt \int_0^T dr \sum_x |\varphi(\epsilon x)| \mathbb{E}_{x,x} \left(|\varphi(\epsilon X(\epsilon^{-2}r))| \mathbb{1}_{\{X(\epsilon^{-2}r)=Y(\epsilon^{-2}r)\}} \right) \\
&\le 2T \|\varphi\|_\infty C_\rho \epsilon^d \sum_x |\varphi(\epsilon x)| \epsilon^2 \mathbb{E}^Z_0 \int_0^{\epsilon^{-2}T} \mathbb{1}_{\{Z(s)=0\}} ds \\
&= 2T \|\varphi\|_\infty C_\rho \epsilon^d \sum_x |\varphi(\epsilon x)| \epsilon^2 \mathbb{E}^Z_0 (l_{\epsilon^{-2}T}(0)).
\end{aligned} \tag{12.182}$$

In the first equality we used stationarity, and in the second orthogonal polynomial duality. By (12.180) we have $\epsilon^2 \mathbb{E}^Z_0(l_{\epsilon^{-2}T}(0)) \to 0$ as $\epsilon \to 0$. □

Remark 12.33

1. Notice that in the proof, the integration over time plays an important role. Indeed, for fixed time $t > 0$, the fluctuation field $\mathcal{Y}_\epsilon(\mathcal{D}(0, 0; \cdot), \varphi; \eta(\epsilon^{-2}t))$ does not converge to zero, but, by stationarity, converges to a normally distributed random variable.

   Next, notice that the essence of the proof of Proposition (12.32) lies in the fact that two dual particles spend $o(\epsilon^{-2})$ time together (or more generally at a fixed distance) in the interval $[0, \epsilon^{-2}]$, or in other words, the expected local time $\mathbb{E}^Z_0(l_{\epsilon^{-2}T}(0)) = o(\epsilon^{-2})$ as $\epsilon \to 0$.

2. Notice further that if the random walk Z is transient then $\mathbb{E}_0^Z(l_{\epsilon^{-2}T}(0))$ remains bounded as $\epsilon \to 0$ and as a consequence the quantity appearing in (12.181), i.e., $\mathbb{E}_{\nu_\rho}\left(\epsilon^{d/2}\sum_x \int_0^T \mathcal{D}(x, x; \eta(\epsilon^{-2}t))\varphi(\epsilon x)dt\right)^2$ is of order ϵ^2 as $\epsilon \to 0$.

The following theorem shows that in the same sense as in the example, fluctuation fields of orthogonal polynomials of order ≥ 2 are negligible. The essence of the proof is once more the fact that dual particles spend $o(\epsilon^{-2})$ at fixed distances in the time interval $[0, \epsilon^{-2}]$. Notice that when we consider orthogonal duality polynomials, then we have that

$$\int \mathcal{D}_\rho(x_1, \ldots, x_n; \eta)\mathcal{D}_\rho(y_1, \ldots, y_n; \eta)\nu_\rho(d\eta) = \tag{12.183}$$

$$\mathbb{1}_{\{\exists \sigma \in S_n : x_{\sigma(1)}, \ldots, x_{\sigma(n)} = (y_1, \ldots, y_n)\}} \int \mathcal{D}_\rho(x_1, \ldots, x_n; \eta)^2 \nu_\rho(d\eta).$$

In other words the polynomials $\mathcal{D}_\rho(x_1, \ldots, x_n; \eta)$ and $\mathcal{D}_\rho(y_1, \ldots, y_n; \eta)$ are orthogonal if and only if the configurations corresponding to the two n tuples are different, i.e., whenever $\sum_{i=1}^n \delta_{x_i} \neq \sum_{i=1}^n \delta_{y_i}$.

Theorem 12.34 (Boltzmann Gibbs Principle) *For all $n \in \mathbb{N}$, $n \geq 2$, $x_1, \ldots, x_n \in \mathbb{Z}^d$, for all Schwartz functions φ and for all $T > 0$ we have*

$$\lim_{\epsilon \to 0} \mathbb{E}_{\nu_\rho}\left(\epsilon^{d/2}\int_0^T \sum_x \mathcal{D}_\rho(x + x_1, \ldots, x + x_n; \eta(\epsilon^{-2}s))\varphi(\epsilon x)dt\right)^2 = 0. \tag{12.184}$$

Proof Fix $T > 0$, $n \in \mathbb{N}$, $x_1, \ldots, x_n \in \mathbb{Z}^d$, a Schwartz function φ. Denote $\|\mathcal{D}_\rho(x + x_1, \ldots, x + x_n; \eta)\|^2_{L^2(\nu_\rho)} = C(\rho)$, where we suppressed the dependence on $(x_1, \ldots, x_n)$ for notational simplicity. Then using stationarity, combined with orthogonal polynomial duality, proceeding as in the proof of Proposition 12.32 we estimate

$$\mathbb{E}_{\nu_\rho}\left(\epsilon^{d/2}\int_0^T \sum_x \mathcal{D}_\rho(x + x_1, \ldots, x + x_n; \eta(\epsilon^{-2}s))\varphi(\epsilon x)dt\right)^2$$

$$\leq \sum_{\sigma \in S_n} 2\epsilon^d C(\rho)\int_0^T dt \int_0^t ds \sum_x |\varphi(\epsilon x)|\mathbb{E}_{x+x_1, \ldots, x+x_n}\Big(|\varphi(\epsilon X_1(\epsilon^{-2}s) - x_{\sigma(1)})|$$

$$\cdot \mathbb{1}_{\{X_1(\epsilon^{-2}s) - x_{\sigma(1)} = \ldots = X_n(\epsilon^{-2}s) - x_{\sigma(n)}\}}\Big)$$

$$\leq 2\epsilon^d C(\rho)\|\varphi\|_\infty T \sum_{\sigma \in S_n}\sum_x |\varphi(\epsilon x)|\epsilon^2$$

$$
\begin{aligned}
&\cdot \int_0^{\epsilon^{-2}T} \mathbb{E}_{x+x_1,\dots,x+x_n}\left(1\!\!1_{\{X_1(s)-x_{\sigma(1)}=\dots=X_n(s)-x_{\sigma(n)}\}}\right) ds \\
&= 2\epsilon^d C(\rho)\|\varphi\|_\infty T \sum_{\sigma\in S_n}\sum_x |\varphi(\epsilon x)|\epsilon^2 \\
&\cdot \int_0^{\epsilon^{-2}T} \mathbb{E}_{x+x_1,\dots,x+x_n}\left(1\!\!1_{\{X_i(s)-X_j(s)=X_{\sigma(i)}(0)-X_{\sigma(j)}(0)\ \forall i,j\in\{1,\dots,n\}\}}\right) ds. \qquad (12.185)
\end{aligned}
$$

Similarly as in the two particle case, one has, for all $\sigma \in S_n$

$$
\lim_{\epsilon\to 0} \epsilon^2 \int_0^{\epsilon^{-2}T} \mathbb{E}_{x+x_1,\dots,x+x_n}\left(1\!\!1_{X_i(s)-X_j(s)=X_{\sigma(i)}(0)-X_{\sigma(j)}(0)\ \forall i,j\in\{1,\dots,n\}} ds\right) = 0. \tag{12.186}
$$

Indeed, from the consistency property, one sees that a randomly selected pair $(X_i(s), X_j(s))$ of $(X_1(s), \dots, X_n(s))$ behaves as two particles, and therefore the expected local time $l_{\epsilon^{-2}T}(z)$ of the difference vector $X_i(s) - X_j(s)$ of that pair at any fixed location z is $o(\epsilon^{-2})$. □

Remark 12.35 If the difference $X_i(s)-X_j(s)$ is transient, then as in Remark 12.33, item 2, the quantity

$$
\mathbb{E}_{\nu_\rho}\left(\epsilon^{d/2}\int_0^T \sum_x \mathcal{D}_\rho(x+x_1,\dots,x+x_n;\eta(\epsilon^{-2}s))\varphi(\epsilon x)dt\right)^2 \tag{12.187}
$$

is of order ϵ^2. In the case of independent random walkers, more quantitative estimates of this quantity can be given, in terms of expected local time of a single random walk. We refer to [230] for more details.

12.6 Additional Notes

The use of duality combined with coupling between n dual particles with n independent particles is the methodology of [71] and was the first method to prove the hydrodynamic limit for the weakly asymmetric exclusion process in [73]. The method developed in [71] is based on the control of the so-called v-functions which are a measure of the deviation from local equilibrium. This method is also reminiscent of correlation function based methods in kinetic theory, e.g. in the proof of the Boltzmann equation.

At present, many methods are available in the proof of hydrodynamic limits, for a complete account of the methodologies developed by Varadhan and co-authors, as well as alternative entropy-based methods such as the method developed by Yau, and methods based on monotonicity see [151].

In this chapter we have highlighted in the simplest possible setting properties which can be obtained via duality which are mainly the time-dependent covariance of the fluctuation field, propagation of local equilibrium, higher-order hydrodynamics and quantitative versions of the Boltzmann-Gibbs principle. In [7] duality is used to obtain results for higher-order fluctuation field. In [6, 49] results are obtained on the scaling limits of the inclusion process in the condensing regime, using scaling limits of two dual particles. The behavior of condensing piles in the scaling limit is related to a system of sticky Brownian motions. The full scaling limit of the (non-stationary) density field in the coarsening process is still an open problem, which amounts to prove the scaling limit for n particles in the condensation limit.

Duality is also used to understand hydrodynamic limits of boundary driven systems, which lead to PDE's such as the heat equation with appropriate boundary conditions (Dirichlet, Robin, Neumann), depending on the strength of the coupling with the reservoirs, see e.g. [101, 207]. The study of hydrodynamic limits has been pioneered in [104] using the entropy method, and has since then been studied by various authors.

Chapter 13
Duality and Integrable Models

In this chapter we study the symmetric harmonic process, and its continuous equivalent, the integrable heat conduction model. These are processes with $\mathfrak{su}(1,1)$ symmetry which in a setting of a chain with left and right reservoirs are exactly solvable, i.e., we have close-form expressions for the moments in the non-equilibrium steady state and obtain an explicit expression of the stationary state as a mixture of product measures. We introduce both models, their algebraic structure, their dualities, and the connection with an integrable quantum spin chain with non-compact spins. The duality functions coincide with those of, respectively, the inclusion process (Chap. 5) and the Brownian energy process (Chap. 6), as they share the same underlying Lie algebra and the same symmetries. We introduce the so-called hidden parameter model which describes the evolution of product measures with equilibrium marginals. We show the Markovian structure of the hidden parameter model and provide a probabilistic proof of the structure of the non-equilibrium steady state of the model on a chain with left and right reservoirs. We also discuss an explicit formula for the rate function describing the large deviations of the density profile in the non-equilibrium steady state.

13.1 Introduction

The models that we have studied so far using duality are usually defined on a general graph $G = (V, E)$, and the proof of duality with factorized (over vertices) duality functions reduces to the corresponding property for the single edge generator. In the literature, exactly solvable models are often available for a chain geometry. E.g. the matrix product ansatz solution of the boundary-driven symmetric (and asymmetric) exclusion process [77] is available for chain geometry. In fact, the one dimensional

C. Giardinà, F. Redig, *Duality for Markov Processes*, Grundlehren der mathematischen Wissenschaften 365, https://doi.org/10.1007/978-3-032-04099-2_13

exclusion process is exactly solvable because it is related to the quantum Heisenberg spin chain, which is integrable.

The exactly solvable process for the $SU(1, 1)$ case is not immediately clear. A natural candidate is the inclusion process, but the corresponding spin chain is not integrable. Therefore, the integrable model with $SU(1, 1)$ symmetry (analogue of the exclusion process with $SU(2)$ symmetry) is not obvious. It is the purpose of this chapter to discuss these models and to analyze them with the Lie algebraic method. These models (which we call *harmonic models*) were introduced in [108] and solved in [106]. Contrary to the case of exclusion process (the quantum Heisenberg spin chain is only integrable for spin $1/2$), these models are integrable for every value of the spin (= label of the representation), and thus provide a one parameter family of exactly solvable models in the chain geometry with left and right reservoirs.

As a second important result, we will show that these harmonic models (and other similar models such as the KMP process) have an extra dual process, called the *hidden parameter model*, which describes how product measures with equilibrium marginals evolve in the course of time. These additional duals are available on any graph $G = (V, E)$. For harmonic models, in the setting of a chain with left and right reservoirs, the hidden parameter model turns out to have a (spatial) Markovian structure, which could be thought of as the probabilistic analog of its "integrability structure". Via this Markovian structure, one can obtain the non-equilibrium steady state in closed form, as a mixture of inhomogeneous product measures. The exact solvability of the harmonic models allows to rigorously establish properties that are rarely available for a non-equilibrium steady state, such as local equilibrium and a large deviation principle for the density profile with an explicit rate function.

13.2 The Symmetric Harmonic Process

We start by defining the symmetric harmonic process, which is central to this chapter. The rates of the process look, at first sight, rather peculiar. The origin of these rates will be clarified in Sect. 13.3 where the integrable spin chain behind the process will be discussed. Here we provide the probabilistic description of this rates and specify their connection to harmonic numbers.

Definition 13.1 (Symmetric Harmonic Process) Let $G = (V, E)$ be a connected graph and let $\alpha > 0$. We define the *symmetric harmonic process* on the graph G as the Markov process $\{\eta(t), t \geq 0\}$ taking values on the set $\mathbb{N}^V$ generated by

$$L = \sum_{(i,j)\in E} L_{i,j}, \tag{13.1}$$

where, for a function $f : \mathbb{N}^V \to \mathbb{R}$ we define the single-edge generator $L_{i,j}$ acting only the variables η_i and η_j as follows

$$L_{i,j} f(\eta) = \sum_{k=1}^{\eta_i} \varphi_\alpha(k, \eta_i)(f(\eta - k\delta_i + k\delta_j) - f(\eta)) \tag{13.2}$$
$$+ \sum_{k=1}^{\eta_j} \varphi_\alpha(k, \eta_j)(f(\eta + k\delta_i - k\delta_j) - f(\eta)).$$

Here the function $\varphi_\alpha : \mathbb{N} \times \mathbb{N} \to \mathbb{R}$ is defined as

$$\varphi_\alpha(k, n) = \frac{1}{k} \frac{\Gamma(n+1)\Gamma(n-k+\alpha)}{\Gamma(n+\alpha)\Gamma(n-k+1)}, \qquad 1 \le k \le n\,, \tag{13.3}$$

while $\varphi_\alpha(k, n) = 0$ when $k > n$. Here we recall that δ_i is the configuration with only one particle at site $i \in V$ and if $k \in \mathbb{N}$ then $k\delta_i$ is the configuration with k particles all at site i.

As usual, $\eta_i(t)$ is interpreted as the number of particles at vertex i at time $t \ge 0$. In the harmonic process, when a site $i \in V$ contains $\eta_i \in \mathbb{N}$ particles, k of them (with $1 \le k \le \eta_i$) are moved from site i to any neighbor $j \in V$ with a rate $\varphi_\alpha(k, \eta_i)$. We call the process "harmonic process" because the harmonic numbers naturally appear in the definition of the model. This is manifest when considering the waiting time in a configuration η before a particle jump occurs, which is an exponential random variable with parameter $\sum_{j:(i,j)\in E} h_\alpha(\eta_i)$ where

$$h_\alpha(n) := \sum_{k=1}^{n} \frac{1}{k+\alpha-1}, \tag{13.4}$$

are the so-called "shifted" harmonic numbers. Indeed, we shall prove below (see Proposition 13.3) that

$$\sum_{k=1}^{n} \varphi_\alpha(k, n) = h_\alpha(n). \tag{13.5}$$

It is useful to look at the simplest instance of the harmonic process, which is obtained by choosing $\alpha = 1$. In this case

$$\varphi_1(k, n) = \frac{1}{k}, \qquad 1 \le k \le n, \tag{13.6}$$

which means that, at any vertex, k out of the available particles jump to a neighboring site at rate $1/k$. Furthermore the holding time in a configuration is proportional to the usual harmonic number

$$\sum_{k=1}^{n} \varphi_1(k,n) = h_1(n) = \sum_{k=1}^{n} \frac{1}{k}\,. \tag{13.7}$$

Remark 13.1 (Generalizations) In Definition 13.1 we have chosen all the edge weights equal to 1 for ease of notation. The definition of the symmetric harmonic process on a weighted graph is then obtained considering the generator

$$L = \sum_{\{i,j\}\in E} p(i,j) L_{i,j}, \tag{13.8}$$

where $p : V \times V \to (0, \infty)$ is a symmetric transition function. Because the duality properties depend only on the structure of the single edge generator, all the duality results discussed in this chapter are also valid for this generalization. Furthermore, we have chosen the parameter α constant for all sites. However, the generalization with inhomogeneities $\boldsymbol{\alpha} = \{\alpha_i, i \in V\}$ is straightforward. Finally, the asymmetric version of the harmonic process on a chain will be discussed in Sect. 13.4.

In the following proposition we show that the rates of the symmetric harmonic process can be expressed in terms of an "improper" Beta-Binomial distribution with success probability given a Beta distribution of which the first parameter approaches zero (hence the name "improper") and the second parameter is fixed to $\alpha > 0$. We recall that a random variable X has a Beta-Binomial distribution with parameters (N, a, b) if

$$P(X = k) = \mathbb{E}\left(\binom{N}{k} P^k (1-P)^{N-k}\right), \qquad k \in \{0, 1, \dots, N\}, \tag{13.9}$$

where the expectation is taken over a random variable P which has a Beta(a, b) distribution, i.e., P takes values in the interval $[0, 1]$ and has a probability density

$$f_P(p) = \frac{p^{a-1}(1-p)^{b-1}}{B(a,b)}, \qquad 0 \le p \le 1, \tag{13.10}$$

where $B(a, b)$ is the Beta function, i.e.

$$B(a,b) = \int_0^1 p^{a-1}(1-p)^{b-1} dp = \frac{\Gamma(a)\Gamma(b)}{\Gamma(a+b)}. \tag{13.11}$$

Proposition 13.2 (Connection to "Improper" Beta-Binomial Distribution) *Let $L_{1,2}$ be the single-edge generator of the symmetric harmonic process*

$$L_{1,2}f(n_1,n_2) = \sum_{k=1}^{n_1} \varphi_\alpha(k,n_1)(f(n_1-k,n_2+k) - f(n_1,n_2)) + \sum_{k=1}^{n_2} \varphi_\alpha(k,n_2)(f(n_1+k,n_2-k) - f(n_1,n_2)), \quad (13.12)$$

and define also

$$L_{1,2}^{(a)} = \mathbb{E}(f(n_1-X_1,n_2+X_1) - f(n_1,n_2)) + \mathbb{E}(f(n_1+X_2,n_2-X_2) - f(n_1,n_2)), \quad (13.13)$$

where X_1 is Beta-Binomial with parameters (n_1, a, α) and X_2 is Beta-Binomial with parameters (n_2, a, α). Then we have

$$L_{1,2} = \lim_{a\to 0+} B(a,\alpha) L_{1,2}^{(a)}, \quad (13.14)$$

where $B(a, \alpha)$ is the Beta function.

Proof We have

$$\begin{aligned} &B(a,\alpha)\mathbb{E}(f(n_1-X_1,n_2+X_1) - f(n_1,n_2)) \\ &= \sum_{k=1}^{n_1} \binom{n_1}{k} \left(\int_0^1 p^{k+a-1}(1-p)^{n_1-k+\alpha-1}dp\right) \big(f(n_1-k,n_2+k) - f(n_1,n_2)\big) \\ &= \sum_{k=1}^{n_1} \binom{n_1}{k} B(k+a, n_1-k+\alpha)\big(f(n_1-k,n_2+k) - f(n_1,n_2)\big) \\ &= \sum_{k=1}^{n_1} \frac{\Gamma(n_1+1)\Gamma(k+a)\Gamma(n_1-k+\alpha)}{\Gamma(k+1)\Gamma(n_1-k+1)\Gamma(n_1+\alpha+a)} \big(f(n_1-k,n_2+k) - f(n_1,n_2)\big). \end{aligned} \quad (13.15)$$

Therefore

$$\lim_{a\to 0+} B(a,\alpha)\mathbb{E}(f(n_1-X_1,n_2+X_1) - f(n_1,n_2)) = \sum_{k=1}^{n_1} \varphi_\alpha(k,n_1)\big(f(n_1-k,n_2+k) - f(n_1,n_2)\big). \quad (13.16)$$

□

In the following proposition we prove the relation between the cumulative rates and the digamma function and, as a corollary, prove the identity (13.5).

Proposition 13.3 (Sum of Rates and Digamma Function) *We have*

$$\sum_{k=1}^{n} \varphi_\alpha(k,n) = \psi(n+\alpha) - \psi(\alpha), \tag{13.17}$$

where ψ *is the digamma function*

$$\psi(z) = \frac{d}{dz}\ln\Gamma(z) = \frac{\Gamma'(z)}{\Gamma(z)}. \tag{13.18}$$

Furthermore (13.17) *implies* (13.5).

Proof This can be shown by employing an integral representation of the digamma function. Indeed, using Proposition 13.2 we get

$$\begin{aligned}\sum_{k=1}^{n} \varphi_\alpha(k,n) &= \lim_{a\to 0+} \mathrm{Beta}(a,\alpha) \sum_{k=1}^{n} \binom{n}{k} \mathbb{E}\Big[P^k(1-P)^{n-k}\Big] \\ &= \lim_{a\to 0+} \mathrm{Beta}(a,\alpha)\mathbb{E}\big[1-(1-P)^n\big] \\ &= \int_0^1 \big(1-(1-p)^n\big)\frac{(1-p)^{\alpha-1}}{p}dp. \end{aligned} \tag{13.19}$$

If we now recall the integral representation

$$\psi(z+1) = -\gamma + \int_0^1 \frac{1-t^z}{1-t}dt\,, \tag{13.20}$$

where γ is the Euler-Mascheroni constant, we obtain the identity (13.17). The identity (13.5) then follows from the recursion relation of the diagamma function

$$\psi(z+1) = \psi(z) + \frac{1}{z}, \tag{13.21}$$

which is in turn implied by the recursion relation of the Gamma function $\Gamma(z+1) = z\Gamma(z)$. □

13.3 Integrable Spin Chains

The harmonic process was introduced in [108] as an integrable model of transport in the setting of a one dimensional chain with left and right reservoirs. To explain the genesis of the harmonic model we need to consider integrable quantum spin chains and their algebraic description. Traditionally, integrability has been used extensively in equilibrium statistical mechanics [13], as the transfer matrices of many 2D systems are related to quantum integrable systems [103]. The connection between integrability and Markovian dynamics in non-equilibrium statistical mechanics is however more recent. For an extended discussion around the concept of stochastic integrability, see [221].

As a prototype of integrable spin chains we consider the Heisenberg XXX_s Hamiltonian with spin s. We shall first recall the case of compact spins and then we discuss the case of non-compact spins. As already mentioned in Chap. 7, the symmetric exclusion process appears as the integrable stochastic jump process associated to the chain with $\mathfrak{su}(2)$ spins when $s = 1/2$. Similarly, the harmonic process introduced in the previous section emerges as the integrable stochastic jump process associated to the chain with non-compact $\mathfrak{su}(1, 1)$ spins. Remarkably, the Heisenberg XXX_s chain with non compact spins can be read as the generator of a Markov process for all spin values $s > 0$. For details about the construction and derivation of integrable quantum spin chains via the Yang-Baxter equation we refer to [90], [215].

Notation In this section, to make a connection with the physics literature about quantum spin chains, we use the so-called *bra-ket* notation. We thus denote by $|n\rangle$, where $n \in \Omega$, a standard orthonormal basis of $\mathbb{R}^\Omega$, being Ω either a finite or a countable set. In particular $\Omega = \{0, 1, \ldots, 2s\}$ with $2s \in \mathbb{N}$ when dealing with a irreducible $(2s + 1)$-dimensional representation of $\mathfrak{su}(2)$ Lie algebra, and $\Omega = \mathbb{N}_0$ for an irreducible representation of the $\mathfrak{su}(1, 1)$ Lie algebra.

In the bra-ket notation, the *ket* $|n\rangle$ is a column vector whose components are all zero expect the $(n + 1)^{\text{th}}$ component which is set to 1. Its transposed (row vector) is the *bra*, denoted by $\langle n|$ and the orthonormality relation reads

$$\langle n|m\rangle = \delta_{n,m}. \tag{13.22}$$

Furthermore an operator A working on functions $f : \Omega \to R$ as

$$(Af)(n) = \sum_{m\in\Omega} A(n, m) f(m) \tag{13.23}$$

is described by its matrix elements via

$$A(n, m) = \langle n|A|m\rangle. \tag{13.24}$$

13.3.1 The Integrable Heisenberg XXX$_s$ Hamiltonian with Compact Spins

We start from the set-up of compact spins associated to the sites of a chain of length N. On each site we consider the spin $\mathbf{S} = (S^x, S^y, S^z)$ satisfying the $\mathfrak{su}(2)$ Lie algebra:

$$[S^x, S^y] = \mathrm{i}S^z, \qquad [S^y, S^z] = \mathrm{i}S^x, \qquad [S^z, S^x] = \mathrm{i}S^y. \tag{13.25}$$

Defining

$$S^+ = S^x + \mathrm{i}S^y, \qquad S^- = S^x - \mathrm{i}S^y \tag{13.26}$$

then we have

$$[S^+, S^-] = 2S^z, \qquad [S^z, S^+] = S^+, \qquad [S^z, S^-] = -S^-, \tag{13.27}$$

which is the more familiar version of $\mathfrak{su}(2)$ algebra encountered in Chap. 7. Irreducible representations are labeled by the value of the spin $s \in \mathbb{N}/2$ and they are $(2s+1)-$dimensional. We choose to enumerate the standard orthonormal base of $\mathbb{R}^{2s+1}$ as $|0\rangle, |1\rangle, \ldots, |2s\rangle$. Then the Casimir operator

$$\mathbf{S}^2 = S^z S^z + \frac{1}{2}\left(S^+S^- + S^-S^+\right) \tag{13.28}$$

is a multiple of the identity

$$\mathbf{S}^2|n\rangle = s(s+1)|n\rangle\,. \tag{13.29}$$

A useful representation is given by

$$\begin{aligned} S^+|n\rangle &= (2s-n)|n+1\rangle, \\ S^-|n\rangle &= n|n-1\rangle, \\ S^z|n\rangle &= (n-s)|n\rangle, \end{aligned} \tag{13.30}$$

with $n \in \{0, 1, \ldots, 2s\}$. As we saw in Chap. 7 this representation allows to recognize the generator of the partial symmetric exclusion process as the co-product of the Casimir (up to a multiple of the identity).

The integrable lattice model describing a chain of N spins with spin value $s \in \mathbb{N}/2$ and free boundary condition is given by the Heisenberg XXX$_s$ Hamiltonian [90]:

$$H^{[s]} = 2\sum_{i=1}^{N-1}\Big[\psi(\mathbb{S}_{i,i+1}+1) - \psi(2s+1)\Big], \tag{13.31}$$

where the $\mathbb{S}_{i,i+1}$ is implicitely defined by the relation

$$\mathbb{S}_{i,i+1}(\mathbb{S}_{i,i+1}+1) = (\mathbf{S}_i + \mathbf{S}_{i+1})^2 . \tag{13.32}$$

Here

$$\psi(z) = \frac{d}{dz} \ln \Gamma(z) \tag{13.33}$$

is the digamma-function, i.e. the logarithmic derivative of the Γ function, which for integer values $n \in \mathbb{N}$ of its argument satisfies

$$\psi(n) = -\gamma + \sum_{k=1}^{n-1} \frac{1}{k}, \tag{13.34}$$

with γ being the Euler-Mascheroni constant.

To identify Markov processes associated with the Hamiltonian (13.31) we need to check if the non-diagonal matrix elements are all of a definite sign. In the tensor product base where for each site we use the representation (13.30) this is not an easy task. It is convenient to switch to the base where the operators $(\mathbf{S}_i + \mathbf{S}_{i+1})^2$ and $(S_i^z + S_{i+1}^z)$ are diagonal, which is obtained by using Clebsch-Gordan coefficients. In this new base, the theory of addition of angular momenta tell us that the operator $(\mathbf{S}_i + \mathbf{S}_{i+1})^2$ will have eigenvalues $\ell(\ell+1)$ with $\ell = 0, 1, \ldots, 2s$. Thus we can rewrite the Hamiltonian in (13.31) as

$$H^{[s]}_{i,i+1} = \sum_{i=1}^{N-1} \left[\alpha_0 \mathbb{1} + \alpha_1 (\mathbf{S}_i + \mathbf{S}_{i+1})^2 + \alpha_2 (\mathbf{S}_i + \mathbf{S}_{i+1})^4 + \ldots + \alpha_{2s} (\mathbf{S}_i + \mathbf{S}_{i+1})^{4s} \right], \tag{13.35}$$

where the coefficient $\alpha_0, \ldots, \alpha_{2s}$ are determined combining together (13.31), (13.32) and (13.35). We illustrate this with two examples.

Remark 13.4 (Spin $s = 1/2$) In this case we have

$$\begin{aligned} 2\big(\psi(1) - \psi(2)\big) &= \alpha_0 \\ 2\big(\psi(2) - \psi(2)\big) &= \alpha_0 + 2\alpha_1 . \end{aligned} \tag{13.36}$$

Using the recursion relation (13.21) we get

$$\begin{aligned} -2 &= \alpha_0 \\ 0 &= \alpha_0 + 2\alpha_1 , \end{aligned} \tag{13.37}$$

which implies $\alpha_0 = -2, \alpha_1 = 1$. Then the integrable Hamiltonian for spin $s = 1/2$ reads:

$$H^{[\frac{1}{2}]} = \sum_{i=1}^{N-1} \Big[-2 + (\mathbf{S}_i + \mathbf{S}_{i+1})^2 \Big] = \sum_{i=1}^{N-1} \Big[2\,\mathbf{S}_i \cdot \mathbf{S}_{i+1} - \frac{1}{2} \Big]. \tag{13.38}$$

To see the spin chain as a stochastic process one can rewrite

$$H^{[\frac{1}{2}]} = \sum_{i=1}^{N-1} \left(S_i^+ S_{i+1}^- + S_i^- S_{i+1}^+ + 2 S_i^z S_{i+1}^z - \frac{1}{2} \right), \tag{13.39}$$

and check the action of the single edge Hamiltonian. Writing $|n_i, n_{i+1}\rangle$ as an abbreviation for the tensor product base $|n_i\rangle \otimes |n_{i+1}\rangle$ and using (13.30), one has

$$\begin{aligned} H_{i,i+1}^{[\frac{1}{2}]} |n_i, n_{i+1}\rangle &= (1 - n_i) n_{i+1} \; |n_i + 1, n_{i+1} - 1\rangle \\ &+ n_i (1 - n_{i+1}) \; |n_i - 1, n_{i+1} + 1\rangle \\ &+ \left(2(n_i - \frac{1}{2})(n_{i+1} - \frac{1}{2}) - \frac{1}{2} \right) \; |n_i, n_{i+1}\rangle. \end{aligned} \tag{13.40}$$

This means that for the corresponding single edge Markov generator, which is the transposed of the single edge Hamiltonian we have the following transition rates:

$$\begin{aligned} |n_i, n_{i+1}\rangle &\to |n_i + 1, n_{i+1} - 1\rangle \quad \text{with rate} \quad (1 - n_i) n_{i+1} \\ |n_i, n_{i+1}\rangle &\to |n_i - 1, n_{i+1} + 1\rangle \quad \text{with rate} \quad n_i (1 - n_{i+1}). \end{aligned} \tag{13.41}$$

The out of diagonal terms of the matrix $H_{i,i+1}^{[\frac{1}{2}]}$ are positive and furthermore the column sums of its elements are zero, so that the transposed is the generator of a Markov process. We retrieved the symmetric exclusion process associated to the integrable spin chain of spin $1/2$.

Remark 13.5 (Spin $s = 1$) In this case we have

$$\begin{aligned} 2\big(\psi(1) - \psi(3)\big) &= \alpha_0 \\ 2\big(\psi(2) - \psi(3)\big) &= \alpha_0 + 2\alpha_1 + 4\alpha_2 \\ 2\big(\psi(3) - \psi(3)\big) &= \alpha_0 + 6\alpha_1 + 36\alpha_2. \end{aligned} \tag{13.42}$$

Using the recursion relation (13.21) we get

$$\begin{aligned} -3 &= \alpha_0 \\ -1 &= \alpha_0 + 2\alpha_1 + 4\alpha_2 \\ 0 &= \alpha_0 + 6\alpha_1 + 36\alpha_2 \end{aligned} \tag{13.43}$$

which implies $\alpha_0 = -2, \alpha_1 = +\frac{5}{4}, \alpha_2 = -\frac{1}{8}$. Then the integrable Heisenberg XXX Hamiltonian of spin value 1 reads:

$$H^{[1]} = \sum_{i=1}^{N-1} \Big[-3 + \frac{5}{4}(\mathbf{S}_i + \mathbf{S}_{i+1})^2 - \frac{1}{8}(\mathbf{S}_i + \mathbf{S}_{i+1})^4 \Big]$$
$$= \frac{1}{2} \sum_{i=1}^{N-1} \Big[\mathbf{S}_i \cdot \mathbf{S}_{i+1} - (\mathbf{S}_i \cdot \mathbf{S}_{i+1})^2 \Big]. \tag{13.44}$$

To our knowledge it is not know how to read the integrable Heisenberg XXX Hamiltonian of spin value 1 as a stochastic process. One can check that, in the tensor product base $|n_j\rangle \otimes |n_{i+1}\rangle$, the hamiltonian density $H^{[1]}_{i,i+1}$ has out-of-diagonal elements of both signs.

13.3.2 *The Heisenberg XXX$_s$ Hamiltonian with Non-compact Spins*

We change now to the set-up of $\mathfrak{su}(1,1)$ spins. On each site we consider a spin $\mathbf{S} = (S^x, S^y, S^z)$ satisfying the $\mathfrak{su}(1,1)$ Lie algebra:

$$[S^x, S^y] = -\mathrm{i}S^z, \qquad [S^y, S^z] = \mathrm{i}S^x, \qquad [S^z, S^x] = \mathrm{i}S^y. \tag{13.45}$$

Defining

$$S^+ = S^x + \mathrm{i}S^y, \qquad S^- = S^x - \mathrm{i}S^y \tag{13.46}$$

then we have

$$[S^+, S^-] = -2S^z, \qquad [S^z, S^+] = S^+, \qquad [S^z, S^-] = -S^-, \tag{13.47}$$

which is the more familiar version of $\mathfrak{su}(1,1)$ algebra encountered in Chap. 5 and in Chap. 6. Representations are labeled by the value of the pseudo-spin $s > 0$ and are infinite dimensional. We choose to enumerate the standard orthonormal base of $\ell_2(\mathbb{N})$ as $|0\rangle, |1\rangle, \ldots$ The Casimir operator

$$\mathbf{S}^2 = S^z S^z - \frac{1}{2}\Big(S^+ S^- + S^- S^+\Big) \tag{13.48}$$

is a multiple of the identity with eigenvalues

$$\mathbf{S}^2 |n\rangle = s(s-1)|n\rangle\,. \tag{13.49}$$

A useful representation is given by

$$\begin{aligned} S^+|n\rangle &= (2s+n)|n+1\rangle \\ S^-|n\rangle &= n|n-1\rangle \\ S^z|n\rangle &= (n+s)|n\rangle, \end{aligned} \tag{13.50}$$

with $n \in \{0, 1, \ldots\}$. As we saw in Chap. 5 this representation is crucial to establish the link between the Hamiltonian with density given by the co-product of the Casimir and the partial symmetric inclusion process.

The integrable lattice model describing a chain of N pseudo-spin with spin value $s > 0$ is given by the Heisenberg XXX$_s$ Hamiltonian [90]:

$$H = 2\sum_{i=1}^{N-1} \Big[\psi(\mathbb{S}_{i,i+1}) - \psi(2s)\Big], \tag{13.51}$$

where $\mathbb{S}_{i,i+1}$ is now defined by the relation

$$\mathbb{S}_{i,i+1}(\mathbb{S}_{i,i+1} - 1) = (\mathbf{S}_i + \mathbf{S}_{i+1})^2 \,. \tag{13.52}$$

As before, here $\psi(x) = \frac{d}{dx}\ln\Gamma(x)$ is the digamma-function.

It turns out that, in the tensor product base where for each site we use the representation (13.50), the transposed of this Hamiltonian is the generator of the symmetric harmonic process introduced in Sect. 13.2. To see this one has to perform a similarity transformation, using the Clebsch-Gordan coefficient, so that in the new base the operators $(\mathbf{S}_i + \mathbf{S}_{i+1})^2$ and $(S_i^z + S_{i+1}^z)$ are diagonal. This is a long computation which is not discussed here. In Sect. 13.5 we will provide an more transparent algebraic description of this Hamiltonian. This alternative form of the Hamiltonian (identified in [106]) is directly given in terms of the algebra generators (13.50) and thus it will be easier to recognize the generator of the symmetric harmonic process.

13.4 The Symmetric Harmonic Process as the Limit of Asymmetric Models

The symmetric harmonic process can be obtained as the limit of asymmetric models previously introduced in relation to the study of the Kardar-Parisi-Zhang universality class. One of them is the multiparticle hopping asymmetric diffusion model (MADM) that was introduced by Sasamoto and Wadati [206], with generator

$$\mathscr{L}^{\mathrm{MADM}} = \sum_{i\in\mathbb{Z}} \mathscr{L}_{i,i+1}, \tag{13.53}$$

with

$$\mathscr{L}_{i,i+1} f(\eta) = \sum_{k=1}^{\eta_i} \frac{1}{[k]_q} \Big[f(\eta - k\delta_i + k\delta_{i+1}) - f(\eta) \Big] \tag{13.54}$$

$$+ \sum_{k=1}^{\eta_{i+1}} \frac{q^k}{[k]_q} \Big[f(\eta + k\delta_i - k\delta_{i+1}) - f(\eta) \Big], \tag{13.55}$$

where, for $0 < q < 1$, the following definition of q-number has been used

$$[k]_q = \frac{1 - q^k}{1 - q}.$$

It is clear that the symmetric harmonic model with $\alpha = 1$ is recovered in the limit $q \to 1$, as the q-numbers reduce to ordinary numbers. A more general asymmetric model is the q-Hahn process, introduced by Barraquand and Corwin [10] as a partial asymmetric model generalizing the totally asymmetric model of Povolotsky [194]. The q-Hahn model is defined by the generator

$$\mathscr{L}^{\text{q-Hahn}} = \sum_{i \in \mathbb{Z}} \mathscr{L}_{i,i+1}, \tag{13.56}$$

with

$$\mathscr{L}_{i,i+1} f(\eta) = \sum_{k=1}^{\infty} \varphi_\alpha^{r,q,\nu}(k, \eta_i)(f(\eta - k\delta_i + k\delta_{i+1}) - f(\eta))$$

$$+ \sum_{k=1}^{\infty} \varphi_\alpha^{\ell,q,\nu}(k, \eta_{i+1})(f(\eta + k\delta_i - k\delta_{i+1}) - f(\eta)). \tag{13.57}$$

Here the functions $\varphi_\alpha^{r/\ell,q,\nu} : \mathbb{N} \times \mathbb{N} \to \mathbb{R}$ are defined as

$$\varphi_\alpha^{r,q,\nu}(k, n) = \frac{1}{[k]_q} \frac{(q;q)_n \, (\nu;q)_{n-k}}{(\nu;q)_n \, (q,q)_{n-k}} \mathbb{1}_{\{1,\dots,n\}}(k) \tag{13.58}$$

$$\varphi_\alpha^{\ell,q,\nu}(k, n) = \frac{\nu^k}{[k]_q} \frac{(q;q)_n \, (\nu;q)_{n-k}}{(\nu;q)_n \, (q,q)_{n-k}} \mathbb{1}_{\{1,\dots,n\}}(k), \tag{13.59}$$

where we have used the Pochhammer symbol

$$(\nu;q)_n = \prod_{j=0}^{n-1} (1 - \nu q^j). \tag{13.60}$$

Putting $\nu = q^\alpha$ and using the fact that

$$\lim_{q\to 1} \frac{(q^\alpha; q)_n}{(1-q)^n} = \frac{\Gamma(\alpha+n)}{\Gamma(\alpha)} \tag{13.61}$$

one obtains

$$\lim_{q\to 1} \varphi_\alpha^{r/\ell,q,q^\alpha}(k,n) = \varphi_\alpha(k,n), \tag{13.62}$$

and thus the symmetric harmonic model with parameter α is recovered from the q-Hahn model in the limit $q \to 1$. We remark that the MADM model, as well as the q-Hahn model, were introduced as models allowing an unbounded number of particles per site for which the coordinate Bethe ansatz can be used to prove properties related to the Kardar-Parisi-Zhang universality class. Their algebraic description was given in [105] in relation to the integrable XXZ$_s$ chain introduced in [37].

13.5 Algebraic Description of the Harmonic Process

Since the generator of the symmetric harmonic process is the transposed of the Hamiltonian (13.51), it follows that the symmetric harmonic process has the $\mathfrak{su}(1,1)$ symmetry. In particular the generator is a function of the co-product of the Casimir, which is a central element. As a consequence, the "total creation operator" $\sum_{i\in V} K_i^+$ and the "total annihilation operator" $\sum_{i\in V} K_i^-$ are hidden symmetries (i.e., commute with the generator), which can be used to produce duality functions.

We here consider an algebraic expression where the generator of the harmonic process on each bond of the graph is decomposed into two parts: one describing the jumps of particle in one direction (which we call right), and the other describing the jumps in the opposite direction (which we call left). This representation was identified in [106].

We recall the operators K^+, K^- and K^0 working on functions $f : \mathbb{N} \to \mathbb{R}$ as:

$$\begin{aligned} K^+ f(n) &= (\alpha + n) f(n+1), \\ K^- f(n) &= n f(n-1), \\ K^0 f(n) &= \left(\tfrac{\alpha}{2} + n\right) f(n). \end{aligned} \tag{13.63}$$

They provide a representation of the conjugate $\mathfrak{su}(1,1)$ Lie algebra. Furthermore we introduce the operator

$$B = \left(K^0 + \frac{\alpha}{2}\right)^{-1} K^+. \tag{13.64}$$

Notice that the operator K^0 is a multiplication operator represented by a diagonal matrix and therefore its inverse is well-defined. The action of B reads

$$Bf(n) = f(n+1). \tag{13.65}$$

Then we have the following result.

Proposition 13.6 (Algebraic Description of the Symmetric Harmonic Process) *Let $L_{1,2}$ be the generator of the symmetric harmonic process on two sites, i.e.*

$$\begin{aligned} L_{1,2}f(n_1,n_2) &= \sum_{k=1}^{n_1} \varphi_\alpha(k,n_1)(f(n_1-k,n_2+k) - f(n_1,n_2)) \\ &+ \sum_{k=1}^{n_2} \varphi_\alpha(k,n_2)(f(n_1+k,n_2-k) - f(n_1,n_2)), \end{aligned} \tag{13.66}$$

with

$$\varphi_\alpha(k,n) = \frac{1}{k}\frac{\Gamma(n+1)\Gamma(n-k+\alpha)}{\Gamma(n+\alpha)\Gamma(n-k+1)}, \qquad 1 \le k \le n\,. \tag{13.67}$$

Then we have

$$\begin{aligned} -L_{1,2} &= e^{K_1^- B_2}\Big(\psi\big(K_1^0 + \frac{\alpha}{2}\big) - \psi(\alpha)\Big)e^{-K_1^- B_2} \\ &+ e^{K_2^- B_1}\Big(\psi\big(K_2^0 + \frac{\alpha}{2}\big) - \psi(\alpha)\Big)e^{-K_2^- B_1}. \end{aligned} \tag{13.68}$$

Proof First, we recall the identity (13.17):

$$h_\alpha(n) = \psi(\alpha+n) - \psi(\alpha). \tag{13.69}$$

Second, using $K^- f(n) = nf(n+1)$, we have for any constant c

$$e^{cK^-} f(n) = \sum_{j=0}^{\infty} \frac{c^j}{j!}\left(K^-\right)^j f(n) = \sum_{j=0}^{n} \frac{c^j}{j!}\frac{n!}{(n-j)!} f(n-j)\,. \tag{13.70}$$

Combining the previous two equations we obtain

$$\begin{aligned} &e^{cK^-}(\psi(K^0 + \frac{\alpha}{2}) - \psi(\alpha))e^{-cK^-} f(n) \\ &= \sum_{l=0}^{\infty}\sum_{j=0}^{\infty} (-1)^l c^{j+l} \frac{n!}{j!l!(n-j-l)!} h_\alpha(n-j) f(n-j-l)\,. \end{aligned} \tag{13.71}$$

Performing the change of variables $k = j + l$ we have

$$e^{cK^-}(\psi(K^0 + \frac{\alpha}{2}) - \psi(\alpha))e^{-cK^-} f(n)$$
$$= \sum_{k=0}^{n} c^k \sum_{j=0}^{k} (-1)^{k-j} \frac{n!}{j!(k-j)!(n-k)!} h_\alpha(n-j) f(n-k) . \quad (13.72)$$

We now separate the term $k = 0$, and use the following identity when $k > 0$

$$\sum_{j=0}^{k} (-1)^{k-j} \frac{n!}{j!(k-j)!(n-k)!} h_\alpha(n-j) = -\varphi_\alpha(k,n) , \quad (13.73)$$

whose proof can be found in [106], Lemma 3.1. This allows to establish the following similarity transformation:

$$e^{cK^-}(\psi(K^0 + \frac{\alpha}{2}) - \psi(\alpha))e^{-cK^-} f(n) = h_\alpha(n) f(n) - \sum_{k=1}^{n} \varphi_\alpha(k,n) c^k f(n-k). \quad (13.74)$$

We then proceed by choosing $c = B_2$, where B_2 is the operator B defined in (13.65) working on the variable n_2. This yields

$$e^{K_1^- B_2}(\psi(K_1^0 + \frac{\alpha}{2}) - \psi(\alpha))e^{-K_1^- B_2} f(n_1, n_2)$$
$$= h_\alpha(n_1) f(n_1, n_2) - \sum_{k=1}^{n_1} \varphi_\alpha(k, n_1)(B_2)^k f(n_1 - k, n_2). \quad (13.75)$$

Using the action of the operator $(B_2)^k$

$$(B_2)^k f(n_1 - k, n_2) = f(n_1 - k, n_2 + k). \quad (13.76)$$

and recalling that $h_\alpha(n_1) = \sum_{k=1}^{n_1} \varphi_\alpha(k, n_1)$ we finally obtain

$$e^{K_1^- B_2}(\psi(K_1^0 + \frac{\alpha}{2}) - \psi(\alpha))e^{-K_1^- B_2} f(n_1, n_2)$$
$$= -\sum_{k=1}^{n_1} \varphi_\alpha(k, n_1)\Big(f(n_1 - k, n_2 + k) - f(n_1, n_2)\Big). \quad (13.77)$$

By an analogous computation, one proves that

$$e^{K_2^- B_1}(\psi(K_2^0 + \frac{\alpha}{2}) - \psi(\alpha))e^{-K_2^- B_1} f(n_1, n_2)$$
$$= -\sum_{k=1}^{n_2} \varphi_\alpha(k, n_2)\Big(f(n_1 + k, n_2 - k) - f(n_1, n_2)\Big). \tag{13.78}$$

Adding up the last two equations the proof is concluded. □

13.6 Reversible Measure and Self-duality of the Harmonic Process

As a consequence of the $\mathfrak{su}(1, 1)$ symmetry, one expects for the symmetric harmonic process the same self-duality results that were established for several other processes sharing the same symmetry. This is indeed the case, as the following theorem shows.

Theorem 13.7 (Reversible Measure and Self-duality) *Consider the symmetric harmonic process with parameter $\alpha > 0$ on a graph $G = (V, E)$, defined by the generator* (13.1)*. Then we have:*

1. *For all $\lambda > 0$ there exists a one-parameter family of reversible distributions, which are the product measure with marginal the (unnormalized) Negative Binomial distribution:*

$$M(\eta) = \prod_{x \in V} m(\eta_i) \qquad \textit{with} \qquad m(n) = \frac{\lambda^n}{n!} \frac{\Gamma(n + \alpha)}{\Gamma(\alpha)}. \tag{13.79}$$

2. *The function*

$$D^{\mathrm{ch}}(\xi, \eta) = \prod_{x \in V} d^{\mathrm{ch}}(\xi_x, \eta_x), \qquad \textit{with} \qquad d^{\mathrm{ch}}(k, n) = \frac{n!\Gamma(\alpha)}{\Gamma(\alpha + n)} \delta_{k,n} \tag{13.80}$$

is a cheap self-duality function.

3. *The function*

$$D(\xi, \eta) = \prod_{i \in V} d(\xi_i, \eta_i) \qquad \textit{with} \qquad d(k, n) = \frac{n!}{(n-k)!} \frac{\Gamma(\alpha)}{\Gamma(\alpha + k)} \mathbb{1}_{\{k \le n\}} \tag{13.81}$$

is a triangular self-duality function.

Proof For the first item, the detailed balance condition is easily verified from the fact that, for all all $n, \ell \in \mathbb{N}$ and for all $1 \le k \le n, \ell$, one has

$$m(n)m(\ell)\varphi_\alpha(k,n) = m(n-k)m(\ell+k)\varphi_\alpha(k,\ell+k)\,. \tag{13.82}$$

As a consequence the cheap self-duality function of the second item follows from Theorem 2.9.

For the last item we provide two proofs. The first is similar to the one of Theorem 5.8. Namely, one verifies that $\sum_{i\in V} K_i^+$ is a symmetry of the generator (see [106] for the computation of the commutator of the generator with $\sum_{i\in V} K_i^+$) and then uses that

$$e^{K^+}[d^{\mathrm{ch}}(\cdot,n)](k) = d(k,n), \tag{13.83}$$

which is part of the proof of Theorem 5.8. An alternative (self-contained) proof of self-duality is obtained by following the idea of a change of representation from the conjugate $\mathfrak{su}(1,1)$ Lie algebra to the $\mathfrak{su}(1,1)$ Lie algebra, as described in Sect. 5.5 for the symmetric inclusion process. Indeed one has

$$\begin{aligned} K^+ &\overset{d^{\mathrm{ch}}}{\longrightarrow} K^- \\ K^- &\overset{d^{\mathrm{ch}}}{\longrightarrow} K^+ \\ K^0 &\overset{d^{\mathrm{ch}}}{\longrightarrow} K^0, \end{aligned} \tag{13.84}$$

which means that

$$\begin{aligned} \left[K^+ d^{\mathrm{ch}}(\cdot,n)\right](k) &= \left[K^- d^{\mathrm{ch}}(k,\cdot)\right](n) \\ \left[K^- d^{\mathrm{ch}}(\cdot,n)\right](k) &= \left[K^+ d^{\mathrm{ch}}(k,\cdot)\right](n) \\ \left[K^0 d^{\mathrm{ch}}(\cdot,n)\right](k) &= \left[K^0 d^{\mathrm{ch}}(k,\cdot)\right](n). \end{aligned} \tag{13.85}$$

As a consequence of this, when we consider the algebraic description (13.68) of the harmonic process generator we have (remember that we need to write the "sequence in the reverse order")

$$\begin{aligned} &e^{K_1^-(K_2^0+\frac{\alpha}{2})^{-1}K_2^+}\Big(\psi\big(K_1^0+\frac{\alpha}{2}\big)-\psi(\alpha)\Big)e^{-K_1^-(K_2^0+\frac{\alpha}{2})^{-1}K_2^+} \\ &+e^{K_2^-(K_1^0+\frac{\alpha}{2})^{-1}K_1^+}\Big(\psi\big(K_2^0+\frac{\alpha}{2}\big)-\psi(\alpha)\Big)e^{-K_2^-(K_1^0+\frac{\alpha}{2})^{-1}K_1^+} \end{aligned}$$

$$\xrightarrow{D^{\mathrm{ch}}} e^{-K_1^+ K_2^- (K_2^0+\frac{\alpha}{2})^{-1}} \Big(\psi\big(K_1^0 + \frac{\alpha}{2}\big) - \psi(\alpha)\Big) e^{K_1^+ K_2^- (K_2^0+\frac{\alpha}{2})^{-1}}$$
$$+ e^{-K_2^+ K_1^- (K_1^0+\frac{\alpha}{2})^{-1}} \Big(\psi\big(K_2^0 + \frac{\alpha}{2}\big) - \psi(\alpha)\Big) e^{K_2^+ K_1^- (K_1^0+\frac{\alpha}{2})^{-1}}. \tag{13.86}$$

The self-duality of the harmonic process with cheap self-duality function is then obtained by observing that the term on the right hand side is again the generator of the symmetric harmonic process, i.e.

$$-L_{1,2} = e^{-K_1^+ K_2^- (K_2^0+\frac{\alpha}{2})^{-1}} \Big(\psi\big(K_1^0 + \frac{\alpha}{2}\big) - \psi(\alpha)\Big) e^{K_1^+ K_2^- (K_2^0+\frac{\alpha}{2})^{-1}}$$
$$+ e^{-K_2^+ K_1^- (K_1^0+\frac{\alpha}{2})^{-1}} \Big(\psi\big(K_2^0 + \frac{\alpha}{2}\big) - \psi(\alpha)\Big) e^{K_2^+ K_1^- (K_1^0+\frac{\alpha}{2})^{-1}}. \tag{13.87}$$

Remark, however, that this identification is the result of a subtle cancellation which originates from the following two equalities:

$$e^{-K_1^+ K_2^- (K_1^0+\frac{\alpha}{2})^{-1}} \Big(\psi\big(K_2^0 + \frac{\alpha}{2}\big) - \psi(\alpha)\Big) e^{K_1^+ K_2^- (K_2^0+\frac{\alpha}{2})^{-1}}$$
$$= -\sum_{k=1}^{n_2} \varphi_\alpha(k, n_2)\Big(f(n_1 + k, n_2 - k) - f(n_1, n_2)\Big)$$
$$+(h_\alpha(n_1) - h_\alpha(n_2)) f(n_1, n_2), \tag{13.88}$$

and

$$e^{-K_2^+ K_1^- (K_1^0+\frac{\alpha}{2})^{-1}} \Big(\psi\big(K_2^0 + \frac{\alpha}{2}\big) - \psi(\alpha)\Big) e^{K_2^+ K_1^- (K_1^0+\frac{\alpha}{2})^{-1}}$$
$$= -\sum_{k=1}^{n_1} \varphi_\alpha(k, n_1)\Big(f(n_1 - k, n_2 + k) - f(n_1, n_2)\Big)$$
$$+(h_\alpha(n_2) - h_\alpha(n_1)) f(n_1, n_2). \tag{13.89}$$

Adding up (13.88) and (13.89) leads to (13.87).

The proof of (13.88) (and similarly of (13.89)) is obtained by a computation that follows the same ideas of Sect. 13.5 and reveals the additional contribution of the diagonal terms. One uses that

$$h_\alpha(n) = \psi(\alpha + n) - \psi(\alpha) \tag{13.90}$$

and

$$e^{cK^+} f(n) = \sum_{j=0}^{\infty} \frac{c^j}{j!} \left(K^+\right)^j f(n) = \sum_{j=0}^{\infty} \frac{c^j}{j!} \frac{\Gamma(n+\alpha+j)}{\Gamma(n+\alpha)} f(n+j) \,. \tag{13.91}$$

Combining together these two equations we get

$$\begin{aligned} & e^{-cK^+} (\psi(K^0 + \frac{\alpha}{2}) - \psi(\alpha)) e^{cK^+} f(n) \\ & = \sum_{l=0}^{\infty} \sum_{j=0}^{\infty} (-1)^l \frac{c^{j+l}}{j! l!} \frac{\Gamma(n+\alpha+j+l)}{\Gamma(n+\alpha)} h_\alpha(n+l) f(n+j+l) \,. \end{aligned} \tag{13.92}$$

Performing the change of variables $k = j + l$ and then changing the order of summation we have

$$\begin{aligned} & e^{-cK^+} (\psi(K^0 + \frac{\alpha}{2}) - \psi(\alpha)) e^{cK^+} f(n) \\ & = \sum_{k=0}^{\infty} c^k \sum_{l=0}^{k} \frac{(-1)^l}{l!(k-l)!} \frac{\Gamma(n+\alpha+k)}{\Gamma(n+\alpha)} h_\alpha(n+l) f(n+k) \,. \end{aligned} \tag{13.93}$$

We now separate the term $k = 0$, and use the following identity when $k > 0$

$$\sum_{l=0}^{k} \frac{(-1)^l}{l!(k-l)!} \frac{\Gamma(n+\alpha+k)}{\Gamma(n+\alpha)} h_\alpha(n+l) = -\frac{1}{k} \,, \tag{13.94}$$

whose proof can be found in [106], Lemma 3.2. Thus we establish the following similarity transformation:

$$e^{-cK^+} \Big(\psi(K^0 + \frac{\alpha}{2}) - \psi(\alpha)\Big) e^{cK^+} f(n) = h_\alpha(n) f(n) - \sum_{k=1}^{n} \frac{c^k}{k} f(n+k). \tag{13.95}$$

Applying this with $c = K_2^-(K_2^0 + \frac{\alpha}{2})^{-1}$ we then have

$$\begin{aligned} & e^{-K_1^+ K_2^- (K_2^0 + \frac{\alpha}{2})^{-1}} \Big(\psi\big(K_1^0 + \frac{\alpha}{2}\big) - \psi(\alpha)\Big) e^{K_1^+ K_2^- (K_2^0 + \frac{\alpha}{2})^{-1}} f(n_1, n_2) \\ & = \sum_{k=1}^{\infty} \frac{1}{k} \Big(K_2^- (K_2^0 + \frac{\alpha}{2})^{-1}\Big)^k f(n_1 + k, n_2) - h_\alpha(n_1) f(n_1, n_2). \end{aligned} \tag{13.96}$$

Observing that

$$\left(K^-(K^0+\frac{\alpha}{2})^{-1}\right) f(n) = \frac{n}{n+\alpha-1} f(n-1), \tag{13.97}$$

it follows

$$\begin{aligned}&\left(K_2^-(K_2^0+\frac{\alpha}{2})^{-1}\right)^k f(n_1+k, n_2)\\ &= \frac{\Gamma(n_2+1)}{\Gamma(n_2-k+1)} \frac{\Gamma(n_2-k+\alpha)}{\Gamma(n_2+\alpha)} f(n_1+k, n_2-k).\end{aligned} \tag{13.98}$$

Inserting this above and recalling the definition of $\varphi_\alpha(k,n) = \frac{1}{k}\frac{\Gamma(n+1)\Gamma(n-k+\alpha)}{\Gamma(n+\alpha)\Gamma(n-k+1)}$, we finally obtain

$$\begin{aligned}&-e^{-K_1^+K_2^-(K_2^0+\frac{\alpha}{2})^{-1}}\left(\psi(K_1^0+\frac{\alpha}{2})-\psi(\alpha)\right)e^{K_1^+K_2^-(K_2^0+\frac{\alpha}{2})^{-1}}\\ &=\sum_{k=1}^{n_2}\varphi_\alpha(k,n_2)\Big(f(n_1+k,n_2-k)-f(n_1,n_2)\Big)\\ &+h_\alpha(n_2)f(n_1,n_2)-h_\alpha(n_1)f(n_1,n_2).\end{aligned} \tag{13.99}$$

Thus the proof of (13.88) is completed. □

13.7 Integrable Heat Conduction Model

In Chap. 6, the Brownian energy process was discussed. The Brownian energy process is a diffusion model of heat conduction, i.e. a continuous quantity (the heat or thermal energy) diffuses and is redistributed among vertices. We proved that we can obtain the Brownian energy process from the symmetric inclusion process by taking a many-particle limit. From the algebraic point of view, this corresponds to considering a representation of non-compact $\mathfrak{su}(1,1)$ spins in terms of differential operators. The change from discrete to continuous representations naturally leads to a duality between the two processes.

Similarly, the harmonic process introduced at the beginning of this chapter does have a many-particle limit, which provides an *integrable* model of heat conduction. However, the generator of the limiting process does not involve only first and second derivatives, but rather contains "derivatives of all orders". In fact we shall see that it is a jump process with infinite activity. In this section we first introduce the process and then we show its connection to the symmetric harmonic process by a many-particle scaling limit. In the next section we shall discuss the relation between the integrable heat conduction process and the harmonic process from the perspective of representation theory.

Definition 13.8 (Integrable Heat Conduction Model) Consider a connected graph $G = (V, E)$ and let $\alpha > 0$. We define the *integrable heat conduction process* on the graph G as the Markov process $\{\zeta(t), t \geq 0\}$ taking values on the set $\mathbb{R}_+^V$ generated by

$$\mathscr{L} = \sum_{\{i,j\} \in E} \mathscr{L}_{i,j} \tag{13.100}$$

where, for smooth functions $f : \mathbb{R}_+^V \to \mathbb{R}$ we have

$$\begin{aligned}\mathscr{L}_{i,j} f(\zeta) &= \int_0^{\zeta_i} \frac{dx}{x} \left(1 - \frac{x}{\zeta_i}\right)^{\alpha-1} \left[f(\zeta - x\delta_i + x\delta_j) - f(\zeta)\right] \\ &+ \int_0^{\zeta_j} \frac{dx}{x} \left(1 - \frac{x}{\zeta_j}\right)^{\alpha-1} \left[f(\zeta + x\delta_i - x\delta_j) - f(\zeta)\right]. \end{aligned} \tag{13.101}$$

We interpret $\zeta_i(t)$ as the energy at vertex i at time $t \geq 0$ and observe that the total energy $|\zeta(t)| = \sum_{i \in V} \zeta_i(t)$ is obviously conserved by the dynamics. The fact that the operator (13.101) generates a Markov process can be established by using Hille-Yosida theory. This approach first defines this operator on polynomials, which form a core. The generator of the process is then obtained as the closure of its action on the set of polynomials. A summary of the Hille-Yosida theory can be found in Appendix A.

Remark 13.9 (Connection to "Improper" Beta Distribution) An alternative form of the generator reads

$$\begin{aligned}\mathscr{L}_{i,j} f(\zeta) &= \int_0^{1} \frac{du}{u} (1-u)^{\alpha-1} \left[f(\zeta - \zeta_i u\delta_i + \zeta_i u\delta_j) - f(\zeta)\right] \\ &+ \int_0^{1} \frac{du}{u} (1-u)^{\alpha-1} \left[f(\zeta + \zeta_j u\delta_i - \zeta_j u\delta_j) - f(\zeta)\right]. \end{aligned} \tag{13.102}$$

We further define

$$\begin{aligned}\mathscr{L}^a_{i,j} f(\zeta) &= \mathbb{E}\left[f(\zeta - \zeta_i U\delta_i + \zeta_i U\delta_j) - f(\zeta)\right] \\ &+ \mathbb{E}\left[f(\zeta + \zeta_j U\delta_i - \zeta_j U\delta_j) - f(\zeta)\right], \end{aligned} \tag{13.103}$$

where U is a random variable with a Beta distribution with parameters (a, α), and $\mathbb{E}$ is expectation w.r.t. U. We then have the analogue of Proposition 13.2, namely

$$\lim_{a \to 0} B(a, \alpha) \mathscr{L}^a_{i,j} = \mathscr{L}_{i,j}. \tag{13.104}$$

This follows from

$$\lim_{a\to 0} B(a,\alpha) f_{a,\alpha}(u) = \frac{1}{u}(1-u)^{\alpha-1}, \tag{13.105}$$

where $B(a,\alpha)$ is the Beta function and where $f_{a,\alpha}(u) = \frac{1}{B(a,\alpha)} u^{a-1}(1-u)^{\alpha-1}$ is the probability density of a Beta random variable with parameters (a,α).

We now show that the integrable heat conduction process is a many-particle limit of the harmonic process.

Proposition 13.10 (Many-particle Limit of the Symmetric Harmonic Process) *Define the process* $\{\zeta^{(N)}(t), t \geq 0\}$ *by*

$$\zeta^{(N)}(t) = \frac{\eta(t)}{N}, \tag{13.106}$$

where $\{\eta(t), t \geq 0\}$ *is the symmetric harmonic process. Assume* $\zeta^{(N)} := \zeta^{(N)}(0)$ *converges to* $\zeta \in \mathbb{R}_+^V$ *as* $N \to \infty$*. Then the sequence of processes* $\{\zeta^{(N)}(t), t \geq 0\}$ *converges to the integrable heat conduction process with generator* (13.100)*, starting from the configuration* $\zeta(0) = \zeta$.

Proof Denote by $L^{(N)}$ the generator of the process $\{\zeta^{(N)}(t), t \geq 0\}$. Then in view of the Trotter-Kurtz theorem, it suffices to prove that for a smooth test function $f : \mathbb{R}_+^V \to \mathbb{R}$ we have the convergence

$$L^N f(\zeta^N) \to \mathscr{L} f(\zeta), \tag{13.107}$$

as $N \to \infty$ where $\mathscr{L}$ is the generator of the integrable heat conduction process, and where ζ^N is a sequence of configurations belonging to $\frac{1}{N}\mathbb{N}^V$ which converges to $\zeta \in [0,\infty)^V$. This in turn reduces to

$$L^N_{i,j} f(\zeta^N) \to \mathscr{L}_{i,j} f(\zeta), \tag{13.108}$$

for all $\{i,j\} \in E$. We give the proof for the case $\alpha = 1$. We have

$$\begin{aligned} L^N_{i,j} f(\zeta^{(N)}) &= \sum_{k=1}^{N\zeta_i^{(N)}} \frac{1}{N}\frac{1}{k/N}\left(f(\zeta^{(N)} - \tfrac{k}{N}\delta_i + \tfrac{k}{N}\delta_j) - f(\zeta^{(N)})\right) \\ &+ \sum_{k=1}^{N\zeta_j^{(N)}} \frac{1}{N}\frac{1}{k/N}\left(f(\zeta^{(N)} - \tfrac{k}{N}\delta_i + \tfrac{k}{N}\delta_j) - f(\zeta^{(N)})\right). \end{aligned} \tag{13.109}$$

The rhs of (13.109) is a Riemann sum approximation of an integral. Using that $\lim_{N\to\infty} \zeta^N = \zeta$ we get

$$\lim_{N\to\infty} L^N_{i,j} f(\zeta^{(N)}) = \int_0^{\zeta_i} \frac{dx}{x} \left[f(\zeta - x\delta_i + x\delta_j) - f(\zeta)\right] + \int_0^{\zeta_j} \frac{dx}{x} \left[f(\zeta - x\delta_j + x\delta_i) - f(\zeta)\right]. \quad (13.110)$$

The general case $\alpha \neq 1$ is proved similarly via Riemann sum approximation and by using the asymptotics

$$\frac{\Gamma(z+\gamma_1)}{\Gamma(z+\gamma_2)} \approx z^{\gamma_1-\gamma_2} \qquad \text{as} \quad z \to \infty. \quad (13.111)$$

□

Remark 13.11 An alternative proof is obtained by using the expression of the rates of the symmetric harmonic process via 'improper' Beta-Binomial distributions of Proposition 13.2 and using the convergence in distribution

$$\frac{X_{N,a,\alpha}}{N} \to B_{a,\alpha} \qquad \text{as} \qquad N \to \infty, \quad (13.112)$$

where $X_{N,a,\alpha}$ is a Beta-Binomial random variable with parameters (N, a, α) and $B_{a,\alpha}$ is a Beta random variable with parameters (a, α).

13.8 Duality of the Integrable Heat Conduction Model

In order to establish duality between the integrable heat conduction model and the symmetric harmonic process, we first provide the algebraic description of the integrable heat conduction model. By considering a representation of the $\mathfrak{su}(1,1)$ Lie algebra with differential operators we show that the integrable heat conduction process is obtained from the Hamiltonian of the integrable XXX$_s$ chain with non-compact spins.

Proposition 13.12 (Algebraic Description of the Integrable Heat Conduction Model) *Let $\mathscr{L}_{1,2}$ be the generator of the harmonic process on two sites, i.e.*

$$\mathscr{L}_{1,2} f(\zeta_1,\zeta_2) = \int_0^{\zeta_1} \frac{dx}{x}\left(1-\frac{x}{\zeta_1}\right)^{\alpha-1} [f(\zeta_1 - x, \zeta_2 + x) - f(\zeta_1,\zeta_2)] + \int_0^{\zeta_2} \frac{dx}{x}\left(1-\frac{x}{\zeta_2}\right)^{\alpha-1} [f(\zeta_1 + x, \zeta_2 - x) - f(\zeta_1,\zeta_2)]. \quad (13.113)$$

Then we have

$$-\mathscr{L}_{1,2} = e^{\mathscr{K}_1^+\mathscr{B}_2}\Big(\psi\big(\mathscr{K}_1^0 + \frac{\alpha}{2}\big) - \psi(\alpha)\Big)e^{-\mathscr{K}_1^+\mathscr{B}_2} + e^{\mathscr{K}_2^+\mathscr{B}_1}\Big(\psi\big(\mathscr{K}_2^0 + \frac{\alpha}{2}\big) - \psi(\alpha)\Big)e^{-\mathscr{K}_2^+\mathscr{B}_1}, \tag{13.114}$$

where the following representation of the $\mathfrak{su}(1,1)$ *Lie algebra is used:*

$$\begin{aligned}\mathscr{K}^+ f(z) &= zf(z)\\ \mathscr{K}^- f(z) &= \left(z\frac{\partial}{\partial z} + \alpha\right)\frac{\partial}{\partial z} f(z)\\ \mathscr{K}^0 f(z) &= \left(z\frac{\partial}{\partial z} + \frac{\alpha}{2}\right) f(z),\end{aligned} \tag{13.115}$$

and the operator $\mathscr{B}$ *is given by*

$$\mathscr{B} = (\mathscr{K}^0 + \frac{\alpha}{2})^{-1}\mathscr{K}^- = \frac{\partial}{\partial z}. \tag{13.116}$$

Proof It is enough to consider the "right" part of the generator, the "left" part is treated in an analogous manner. Using (13.115) and (13.116) we have

$$\begin{aligned}&e^{\mathscr{K}_1^+\mathscr{B}_2}\Big(\psi\big(\mathscr{K}_1^0 + \frac{\alpha}{2}\big) - \psi(\alpha)\Big)e^{-\mathscr{K}_1^+\mathscr{B}_2}\\ &= e^{\zeta_1\frac{\partial}{\partial\zeta_2}}\Big(\psi\big(\zeta_1\frac{\partial}{\partial\zeta_1} + \alpha\big) - \psi(\alpha)\Big)e^{-\zeta_1\frac{\partial}{\partial\zeta_2}}.\end{aligned} \tag{13.117}$$

By using the Euler integral representation of the digamma function (see (13.20)) we write

$$\psi(z+\alpha) - \psi(\alpha) = \int_0^1 d\beta\, \beta^{\alpha-1}\frac{1-\beta^z}{1-\beta}, \tag{13.118}$$

so that we obtain

$$\begin{aligned}&e^{\mathscr{K}_1^+\mathscr{B}_2}\Big(\psi\big(\mathscr{K}_1^0 + \frac{\alpha}{2}\big) - \psi(\alpha)\Big)e^{-\mathscr{K}_1^+\mathscr{B}_2}\\ &= e^{\zeta_1\frac{\partial}{\partial\zeta_2}}\int_0^1 \frac{d\beta}{1-\beta}\beta^{\alpha-1}\Big(1 - \beta^{\zeta_1\frac{\partial}{\partial\zeta_1}}\Big)e^{-\zeta_1\frac{\partial}{\partial\zeta_2}}.\end{aligned} \tag{13.119}$$

Moreover we have the following expressions of translations and dilatations

$$f(z+c) = e^{c\frac{\partial}{\partial z}} f(z), \tag{13.120}$$

$$f(cz) = c^{z\frac{\partial}{\partial z}} f(z). \tag{13.121}$$

By this "formal" notation we mean the following. For a first order differential operator A we define

$$e^{tA} f(z) = g(z,t), \tag{13.122}$$

where $g(z,t)$ is the solution of

$$\frac{\partial g(z,t)}{\partial t} = Ag(\cdot,t)(z), \tag{13.123}$$

with initial condition $g(z,0) = f(z)$. For the operator $A = \frac{\partial}{\partial z}$ we get $g(z,t) = f(z+t)$ and for $A = z\frac{\partial}{\partial z}$ we get $g(z,t) = f(ze^t)$, which gives (13.120) and (13.121) (in the latter $t = \ln(c)$).

Using this we obtain

$$\begin{aligned} &e^{\mathscr{K}_1^+ \mathscr{B}_2}\Big(\psi\big(\mathscr{K}_1^0 + \frac{\alpha}{2}\big) - \psi(\alpha)\Big)e^{-\mathscr{K}_1^+ \mathscr{B}_2} f(\zeta_1,\zeta_2) \\ &= \int_0^1 \frac{d\beta}{1-\beta}\beta^{\alpha-1}\Big(f(\zeta_1,\zeta_2) - f(\beta\zeta_1, \zeta_2 + (1-\beta)\zeta_1)\Big). \end{aligned} \tag{13.124}$$

Changing variable by defineing $x = (1-\beta)\zeta_1$ we have

$$\begin{aligned} &e^{\mathscr{K}_1^+ \mathscr{B}_2}\Big(\psi\big(\mathscr{K}_1^0 + \frac{\alpha}{2}\big) - \psi(\alpha)\Big)e^{-\mathscr{K}_1^+ \mathscr{B}_2} f(\zeta_1,\zeta_2) \\ &= -\int_0^{\zeta_1} \frac{dx}{x}\Big(1-\frac{x}{\zeta_1}\Big)^{\alpha-1}\Big(f(\zeta_1 - x, \zeta_2 + x) - f(\zeta_1,\zeta_2)\Big). \end{aligned} \tag{13.125}$$

By an analogous computation, one proves that

$$\begin{aligned} &e^{\mathscr{K}_2^+ \mathscr{B}_1}\Big(\psi\big(\mathscr{K}_2^0 + \frac{\alpha}{2}\big) - \psi(\alpha)\Big)e^{-\mathscr{K}_2^+ \mathscr{B}_1} f(\zeta_1,\zeta_2) \\ &= -\int_0^{\zeta_2} \frac{dx}{x}\Big(1-\frac{x}{\zeta_2}\Big)^{\alpha-1}\Big(f(\zeta_1 + x, \zeta_2 - x) - f(\zeta_1,\zeta_2)\Big). \end{aligned} \tag{13.126}$$

Adding up the last two equations the proof is concluded. □

Having established the algebraic description (13.114), it is an easy consequence to establish duality.

Theorem 13.13 (Duality Between Integrable Heat Conduction Process and Symmetric Harmonic Process) *The integrable heat conduction process with parameter $\alpha > 0$ on a graph $G = (V, E)$, defined by the generator* (13.100) *is dual to the symmetric harmonic process with duality functions given by*

$$D(\xi, \zeta) = \prod_{i \in V} d(\xi_i, \zeta_i) \qquad \text{with} \qquad d(k, z) = z^k \frac{\Gamma(\alpha)}{\Gamma(\alpha + k)} . \tag{13.127}$$

Proof The proof is analogous to the one of Theorem 6.2, where the following basic duality for the $\mathfrak{su}(1,1)$ Lie algebra was first obtained:

$$\begin{aligned} K^+ &\xrightarrow{d} \mathscr{K}^+ \\ K^- &\xrightarrow{d} \mathscr{K}^- \\ K^0 &\xrightarrow{d} \mathscr{K}^0 . \end{aligned} \tag{13.128}$$

The duality relation

$$\mathscr{L}_{1,2} \xrightarrow{D} L_{1,2} \tag{13.129}$$

then follows combining Eq. (13.114) (which provides the algebraic description of the integrable heat conduction model in terms of $\mathscr{K}$ operators) and Eq. (13.87) (which provides the algebraic description of the symmetric harmonic process in terms of K operators). More explicitly

$$\begin{aligned} & e^{\mathscr{K}_1^+ (\mathscr{K}_2^0 + \frac{\alpha}{2})^{-1} \mathscr{K}_2^-} \Big(\psi\big(\mathscr{K}_1^0 + \frac{\alpha}{2}\big) - \psi(\alpha) \Big) e^{-\mathscr{K}_1^+ (\mathscr{K}_2^0 + \frac{\alpha}{2})^{-1} \mathscr{K}_2^-} \\ & + e^{\mathscr{K}_2^+ (\mathscr{K}_1^0 + \frac{\alpha}{2})^{-1} \mathscr{K}_1^-} \Big(\psi\big(\mathscr{K}_2^0 + \frac{\alpha}{2}\big) - \psi(\alpha) \Big) e^{-\mathscr{K}_2^+ (\mathscr{K}_1^0 + \frac{\alpha}{2})^{-1} \mathscr{K}_1^-} \\ & \quad \xrightarrow{D} e^{-K_1^+ K_2^- (K_2^0 + \frac{\alpha}{2})^{-1}} \Big(\psi\big(K_1^0 + \frac{\alpha}{2}\big) - \psi(\alpha) \Big) e^{K_1^+ K_2^- (K_2^0 + \frac{\alpha}{2})^{-1}} \\ & \qquad + e^{-K_2^+ K_1^- (K_1^0 + \frac{\alpha}{2})^{-1}} \Big(\psi\big(K_2^0 + \frac{\alpha}{2}\big) - \psi(\alpha) \Big) e^{K_2^+ K_1^- (K_1^0 + \frac{\alpha}{2})^{-1}} . \end{aligned} \tag{13.130}$$

□

Remark 13.14 (Poisson Intertwining) As a consequence of the duality between the symmetric harmonic process and the integrable heat conduction process we have the Poisson intertwiner, as in Theorem 6.8. More precisely,

$$\Lambda L = \mathscr{L} \Lambda, \tag{13.131}$$

with

$$(\Lambda f)(\zeta) = \sum_{\eta \in \mathbb{N}^V} f(\eta) \prod_{x \in V} \frac{\zeta_x^{\eta_x}}{\eta_x!} e^{-\zeta_x}. \tag{13.132}$$

Therefore evolving a Poisson product measure with parameters ζ_i at site i under the symmetric harmonic process yields at time $t > 0$ a mixture of Poisson product measures where the mixing measure is given by the probability distribution of the integrable heat conduction model at time t started from $\zeta = \{\zeta_i, i \in V\}$.

13.9 The Hidden Parameter Model and Propagation of Mixtures

In this section we will prove that, besides the heat conduction model of Sect. 13.7, the harmonic model has another dual process, the *hidden parameter model*. This novel process will be associated to the time evolution of Negative Binomial product measures and it will be particularly useful in the non-equilibrium set-up discussed in Sect. 13.10.

Definition 13.15 (Hidden Parameter Process) Consider a connected graph $G = (V, E)$ and let $\alpha > 0$. We define the *hidden parameter process* on the graph G as the Markov process $\{\theta(t), t \geq 0\}$ taking values on the set $\mathbb{R}_+^V$ generated by

$$\mathcal{L} = \sum_{\{i,j\} \in E} \mathcal{L}_{i,j}, \tag{13.133}$$

where, for functions $f : \mathbb{R}_+^V \to \mathbb{R}$ we set

$$\begin{aligned} \mathcal{L}_{i,j} f(\theta) &= \int_0^1 \frac{du}{u} (1-u)^{\alpha-1} \left[f(\theta - u\theta_i \delta_i + u\theta_j \delta_i) - f(\theta) \right] \\ &+ \int_0^1 \frac{du}{u} (1-u)^{\alpha-1} \left[f(\theta - u\theta_j \delta_j + u\theta_i \delta_j) - f(\theta) \right]. \end{aligned} \tag{13.134}$$

Proposition 13.16 (Algebraic Description of the Hidden Parameter Model) *Let $\mathcal{L}_{1,2}$ be the generator of the hidden parameter process on two sites, i.e.*

$$\begin{aligned} \mathcal{L}_{1,2} f(\theta_1, \theta_2) &= \int_0^1 \frac{du}{u} (1-u)^{\alpha-1} \left[f((1-u)\theta_1 + u\theta_2, \theta_2) - f(\theta_1, \theta_2) \right] \\ &+ \int_0^1 \frac{du}{u} (1-u)^{\alpha-1} \left[f(\theta_1, u\theta_1 + (1-u)\theta_2) - f(\theta_1, \theta_2) \right]. \end{aligned} \tag{13.135}$$

Then we have

$$-\mathcal{L}_{1,2} = e^{-\mathcal{K}_1^-\mathcal{K}_2^+(\mathcal{K}_2^0+\frac{\alpha}{2})^{-1}}\Big(\psi\big(\mathcal{K}_1^0+\frac{\alpha}{2}\big)-\psi(\alpha)\Big)e^{\mathcal{K}_1^-\mathcal{K}_2^+(\mathcal{K}_2^0+\frac{\alpha}{2})^{-1}}$$
$$+\,e^{-\mathcal{K}_2^-\mathcal{K}_1^+(\mathcal{K}_1^0+\frac{\alpha}{2})^{-1}}\Big(\psi\big(\mathcal{K}_2^0+\frac{\alpha}{2}\big)-\psi(\alpha)\Big)e^{\mathcal{K}_2^-\mathcal{K}_1^+(\mathcal{K}_1^0+\frac{\alpha}{2})^{-1}}, \tag{13.136}$$

where the following representation of the $\mathfrak{su}(1,1)$ *Lie algebra is used:*

$$\mathcal{K}^+ f(\theta) = \theta\left(\theta\frac{\partial}{\partial\theta}+\alpha\right)f(\theta)$$
$$\mathcal{K}^- f(\theta) = \frac{\partial}{\partial\theta}f(\theta)$$
$$\mathcal{K}^0 f(\theta) = \left(\theta\frac{\partial}{\partial\theta}+\frac{\alpha}{2}\right)f(\theta). \tag{13.137}$$

Proof It is enough to consider the "right" part of the generator, the "left" part is treated in an analogous manner. Using (13.137) we have

$$\mathcal{K}^+(\mathcal{K}^0+\frac{\alpha}{2})^{-1}f(\theta) = \theta f(\theta). \tag{13.138}$$

It follows that

$$e^{-\mathcal{K}_1^-\mathcal{K}_2^+(\mathcal{K}_2^0+\frac{\alpha}{2})^{-1}}\Big(\psi\big(\mathcal{K}_1^0+\frac{\alpha}{2}\big)-\psi(\alpha)\Big)e^{\mathcal{K}_1^-\mathcal{K}_2^+(\mathcal{K}_2^0+\frac{\alpha}{2})^{-1}}f(\theta_1,\theta_2)$$
$$= e^{-\theta_2\frac{\partial}{\partial\theta_1}}\Big(\psi\big(\theta_1\frac{\partial}{\partial\theta_1}+\alpha\big)-\psi(\alpha)\Big)e^{\theta_2\frac{\partial}{\partial\theta_1}}f(\theta_1,\theta_2). \tag{13.139}$$

By employing the Euler integral representation of the digamma function (13.118) we obtain

$$e^{-\mathcal{K}_1^-\mathcal{K}_2^+(\mathcal{K}_2^0+\frac{\alpha}{2})^{-1}}\Big(\psi\big(\mathcal{K}_1^0+\frac{\alpha}{2}\big)-\psi(\alpha)\Big)e^{\mathcal{K}_1^-\mathcal{K}_2^+(\mathcal{K}_2^0+\frac{\alpha}{2})^{-1}}f(\theta_1,\theta_2)$$
$$= e^{-\theta_2\frac{\partial}{\partial\theta_1}}\int_0^1\frac{d\beta}{1-\beta}\beta^{\alpha-1}\Big(1-\beta^{\theta_1\frac{\partial}{\partial\theta_1}}\Big)e^{\theta_2\frac{\partial}{\partial\theta_1}}f(\theta_1,\theta_2). \tag{13.140}$$

Moreover the formal rewriting of translations and dilatations

$$f(\theta+c) = e^{c\frac{\partial}{\partial\theta}}f(\theta) \tag{13.141}$$

$$f(c\theta) = c^{\theta\frac{\partial}{\partial\theta}}f(\theta) \tag{13.142}$$

implies that

$$e^{-\mathcal{K}_1^-\mathcal{K}_2^+(\mathcal{K}_2^0+\frac{\alpha}{2})^{-1}}\Big(\psi\big(\mathcal{K}_1^0+\frac{\alpha}{2}\big)-\psi(\alpha)\Big)e^{\mathcal{K}_1^-\mathcal{K}_2^+(\mathcal{K}_2^0+\frac{\alpha}{2})^{-1}}f(\theta_1,\theta_2)$$
$$=\int_0^1\frac{d\beta}{1-\beta}\beta^{\alpha-1}\Big(f(\theta_1,\theta_2)-f(\beta\theta_1+(1-\beta)\theta_2,\theta_2)\Big). \quad (13.143)$$

Changing variable by defining $u=1-\beta$ we have

$$e^{-\mathcal{K}_1^-\mathcal{K}_2^+(\mathcal{K}_2^0+\frac{\alpha}{2})^{-1}}\Big(\psi\big(\mathcal{K}_1^0+\frac{\alpha}{2}\big)-\psi(\alpha)\Big)e^{\mathcal{K}_1^-\mathcal{K}_2^+(\mathcal{K}_2^0+\frac{\alpha}{2})^{-1}}f(\theta_1,\theta_2)$$
$$=\int_0^1\frac{du}{u}(1-u)^{\alpha-1}\Big(f(\theta_1,\theta_2)-f((1-u)\theta_1+u\theta_2,\theta_2)\Big). \quad (13.144)$$

By an analogous computation, one proves that

$$e^{-\mathcal{K}_2^-\mathcal{K}_1^+(\mathcal{K}_1^0+\frac{\alpha}{2})^{-1}}\Big(\psi\big(\mathcal{K}_2^0+\frac{\alpha}{2}\big)-\psi(\alpha)\Big)e^{\mathcal{K}_2^-\mathcal{K}_1^+(\mathcal{K}_1^0+\frac{\alpha}{2})^{-1}}f(\theta_1,\theta_2)$$
$$=\int_0^1\frac{du}{u}(1-u)^{\alpha-1}\Big(f(\theta_1,\theta_2)-f(\theta_1,(1-u)\theta_2+u\theta_1)\Big). \quad (13.145)$$

Adding up the last two equations the proof is concluded. □

From the algebraic description of the hidden parameter process we now derive a duality relation.

Theorem 13.17 (Duality Hidden Parameter Process and Symmetric Harmonic Process) *The hidden parameter process on a graph $G=(V,E)$, defined by the generator* (13.133) *is dual to the symmetric harmonic process with duality functions given by*

$$D(\xi,\theta)=\prod_{i\in V}d(\xi_i,\theta_i) \qquad \textit{with} \qquad d(k,\Theta)=\Theta^k. \quad (13.146)$$

Proof The proof is a consequence of the following basic duality for the $\mathfrak{su}(1,1)$ Lie algebra:

$$K^+\xrightarrow{d}\mathcal{K}^+$$
$$K^-\xrightarrow{d}\mathcal{K}^-$$
$$K^0\xrightarrow{d}\mathcal{K}^0, \quad (13.147)$$

which can be easily checked. Then the duality relation

$$\mathcal{L}_{1,2}\xrightarrow{D}L_{1,2} \quad (13.148)$$

follows combining Eq. (13.136) (which provides the algebraic description of the hidden parameter model in terms of $\mathcal{K}$ operators) and Eq. (13.68) (which provides the algebraic description of the symmetric harmonic process in terms of K operators):

$$
\begin{aligned}
& e^{-\mathcal{K}_1^-\mathcal{K}_2^+(\mathcal{K}_2^0+\frac{\alpha}{2})^{-1}}\Big(\psi\big(\mathcal{K}_1^0+\frac{\alpha}{2}\big)-\psi(\alpha)\Big)e^{\mathcal{K}_1^-\mathcal{K}_2^+(\mathcal{K}_2^0+\frac{\alpha}{2})^{-1}} \\
& +e^{-\mathcal{K}_2^-\mathcal{K}_1^+(\mathcal{K}_1^0+\frac{\alpha}{2})^{-1}}\Big(\psi\big(\mathcal{K}_2^0+\frac{\alpha}{2}\big)-\psi(\alpha)\Big)e^{\mathcal{K}_2^-\mathcal{K}_1^+(\mathcal{K}_1^0+\frac{\alpha}{2})^{-1}} \\
& \quad \xrightarrow{D} e^{K_1^-\left(K_2^0+\frac{\alpha}{2}\right)^{-1}K_2^+}\Big(\psi\big(K_1^0+\frac{\alpha}{2}\big)-\psi(\alpha)\Big)e^{-K_1^-\left(K_2^0+\frac{\alpha}{2}\right)^{-1}K_2^+} \\
& \qquad +e^{K_2^-\left(K_1^0+\frac{\alpha}{2}\right)^{-1}K_1^+}\Big(\psi\big(K_2^0+\frac{\alpha}{2}\big)-\psi(\alpha)\Big)e^{-K_2^-\left(K_1^0+\frac{\alpha}{2}\right)^{-1}K_1^+}.
\end{aligned}
\tag{13.149}
$$

□

As a consequence of the duality between the symmetric harmonic process and the hidden parameter process we have an additional intertwining relation. Namely, if we define the intertwiner

$$\Lambda f(\theta)=\sum_{\eta\in\mathbb{N}^V}\prod_{i\in V}\mu_{\theta_i}(\eta_i)f(\eta), \tag{13.150}$$

with

$$\mu_\Theta(n)=\frac{1}{n!}\frac{\Gamma(\alpha+n)}{\Gamma(\alpha)}\left(\frac{\Theta}{1+\Theta}\right)^n\left(\frac{1}{1+\Theta}\right)^\alpha, \qquad \Theta>0 \tag{13.151}$$

then, using Theorem 2.29, we have that the duality of Theorem 13.17 implies

$$\Lambda L=\mathcal{L}\Lambda. \tag{13.152}$$

As we show in Proposition 13.18 below, this implies that evolving a Negative Binomial product measure with parameter θ_i at site i under the symmetric harmonic process produces at time $t>0$ a mixture of Negative Binomial product measures where the mixing measure is given by the probability distribution of the hidden parameter model at time t started from $\theta=\{\theta_i, i\in V\}$. As a further consequence, this implies that mixtures of Negative Binomial product distributions form an invariant set of measures for the discrete harmonic process.

Proposition 13.18 (Negative Binomial Intertwining and Propagation of Mixtures) *Consider the symmetric harmonic process on a graph $G=(V,E)$ initialized from the inhomogeneous product measure*

$$\mu^{(0)}=\otimes_{i\in V}\mu_{\theta_i}, \tag{13.153}$$

where μ_{θ_i} is the Negative Binomial distribution defined in (13.151) *with parameter* $\Theta = \theta_i > 0$. *Then, at time* $t > 0$, *the evolved measure is*

$$\mu^{(t)} = \mathbb{E}_\theta\Big(\otimes_{i\in V} \mu_{\theta_i(t)}\Big), \tag{13.154}$$

where $\mathbb{E}_\theta$ *denotes expectation with respect to the hidden parameter process* $\{\theta(t)\,,\ t \geq 0\}$ *initialized from the configuration* $\theta = (\theta_i)_{i\in V}$.

Proof Denote by $S(t)$ the semigroup of the symmetric harmonic process and by $\mathcal{S}(t)$ the semigroup of the hidden parameter process. For a bounded function $f : \mathbb{N}^V \to \mathbb{R}$ we compute

$$\begin{aligned}\sum_{\eta\in\mathbb{N}^V} \mu^{(0)}(\eta)(S(t)f(\eta)) &= (\Lambda S(t) f)(\theta)\\ &= (\mathcal{S}(t)\Lambda f)(\theta)\\ &= \mathbb{E}_\theta\Big(\Lambda f(\theta(t))\Big)\\ &= \sum_{\eta\in\mathbb{N}^V} \mathbb{E}_\theta\Big(\prod_{i\in V} \mu_{\theta_i(t)}(\eta)\Big) f(\eta),\end{aligned} \tag{13.155}$$

where we have used (13.150) and (13.152). □

Remark 13.19 The propagation of mixed Negative Binomial into mixed Negative Binomial holds for the symmetric harmonic process on any graph, even in the presence of reservoirs. In the next section we will exploit this property for a chain with two boundary reservoirs and thus obtain a closed formula for the non-equilibrium steady state.

13.10 Boundary-driven Harmonic Chain

In this last section we consider the harmonic model in the boundary-driven set up and show how the combination of duality and quantum inverse scattering method allows to identify the stationary measure, i.e. the non-equilibrium steady state, in closed form. We give here a concise account with some hints of proofs, for details we refer to the works [108], [107], [106], [102], [51], [50], [118].

Definition 13.20 (Boundary-driven Symmetric Harmonic Model) The boundary-driven symmetric harmonic model is the process $\{\eta(t) = (\eta_1(t), \ldots \eta_N(t))$, $t \geq 0\}$ taking values in $\mathbb{N}^N$ defined by the following generator:

$$L = L_1 + \left(\sum_{i=1}^{N-1} L_{i,i+1}\right) + L_N, \tag{13.156}$$

where

$$L_{i,i+1}f(\eta) = \sum_{k=1}^{\eta_i} \varphi_\alpha(k,\eta_i)\Big[f(\eta - k\delta_i + k\delta_{i+1}) - f(\eta)\Big]$$
$$+ \sum_{k=1}^{\eta_{i+1}} \varphi_\alpha(k,\eta_{i+1})\Big[f(\eta + k\delta_i - k\delta_{i+1}) - f(\eta)\Big], \quad (13.157)$$

and

$$L_1 f(\eta) = \sum_{k=1}^{\eta_1} \varphi_\alpha(k,\eta_i)\Big[f(\eta - k\delta_1) - f(\eta)\Big]$$
$$+ \sum_{k=1}^{\infty} \frac{1}{k}\left(\frac{\theta_L}{1+\theta_L}\right)^k \Big[f(\eta + k\delta_1) - f(\eta)\Big], \quad (13.158)$$

and

$$L_N f(\eta) = \sum_{k=1}^{\eta_N} \varphi_\alpha(k,\eta_i)\Big[f(\eta - k\delta_N) - f(\eta)\Big]$$
$$+ \sum_{k=1}^{\infty} \frac{1}{k}\left(\frac{\theta_R}{1+\theta_R}\right)^k \Big[f(\eta + k\delta_N) - f(\eta)\Big]. \quad (13.159)$$

Here θ_L and $\theta_R > 0$ are two parameters that are used to tune to densities of particles at the left and right boundaries. In particular, when $\theta_L = \theta_R = \Theta$ one can check that the product measure $\prod_{i=1}^N \mu_\Theta$ is reversible, where

$$\mu_\Theta(n) = \frac{1}{n!}\frac{\Gamma(\alpha+n)}{\Gamma(\alpha)}\left(\frac{\Theta}{1+\Theta}\right)^n\left(\frac{1}{1+\Theta}\right)^\alpha, \qquad \Theta > 0. \quad (13.160)$$

Thus, in equilibrium, the numbers of particles at each site are i.i.d. random variables with Negative Binomial distribution of mean $\alpha\Theta$.

The boundary generators L_1 and L_N can be recognized to be of the form discussed in Chap. 11 as the following proposition shows.

Proposition 13.21 (Reservoirs of the Harmonic Process) *Let*

$$L^\Theta f(n) = \sum_{k=1}^{n} \varphi_\alpha(k,n)\Big[f(n-k) - f(n)\Big]$$
$$+ \sum_{k=1}^{\infty} \frac{1}{k}\left(\frac{\Theta}{1+\Theta}\right)^k \Big[f(n+k) - f(n)\Big], \quad (13.161)$$

and

$$L^{res,\Theta} f(n) = \sum_{k=1}^{n} \varphi_\alpha(k,n)\Big[f(n-k) - f(n)\Big]$$
$$+ \mathbb{E}_{N_\Theta}\Big[\sum_{k=1}^{N_\Theta} \varphi_\alpha(k, N_\Theta)\Big]\Big[f(n+k) - f(n)\Big], \qquad (13.162)$$

where $\mathbb{E}_{N_\Theta}$ denotes expectation w.r.t. a random variable N_Θ with Negative binomial distribution of mean $\alpha\Theta$. Then we have

$$L^{res,\Theta} = L^{\Theta}. \qquad (13.163)$$

Proof We compute

$$\mathbb{E}_{N_\Theta}\Big[\sum_{k=1}^{N_\Theta} \varphi_\alpha(k, N_\Theta)\Big] = \sum_{m=1}^{\infty}\sum_{k=1}^{m} \varphi_\alpha(k,m)\mu_\Theta(m)$$
$$= \sum_{k=1}^{\infty}\sum_{m=k}^{\infty} \varphi_\alpha(k,m)\mu_\Theta(m). \qquad (13.164)$$

We now insert the expression of $\varphi_\alpha(k,m)$ and of $\mu_\Theta(m)$ to find

$$\mathbb{E}_{N_\Theta}\Big[\sum_{k=1}^{N_\Theta} \varphi_\alpha(k, N_\Theta)\Big]$$
$$= \sum_{k=1}^{\infty}\sum_{m=k}^{\infty} \frac{1}{k}\frac{\Gamma(m+1)\Gamma(m-k+\alpha)}{\Gamma(m+\alpha)\Gamma(m-k+1)} \cdot \frac{1}{m!}\frac{\Gamma(\alpha+m)}{\Gamma(\alpha)}\left(\frac{\Theta}{1+\Theta}\right)^{m}\left(\frac{1}{1+\Theta}\right)^{\alpha}$$
$$= \sum_{k=1}^{\infty}\frac{1}{k}\sum_{m=k}^{\infty} \frac{\Gamma(m-k+\alpha)}{(m-k)!\Gamma(\alpha)}\left(\frac{\Theta}{1+\Theta}\right)^{m}\left(\frac{1}{1+\Theta}\right)^{\alpha}$$
$$= \sum_{k=1}^{\infty}\frac{1}{k}\left(\frac{\Theta}{1+\Theta}\right)^{k}, \qquad (13.165)$$

where in the last step we used

$$\sum_{m=k}^{\infty} \frac{\Gamma(m-k+\alpha)}{(m-k)!\Gamma(\alpha)}\left(\frac{\Theta}{1+\Theta}\right)^{m-k}\left(\frac{1}{1+\Theta}\right)^{\alpha} = 1.$$

□

When $\theta_L \neq \theta_R$ the boundary driven symmetric harmonic process, has a unique stationary measure which is called the non-equilibrium steady state, which is not reversible and not product. The following theorem [106, 107] fully characterizes this non-equilibrium steady state by giving closed-form formulas of its moments. We will give a partial proof of this result, referring to [106, 107] to details on the computation of the absorption probabilities via the quantum inverse scattering method. An alternative self-contained proof, based on spatial Markovianity of the hidden parameter process is provided in Sect. 13.11.

Theorem 13.22 (Factorial Moments) *Consider the boundary-driven harmonic model with parameters θ_L, θ_R. Denote by $\mathbb{E}$ expectation with respect to the stationary measure and call $(Y_1, \ldots, Y_N)$ the random variables with this joint distribution. Then, for $\xi = (\xi_1, \ldots, \xi_N) \in \mathbb{N}_0^N$, the factorial moments of order ξ are given by*

$$\mathbb{E}\left[\prod_{i=1}^{N} \frac{Y_i!}{(Y_i - \xi_i)!}\right] = \left[\prod_{i=1}^{N} \frac{\Gamma(\xi_i + \alpha)}{\Gamma(\alpha)}\right] \sum_{n=0}^{|\xi|} \theta_R^{|\xi|-n} (\theta_L - \theta_R)^n g_\xi(n), \tag{13.166}$$

with

$$g_\xi(n) = \sum_{\substack{n_1,\ldots,n_N \\ \sum_i n_i = n}} \prod_{i=1}^{N} \binom{\xi_i}{n_i} \prod_{j=1}^{\alpha} \frac{\alpha(N+2-i) - j}{\alpha(N+2-i) - j + \sum_{k=i}^{N} n_k}. \tag{13.167}$$

Proof The proof is obtained in two steps. The first step is to use duality to reduce the computation of the factorial moment of order ξ to a problem with $|\xi| = \sum_{i=1}^{N} \xi_i$ dual particles. In particular we have duality of the boundary-driven harmonic model with an absorbing dual process $\{\tilde{\xi}(t) = (\tilde{\xi}_0(t), \tilde{\xi}_1(t), \ldots \tilde{\xi}_N(t), \tilde{\xi}_{N+1}(t)),\ t \geq 0\}$ with generator:

$$L^{\text{dual}} = L_1^{\text{dual}} + \left(\sum_{i=1}^{N-1} L_{i,i+1}\right) + L_N^{\text{dual}} \tag{13.168}$$

where

$$L_1^{\text{dual}} f(\tilde{\xi}) = \sum_{k=1}^{\tilde{\xi}_1} \varphi_\alpha(k, \tilde{\xi}_i)\Big[f(\tilde{\xi} - k\delta_1 + k\delta_0) - f(\tilde{\xi})\Big], \tag{13.169}$$

and

$$L_N^{\text{dual}} f(\tilde{\xi}) = \sum_{k=1}^{\tilde{\xi}_N} \varphi_\alpha(k, \tilde{\xi}_i)\Big[f(\tilde{\xi} - k\delta_N + k\delta_{N+1}) - f(\tilde{\xi})\Big]. \tag{13.170}$$

One can check that

$$\mathbb{E}_\eta\Big[D(\eta(t),\tilde{\xi})\Big]=\mathbb{E}_{\tilde{\xi}}\Big[D(\eta,\tilde{\xi}(t))\Big], \tag{13.171}$$

with duality function

$$D(\tilde{\xi},\eta)=\theta_L^{\tilde{\xi}_0}\left[\prod_{i=1}^{N}\frac{\eta_i!}{(\eta_i-\tilde{\xi}_i)!}\frac{\Gamma(\alpha)}{\Gamma(\tilde{\xi}_i+\alpha)}\right]\theta_R^{\tilde{\xi}_{N+1}}. \tag{13.172}$$

Taking the limit $t\to\infty$ of the duality relation with the dual process initialized from the configuration $\tilde{\xi}=\sum_{i=1}^{N}\xi_i\delta_i$ one gets

$$\mathbb{E}\left[\prod_{i=1}^{N}\frac{Y_i!}{(Y_i-\xi_i)!}\frac{\Gamma(\alpha)}{\Gamma(\xi_i+\alpha)}\right]=\sum_{k=0}^{|\xi|}\theta_L^k\,\theta_R^{|\xi|-k}\;p_\xi(k), \tag{13.173}$$

where

$$p_\xi(k)=\mathbb{P}\left[\xi(\infty)=k\delta_0+(|\xi|-k)\delta_{N+1}\,\Big|\,\xi(0)=\sum_{i=1}^{N}\xi_i\delta_i\right]. \tag{13.174}$$

Thus the factorial moments are expressed as polynomials in the parameters θ_L,θ_R whose coefficients are the probabilities $p_\xi(k)$ that starting the dual process from the configuration $\sum_{i=1}^{N}\xi_i\delta_i$, one eventually has k particles absorbed at the extra site $\{0\}$ and the remaining ones absorbed at the extra site $\{N+1\}$.

The second step amounts to the explicit computation of these absorption probabilities. For this we refer to the original paper [106], where the computation is performed by using a non-trivial symmetry of the boundary-driven chain, which is in turn derived from the application of the quantum inverse scattering method. □

From the knowledge of the factorial moments one can reconstruct the stationary measure of the open harmonic process with parameters θ_L,θ_R by using the inversion formula

$$\mathbb{P}(Y=\eta)=\sum_{\xi\geq\eta}\mathbb{E}\left[\prod_{i=1}^{N}\frac{Y_i!}{(Y_i-\xi_i)!}\frac{(-1)^{\xi_i-\eta_i}}{\xi_i!}\binom{\xi_i}{\eta_i}\right]. \tag{13.175}$$

In a next step [50, 51] an integral representation of the stationary measure has been found. From this representation it can be recognized the structure of a mixed measure.

Theorem 13.23 (Non-equilibrium Steady State) *The stationary measure of the boundary-driven harmonic model with parameters $\theta_L < \theta_R$ can be written as a mixture of product measures as follows:*

$$\mu^{\theta_L,\theta_R}(\eta) = \mathbb{E}\left[\prod_{i=1}^{N} \mu_{\Theta_i}(\eta_i)\right], \tag{13.176}$$

where μ_Θ is the Negative Binomial distribution defined in (13.160) *and the components of the random vector $\Theta = (\Theta_1, \ldots, \Theta_N)$ are*

$$\Theta_i = \theta_L + (\theta_R - \theta_L)\sum_{j=1}^{i} R_i. \tag{13.177}$$

Here the random vector $R = (R_1, R_2, \ldots, R_{N+1})$ has the symmetric Dirichlet distribution with parameter α, taking values on the $(N+1)-$dimensional simplex $\sum_{i=1}^{N+1} r_i = 1$ and having probability density

$$f(r_1, \ldots, r_{N+1}) = \frac{\Gamma(\alpha(N+1))}{\Gamma(\alpha)^{N+1}} \prod_{i=1}^{N+1} r_i^{\alpha-1} \mathbb{1}_{\{\sum_{i=1}^{N+1} r_i = 1\}}. \tag{13.178}$$

As a consequence, defining $\theta_0 = \theta_L$ and $\theta_{N+1} = \theta_R$, one has the closed formula

$$\mu^{\theta_L,\theta_R}(\eta) = \frac{\Gamma(\alpha(N+1))}{\Gamma(\alpha)^{N+1}(\theta_R - \theta_L)^{\alpha(N+1)-1}} \int_{\theta_L}^{\theta_R} d\theta_1 \int_{\theta_1}^{\theta_R} d\theta_2 \cdots \int_{\theta_{N-1}}^{\theta_R} d\theta_N$$
$$\left[\prod_{i=1}^{N+1} (\theta_i - \theta_{i-1})^{\alpha-1}\right] \cdot \left[\prod_{i=1}^{N} \frac{1}{\eta_i!} \frac{\Gamma(\alpha+\eta_i)}{\Gamma(\alpha)} \left(\frac{\theta_i}{1+\theta_i}\right)^{\eta_i} \left(\frac{1}{1+\theta_i}\right)^{\alpha}\right]. \tag{13.179}$$

Proof It is convenient to work with generating functions. One starts by recalling that for a random variable X with Negative Binomial distribution (13.160) the generating function is given by

$$\mathbb{E}(e^{hX}) = \left(\frac{1}{1+(1-e^h)\Theta}\right)^{\alpha}. \tag{13.180}$$

Using the factorial moments of Theorem 13.22, one can compute the generating function of the random variables $(Y_1, \ldots, Y_N)$, for details see [50]. The result is

$$\mathbb{E}\Big[e^{\sum_{i=1}^N h_i Y_i}\Big]$$
$$= \frac{\Gamma(\alpha(N+1))}{\Gamma(\alpha)^{N+1}} \int_0^1 dv_1 \int_{v_1}^1 dv_2 \cdots \int_{v_{N-1}}^1 dv_N \Bigg(\prod_{i=1}^{N+1} (v_i - v_{i-1})^{\alpha-1}$$
$$\prod_{i=1}^N \Big(\frac{1}{1+(1-e^{h_i})\big(\theta_L + (\theta_R - \theta_L)v_i\big)}\Big)^\alpha\Bigg), \tag{13.181}$$

with the convention $v_0 = 0$ and $v_{N+1} = 1$. From this expression one recognizes the structure of a mixed measure, i.e.

$$\mathbb{E}\Big[e^{\sum_{i=1}^N h_i Y_i}\Big] = \mathbb{E}\Big[\mathbb{E}\Big[\prod_{i=1}^N e^{h_i Y_i} \mid \Theta\Big]\Big], \tag{13.182}$$

where

$$\Theta_i = \theta_L + (\theta_R - \theta_L)V_i, \tag{13.183}$$

and $V = (V_1, \ldots, V_N)$ is the vector with probability density

$$f_{V_1,\ldots,V_n}(v_1, \ldots, v_N) = \frac{\Gamma(\alpha(N+1))}{\Gamma(\alpha)^{N+1}} \prod_{i=1}^{N+1} (v_i - v_{i-1})^{\alpha-1} \mathbb{1}_{\{0 \le v_1 \le \ldots \le v_N \le 1\}}. \tag{13.184}$$

□

Remark 13.24 (Non-equilibrium Steady State of the Heat Conduction Model with Reservoirs) Using duality, it follows from Theorem 13.23 that one can also characterize the non-equilibrium steady state of the boundary-driven integrable heat conduction model as a mixture [102]. More precisely, one get a mixture of product Gamma distributions whose scale parameters are given by the ordered Dirichlet distribution.

The knowledge of the stationary measure established in Theorem 13.23 can be used to establish the large deviation principle for the empirical density profile, with the large deviation function predicted by the Macroscopic Fluctuation Theory[1]

[1] The Macroscopic Fluctuation Theory (MFT) can strictly speaking not be applied to the harmonic process, because MFT is based on a *dynamical* large deviation principle which is currently not available for the harmonic process. The reason is that the stationary measures have exponential tails, and the proof of the dynamical large deviation principle, based on super-exponential replacement lemmas, requires super-exponential tails of the stationary measures.

[27, 28]. Furthermore, the Markov property of the ordered Dirichlet distribution implies an additivity formula for the large deviation function, or equivalently for the pressure. A similar additivity formula was first established for the boundary-driven symmetric exclusion process [78], although it was not possible to associate it with a direct physical meaning, as it was formulated as a variational problem with a minimization procedure of the pressure. For the harmonic process, instead, one gets a maximization procedure, which in turn can be explained by a contraction principle.

We will briefly mention this result, together with the ideas of proof, referring to [50] for full details. In what follows we use the informal notation

$$\mathbb{P}(X_N \approx x) \sim e^{-NI(x)}, \tag{13.185}$$

to indicate that the sequence of random variables $\{X_N, N \in \mathbb{N}\}$ taking values in a space χ satisfies the large deviation principle with rate N and rate function $I : \chi \to [0, \infty]$, which means

$$\limsup_{N\to\infty} \frac{1}{N} \log \mathbb{P}(X_N \in F) \leq - \inf_{x\in F} I(x), \tag{13.186}$$

for all closed sets $F \subset \chi$, and

$$\liminf_{N\to\infty} \frac{1}{N} \log \mathbb{P}(X_N \in G) \geq - \inf_{x\in G} I(x), \tag{13.187}$$

for all open sets $G \subset \chi$.

Theorem 13.25 (Large Deviation Principle and Additivity Formula) *Consider the boundary-driven harmonic model with parameters $\theta_L < \theta_R$. Denote*

$$\mathcal{M}_{\theta_L,\theta_R} = \{\theta : [0,1] \to \mathbb{R} : \theta(0) = \theta_L, \theta(1) = \theta_R, \text{ non-decreasing } \} \tag{13.188}$$

Then we have the following results.

1. The empirical profile

$$\pi_N = \frac{1}{N} \sum_{i=1}^{N} \eta_i \delta_{\frac{i}{N}} \tag{13.189}$$

satisfies a large deviation principle

$$\mathbb{P}\Big[\pi_N(dx) \approx \rho(x)dx\Big] \sim e^{-NI(\rho)}, \tag{13.190}$$

with rate function

$$I(\rho) = \inf_{\theta \in \mathcal{M}_{\theta_L,\theta_R}} \alpha \int_0^1 dx \Big[\frac{\rho(x)}{\alpha} \ln \frac{\frac{\rho(x)}{\alpha}}{\theta(x)} + (1 + \frac{\rho(x)}{\alpha}) \ln \frac{1+\theta(x)}{1+\frac{\rho(x)}{\alpha}} - \ln \frac{\theta'(x)}{\theta_R - \theta_L} \Big]. \tag{13.191}$$

2. *The pressure*

$$P(h) = \lim_{N\to\infty} \frac{1}{N} \log \mathbb{E}\Big[e^{N\langle \pi_N, h\rangle} \Big] \tag{13.192}$$

satisfies the following variational problem

$$P(h) = \sup_{\theta \in \mathcal{M}_{\theta_L,\theta_R}} \alpha \int_0^1 dx \Big[\ln \Big(\frac{1}{1+(1-e^{h(x)})\theta(x)} \Big) + \ln \frac{\theta'(x)}{\theta_R - \theta_L} \Big]. \tag{13.193}$$

3. *Furthermore the pressure satisfies the following additivity principle:*

$$\widetilde{P}^{[0,1]}_{\theta_L,\theta_R}(h) = \sup_{\theta_L \le \theta_m \le \theta_R} \Big[\widetilde{P}^{[0,x]}_{\theta_L,\theta_m}(h_1) + \widetilde{P}^{[x,1]}_{\theta_m,\theta_R}(h_2) \Big] \qquad 0 < x < 1, \tag{13.194}$$

where

$$\widetilde{P}^{[a,b]}_{\theta_a,\theta_b}(h) := P^{[a,b]}_{\theta_a,\theta_b}(h) + \alpha(b-a)\log\Big(\frac{\theta_R - \theta_L}{b-a}\Big), \tag{13.195}$$

and

$$P^{[a,b]}_{\theta_a,\theta_b}(h) = \sup_{\theta} \alpha \int_a^b \Big[\ln \Big(\frac{1}{1+(1-e^{h(x)})\theta(x)} \Big) + \ln \Big(\frac{(b-a)\theta'(x)}{\theta_b - \theta_a} \Big) \Big] dx. \tag{13.196}$$

Proof The full proof can be found in [50]. We state in words the main arguments. Items 1 is obtained by a contraction principle that can be applied to the joint large deviations of the empirical density profile of an inhomogeneous product measure and of the empirical profile of the ordered Dirichlet distribution. The former can be analyzed by standard techniques [134], the latter is also known (see e.g. [82] for large deviations of the order statistics of uniform i.i.d.). Item 2 is obtained as an application of Varadhan's lemma. Item 3 follows by exploiting the spatial Markov properties of the ordered Dirichlet distribution[5]. □

13.11 Spatial Markovian Structure of the Hidden Parameter Process

In this section we show that the hidden parameter process on a chain with left and right reservoirs has a spatial Markovian structure. This allows us to give an alternative, self-contained and more probabilistic proof of the structure of the non-equilibrium steady state as a mixture of products of negative binomials where the mixing measure is the Dirichlet process.

First we state the analogue of the intertwining result for the boundary driven case. Similarly to Proposition 13.18, we can obtain the following intertwining relation (see [118])

$$\Lambda L = \mathcal{L}^{N,\theta_L,\theta_R} \Lambda, \tag{13.197}$$

where L is the generator of the boundary driven symmetric harmonic process (13.156), Λ is the Negative Binomial intertwiner defined in (13.152) and $\mathcal{L}^{N,\theta_L,\theta_R}$ is the generator of a hidden parameter process $\theta(t) = (\theta_0(t), \ldots, \theta_{N+1}(t))$ where $\theta_0(t) = \theta_L$ and $\theta_{N+1}(t) = \theta_R$ for all $t \geq 0$. The generator $\mathcal{L}^{N,\theta_L,\theta_R}$ then reads

$$\mathcal{L}^{N,\theta_L,\theta_R} = \mathcal{L}^{\theta_L} + \left(\sum_{i=1}^{N-1} \mathcal{L}_{i,i+1}\right) + \mathcal{L}^{\theta_R}, \tag{13.198}$$

with

$$\mathcal{L}^{\theta_L} f(\theta) = \int_0^1 \frac{du}{u} (1-u)^{\alpha-1} \left[f(\theta - u\theta_1\delta_1 + u\theta_L\delta_1) - f(\theta)\right], \tag{13.199}$$

and

$$\begin{aligned}\mathcal{L}_{i,i+1} f(\theta) = &\int_0^1 \frac{du}{u} (1-u)^{\alpha-1} \left[f(\theta - u\theta_i\delta_i + u\theta_{i+1}\delta_i) - f(\theta)\right] \\ &+ \int_0^1 \frac{du}{u} (1-u)^{\alpha-1} \left[f(\theta - u\theta_{i+1}\delta_{i+1} + u\theta_i\delta_{i+1}) - f(\theta)\right],\end{aligned} \tag{13.200}$$

and

$$\mathcal{L}^{\theta_R} f(\theta) = \int_0^1 \frac{du}{u} (1-u)^{\alpha-1} \left[f(\theta - u\theta_N\delta_N + u\theta_R\delta_N) - f(\theta)\right]. \tag{13.201}$$

Here we make the convention that the empty sum is zero, i.e., for $N = 1$, the generator of a one-site system with left reservoir parameter θ_L and right reservoir parameter θ_R reads

$$\mathcal{L}^{1,\theta_L,\theta_R} = \mathcal{L}^{\theta_L} + \mathcal{L}^{\theta_R}. \tag{13.202}$$

To prove the intertwining (13.197), it suffices to see that for the bulk term we can use (13.152), whereas for the boundary terms, the intertwining follows from an simple explicit computation, see [118] for full details. First we identify the invariant measure of the one-site system with two reservoirs.

Lemma 13.26 *The density of the unique invariant measure of the process with generator* (13.202) *is given by*

$$g^{1,\theta_L,\theta_R}(\theta_1) = \frac{1}{B(\alpha,\alpha)} \frac{(\theta_1-\theta_L)^{\alpha-1}(\theta_R-\theta_1)^{\alpha-1}}{(\theta_R-\theta_L)^{2\alpha-1}}. \tag{13.203}$$

Proof We prove the result for $\theta_L = 0, \theta_R = 1$ without loss of generality. Then it is easy to see that the generator (13.202) is the same as the generator of the process $\zeta_1(t)$ where $(\zeta_1(t), \zeta_2(t))$ evolves according to the two-site integrable heat conduction model, i.e., the process with generator (13.113), with initial condition $\zeta_1(0) + \zeta_2(0) = 1 = \zeta_1(t) + \zeta_2(t)$ (recall that the process with generator (13.113) conserves $\zeta_1 + \zeta_2$). Since this process has as invariant measure the product of gamma distributions with identical shape parameters α and identical scale parameters, the conditional density given $\zeta_1 + \zeta_2 = 1$ is precisely the Beta distribution with parameters (α, α). □

The following result is a useful decomposition of the generator. It states that the action of the generator (13.198) on the variable θ_i is the same as the action of a one-site systems with left reservoir parameter θ_{i-1} and right reservoir parameter θ_{i+1} (with the convention $\theta_0 = \theta_L, \theta_{N+1} = \theta_R$.

Lemma 13.27 *The generator of the boundary driven hidden parameter process on the chain can be decomposed as follows.*

$$\mathcal{L}^{N,\theta_L,\theta_R} = \sum_{i=1}^{N} \mathcal{L}_i^{1,\theta_{i-1},\theta_{i+1}}, \tag{13.204}$$

where by $\mathcal{L}_i^{1,\theta_{i-1},\theta_{i+1}}$ *we denote the generator* (13.202) *acting on the variable* θ_i, *and where* $\theta_0 = \theta_L$, $\theta_{N+1} = \theta_R$.

The next proposition then shows how to obtain the invariant measure of the N-site system from the invariant measure of the 1-site system. Let us denote by $g^{N,\theta_L,\theta_R}(\theta_1, \ldots, \theta_N)$ a probability density for the N-site system with left reservoir parameter θ_L and right reservoir parameter θ_R. Let us also denote by $g^{N,\theta_L,\theta_R}(\theta_i|\theta_{\neq i})$ the corresponding conditional density of θ_i, given $\theta_{\neq i} := (\theta_1, \dots, \theta_{i-1}, \theta_{i+1}, \dots, \theta_N)$. Then we have the following result.

Proposition 13.28 *Assume that for all $i = 1, \dots, N$ we have*

$$g^{N,\theta_L,\theta_R}(\theta_i|\theta_{\neq i}) = g^{1,\theta_{i-1},\theta_{i+1}}(\theta_i), \tag{13.205}$$

where $g^{1,\theta_{i-1},\theta_{i+1}}$ is given in (13.203). *Then $g^{N,\theta_L,\theta_R}(\theta_1, \dots, \theta_N)$ is the probability density of the invariant measure for the N-site system with left reservoir parameter θ_L and right reservoir parameter θ_R.*

Proof Let us denote by $h_i(\theta_{\neq i})$ the marginal density corresponding to the given probability density $g^{N,\theta_L,\theta_R}(\theta_1, \dots, \theta_N)$. Then we have for $f : \mathbb{N}^N \to \mathbb{R}$,

$$\int g^{N,\theta_L,\theta_R}(\theta_1, \dots, \theta_N)\mathcal{L}_i^{1,\theta_{i-1},\theta_{i+1}} f(\theta_1, \dots, \theta_N) \prod_{j=1}^{n} d\theta_j$$
$$= \int h_i(\theta_{\neq i}) \prod_{j\neq i} d\theta_j \left(\int g^{N,\theta_L,\theta_R}(\theta_i|\theta_{\neq i})(\mathcal{L}_i^{1,\theta_{i-1},\theta_{i+1}} f)d\theta_i \right) = 0. \tag{13.206}$$

Here in the last step we used that by the assumption (13.205) we have

$$\left(\int g^{N,\theta_L,\theta_R}(\theta_i|\theta_{\neq i})(\mathcal{L}_i^{1,\theta_{i-1},\theta_{i+1}} f)d\theta_i \right) = 0, \tag{13.207}$$

combined with Lemma 13.27.

Then, summing over i and by using Lemma 13.27 again we conclude

$$\int g^{N,\theta_L,\theta_R}(\theta_1, \dots, \theta_N) \left(\mathcal{L}^{N,\theta_L,\theta_R} f \right) \prod_{j=1}^{n} d\theta_j = 0, \tag{13.208}$$

which shows that $g^{N,\theta_L,\theta_R}(\theta_1, \dots, \theta_N)$ is the density of the invariant measure. □

Proposition 13.28 characterizes the density of the invariant measure implicitly by specifying its one-site conditional densities $g^{N,\theta_L,\theta_R}(\theta_i|\theta_{\neq i})$. These one-site conditional densities are completely determined by the one-site system with reservoirs on left and right end. Because the conditional density of θ_i given $\theta_{\neq i}$ is only depending on $\theta_{i-1}, \theta_{i+1}$, we see that we have a two-sided Markov property (in space) of the invariant measure. Also we can explicitly find this joint density of the invariant measure $g^{N,\theta_L,\theta_R}(\theta_1, \dots, \theta_N)$ by recursively constructing from the conditional densities $g^{N,\theta_L,\theta_R}(\theta_i|\theta_{\neq i})$. On the other hand, it is also easy to see that the density appearing in Theorem 13.23, namely, the density

$$g^{N,\theta_L,\theta_R}(\theta_1,\ldots,\theta_N) \tag{13.209}$$
$$= \frac{\Gamma(\alpha(N+1))}{\Gamma(\alpha)^{N+1}(\theta_R-\theta_L)^{\alpha(N+1)-1}}\left[\prod_{i=1}^{N+1}(\theta_i-\theta_{i-1})^{\alpha-1}\right]\mathbb{1}_{\{\theta_L\leq\theta_1\leq\ldots\leq\theta_N\leq\theta_R\}}$$

has indeed the required conditional densities. More precisely via a simple computation, from (13.209) we see

$$g^{N,\theta_L,\theta_R}(\theta_i|\theta_{\neq i}) = g^{1,\theta_{i-1},\theta_{i+1}}(\theta_i) \tag{13.210}$$

and therefore, $g^{N,\theta_L,\theta_R}(\theta_1,\ldots,\theta_N)$ is indeed the invariant density of the hidden parameter model.

Remark 13.29 One can ask how a function of the form $g^{1,\theta_{i-1},\theta_{i+1}}(\theta_i)$ determines a joint density $g^{N,\theta_L,\theta_R}(\theta_1,\ldots,\theta_N)$ via the specification (13.205). This is naturally related to the question for which systems the one-site system with two reservoirs determines the invariant densities for the N-site system with two reservoirs. This question is studied in the recent preprint [199], where under natural scaling and shift invariance properties, it is shown that densities of the form (11.24) are the only joint densities.

13.12 Additional Notes

The integrable XXX_s spin chain can be found in [90]. In high-energy physics, the version with non-compact spins has been studied from different perspectives, especially those related to AdS/CFT correspondence and high-energy QCD [14, 34, 91, 154, 175]. The fact that the integrable XXX_s Hamiltonian with non-compact spins can be interpreted as a Markov process is more recent [107, 108]. As already remarked above (see Sect. 13.4), the asymmetric version of the harmonic process was already considered by Sasamoto-Wadati [206] for parameter $\alpha = 1$ (the so-called MADM process) and by Povolotsky [194] and Barraquand-Corwin [10] for generic values of α (the so-called q-Hahn process). Remarkably these asymmetric models were studied in relation to KPZ universality class just constructing them in the framework of Bethe ansatz, without making an explicit connection with integrable spin chains. This connection was established in [105], where it was show that they arise from the integrable XXZ_s Hamiltonian with non-compact spins.

The boundary driven set-up of the symmetric harmonic model was introduced in [107, 108] by using duality to identify integrable boundaries for the open spin chain. The solution for the stationary measure [106], [51], [50] opens up a new perspective on non-equilibrium steady states as a mixture of local Gibbs measures. From the point of duality, this is related to the identification of a proper intertwiner which then yields propagation of mixed local Gibbs measure over time.

Remarkably, the same closure property of the dynamics has been noticed for the Kipnis-Marchioro-Presutti model [69]. The corresponding hidden parameter process is given by a random version of the Aldous averaging process, where the averaging at neighbouring sites is done using a uniform random variable. However, for the KMP process, the stationary measure of the hidden parameter process it is not known and thus also the KMP non-equilibrium steady state is not currently known.

Appendix A
Markov Processes

In this appendix we provide some basic background material on Markov processes, in view of what is needed in the book. Most of this material can be found in standard books, but we have collected here the basic notions which we need in the book in a spirit of "pragmatism" and self-consistency, i.e., avoiding unnecessary technicalities of measure theoretic or functional analytic nature. For more background and detailed accounts on Markov process theory we refer to [89, 146, 171, 172]. For more background on random walks we refer to [162, 218], and on Brownian motion [166, 183]. For basics on probability theory and its measure theoretic background we refer to [130, 236]. For more background on ergodic theory and analytic aspects of transition operators and semigroups, we refer to [155, 234]. For basics on functional analytic background on semigroups we refer to [84], and for general analytic background to [203]. For the material on Hille-Yosida theory we follow closely chapter 1 of [172], and [84].

A.1 Discrete-time Markov Chains

We start with the simplest case of a discrete-time Markov chain on a countable set S. In doing so, we will set up notation and provide proofs which can be easily transferred to the setting of a Markov process in discrete time on more much general state spaces.

A.1.1 Path Space

The space of discrete-time trajectories, also called (discrete-time) path space, is $S^{\mathbb{N}}$, with $\mathbb{N} = \{0, 1, 2, \ldots\}$. Elements of $S^{\mathbb{N}}$ are denoted by Greek letters ω, ζ. For

C. Giardinà, F. Redig, *Duality for Markov Processes*, Grundlehren der mathematischen Wissenschaften 365, https://doi.org/10.1007/978-3-032-04099-2

$\omega \in S^{\mathbb{N}}$ a trajectory, we denote $\omega(t) \in S$ its state at time $t \in \mathbb{N}$. We equip this path space $S^{\mathbb{N}}$ with the canonical σ-algebra $\mathcal{F}$ generated by cylinders, i.e., the smallest σ-algebra which makes the projections

$$\pi_t : S^{\mathbb{N}} \to S : \omega \to \omega(t) \tag{A.1}$$

measurable. We also denote by $\mathcal{F}_n$ the σ-algebra generated by the projections $\pi_t, 0 \le t \le n$. On path space we define the shift θ_n via $(\theta_n \omega)(t) = \omega(t+n)$. A measurable subset $A \in \mathcal{F}$ is called shift invariant if $\theta_n A = A$ for all $n \in \mathbb{N}$. We will typically denote by $\mathbb{P}$ a probability measure on the discrete path space $(\Omega, \mathcal{F})$.

Definition A.1 (Stationarity and Ergodicity)

1. A probability measure $\mathbb{P}$ on $(\Omega, \mathcal{F})$ is called stationary if $\mathbb{P} \circ \theta_n = \mathbb{P}$ for all $n \in \mathbb{N}$.
2. A stationary probability measure $\mathbb{P}$ on Ω is called ergodic if for all shift-invariant sets $A \in \mathcal{F}$, $\mathbb{P}(A) = 0$ or $\mathbb{P}(A) = 1$.

One of the equivalent and useful characterisations of ergodicity is given in the following proposition.

Proposition A.2 *A stationary probability measure $\mathbb{P}$ on Ω is ergodic if and only if for all $f, g \in L^2(\Omega, \mathcal{F}, \mathbb{P})$ one has*

$$\lim_{N\to\infty} \int \left(\frac{1}{N} \sum_{n=1}^{N} f \circ \theta_n \right) g d\mathbb{P} = \int f d\mathbb{P} \int g d\mathbb{P}. \tag{A.2}$$

Proof See [145]. □

A.1.2 Transition Matrix, Markov Property

A transition probability matrix $P = (P_{xy})_{x,y \in S}$, is a matrix indexed by elements of S such that

$$\begin{aligned} &0 \le P_{xy} \le 1 \text{ for all } x, y \in S, \\ &\sum_{y \in S} P_{xy} = 1 \text{ for all } x \in S. \end{aligned} \tag{A.3}$$

The transition matrix encodes the one-step transition probabilities.

The powers of the transition matrix P are defined via the usual matrix multiplication

$$(P^n)_{xy} = \sum_{z \in S} P_{xz} P^{n-1}_{zy}, \tag{A.4}$$

where by definition $P^0_{xy} = \delta_{x,y}$. We call the matrix irreducible if for all $x, y \in S$ there exist $n \in \mathbb{N}$ such that $P^n_{xy} > 0$.

With the transition matrix we build the discrete-time Markov process via

$$\mathbb{P}(X_{t_1} = x_1, \ldots, X_{t_n} = x_n | X_0 = x_0) = P^{t_1}_{x_0,x_1} P^{t_2-t_1}_{x_1,x_2} \ldots P^{t_n-t_{n-1}}_{x_{n-1},x_n}. \tag{A.5}$$

This defines a unique collection of probability measures $\mathbb{P}_x$ on the discrete-time path space $S^{\mathbb{N}}$, endowed with the σ-algebra of cylinder events, indexed by the starting point $x \in S$, and correspondingly a unique discrete-time stochastic process $X_n, n \in \mathbb{N}$ with values in S. We denote by $\mathbb{P}_x$ the joint distribution of $X_0, X_1, \ldots, X_n, \ldots$ with starting point x, i.e., conditioned on $X_0 = x$, in particular $\mathbb{P}_x(X_0 = x) = 1$, and $\mathbb{E}_x$ denotes the corresponding expectation. More generally if μ is a probability measure on S, i.e., a collection of numbers $\mu(x) \geq 0$ with $\sum_{x \in S} \mu(x) = 1$ then $\mathbb{P}_\mu := \sum_x \mu(x)\mathbb{P}_x$ is the trajectory measure when the starting point X_0 is distributed according to μ, and we denote by $\mathbb{E}_\mu$ the corresponding expectation.

As a consequence of the way in which we have defined the finite dimensional marginals of the process, namely via (A.5), we obtain the Markov property: the conditional distribution of the trajectory after time n, i.e., the joint distribution of $X_{n+s}, s \in \mathbb{N}$ conditional on the past $\{X_0, \ldots, X_n\}$ depends only on the past via the "current state" X_n. Let us formalize this property. We call a function $f : S^{\mathbb{N}} \to \mathbb{R}$ cylindrical if it depends only on a finite number of coordinates, i.e., if there exists a finite subset $A \subset \mathbb{N}$ such that if $\omega, \omega' \in S^N$ satisfy $\omega_A = \omega'_A$ then $f(\omega) = f(\omega')$. For a trajectory $\omega \in S^{\mathbb{N}}$ we denote its shift over time n via $(\theta_n \omega)(t) = \omega(t+n), t \in \mathbb{N}$. The Markov property is then expressed as follows. For all bounded cylindrical functions $f : S^{\mathbb{N}} \to \mathbb{R}$ and for all $n \in \mathbb{N}$:

$$\mathbb{E}(f \circ \theta_n | X_0, \ldots, X_n) = \mathbb{E}(f \circ \theta_n | X_n) = \int f(\omega) d\mathbb{P}_{X_n}(\omega). \tag{A.6}$$

The first equality expresses the Markov property, i.e., the distribution of the trajectory after time n only depends on the state at time n and not on the further history. The second equality expresses the time-homogeneity of the process, i.e., the transition probabilities for a single step in the process do not depend on the time at which the step is taken. More explicitly, (A.6) can be restated in terms of conditional probabilities

$$\begin{aligned} &\mathbb{P}(X_{n+t_1} = y_1, \ldots, X_{n+t_k} = y_k | X_0 = x_0, \ldots, X_n = x_n) \\ &= \mathbb{P}(X_{t_1} = y_1, \ldots, X_{t_k} = y_k | X_0 = x_n), \end{aligned} \tag{A.7}$$

for all $k, n \in \mathbb{N}$, $t_1, \ldots, t_k \in \mathbb{N}$, $y_1, \ldots, y_k \in S$ and $x_0, \ldots, x_n \in S$. This expresses both the Markov property, i.e., the conditional probability in the lhs of (A.7) only depends on the last time in the conditioning, i.e., the event $X_n = x_n$, and on the time difference, i.e., not explicitly on n.

A.1.3 Transition Operator, Invariant Measures

Given the transition probabilities, we define the transition operator acting on bounded $f : S \to \mathbb{R}$

$$Pf(x) = \sum_{y \in S} P_{xy} f(y) = \mathbb{E}(f(X_1)|X_0 = x). \tag{A.8}$$

In this setting, a function can be thought of as a column vector, and the action of the transition operator is obtained by multiplying this column (on the left) with the transition matrix P. The basic properties satisfied by the transition operator P are the following.

1. Normalization: $P1 = 1$.
2. Positivity: if $f \geq 0$, then $Pf \geq 0$.
3. Contraction in the sup-norm: $\|Pf\|_\infty \leq \|f\|_\infty$. Here $\|f\|_\infty = \sup_{x \in S} |f(x)|$.

We can then also define the action of the transition operator on probability measures μ on S via

$$\int Pf d\mu = \sum_x Pf(x)\mu(x) =: \int f d\mu P, \tag{A.9}$$

where

$$\mu P(y) = \sum_x \mu(x) P_{xy}. \tag{A.10}$$

Thus μP represents the distribution at time 1 when the process at time zero is started from a state distributed according to μ, i.e.,

$$\int f d\mu P = \mathbb{E}_\mu f(X_1) = \int \mathbb{E}_x f(X_1) d\mu(x). \tag{A.11}$$

In the matrix notation, (A.10) expresses the fact that a probability measure can be viewed as a row vector, which after one time step evolves by multiplying (on the right) with the transition matrix.

Definition A.3 (Invariant, Reversible and Ergodic Measures)

1. A probability measure μ is called invariant or stationary if $\mu P = \mu$, or equivalently for all $f : S \to \mathbb{R}$ bounded $\int P f d\mu = \int f d\mu$.
2. A probability measure μ is called reversible if for all $f, g : S \to \mathbb{R}$ bounded functions

$$\int f(Pg) d\mu = \int (Pf) g d\mu. \tag{A.12}$$

3. An invariant probability measure μ is called ergodic if $Pf = f$ implies $f = \int f d\mu$, μ-almost surely.

The following proposition shows that this notion of ergodicity is consistent with the previously defined notion of ergodicity on path space.

Proposition A.4 *Let μ be an invariant probability measure. Then we have that μ is ergodic if and only if $\mathbb{P}_\mu$ is ergodic in the sense of Definition A.1*

Proof Let μ be ergodic, and let A be a shift invariant set. Our aim is to show that $\mathbb{P}_\mu(A) \in \{0, 1\}$. Denote

$$f_A(x) = \mathbb{E}_\mu(1_A|\mathcal{F}_0)(x). \tag{A.13}$$

Then by shift invariance of A, we obtain $Pf_A(x) = f_A(x)$, which implies, by ergodicity of μ that

$$f_A(x) = \int f_A(x)d\mu(x) = \mathbb{P}_\mu(A). \tag{A.14}$$

Similarly, denote

$$f_A^n(x_1, \ldots, x_n) = \mathbb{E}_\mu(1_A|\mathcal{F}_n)(x_1, \ldots, x_n). \tag{A.15}$$

By shift invariance of A, combined with the Markov property (A.6) and with (A.14), we get

$$\begin{aligned} \mathbb{E}_\mu(1_A|\mathcal{F}_n)(x_1, \ldots, x_n) &= \mathbb{E}_\mu(1_A \circ \theta_n|\mathcal{F}_n)(x_1, \ldots, x_n) \\ &= \mathbb{E}_\mu(1_A|\mathcal{F}_0)(x_n) \\ &= \mathbb{P}_\mu(A), \qquad \mathbb{P}_\mu\text{-almost surely.} \end{aligned} \tag{A.16}$$

Because $\mathcal{F}_n \uparrow \mathcal{F}$ as $n \to \infty$, $\mathbb{E}_\mu(1_A|\mathcal{F}_n) \to \mathbb{E}_\mu(1_A|\mathcal{F}) = 1_A$ as $n \to \infty$. This, combined with (A.16), yields

$$1_A = \mathbb{P}_\mu(A), \qquad \mathbb{P}_\mu\text{-almost surely,} \tag{A.17}$$

which shows that $\mathbb{P}_\mu(A) \in \{0, 1\}$.

Conversely, if $\mathbb{P}_\mu$ is ergodic and f is a bounded function such that $Pf = f$, then $\mathbb{P}_\mu$-almost surely we have

$$\mathbb{E}_\mu(f \circ \pi_1|\mathcal{F}_0) = f(\omega(0)), \tag{A.18}$$

which implies that for all $g : S \to \mathbb{R}$ bounded

$$\int f(\omega(1))g(\omega(0))d\mathbb{P}_\mu(\omega) = \int f(\omega(0))g(\omega(0))d\mathbb{P}_\mu(\omega). \tag{A.19}$$

Similarly, because $P^n f = f$, we obtain for all $n \in \mathbb{N}$:

$$\int f(\omega(n))g(\omega(0))d\mathbb{P}_\mu(\omega) = \int f(\omega(0))g(\omega(0))d\mathbb{P}_\mu(\omega) \tag{A.20}$$

and averaging over n gives

$$\lim_{N\to\infty} \int \frac{1}{N}\sum_{n=1}^{N} f(\omega(n))g(\omega(0))d\mathbb{P}_\mu(\omega)$$
$$= \int f(\omega(0))g(\omega(0))d\mathbb{P}_\mu(\omega) = \int fgd\mu. \tag{A.21}$$

By the ergodicity of $\mathbb{P}_\mu$ we have, using Proposition A.2,

$$\lim_{N\to\infty} \int \frac{1}{N}\sum_{n=1}^{N} f(\omega(n))g(\omega(0))d\mathbb{P}_\mu(\omega) = \int gd\mu \int fd\mu. \tag{A.22}$$

Combining (A.22) and (A.21) gives

$$\int fgd\mu = \int gd\mu \int fd\mu$$

for all g bounded. Then, by choosing $f = g$ we obtain $f = \int fd\mu$, μ-almost surely. □

The following lemma shows that different ergodic measures are singular.

Lemma A.5 *If μ and ν are two different ergodic measures, then μ and ν are mutually singular.*

Proof Assume the contrary, i.e., assume that μ is ergodic and

$$\nu(A) = \int_A fd\mu \tag{A.23}$$

with $f \geq 0$ and $\int fd\mu = 1$, and that also ν is ergodic. Then by invariance of both ν and μ we obtain

$$\int (P^n g)d\nu = \int fP^n gd\mu = \int gd\nu = \int fgd\mu \tag{A.24}$$

for all $n \in \mathbb{N}$ and $g : S \to \mathbb{R}$ bounded. Averaging and using that $\frac{1}{N}\sum_{n=1}^{N} P^n g \to \int g d\mu$, in μ probability, combined with dominated convergence (using that g is bounded), we obtain

$$\int f d\mu \int g d\mu = \int f g d\mu. \tag{A.25}$$

This holds for all g bounded measurable. As a consequence f is constant μ-almost surely, i.e. $f = \int f d\mu\ \mu$. This in turn gives that $\mu = \nu$. □

We collect in the next proposition some basic properties of stationary, ergodic and reversible measures.

Proposition A.6 *We have the following:*

1. *If S is finite then there exists a stationary distribution.*
2. *If S is finite then the set of invariant measures $\mathcal{I}$ is a non-empty convex set and the ergodic measures are the extreme points of $\mathcal{I}$.*
3. *If μ is stationary, then the transition operator is a contraction on $L^p(\mu)$ for all $1 \le p \le \infty$.*
4. *If S is finite and P is irreducible then there exists a unique stationary measure.*
5. *Reversibility implies stationarity.*
6. *Reversibility of a probability measure μ is equivalent with the detailed balance condition*

$$\mu(x)P_{xy} = \mu(y)P_{yx} \tag{A.26}$$

for all $x, y \in S$.

Proof We prove the statements one by one:

1. For item 1, we provide a well-known and standard proof which is based on the so-called Bogoliubov-Krylov argument and only uses that the set of probability measures on S, denoted $\mathcal{P}(S)$ is compact (under the weak topology). As a consequence, this argument is valid on general state spaces as long as $\mathcal{P}(S)$ is compact (e.g. when S is a compact metric space and $\mathcal{P}(S)$ is equipped with the weak topology). Notice that in the context of countable state spaces, the statement can also be proved via the Perron-Frobenius theorem, but this proof method is then restricted to that context where the transition operator is a matrix, whereas the Bogoliubov-Krylov argument is more general.

   Consider $\mu \in \mathcal{P}(S)$ and denote

$$\Gamma_N := \frac{1}{N}\sum_{k=1}^{N} \mu P^k. \tag{A.27}$$

We estimate, for $f : S \to \mathbb{R}$ bounded

$$\left|\int f d\Gamma_N - \int f d\Gamma_N P\right| \le \frac{1}{N}\left(\int |f| d\mu + \int |f| d\mu P^{N+1}\right) \le \frac{2\|f\|_\infty}{N}, \tag{A.28}$$

which implies that

$$\lim_{N\to\infty}\left|\int f d\Gamma_N - \int f d\Gamma_N P\right| = 0. \tag{A.29}$$

By compactness, there exists a subsequence $\Gamma_{N_k} \to \nu$ as $k \to \infty$. By (A.29) we then have

$$\int f d\nu = \int f d\nu P, \tag{A.30}$$

which shows that ν is invariant.

2. Non-emptyness has already been obtained, and convexity of the set is clear because the defining relation $\int P f d\mu = \int f d\mu$ for all $f : S \to \mathbb{R}$ bounded is a linear relation. It suffices to see that the ergodic measures are precisely the extreme points. Assume that μ is not ergodic, then there exists $A \subset S$ such that $0 < \mu(A) < 1$ and such that $P1_A = 1_A$, μ-almost surely. We first show that this implies that $\mu_A = \mu(\cdot|A)$ is invariant. Notice that $P1_A = 1_A$, μ-almost surely, implies for all $g : S \to \mathbb{R}$ bounded

$$\int 1_A(\omega(0))g(\omega(0))d\mathbb{P}_\mu = \int 1_A(\omega(1))g(\omega(0))d\mathbb{P}_\mu. \tag{A.31}$$

Choosing $g = 1_A$ gives

$$\int 1_A(\omega(0))1_A(\omega(0))d\mathbb{P}_\mu = \int 1_A(\omega(1))1_A(\omega(0))d\mathbb{P}_\mu, \tag{A.32}$$

which combined with stationarity of $\mathbb{P}_\mu$ gives

$$\int (1_A(\omega(0)) - 1_A(\omega(1)))^2 d\mathbb{P}_\mu = 0, \tag{A.33}$$

which implies $1_A(\omega(1)) = 1_A(\omega(0))$ $\mathbb{P}_\mu$-almost surely. This in turn yields

$$\begin{aligned}\int (Pg)1_A d\mu &= \int g(\omega(1))1_A(\omega(0))d\mathbb{P}_\mu \\ &= \int g(\omega(1))1_A(\omega(1))d\mathbb{P}_\mu = \int g1_A d\mu,\end{aligned} \tag{A.34}$$

which implies the invariance of μ_A. Then we have

$$\mu = \mu(A)\mu_A + (1 - \mu(A))\mu_{A^c}, \tag{A.35}$$

which is a non-trivial convex decomposition of μ in the set $\mathcal{I}$, showing that μ is not extreme in $\mathcal{I}$. Assume now that $\mu \in \mathcal{I}$ is not extreme and ergodic. We show that this leads to a contradiction. By the assumed non-extremality there exists $0 < \lambda < 1$, and $\mu_1 \neq \mu_2$ such that

$$\mu = \lambda\mu_1 + (1 - \lambda)\mu_2. \tag{A.36}$$

This implies that for the corresponding path space measures

$$\mathbb{P}_\mu = \lambda\mathbb{P}_{\mu_1} + (1 - \lambda)\mathbb{P}_{\mu_2}. \tag{A.37}$$

This in turn implies $\mathbb{P}_{\mu_1} << \mathbb{P}_\mu$. Put $F = \frac{d\mathbb{P}_{\mu_1}}{d\mathbb{P}_\mu}$. Because both $\mathbb{P}_\mu$ and $\mathbb{P}_{\mu_1}$ are shift invariant, it holds that F is shift invariant. By ergodicity of μ, this implies via Proposition A.4 that $\mathbb{P}_\mu$ is ergodic, and therefore combining this with translation invariance of F, we conclude $F = \int F d\mathbb{P}_\mu = 1$ $\mathbb{P}_\mu$-almost surely. This implies $\mathbb{P}_{\mu_1} = \mathbb{P}_\mu$ and hence $\mu_1 = \mu = \mu_2$, which is a contradiction.

3. If μ is invariant, then we have

$$\begin{aligned} \int |Pf|^p d\mu &= \int |\mathbb{E}_x f(X_1)|^p d\mu(x) \\ &\leq \int \mathbb{E}_x(|f|^p(X_1))d\mu(x) = \int |f|^p(x)d\mu(x), \end{aligned} \tag{A.38}$$

where we used Jensen's inequality combined with the stationarity of μ.

4. Assume the contrary, then there exists two different ergodic measures μ, ν. These measures are mutually singular. So there exists A with $\mu(A) = 0$ and $\nu(A) = 1$. Choose $x \in A$ such that $\mu(x) > 0$ and $y \in A^c$. Then choose n such that $P^n_{x,y} > 0$ which is possible by irreducibility. Then we have

$$\mathbb{P}_\mu(\omega(n) \in A^c) = \mu(A^c) = 0. \tag{A.39}$$

On the other hand

$$\mathbb{P}_\mu(\omega(n) \in A^c) \geq \mu(x)P^n_{x,y} > 0, \tag{A.40}$$

which is a contradiction.

5. If μ is reversible then choosing $g = 1$ in (A.12) gives, using $P1 = 1$, that for all f bounded $\int Pf d\mu = \int f d\mu$, which is invariance of μ.

6. If μ is reversible then we have, by definition,

$$\int fPgd\mu = \int gPfd\mu. \tag{A.41}$$

Choose $f(z) = \delta_{x,z}$, $g(z) = \delta_{y,z}$, then we obtain

$$\mu(x)P_{x,y} = \mu(y)P_{y,x}. \tag{A.42}$$

Conversely, if detailed balance holds, then

$$\int fPgd\mu = \sum_{x,y} \mu(x)f(x)P_{x,y}g(y) = \sum_{x,y} \mu(y)P_{y,x}f(x)g(y) = \int gPfd\mu. \tag{A.43}$$

□

A.2 Continuous-time Jump Processes

In continuous time, we will first restrict to the simplest case of a countable state space S. We will find back many of the results from the discrete time setting. The continuous-time setting, in its simplest form, already provides the notion of semigroup, generator, and Dynkin martingale which are important in the general theory.

A.2.1 Path Space

In continuous-time setting, the path space is $\Omega := D([0,\infty), S)$ which is the space of trajectories $\omega : [0,\infty) \to S$ which are right-continuous at any $t \geq 0$ and have left limits at any $t > 0$. We call such paths cadlag, from the French "continu à droite, limité à gauche". We denote by $\mathcal{F}_t$ the σ-algebra generated by the projections $\pi_s : \omega \to \omega(s)$, $s \leq t$ and $\mathcal{F} = \sigma\left(\cup_{t\geq 0}\mathcal{F}_t\right)$. On the path space we have the shift $\theta_t(\omega)(s) = \omega(s+t)$. We call a measure $\mathbb{P}$ on path space stationary if $\mathbb{P} \circ \theta_t = \mathbb{P}$ for all $t > 0$. We call a set $A \in \mathcal{F}$ invariant if $\theta_t(A) = A$ for all $t > 0$, and we call a measure on path space ergodic under time shifts if every invariant set has measure zero or one.

A.2.2 Markov Processes, Transition Rates

In continuous-time jump processes, starting from a state $x \in S$, the process waits an exponential time with parameter λ_x after which it jumps with probability $p(x, y)$

to a new state y. The jump times have to be exponentially distributed to ensure the Markov property via the memoryless property of the exponential distribution. First we formulate the Markov property. As before, to the process starting from $x \in S$ is associated a probability measure $\mathbb{P}_x$ on path space Ω. We denote by $\mathbb{E}_x$ expectation w.r.t. this probability measure. We say that the process $\{X_t, t \geq 0\}$ is time-homogeneous Markov if for all bounded cylindrical functions $f : \Omega \to \mathbb{R}$ we have

$$\mathbb{E}_x (f \circ \theta_t | \mathcal{F}_t) = \mathbb{E}_{X_t}(f). \tag{A.44}$$

As in the discrete case, this expresses the fact that conditional on the past up to time $t > 0$, the distribution of the future, i.e., of $\{X_{t+s}, s \geq 0\}$ is determined by X_t only, and not by any information about the further past. The time homogeneity is expressed by the fact that this distribution of the future does not explicitly depend on t.

In the following lemma we show that this implies that jump times have to be exponentially distributed.

Lemma A.7 *Denote* $T_x := \inf\{t > 0 : X_t \neq x\}$ *then there exists* $\lambda_x > 0$ *such that*

$$\mathbb{P}_x(T_x > t) = e^{-\lambda_x t} \tag{A.45}$$

Proof By the Markov property we have

$$\mathbb{P}_x(T_x > t + s | T_x > t) = \mathbb{P}_x(T_x > t + s | X_s = x, \ \forall\, 0 \leq s \leq t) = \mathbb{P}_x(T_x > s). \tag{A.46}$$

This implies $\mathbb{P}_x(T_x > t + s) = \mathbb{P}_x(T_x > t)\mathbb{P}_x(T_x > s)$, which is a characterization of the exponential distribution. □

The process $\{X_t, t \geq 0\}$ can then be defined via transition rates $c(x, y) \geq 0$ for $y \neq x$, which can be thought of as "transition probability per unit of time". We also make the convention that $c(x, x) = 0$. When starting from $x \in S$ the process waits an exponential time with parameter $\lambda_x = \sum_{z \neq x} c(x, z)$, after which it jumps to a new state $y \in S$ with probability $p(x, y) = c(x, y)/\lambda_x$. In order to ensure that the process is well-defined we assume

$$\sup_{x \in S} \lambda_x < \infty. \tag{A.47}$$

This is not necessary but sufficient to ensure that the process does not escape in finite time to "infinity", and that it admits a version with cadlag paths. We call the process irreducible if for all $x, y \in S$ there exists $x_1, \ldots, x_n$ with $x_1 = x, x_n = y$ and $\prod_{i=1}^{n-1} c(x_i, x_{i+1}) > 0$.

The process $\{X_t, t \geq 0\}$ admits the following graphical representation. For each pair $(x, y) \in S \times S$, we consider a Poisson process $\{N_t^{xy}, t \geq 0\}$ with rate

$c(x, y)$. For different pairs, these Poisson processes are independent. At each event time of the Poisson process we draw an arrow from x to y. The process, with given initial point x is then constructed by "following the arrows". This graphical representation then provides a coupling of the processes $\{X_t^x, t \geq 0\}$ with different initial conditions $X_0^x = x$. I.e., given the realization of all the Poisson processes $\{N_t^{xy}, t \geq 0\}$, with $(x, y) \in S \times S$, the processes $\{X_t^x, t \geq 0\}$ with $x \in S$ are deterministic functions of this realization.

A.2.3 Semigroups, Generators and Invariant Measures

In this section we review the connection generator-semigroup-Markov process in the simple setting of Markovian jump processes of the previous section. As we did before in the discrete case, in doing so, we set up notation and provide proofs which can be used in much wider context without too much adaptation.

Let us denote $\mathcal{B}(S)$ the set of bounded functions $f : S \to \mathbb{R}$. To the Markov process $\{X_t, t \geq 0\}$ we associate the collection of operators

$$S(t)f(x) = \mathbb{E}_x f(X_t), \tag{A.48}$$

acting on $f \in \mathcal{B}(S)$. Notice that because S is discrete, this space coincides of course with the space of bounded continuous functions. The following proposition collects the basic properties of $\{S(t), t \geq 0\}$.

Proposition A.8 *The collection of operators $\{S(t) : t \geq 0\}$ satisfies the following properties.*

1. *$S(0) = I$, where I denotes the identity.*
2. *Normalization: $S(t)1 = 1$.*
3. *Positivity: $f \geq 0$ implies $S(t)f \geq 0$.*
4. *Contraction in sup-norm: $\|S(t)f\|_\infty \leq \|f\|_\infty$.*
5. *Semigroup property: $S(t+s) = S(t)S(s)$ for all $t, s \geq 0$.*
6. *Right continuity in the sup-norm*

$$\lim_{t \to 0} \|S(t)f(x) - f(x)\|_\infty = 0. \tag{A.49}$$

Proof The first four properties are immediate from the definition. We prove the fifth property, which is intimately connected to the Markov property:

$$\begin{aligned} S(t+s)f(x) &= \mathbb{E}_x(f(X_{t+s})) \\ &= \mathbb{E}_x\left(\mathbb{E}_x(f(X_{t+s})|\mathcal{F}_t)\right) \\ &= \mathbb{E}_x\left(\mathbb{E}_{X_t}(f(X_s))\right) \end{aligned}$$

$$= \mathbb{E}_x(S(s)f(X_t))$$
$$= (S(t)(S(s)f))(x). \tag{A.50}$$

Finally, to prove right continuity, notice that

$$|S(t)f(x) - f(x)| \le 2\|f\|_\infty \mathbb{P}_x(T_x \le t) = 2\|f\|_\infty (1 - e^{-\lambda_x t}). \tag{A.51}$$

Then use (A.47) to conclude. □

Remark A.9 The semigroup $S(t)$ can be viewed as a matrix indexed by elements of the countable state space S. The matrix element $S(t)_{xy}$ is simply the transition probability to move from x to y in a time t:

$$S(t)_{xy} = p_t(x, y) = \mathbb{E}_x(\mathbb{1}_{\{X_t = y\}}). \tag{A.52}$$

The semigroup property is equivalent with the Chapman-Kolmogorov equation

$$p_{t+s}(x, y) = \sum_z p_t(x, z) p_s(z, y). \tag{A.53}$$

We can now introduce the generator. Notice that because we are restricting here to the case $\sup_x \lambda_x < \infty$, the generator is a bounded operator (on the set of bounded functions $f : S \to \mathbb{R}$) so we will not (yet) have to deal with domain problems, or problems on how to define the exponential of the generator.

Definition A.10 (Generator) The generator of the semigroup is defined as the limit

$$Lf = \lim_{t \to 0} \frac{S(t)f - f}{t}, \tag{A.54}$$

where the limit is in the uniform topology, i.e., such that

$$\lim_{t \to 0} \left\| \frac{S(t)f - f}{t} - Lf \right\|_\infty = 0. \tag{A.55}$$

The domain of the generator is

$$\mathcal{D}(L) = \left\{ f : \lim_{t \to 0} \frac{S(t)f - f}{t} \text{ exists} \right\}. \tag{A.56}$$

Proposition A.11 *Under condition* (A.47), *L is a bounded operator, i.e., $\mathcal{D}(L) = \mathcal{B}(S)$, and we have the "exponentiation" relation between the generator and the semigroup*

$$S(t) = e^{tL} = \sum_{n=0}^{\infty} \frac{t^n L^n}{n!}, \tag{A.57}$$

where the sum converges in the uniform operator topology. The generator is explicitly given by

$$Lf(x) = \sum_y c(x,y)(f(y) - f(x)). \tag{A.58}$$

Proof Assume the process starts from $X_0 = x$. Let us denote N_t the total number of jumps of the process in time $[0, t]$. Then, reminding that $\sup_x \lambda_x < \infty$, we have $\mathbb{P}_x(N_t \geq 2) = o(t)$, i.e, uniformly in x, $\lim_{t\to 0} \mathbb{P}_x(N_t \geq 2)/t = 0$. Therefore, we can write

$$\begin{aligned} \mathbb{E}_x f(X_t) &= \mathbb{E}_x(f(X_t)\mathbb{1}_{\{N_t=0\}}) + \mathbb{E}_x(f(X_t)\mathbb{1}_{\{N_t=1\}}) + o(t) \\ &= f(x)e^{-\lambda_x t} + \sum_{y\neq x} p(x,y)t\lambda_x e^{-\lambda_x t} f(y) + o(t). \end{aligned} \tag{A.59}$$

As a consequence, using that $c(x, y) = p(x, y)\lambda_x$ when $y \neq x$ and $\sum_y p(x, y) = 1$ we obtain

$$\lim_{t\to 0} \frac{\mathbb{E}_x f(X_t) - f(x)}{t} = \sum_{y\neq x} c(x,y)(f(y) - f(x)). \tag{A.60}$$

The fact that the limit is uniform in x follows from (A.47). From the semigroup property combined with the fact $L = \lim_{t\to 0} \frac{S(t)-I}{t}$ we obtain

$$\frac{d}{dt}S(t) = LS(t) = S(t)L, \tag{A.61}$$

which leads to $S(t) = e^{tL}$. □

In the following we collect some basic properties of the generator. The simple proof is left to the reader. Notice that in matrix notation, the generator has as off-diagonal elements L_{xy} the jump rate $c(x, y)$ and on the diagonal $L_{xx} = -\lambda_x$, which implies that the row sums are zero, which corresponds to the normalization property of the semigroup ($S(t)1 = 1$).

Proposition A.12 *The generator satisfies the following properties:*

1. $L1 = 0$
2. *L satisfies the maximum principle, i.e., if $f(x) = \max_y f(y)$ then $Lf(x) \leq 0$.*

We can then define the Kolmogorov forward and backward equations. In the physics literature, the Kolmogorov forward equation is called the master equation. For a probability measure $\mu \in \mathcal{P}(S)$ we define the measure at time $t > 0$, which we denote $\mu_t = \mu S(t)$ via

$$\int S(t)f d\mu = \int f d\mu S(t). \tag{A.62}$$

In matrix notation, the measure at time t is obtained by multiplying the row matrix μ (on the right) with the matrix $S(t)$. As a consequence we have the following differential equations for μ_t and $S(t)f$, which we give without proof, because the statements follow immediately from $S(t) = e^{tL}$, and as a consequence

$$\frac{dS(t)f}{dt} = LS(t)f = S(t)Lf, \tag{A.63}$$

for all f.

Theorem A.13 *We have the following:*

1. *Kolmogorov backwards equation. The function $f_t(x) = S(t)f(x)$ is the unique solution of the differential equation*

$$\frac{df_t(x)}{dt} = \sum_y c(x,y)(f_t(y) - f_t(x)), \tag{A.64}$$

with initial condition $f_0(x) = f(x)$. This equation is called the Kolmogorov backwards equation.

2. *Kolmogorov forward equation or master equation. The function $\mu_t(x) = \mu S(t)(\{x\})$ is the unique solution of the differential equation*

$$\frac{d\mu_t(x)}{dt} = \sum_y (\mu_t(y)c(y,x) - \mu_t(x)c(x,y)). \tag{A.65}$$

Next, we define the notions of invariant, reversible, ergodic measures in this context of continuous-time Markov jump processes.

Definition A.14 (Invariant, Ergodic and Reversible Measure)

1. A probability measure μ is called invariant if $\mu_t = \mu$ for all $t > 0$, where μ_t is defined via

$$\int S(t)f d\mu = \int f d\mu_t, \tag{A.66}$$

for all $f : S \to \mathbb{R}$ bounded. The set of invariant measures is denoted $\mathcal{I}$.

2. An element of $\mathcal{I}$ is called ergodic if the corresponding process measure $\mathbb{P}_\mu = \int \mathbb{P}_x d\mu(x)$ is ergodic under time shifts.
3. A probability measure μ is called reversible if

$$\int (S(t)f)g d\mu = \int f(S(t)g)d\mu, \tag{A.67}$$

for all $t \geq 0$, $f, g : S \to \mathbb{R}$ bounded.

It follows from (A.67) that if μ is reversible, then both the semigroup $S(t)$ and the generator L are self-adjoint operators in $L^2(\mu)$. We have the following further properties of invariant measures, which are the analogues of the Properties listed in proposition A.6 in the continuous-time setting.

Proposition A.15 *We have the following:*

1. $\mu \in \mathcal{I}$ *if and only if* $\int Lf d\mu = 0$ *for all* $f \in \mathcal{D}(L)$.
2. μ *is ergodic if and only if* $S(t)f = f$ *for all* $t > 0$ *implies* $f = \int f d\mu$, μ*-a.s.*
3. *The set of ergodic measures is the set of extreme points of the convex set* $\mathcal{I}$.
4. *If* S *is finite, then* $\mathcal{I}$ *is non-empty.*
5. *If* $\mu \in \mathcal{P}(S)$, *then* $S(t)$ *is a semigroup of contractions in* $L^p(\mu)$ *for all* $p \in [1, \infty]$.
6. *If* μ *is reversible, then* μ *is invariant.*
7. μ *is reversible if and only if* μ *satisfies the detailed balance relation*

$$\mu(x)c(x, y) = \mu(y)c(y, x) \tag{A.68}$$

for all $x, y \in S$.

Proof The proofs of items 2–7 are the obvious modification of the proofs of the corresponding discrete time statements in Proposition A.6 (i.e. replacing discrete time averages by integrals). For item 1, if $\mu \in \mathcal{I}$

$$\int Lf d\mu = \int \lim_{t\to 0} \frac{S(t)f - f}{t} d\mu = \lim_{t\to 0} \frac{1}{t} \int (S(t)f - f) d\mu = 0, \tag{A.69}$$

where we are allowed to pull the limit out of the integral because of uniformity of the limit. Conversely, if $\int Lf d\mu = 0$ for all $f \in \mathcal{D}(L)$, then for all such f and for all $t > 0$, using (A.63) we obtain

$$\frac{d}{dt} \int S(t) f d\mu = 0, \tag{A.70}$$

which implies that $\int S(t) f d\mu = \int S(0) f d\mu = \int f d\mu$ for all $t > 0$. □

In the following proposition we show how from the generator natural martingales can be generated.

Proposition A.16 *For all* $f : S \to \mathbb{R}$ *bounded, the process*

$$M_t = f(X_t) - f(X_0) - \int_0^t Lf(X_s) ds \tag{A.71}$$

is a martingale. The predictable quadratic variation of this martingale is given by

$$[M, M]_t = \int_0^t \Gamma(f)(X_s) ds, \tag{A.72}$$

where the quadratic form $\Gamma(f)$ *is given by*

$$\Gamma(f)(x) = (L(f^2))(x) - 2f(x)Lf(x) = \sum_y c(x,y)(f(y) - f(x))^2. \tag{A.73}$$

Proof We prove that $\mathbb{E}(M_t - M_s|\mathcal{F}_s) = 0$ for all $0 \le s < t$. We compute, using the Markov property, combined with the definition of the semigroup

$$\begin{aligned}\mathbb{E}(M_t - M_s|\mathcal{F}_s) &= \mathbb{E}\left(f(X_t) - f(X_s) - \int_s^t Lf(X_r)dr|\mathcal{F}_s\right) \\ &= S(t-s)f(X_s) - f(X_s) - \int_s^t LS(r-s)f(X_s)dr \\ &= S(t-s)f(X_s) - f(X_s) - \int_s^t \frac{d}{dr}S(r-s)f(X_s)dr = 0.\end{aligned} \tag{A.74}$$

To prove the formula for the quadratic variation, it suffices to prove that

$$\mathbb{E}\left((M_t - M_s)^2 - \int_s^t \Gamma(f)(X_r)dr\Big|\mathcal{F}_s\right) = o(t-s) \tag{A.75}$$

as $t - s \to 0$. We put $s = 0$, the case $s > 0$ is analogous. We then have to prove

$$\mathbb{E}_x\left(M_t^2 - \int_0^t \Gamma(f)(X_r)dr\right) = o(t) \tag{A.76}$$

as $t \to 0$. Working out the square and neglecting obvious $o(t)$ terms, we have

$$\begin{aligned}&\mathbb{E}_x\left(M_t^2 - \int_0^t \Gamma(f)(X_r)dr\right) \\ &= \mathbb{E}_x\Bigg(f^2(X_t) - 2f(X_t)f(x) + f(x)^2 - \int_0^t Lf^2(X_r)dr \\ &+ 2\int_0^t f(X_r)Lf(X_r)dr\Bigg) + o(t) \\ &= \mathbb{E}_x\Bigg(f^2(x) + \int_0^t Lf^2(X_r)dr - f(x)^2 - 2\int_0^t f(x)Lf(X_r) - \int_0^t Lf^2(X_r)dr \\ &+ 2\int_0^t f(X_r)Lf(X_r)dr\Bigg) + o(t) \\ &= 2\mathbb{E}_x\int_0^t Lf(X_r)(f(X_r) - f(x)) + o(t).\end{aligned} \tag{A.77}$$

The last expression is clearly $o(t)$ and therefore (A.76) follows. Finally, the expression (A.73) follows by computing explicitly $Lf^2 - 2fLf$ for the generator (A.58). □

Remark A.17 The martingale (A.71) is called the Dynkin martingale (associated to f), whereas the quadratic form $\Gamma(f) = Lf^2 - 2fLf$ is called the "carré du champ" operator.

A.2.4 Examples

Example 1: Two State Markov Chain

The simplest example of a continuous-time Markov chain is the chain with two states $1, 2$, hopping from 1 to 2 at rate $\alpha > 0$ and from 2 to 1 at rate β. Its generator is given by the two by two matrix

$$L = \begin{pmatrix} -\alpha & \alpha \\ \beta & -\beta \end{pmatrix} \tag{A.78}$$

The semigroup, which is the exponential of this matrix can be obtained explicitly by diagonalization and is given by

$$S(t) = e^{tL} = \begin{pmatrix} \frac{\beta}{\alpha+\beta} & \frac{\alpha}{\alpha+\beta} \\ \frac{\beta}{\alpha+\beta} & \frac{\alpha}{\alpha+\beta} \end{pmatrix} + \frac{e^{-(\alpha+\beta)t}}{\alpha+\beta} \begin{pmatrix} \alpha & -\alpha \\ -\beta & \beta \end{pmatrix} \tag{A.79}$$

From this we see that the unique invariant and reversible measure is given by $\nu = (\frac{\beta}{\alpha+\beta}, \frac{\alpha}{\alpha+\beta})$, and for each $f : \{1, 2\} \to \mathbb{R}$, $S(t)f$ converges exponentially fast to $\int f d\nu$. The fact that the invariant measure is automatically reversible is because we have two states. As soon as the number of states is 3 or larger, invariant measures can be non-reversible.

Example 2: Poisson Process with Intensity λ

The Poisson process is the process $\{N(t) : t \geq 0\}$ on $\mathbb{N}$ characterized by

1. Independent Poissonian increments: for $0 = t_0 < t_1 < t_2 \ldots < t_n$ the random variables $N(t_i) - N(t_{i-1}), i = 1, \ldots, n$ are independent Poisson distributed with parameter $\lambda(t_i - t_{i-1})$, i.e., for all $n = 0, 1, 2, \ldots$

$$\mathbb{P}\left(N(t_i) - N(t_{i-1}) = n\right) = \frac{(\lambda(t_i - t_{i-1}))^n}{n!} e^{-\lambda(t_i - t_{i-1})}. \tag{A.80}$$

2. $N_0 = 0$.

Equivalently, the Poisson process can be described starting from a family $\{T_n, n = 1, 2, \ldots\}$ of independent exponential (with parameter λ) random variables via

$$N(t) = \sup\left\{k \in \mathbb{N} : \sum_{i=1}^{k} T_i \le t\right\}. \tag{A.81}$$

The Poisson process starting from $n \in \mathbb{N}$ is then defined as $\{N_t + n, t \ge 0\}$. The Poisson process is a Markov process with generator defined on bounded functions $f : \mathbb{N} \to \mathbb{R}$ by

$$Lf(n) = \lambda(f(n+1) - f(n)). \tag{A.82}$$

The corresponding semigroup is given by

$$S(t)f(n) = \sum_{k=0}^{\infty} \frac{(\lambda t)^k}{k!} f(n+k). \tag{A.83}$$

From the independent Poissonian increments property, one derives that the following processes are martingales w.r.t. the natural filtration $\mathcal{F}_t = \sigma\{N(s) : 0 \le s \le t\}$

1. The compensated Poisson process: $N(t) - \lambda t$.
2. The Poissonian exponential martingale: $X_\alpha(t) = e^{\alpha N(t) - \lambda(e^\alpha - 1)t}$.
3. The compensated variance: $(N(t) - \lambda t)^2 - \lambda t$.

Notice that the martingales (1) and (3) can be obtained (formally) as a Dynkin martingale (cf. proposition A.16) via

$$L(n) = \lambda(n + 1 - n) = \lambda \tag{A.84}$$

which gives that $N(t) - \lambda t$ is a martingale with quadratic variation computed via the carré du champ

$$L(n^2) - 2nL(n) = \lambda((n+1)^2 - n^2) - 2n\lambda = \lambda \tag{A.85}$$

i.e., the quadratic variation of $N(t) - \lambda t$ equals λt and therefore $(N(t) - \lambda t)^2 - \lambda t$ is a martingale. Notice that the martingales (1) and (3) can be obtained by taking derivatives of $X_\alpha(t)$ w.r.t. α, and then putting $\alpha = 0$.

Example 3: Continuous-time Simple Random Walk on $\mathbb{Z}^d$

The continuous-time simple random walk with edge rate λ on $\mathbb{Z}^d$ is described by the generator

$$Lf(x) = \sum_{e:|e|=1} \lambda(f(x+e) - f(x)). \tag{A.86}$$

The process, when started at x waits an exponential time with parameter $\lambda 2d$, after which it jumps with equal probability to one of the $2d$ neighbors of x. This can alternatively been described as follows: to each oriented nearest-neighbor edge we associate a Poisson process with rate λ, where the processes are independent for different edges. At each event time (=jump time) of the Poisson process associated to the edge (xy) we put an arrow from x to y. The random walk $X^x(t)$ starting from x is then obtained by "following the outgoing arrows". This construction of the random walk based on Poisson processes has the advantage that it provides a joint coupling of $X^x(t)$ for different starting points $x \in \mathbb{Z}^d$. The process $X(t) = (X_1(t), \ldots, X_d(t))$ has independent components which evolve according to the continuous-time simple random walk on $\mathbb{Z}$. If we have two independent copies $X(t), Y(t)$ of the process with generator L, then the difference $Z(t) = X(t) - Y(t)$ is a continuous-time simple random walk with edge rate 2λ on $\mathbb{Z}^d$. A simple representation in $d = 1$ of the process $X(t)$ starting from zero is given by

$$X(t) = N^+(t) - N^-(t), \tag{A.87}$$

where $N^+(t)$ and $N^-(t)$ are two independent Poisson processes of rate λ. From this one obtains the generating function

$$\mathbb{E}_0 e^{\alpha X(t)} = e^{\lambda t(e^\alpha + e^{-\alpha} - 2)}, \tag{A.88}$$

as well as the characteristic function

$$\mathbb{E}_0 e^{iqX(t)} = e^{-4\lambda t \sin^2(q/2)}. \tag{A.89}$$

Example 4: Symmetric Exclusion Process on a Finite Set

Let (V, E) be a finite unoriented and connected graph with vertices V and edges E. A configuration of the exclusion process is an element of $\Omega = \{0, 1\}^V$ where for $\eta \in \Omega$, $i \in V$, $\eta_i \in \{0, 1\}$ is interpreted as the number of particles at i in the configuration η. Furthermore, by $\eta^{x,y}$ we denote the configuration where occupancies at x and y are interchanged, i.e.,

$$\eta_z^{x,y} = \begin{cases} \eta_x & \text{if } z = y \\ \eta_y & \text{if } z = x \\ \eta_z & \text{otherwise.} \end{cases} \tag{A.90}$$

The generator of the symmetric exclusion process $\{\eta(t) : t \geq 0\}$ on (V, E) is given by

$$Lf(\eta) = \sum_{(xy)\in E} (f(\eta^{x,y}) - f(\eta)). \tag{A.91}$$

More generally, if $\kappa : E \to (0, \infty)$, we call the symmetric exclusion process with edge rate κ the process with generator

$$Lf(\eta) = \sum_{(xy)\in E} \kappa(x, y)(f(\eta^{x,y}) - f(\eta)). \tag{A.92}$$

Notice that we associate the weight to the unoriented edges. Denote by ν_ρ the Bernoulli product measure on Ω with $\nu(\eta_x = 1) = \rho$ for all $x \in V$. Then it holds, $\nu_\rho(\eta^{x,y}) = \nu_\rho(\eta)$ for all η, and $x, y \in V$. As a consequence, the measures ν_ρ are reversible, for any choice of (connected) graph (V, E) and edge rates κ. Because the total number of particles $|\eta| = \sum_{x\in V} \eta_x$ is conserved, these measures are not ergodic. Instead, the "canonical measures" on the set of configurations with n particles given by conditioning the Bernoulli measure on having n particles are both reversible and ergodic. These measures are given by

$$\nu^{(n)}(\eta) = \nu_\rho(\eta||\eta| = n) = \begin{cases} \frac{1}{Z_n} & \text{if } \sum_{x\in V} \eta_x = n \\ 0 & \text{otherwise,} \end{cases} \tag{A.93}$$

with $Z_n = \binom{|V|}{n}$. Notice that when V is infinite, the situation changes drastically, because the "total number of particles" can be infinite. In that case, the Bernoulli product measures are reversible and ergodic under reasonable conditions on the graph (and or the edge rates). See [172], Chapter 8 for a detailed study.

A.3 Generators and Semigroups: General Case

In general setting, Markov semigroups and their generators are still related by "exponentiation" but this exponentiation is no longer defined via the sum of the classical Taylor series, because the generator is in general an unbounded operator. In a suitable function space, which is depending on the process under consideration, the Markov semigroup acts as a semigroup of positive contractions, and therefore, one can use Hille-Yosida theory to define the exponential of the generator via the resolvents (Laplace transform), i.e., using the so-called Yosida-approximants. In general, it is not easy to explicitly characterize the domain of the generator of a Markov semigroup. The strategy is usually to start from a "pregenerator", i.e., an operator which is defined on a smaller set of functions (e.g. smooth functions in diffusion process context or local functions in the context of interacting particle systems) and has "all the properties of a Markov generator", except for not being closed. The challenge is then to prove that the closure of this operator generates a Markov semigroup.

A.3.1 Hille Yosida Theory of Markov Semigroups

Following [172], Chapter 1, we give here an overview of this theory in the case where the state space of the Markov process is a compact metric space Ω, and the corresponding function space on which the Markov semigroup acts is the space $\mathscr{C}(\Omega)$ of continuous functions equipped with the supremum norm.

Definition A.18 A collection of operators $\{S(t) : t \geq 0\}$ is called a Markov semigroup (also called Feller semigroup) when the following properties are satisfied.

1. Contraction: $S(t) : \mathscr{C}(\Omega) \to \mathscr{C}(\Omega)$ satisfies $\|S(t)f\|_\infty \leq \|f\|_\infty$.
2. Normalization $S(t)1 = 1$.
3. Positivity: $f \geq 0$ implies $S(t)f \geq 0$.
4. Right continuity: for all $f \in \mathscr{C}(\Omega)$: $\lim_{t\to 0} \|S(t)f - f\|_\infty = 0$.
5. Semigroup property: $S(0) = I$ and $S(t+s) = S(t)S(s)$ for all $s, t \geq 0$.

To a Markov semigroup corresponds a unique Markov process in the following sense.

Proposition A.19 *Let $S(t), t \geq 0$ be a Markov semigroup then there exists a unique process $\{X_t, t \geq 0\}$ on Ω, such that*

1. *The process has a version with paths in $D([0,\infty), \Omega)$, i.e., the joint distribution of $\{X_t, t \geq 0\}$ conditioned on $X_0 = x$ is a probability measure $\mathbb{P}_x$ on $D([0,\infty), \Omega)$.*
2. *The relation between the Markov semigroup and the process is given by*

$$S(t)f(x) = \mathbb{E}_x(f(X_t)) = \int f \circ \pi_t(\omega) d\mathbb{P}_x(\omega), \tag{A.94}$$

where $\pi_t(\omega) = \omega(t)$ is the projection at time t on path space.

Remark A.20 For the uniqueness it is important that we specify that we choose the cadlag version, i.e., such that $\{X_t, t \geq 0\}$ has trajectories in $D([0,\infty), \Omega)$. The semigroup determines the finite dimensional distributions via the Markov property. By Kolomogorov theorem this leads, for a given starting point $x \in \Omega$ to a unique probability measure $\mathscr{P}_x$ on the product space $\Omega^{[0,\infty)}$ equipped with the cylinder σ-algebra. It is then via the right-continuity of the semigroup that one proves that this further leads to a cadlag process $\mathbb{P}_x$ satisfying (A.94). See [89], for full details.

We now discuss the relation between the semigroup and the generator, which is an appropriate generalization of the simple relation $S(t) = e^{tL}$ which we encountered in the finite or countable state space case.

Following [172], chapter 1, we define the notion of a pregenerator.

Definition A.21 (Pregenerator) An operator L with domain $D(L) \subset \mathscr{C}(\Omega)$ is called a pregenerator if

(a) $1 \in D(L)$, and $L1 = 0$.

(b) $D(L)$ is dense in $\mathscr{C}(\Omega)$ with the supremum norm.
(c) If $f \in D(L)$, $\lambda \geq 0$ and $(I - \lambda L)f = g$ then

$$\min_{\eta \in \Omega} f(\eta) \geq \min_{\eta \in \Omega} g(\eta). \tag{A.95}$$

Remark A.22 As a consequence of item (c), applying it to f and $-f$, and using $\|f\| = \max\{-\min_\eta f(\eta), -\min_\eta(-f(\eta))\}$, we obtain the following: if for some $\lambda \geq 0$, $(I - \lambda L)f = g$, then $\|f\| \leq \|g\|$. This implies that $(I - \lambda L)^{-1}$ acts as a contraction when it is defined.

In particular, f is uniquely determined by $(I - \lambda L)f = g$. Indeed, if $(I - \lambda L)f_1 = (I - \lambda L)f_2 = g$, then $(I - \lambda L)(f_1 - f_2) = 0$ and therefore $\|f_1 - f_2\| = 0$, which gives $f_1 = f_2$.

The following proposition shows that item (c) from Definition A.21 follows from the so-called maximum principle.

Proposition A.23 *Suppose L is a linear operator with domain $D(L)$ such that for $f \in D(L)$, we have that if $f(\eta) = \min_{\xi \in \Omega} f(\xi)$, then*

$$Lf(\eta) \geq 0. \tag{A.96}$$

Then L satisfies item (c) of Definition A.21

Proof Put $f \in \mathscr{C}(\Omega)$, $\lambda \geq 0$ and $g = (I - \lambda L)f$. Let η be such that $f(\eta) = \min_{\zeta \in \Omega} f(\zeta)$ (such η exists by compactness of Ω and continuity of f). Then $\lambda Lf(\eta) \geq 0$ by assumption, and hence

$$\min_{\zeta \in \Omega} f(\zeta) \geq f(\eta) - \lambda Lf(\eta) = g(\eta) \geq \min_{\zeta \in \Omega} g(\zeta). \tag{A.97}$$

□

Remark A.24 The following remarks are in place:

1. Coming back to the case of a finite state space, where $S_t = e^{tL}$. We then have

$$\begin{aligned} (I - \lambda L)^{-1} &= \int_0^\infty e^{-t(I - \lambda L)} dt \\ &= \int_0^\infty \frac{1}{\lambda} e^{-\frac{t}{\lambda}} S(\lambda t) dt \\ &= \mathbb{E}(S(\tau)), \end{aligned} \tag{A.98}$$

where in the last equality τ is a random variable which is exponentially distributed with expectation λ. As a consequence if $(I - \lambda L) f = g$ we have

$$f(x) = \int_0^\infty e^{-t} S(\lambda t) g(x)\, dt \geq \min_x g(x) \int_0^\infty e^{-t} S(\lambda t) 1\, dt = \min_x g(x), \tag{A.99}$$

and therefore item (c) of Definition A.21 automatically follows. So we see that this item is related to positivity and normalization of the semigroup.
2. The maximum principle (A.96) in the finite case simply follows from $S_t f(\eta) \geq \min_\zeta f(\zeta) = f(\eta)$ therefore $Lf(\eta) = \lim_{t\to 0}(S_t f(\eta) - f(\eta))/t \geq 0$. In the case of processes of the type Brownian motion (see Sect. A.3.2 below), the generator is of the type "second derivative with positive coefficients", which is positive where the function attains a minimum.

Definition A.25 (Closability)

1. A linear operator L with domain $D(L) \subset \mathscr{C}(\Omega)$ is called closed if the graph

$$\mathcal{G}(L) = \{(f, Lf) : f \in D(L)\}$$

is a closed subset of $\mathscr{C}(\Omega) \times \mathscr{C}(\Omega)$.
2. For a linear operator L with domain $D(L)$ we call the operator closable if for any sequence such that $f_n \to 0$, $f_n \in D(L)$, $Lf_n \to h$, it holds that $h = 0$.

Closability implies that if $f_n \to f$, $f'_n \to f$, $Lf_n \to h$, $Lf'_n \to h'$ then $h - h' = 0$ and so we can define the closure of the operator via $\bar{L} f = h$. In other words, an operator is closable if the closure of the graph $\{(f, Lf) : f \in D(L)\}$ is the graph of an operator. This is then the smallest closed extension of the operator. An operator is called closed if it is equal to its closure. It is a general property of generators of right continuous contraction semigroups to be closed, as we will prove below. Therefore, starting from a pregenerator, we need to consider its closure in order to have a candidate generator of a contraction semigroup. The following proposition shows that the closure of a Markov pregenerator is again a Markov pregenerator.

Proposition A.26 *A Markov pregenenerator L is closable and its closure $\bar{L}$ is again a Markov pregenerator.*

Proof First we show that L is closable. Let $f_n \in D(L)$ and $f_n \to 0$, $Lf_n \to h$. Choose $g \in D(L)$ then we have

$$\|(I - \lambda L)(f_n + \lambda g)\| \geq \|f_n + \lambda g\|. \tag{A.100}$$

Taking the limit $n \to \infty$ gives

$$\|\lambda g - \lambda h - \lambda^2 Lg\| \geq \|\lambda g\|, \tag{A.101}$$

which leads to

$$\|g - h\| \geq \|g\|. \tag{A.102}$$

Because this holds for all $g \in D(L)$ and $D(L)$ is dense by assumption, we conclude that $h = 0$. To show that $\bar{L}$ is a pregenerator, we only have to verify item (c) of Definition A.21 for $\bar{L}$. So assume

$$f - \lambda \bar{L} f = g. \tag{A.103}$$

Choose a sequence $f_n \in D(L)$ such that $f_n \to f$, $Lf_n \to \bar{L}f$. Define

$$f_n - \lambda L f_n = g_n, \tag{A.104}$$

then because L is a pregenerator

$$\min_\zeta f_n(\zeta) \geq \min g_n(\zeta). \tag{A.105}$$

Since f_n, g_n both uniformly converge to f, g we can take the limit and obtain

$$\min_\zeta f(\zeta) \geq \min_\zeta g(\zeta). \tag{A.106}$$

□

Proposition A.27 *Let L be a closed Markov pregenerator. Then for all $\lambda \geq 0$. The range $\mathcal{R}(I - \lambda L)$ is closed.*

Proof Assume

$$f_n - \lambda L f_n = g_n, \tag{A.107}$$

and $g_n \to g$. Then

$$f_n - f_m - \lambda L(f_n - f_m) = g_n - g_m. \tag{A.108}$$

Because $g_n \to g$, g_n is a Cauchy sequence and therefore, $\|g_n - g_m\| \to 0$ as $n, m \to \infty$. We then obtain from Remark A.22 that $\|f_n - f_m\| \to 0$ as $n, m \to \infty$, hence f_n is a Cauchy sequence converging to some f (because $\mathscr{C}(\Omega)$ is complete, i.e., every Cauchy sequence converges), and from (A.107) we then obtain that Lf_n also converges to -say- h. As a consequence, since L is a closed operator by assumption, $h = Lf$ and we obtain $f - Lf = g$ showing that $g \in \mathcal{R}(I - \lambda L)$. □

Definition A.28 A Markov generator is a closed Markov pregenerator which satisfies $\mathcal{R}(I - \lambda L) = \mathscr{C}(\Omega)$ for all λ sufficiently small.

For the next proposition, let us remind that we call an operator A bounded if $\|Af\| \leq C\|f\|$ for all f, and then we define the norm of the operator $\|A\| = \sup_{f:\|f\|=1} \|Af\|$.

Proposition A.29 *We have the following:*

(a) A bounded Markov pregenerator is a Markov generator.
(b) A Markov generator satisfies.

$$\mathcal{R}(I - \lambda L) = \mathcal{C}(\Omega) \tag{A.109}$$

for all $\lambda \geq 0$.

Proof

(a) For $\lambda < \|L\|^{-1}$

$$(I - \lambda L)^{-1} = \sum_{n=0}^{\infty} \lambda^n L^n. \tag{A.110}$$

(b) Suppose we know that $\mathcal{R}(I-\lambda L) = \mathcal{C}$, we then have to show that for all $\gamma > \lambda$: $\mathcal{R}(I - \gamma L) = \mathcal{C}$. I.e., for $g \in \mathcal{C}$ we have to solve

$$(I - \gamma L)f = g, \tag{A.111}$$

which is equivalent with

$$(I - \lambda L)f = \frac{\lambda}{\gamma}g + \frac{\gamma - \lambda}{\gamma}f, \tag{A.112}$$

or

$$f = \frac{\lambda}{\gamma}(I - \lambda L)^{-1}g + \frac{\gamma - \lambda}{\gamma}(I - \lambda L)^{-1}f. \tag{A.113}$$

To see that such f exists, consider the map T which maps h to

$$Th = \frac{\lambda}{\gamma}(I - \lambda L)^{-1}g + \frac{\gamma - \lambda}{\gamma}(I - \lambda L)^{-1}h. \tag{A.114}$$

Then (A.113) reads $Tf = f$, i.e., we have to show that f has a unique fixed point. We have

$$\|Th_1 - Th_2\| = \frac{\gamma - \lambda}{\gamma}\|(I-\lambda L)^{-1}(h_1 - h_2)\| \leq \frac{\gamma - \lambda}{\gamma}\|h_1 - h_2\|. \tag{A.115}$$

Therefore, T is a contraction and hence has a unique fixed point.

□

For $\lambda > 0$ the operator $R(\lambda, L) = (\lambda - L)^{-1}$ is called the resolvent: it is well defined for a Markov generator, and we have $\lambda R(\lambda, L) = (I - \lambda^{-1}L)^{-1}$ is a contraction, hence $\|\lambda R(\lambda, L)\| \leq 1$. This is the basic ingredient of the Hille Yosida theorem which defines e^{tL} -the semigroup associated to the generator L- via these resolvents. In fact we have the following basic theorem of Hille and Yosida. For a general linear operator A with domain $D(A)$ we denote $\rho(A)$ the set of $\lambda \in \mathbb{C}$ such that $(\lambda - A)$ has a bounded inverse, i.e., such that $(\lambda - A)^{-1}$ is a well-defined and bounded operator. We can then state the Hille-Yosida theorem. Notice that to distinguish this general theorem from the context of Markov processes, we slightly changed notation, calling A the generator, and $\{T(t) : t \geq 0\}$ its associated semigroup. For this theorem we follow the proof of [84].

Theorem A.30 *For an operator A with domain $D(A)$ the following are equivalent*

(a) A is the generator of a strongly continuous contraction semigroup.
(b) A is closed, $D(A)$ is dense and for every $\lambda > 0$, $\lambda \in \rho(A)$, $R(\lambda, A) = (\lambda - A)^{-1}$ is well defined and satisfies $\|\lambda R(\lambda, A)\| \leq 1$.

Proof First we will show that $D(A)$ is dense and A is closed whenever A is the generator of a contraction semigroup. First to see that $D(A)$ is dense, define

$$\psi(t, f) = \int_0^t T(s) f. \tag{A.116}$$

We will show that $\psi(t, f)$ is in the domain of the generator, and $A\psi(t, f) = T(t)f - f$. By right-continuity, we have that

$$\frac{\psi(t, f)}{t} \to f, \tag{A.117}$$

as $t \to 0$ and therefore, $D(A)$ is dense. To see that $\psi(t, f) \in D(A)$, notice that

$$\frac{T(\epsilon)\psi(t, f) - \psi(t, f)}{\epsilon} = \frac{\int_t^{t+\epsilon} T(s)f - \int_0^{\epsilon} T(s)f}{\epsilon}. \tag{A.118}$$

This, using the right-continuity of the semigroup converges to $T(t)f - f$ as $\epsilon \to 0$. Therefore, we indeed obtain that $\psi(t, f)$ is in the domain of the generator, and $A\psi(t, f) = T(t)f - f$. To see that A is a closed operator, let $f_n \to f$ and $Af_n \to g$ then we have to show that $f \in D(A)$ and $Af = g$. To see this write

$$T(t) f_n - f_n = \int_0^t T(s) A f_n, \tag{A.119}$$

and take the limit $n \to \infty$, using the assumption $f_n \to f$ and $Af_n \to g$, together with the fact that $T(s)$ is a contraction, to obtain

$$T(t)f - f = \int_0^t T(s)g, \tag{A.120}$$

which shows that $f \in D(A)$ and $Af = g$. Indeed, by right continuity

$$\frac{1}{t}\left(\int_0^t T(s)g\right) \to g \text{ as } t \to 0. \tag{A.121}$$

Next, we will not give the full proof of the rest of the theorem (see e.g. [84] for this which we follow here partly) but we will explain how the resolvents are used to construct the semigroup, under the assumption $\|\lambda R(\lambda, A)\| \leq 1$. The clue is the Yosida approximation of the generator, i.e., we consider

$$A_n := nAR(n, A) = n^2 R(n, A) - nI. \tag{A.122}$$

Here the second equality follows from the general fact

$$AR(\lambda, A) = A(\lambda - A)^{-1} = -(\lambda - A)(\lambda - A)^{-1} + \lambda R(\lambda, A) = -I + \lambda R(\lambda, A). \tag{A.123}$$

Moreover, for all $f \in D(A)$ we have

$$\lambda AR(\lambda, A)f \to Af, \qquad \text{as } \lambda \to \infty. \tag{A.124}$$

We see from (A.122) that A_n is a bounded operator, and therefore we can define

$$T_n(t) = e^{tA_n}. \tag{A.125}$$

This is a contraction semigroup because

$$\|T_n(t)\| = \|e^{-nt+n^2 tR(n,A)}\| \leq e^{-nt}\|e^{n^2R(n,A)t}\| \leq e^{-nt}e^{\|n^2R(n,A)\|} \leq e^{-nt}e^{nt} = 1, \tag{A.126}$$

where in the last step we used the assumption $\|\lambda R(\lambda, A)\| \leq 1$.

The next step is to show that the sequence of semigroups $\{T_n(t) : t \geq 0\}$ converges, as $n \to \infty$, to a limiting semigroup. To prove this, one shows that for all $f \in D(A)$, $t \geq 0$ $T_n(t)f$ form a Cauchy sequence. We have

$$\begin{aligned} T_n(t)f - T_m(t)f &= \int_0^t \frac{d}{ds}(T_m(t-s)T_n(s))\,ds \\ &= \int_0^t T_m(t-s)T_n(s)(A_n f - A_m f). \end{aligned} \tag{A.127}$$

Therefore, using that $T_n(t)$ are contractions:

$$\|T_n(t)f - T_m(t)\| \leq t\|A_n f - A_m f\|. \tag{A.128}$$

Remember now that $A_n f \to Af$ for $f \in D(A)$. Therefore, for $f \in D(A)$ we have that $T_n(t)f$ is a Cauchy sequence which has a limit, which we then call $T(t)f$. Since both $T_n(t)$ and $T(t)$ are contraction semigroups, this convergence $T_n(t)f \to T(t)f$ can then be extended to the whole space $\mathscr{C}(\Omega)$ by approximation with elements from $D(A)$. Then it is not very hard to show that $T(t)$ has generator A, see [84] for full details. □

The application of the Hille Yosida theorem to the context of Markov generators then gives the following.

Theorem A.31 *There is a one-to-one correspondence between a Markov generator L and a Markov semigroup $\{S(t), t \geq 0\}$ via*

(a) The domain $D(L)$ is given by

$$D(L) = \left\{ f : \lim_{t\to 0} \frac{S(t)f - f}{t} \textit{ exists} \right\}, \tag{A.129}$$

and for $f \in D(L)$

$$Lf = \lim_{t\to 0} \frac{S(t)f - f}{t}. \tag{A.130}$$

(b) The semigroup is given by

$$S(t) = \lim_{n\to\infty} \left(I - \frac{t}{n} L \right)^{-n}. \tag{A.131}$$

(c) For $f \in D(L)$, $S(t)f \in D(L)$ and

$$\frac{d}{dt} S(t)f = S(t)Lf = LS(t)f. \tag{A.132}$$

Moreover, $S(t)f$ is the unique solution of the differential equation

$$\frac{d\psi_t}{dt} = L\psi_t, \tag{A.133}$$

with initial condition $\psi_0 = f$.

Remark A.32 Remark that

$$(I - tL)^{-1} = \int_0^\infty \frac{1}{t} e^{-u/t} S(u)du = \mathbb{E}S(U(t,1)) = \mathbb{E}(S(U(t,1)), \tag{A.134}$$

where $U(t, 1)$ is an exponential random variable with parameter $1/t$, and $\mathbb{E}$ denotes expectation w.r.t. this random variable. Similarly

$$\left(I - \frac{t}{n}L\right)^{-n} = \mathbb{E}\left(S\left(\frac{1}{n}\sum_{i=1}^{n} U_i(t, 1)\right)\right), \tag{A.135}$$

where $U_i(t, 1)$ are i.i.d. exponential random variables with parameter $1/t$. As a consequence of the strong law of large numbers $\frac{1}{n}\sum_{i=1}^{n} U_i(t, 1) \to t$ as $n \to \infty$, which provides an intuitive explanation for the convergence of $\left(I - \frac{t}{n}L\right)^{-n}$ to $S(t)$.

The logic of the general functional analytic construction of a process from a "candidate generator" in the area of interacting particle systems, i.e., as in [172] runs then schematically as follows.

1. One starts from *a pregenerator* L, and then to shows that for λ small enough the range $\mathcal{R}(I - \lambda L)$ is dense in $\mathcal{C}(\Omega)$. This is sufficient to conclude that $\bar{L}$ is a Markov generator. This is also the difficult step which is needed to show that the closure of L is a Markov semigroup.
2. To perform this step, one introduces an auxiliary space $\mathfrak{D}(\Omega) \subset D(L)$ of "smooth functions" with stronger norm $\|.\|_S$ (called the "triple norm" in [172]) such that for $g \in \mathfrak{D}$, and $(f - \lambda Lf) = g$, one has a priori estimates of $\|f\|_S$.
3. One then proceeds by approximating L by bounded pregenerators L_n (usually this is naturally done by finite-volume approximations). One then defines, for $g \in \mathfrak{D}$

$$f_n - \lambda L_n f_n = g, \tag{A.136}$$

shows that $f_n \in \mathfrak{D}$ and defines

$$g_n = (I - \lambda L) f_n = (I - \lambda L)(I - \lambda L_n)^{-1} g, \tag{A.137}$$

and shows that $g_n \to g$. This then shows that $\mathfrak{D} \subset \overline{\mathcal{R}(I - \lambda L)}$, which implies $\mathcal{R}(I - \lambda \bar{L}) = \mathcal{C}(\Omega)$, and hence that $\bar{L}$ is a Markov generator.
4. An important notion is "a core" which is a smaller domain $\mathfrak{D} \subset D(L)$ (smaller than the full domain $D(L)$) of the generator L but is such that the closure of the generator L restricted to $\mathfrak{D}$ is still the full generator. In the context of interacting particle systems, where $\Omega = E^{\mathbb{Z}^d}$ with E a finite set, this core is in most cases the set of local functions, i.e., functions depending only on a finite number of coordinates.

Let us give two simple examples from interacting particle systems.

1. **Independent spin-flip dynamics.** $\Omega = \{-1, 1\}^{\mathbb{Z}^d}$. The process $\{\sigma(t), t \geq 0\}$ is defined by letting the symbols $\{\sigma_x(t) : t \geq 0\}$ flip independently for different x

at the event times of a Poisson process of rate $\lambda > 0$. This is obviously a Markov process on Ω. A core for the generator is the set of local functions

$$\mathcal{D} = \{f : \exists A \subset \mathbb{Z}^d, |A| < \infty \text{ such that } \sigma_A = \eta_A \text{ implies} f(\sigma) = f(\eta)\}, \tag{A.138}$$

and the generator reads, for $f \in \mathcal{D}$

$$Lf(\sigma) = \sum_{x \in \mathbb{Z}^d} \lambda(f(\sigma^x) - f(\sigma)). \tag{A.139}$$

Notice that L is not a bounded operator, as can be seen e.g. from

$$L\left(\prod_{i \in A} \sigma_i\right) = -2|A| \prod_{i \in A} \sigma_i, \tag{A.140}$$

which implies that $\|L \prod_{i \in A} \sigma_i\|_\infty$ diverges when $A \uparrow \mathbb{Z}^d$, while for all A finite subset of $\mathbb{Z}^d$ we have $\|\prod_{i \in A} \sigma_i\|_\infty = 1$.

2. **Local spin-flip dynamics.** $\Omega = \{-1, 1\}^{\mathbb{Z}^d}$. The process $\{\sigma(t), t \geq 0\}$ is defined by letting the symbols $\{\sigma_x(t) : t \geq 0\}$ flip at rate $c(x, \sigma)$, which depends on σ locally around x (e.g. on σ_y for y neighbor of x). More formally, one defines

$$\Gamma_{x,y} := \sup \sigma |c(x, \sigma^y) - c(x, \sigma)|, \tag{A.141}$$

and requires $\sup_x \sum_y \Gamma_{xy} < \infty$. A core for the generator is again the set of local functions

$$\mathcal{D} = \{f : \exists A \subset \mathbb{Z}^d, |A| < \infty \text{ such that } \sigma_A = \eta_A \text{ implies} f(\sigma) = f(\eta)\}, \tag{A.142}$$

and the generator reads, for $f \in \mathcal{D}$

$$Lf(\sigma) = \sum_{x \in \mathbb{Z}^d} c(x, \sigma)(f(\sigma^x) - f(\sigma)). \tag{A.143}$$

An alterative core for the generator is the functions of finite "triple" seminorm, defined via $\|f\|_S = \sum_{x \in \mathbb{Z}^d} \delta_x f$ with $\delta_x f(\sigma) = \sup_{\sigma \in \Omega} f(\sigma^x) - f(\sigma)$ where σ^x is the configuration obtained from σ by flipping the symbol at x and leaving all other coordinates unchanged:

$$\mathcal{D}' = \{f : \|f\|_S < \infty\}. \tag{A.144}$$

This set is also dense in the set of continuous functions (as is the set of local functions) but it also is closed under the action of the semigroup, i.e., $S(t)\mathcal{D}' \subset$

$\mathcal{D}'$ (this is not the case for the set of local functions), a property which in fact implies that $\mathcal{D}'$ is a core, see e.g. [84] for a proof.

A.3.2 Brownian Motion and Diffusion Processes

Here we shortly review the expressions for the generators of (Markovian) solutions of stochastic differential equations. For more background on this subject, see [222], [12].

When the state space $\Omega = \mathbb{R}^d$ is locally compact (but not compact), one has to choose other function spaces (than $\mathscr{C}(\Omega)$) on which the Markov semigroup acts.

Brownian Motion

We denote by $\{W(t) : t \geq 0\}$ Brownian motion and denote $\mathbb{E}_x$ the expectation under the path space measure of $\{x + W(t), t \geq 0\}$ (which is "Brownian motion starting from x").

For the reader's convenience, we recall the definition of Brownian motion.

Definition A.33 (Brownian Motion) A process $\{W(t) : t \geq 0\}$ is called Brownian motion if

1. Starting point zero: $W(0) = 0$.
2. Normally distributed increments: for $0 \leq s \leq t$ the increment $W(t) - W(s)$, is normally distributed with expectation zero and variance $t - s$. We denote this by $W(t) - W(s) \simeq N(0, t - s)$.
3. Independent increments: for $0 \leq t_1 < t_2 \ldots < t_n$, the increments $W(t_i) - W(t_{i-1})$ are jointly independent.
4. Continuous trajectories: the map $t \mapsto W(t)$ is almost surely continuous.

Brownian motion started from x is then defined as $W(t) + x$. Brownian motion is a Markov process, and at the same time a Gaussian process with covariance function $\mathrm{cov}(W(t), W(s)) = \min\{t, s\}$. Brownian motion on $\mathbb{R}^d$ is then defined as $(W^{(1)}(t), \ldots, W^{(d)}(t))$, where $W^{(i)}(t), i = 1, \ldots, d$ are independent one dimensional Brownian motions.

For Brownian motion, the state space $\Omega = \mathbb{R}$ is locally compact, and the function space on which the semigroup acts is $\mathscr{C}_0(\mathbb{R})$ the continuous functions vanishing at infinity. The smooth test functions $\mathcal{D}$ are defined as the $\mathscr{C}^\infty$ functions with compact support. Another class of smooth test functions is $\mathcal{S}$, the set of Schwartz functions, defined as those functions of which all derivatives vanish at infinity faster than any polynomial. More explicitly,

$$\mathcal{S} = \{f : \mathbb{R} \to \mathbb{R} : \forall n \in \mathbb{N}, \alpha > 0 \lim_{|x| \to \infty} |x|^\alpha f(x) = 0\}. \tag{A.145}$$

The semigroup of Brownian motion is

$$S(t)f(x) = \mathbb{E}_x f(X(t)) = \mathbb{E}(f(x + W(t))) = \int \frac{1}{\sqrt{2\pi t}} e^{-(x-y)^2/2t} f(y)dy. \tag{A.146}$$

This semigroup leaves the Schwartz space invariant, therefore the Schwartz space is a core for the generator (as we saw earlier, a set of functions in the domain of the generator and closed under the action of the semigroup is automatically a core), and for $f \in \mathcal{S}$

$$Lf = \frac{1}{2} f''(x). \tag{A.147}$$

The fact that the action of the generator is $\frac{1}{2} f''(x)$ can be obtained from the simple computation

$$S(t)f(x) - f(x) = \mathbb{E}f(x + N(0,t)) - f(x) \tag{A.148}$$

$$= f'(x)\mathbb{E}(N(0,t)) + \frac{1}{2} f''(x)\mathbb{E}N^2(0,t) + o(t)$$

$$= \frac{1}{2} f''(x) + o(t), \tag{A.149}$$

where, as before, we denoted $N(0,t)$ a normal random variable with mean zero and variance t. In this case we can explicitly define the domain of the generator

$$D(L) = \{f : f', f'' \in \mathscr{C}_0(\mathbb{R})\}, \tag{A.150}$$

and the action of L is given by (A.147). Similarly, on $\mathbb{R}^d$, the generator of d-dimensional Brownian motion is given by

$$Lf(x) = \frac{1}{2} \Delta f(x) = \sum_{i=1}^{d} \frac{\partial^2 f}{\partial^2 x_i}(x), \tag{A.151}$$

for $f \in \mathscr{C}_0(\mathbb{R}^d)$ twice continuously differentiable.

Markov Diffusion Processes

If a Markovian diffusion process $X(t)$ on $\mathbb{R}$ is the solution of the stochastic differential equation

$$dX(t) = b(X(t))dt + \sigma(X(t))dW(t), \tag{A.152}$$

where b, σ are smooth real-valued functions, and $\sigma > 0$, then on smooth functions the generator is given by

$$Lf(x) = b(x)f'(x) + \frac{1}{2}\sigma^2(x)f''(x). \tag{A.153}$$

Notice that this form of the generator on smooth functions can be derived via Itô's formula which yields

$$f(X(t)) - f(X(0)) - \int_0^t b(X(s))f'(X(s)) - \frac{1}{2}\int_0^t \sigma^2(X(s))f''(X(s))$$

$$= \int_0^t f'(X(s))\sigma(X(s))dW(s), \tag{A.154}$$

which gives that

$$f(X(t)) - f(X(0)) - \int_0^t Lf(X(s)) = M_t, \tag{A.155}$$

with $\{M_t, t \geq 0\}$ a martingale.

More generally, if we have a Markovian diffusion process $X(t)$ on $\mathbb{R}^d$ which solves the SDE

$$dX(t) = b(X(t))dt + \sqrt{a(X(t))}dW(t), \tag{A.156}$$

with now $b : \mathbb{R}^d \to \mathbb{R}^d$ and $a : R^d \to M_n^+$ smooth functions, with M_n^+ the set of $n \times n$ positive definite matrices. Via the d-dimensional Itô's formula one can derive the action of the generator on smooth functions

$$Lf(x) = \sum_{i=1}^{d} b_i(x)\frac{\partial f}{\partial x_i}(x) + \frac{1}{2}\sum_{ij} a_{ij}(x)\frac{\partial^2 f}{\partial x_i \partial x_j}. \tag{A.157}$$

Further Examples of Diffusion Processes and Their Generators

1. **Ornstein-Uhlenbeck process.** This is the process on $\mathbb{R}$ which solves the stochastic differential equation

$$dX(t) = -\kappa X(t)dt + dW(t), \tag{A.158}$$

with $\kappa > 0$. The solution when $X(0) = x$ can be written explicitly as

$$X^x(t) = xe^{-\kappa t} + \int_0^t e^{-\kappa(t-s)}dW(s). \tag{A.159}$$

From this one sees that $X^x(t)$ is normally distributed with expectation $xe^{-\kappa t}$ and variance $\int_0^t e^{-2\kappa(t-s)}ds = \frac{1-e^{-2\kappa t}}{2\kappa}$. The semigroup is therefore given explicitly by

$$S(t)f(x) = \mathbb{E}f(X^x(t)) = \int_{\mathbb{R}} f\left(e^{-\kappa t}x + \sqrt{\frac{1-e^{-2\kappa t}}{2\kappa}}y\right)\frac{e^{-y^2/2}}{\sqrt{2\pi}}dy. \tag{A.160}$$

This formula for the semigroup is called "the Mehler formula" and plays a crucial role in the harmonic analysis of the Ornstein Uhlenbeck generator. We see from (A.160) that the Ornstein-Uhlenbeck process has as its unique invariant (reversible) measure the normal $N(0, \frac{1}{2\kappa})$ We can also derive from (A.160) that for smooth functions f the action of the generator of the process $\{X(t) : t \geq 0\}$ is given by

$$Lf(x) = -\kappa x\frac{df}{dx} + \frac{1}{2}\frac{d^2f}{dx^2}. \tag{A.161}$$

The process $\{X(t) : t \geq 0\}$ provides us with a simple example of so-called Laplace duality. Consider $D(y,x) = e^{xy}$, then we have that D satisfies

$$LD(y,\cdot)(x) = \mathcal{L}D(\cdot,x)(y), \tag{A.162}$$

where

$$\mathcal{L} = -\kappa y\frac{d}{dy} + \frac{1}{2}y^2. \tag{A.163}$$

The semigroup associated to $\mathcal{L}$ can be computed via the Feynman-Kac formula as follows

$$e^{t\mathcal{L}}f(y) = e^{\int_0^t \frac{1}{2}Y^y(s)^2ds}f(Y^y(s)), \tag{A.164}$$

where $Y^y(t)$ is the deterministic process generated by $-\kappa yd/dy$, i.e., $Y^y(t) = ye^{-\kappa t}$. This gives the "duality relation"

$$\mathbb{E}_xD(X(t),y) = \mathbb{E}_xe^{X(t)y} = \exp\left(\frac{y^2}{2}\frac{1-e^{-2\kappa t}}{2\kappa}\right)e^{yxe^{-\kappa t}} = (e^{t\mathcal{L}}e^{x\cdot})(y). \tag{A.165}$$

2. **Brownian motion on** $[0,\infty)$. In domains $D \subset \mathbb{R}^d$, the generator of Brownian motion depends on the boundary behavior. This boundary behavior is encoded in the domain of the generator. It shows that not only the action of the generator determines the process, but in the case of domains also the boundary conditions,

encoded in the domain of the generator. We illustrate this for the domain $[0, \infty) \subset \mathbb{R}$.

(a) Brownian motion absorbed at zero, i.e., once it hits zero it stays at zero has as a generator

$$L = \frac{1}{2}\frac{d^2}{dx^2}, \tag{A.166}$$

on the domain

$$\mathcal{D}(L) = \{f \in \mathscr{C}_0([0, \infty)) : f', f'' \in \mathscr{C}_0([0, \infty)), f''(0) = 0\}. \tag{A.167}$$

It is the boundary condition $f''(0) = 0$ which corresponds to the "absorbing behavior". It can be understood easily that $Lf(0) = 0$ is a necessary condition to be in the domain of the generator, because $S(t)f(0) = f(0)$ by the absorbing property of the Brownian motion, which implies $Lf(0) = 0$.

(b) Brownian motion reflected at zero has as a generator

$$L = \frac{1}{2}\frac{d^2}{dx^2}, \tag{A.168}$$

on the domain

$$\mathcal{D}(L) = \{f \in \mathscr{C}_0([0, \infty)) : f', f'' \in \mathscr{C}_0([0, \infty)), f'(0) = 0\}. \tag{A.169}$$

It is the Neumann boundary condition $f'(0) = 0$ which corresponds to the "reflecting behavior".

(c) Brownian motion killed at zero. The semigroup is given by

$$S(t)f(x) = \mathbb{E}_x f(X^x(t \wedge \tau)I(\tau > t)), \tag{A.170}$$

where τ denotes the hitting time of zero and where $X^x(t)$ denotes Brownian motion starting from x. This process is not "mass-preserving", i.e., $S(t)1 \neq 1$, in this sense, $S(t)$ is a semigroup but not a Markov semigroup as we defined before. The semigroup (A.170) has as a generator

$$L = \frac{1}{2}\frac{d^2}{dx^2}, \tag{A.171}$$

on the domain

$$\mathcal{D}(L) = \{f \in \mathscr{C}_0([0, \infty)) : f', f'' \in \mathscr{C}_0([0, \infty)), f(0) = 0\}. \tag{A.172}$$

It is the Dirichlet boundary condition $f(0) = 0$ which corresponds to the "killing when hitting the boundary". From (A.170) one see that $S(t)f(0) =$

0 which implies that in order for f to belong to the domain of the generator, $f(0) = 0$ is a necessary condition.

A.4 Coupling

Coupling is a widely applied technique in Markov process theory, interacting particle systems and probability theory in general. Here we briefly review some essential elements of this theory, focusing on the results needed in the book. Excellent books on coupling are [174, 228] and Chapter 2 of [172]. For X, Y two random variables with values in E, resp. E', we call a coupling of X and Y a $E \times E'$-valued random variable (X_1, X_2), such that the distribution of X_1 equals the distribution of X and the distribution of X_2 equals that of Y. The simplest coupling is given by the independent joining of X and Y, but many more interesting coupling are possible. As a starting example consider two random variables X, Y with values in $\mathbb{R}$, and with distribution functions F_X, F_Y. Then whenever U_1, U_2 are two uniform random variables on $[0, 1]$, $X_1 = F_X^{-1}(U_1), X_2 = F_Y^{-1}(U_2)$ provides a coupling of X and Y. Choosing $U_1 = U_2 = U$ is the co-called maximal coupling which minimizes the probability of mismatch $\mathbb{P}(X_1 \neq X_2)$.

If the distributions of X and Y are ordered in the sense $F_X(x) \le F_Y(x)$ for all $x \in \mathbb{R}$, then for the maximal coupling $X_1 \le X_2$ with probability one. For a simple example: if $X = \text{Ber}(p)$ and $Y = \text{Ber}(p')$ with $p < p'$, the maximal coupling is given by $\mathbb{P}(X_1 = 1, X_2 = 1) = p, \mathbb{P}(X_1 = 0, X_2 = 1) = p' - p, \mathbb{P}(X_1 = 1, X_2 = 0) = 0, \mathbb{P}(X_1 = 0, X_2 = 0) = 1 - p'$

Let $\{X(t), t \ge 0\}$ and $\{Y(t) : t \ge 0\}$ denote two Markov processes with states spaces Ω, resp. Ω'. A coupling is a process $\{(X_1(t), X_2(t) : t \ge 0\}$ on the product space $\Omega \times \Omega'$ such that $\{X(t), t \ge 0\}$ is equal to $\{X_1(t), t \ge 0\}$ in distribution, and $\{Y(t) : t \ge 0\}$ is equal to $\{X_2(t), t \ge 0\}$ in distribution. A coupling always exist because one can consider the independent joining of the processes $\{X(t), t \ge 0\}$ and $\{Y(t) : t \ge 0\}$, which is the so-called product coupling. The coupling time is defined as

$$\tau = \inf\{t > 0 : X_1(s) = X_2(s), \ \text{for all } s \ge t\}, \tag{A.173}$$

with the convention $\inf \emptyset = +\infty$. The coupling is called successful if $\tau < \infty$ with probability one.

Definition A.34 (Successfull Coupling) Let $\{X(t), t \ge 0\}$ be a Markov process on a state space Ω and denote by $\{X^x(t) : t \ge 0\}, x \in \Omega$ the process with starting point x. We then say that the Markov process admits a successful coupling if for all $x, y \in \Omega$ there exists a successful coupling for the processes $\{X^x(t), t \ge 0\}$ and $\{X^y(t) : t \ge 0\}$.

The following proposition is from [172] Chapter 2, and shows that existence of a successful coupling implies that bounded harmonic functions are constant.

Proposition A.35 *Let $\{X(t), t \geq 0\}$ be a Markov process on a state space Ω, and assume there exists a successful coupling. Let f be a bounded measurable function such that $S(t)f(x) = f(x)$ for all $x \in \Omega, t \geq 0$ (this is what we call a harmonic function). Then f is a constant.*

Proof Let $x, y \in \Omega$, we will prove that $f(x) = f(y)$. Denote by $(X_1(t), X_2(t)$ the successful coupling of $\{X^x(t), t \geq 0\}$ and $\{X^y(t) : t \geq\}$, with corresponding coupling time τ.

$$\begin{aligned}|f(x) - f(y)| &= |S(t)f(x) - S(t)f(y)| = |\mathbb{E}(f(X^x(t)) - f(X^y(t)))| \\ &= |\mathbb{E}(f(X_1(t)) - f(X_2(t)))| \\ &= |\mathbb{E}(f(X_1(t)) - f(X_2(t))I(\tau > t))| \\ &\leq 2\|f\|_\infty \mathbb{P}(\tau > t). \end{aligned} \tag{A.174}$$

The proof is then concluded by letting $t \to \infty$. □

In the next proposition we show that the existence of a successful coupling implies that there exists at most one stationary (probability) measure. Moreover, the decay of the coupling time is a measure of the speed of convergence to this unique stationary probability measure (if it exists).

Proposition A.36 *Let $\{X(t), t \geq 0\}$ be a Markov process on a state space Ω, and assume there exists a successful coupling. Then there exists at most one invariant probability measure.*

Proof Let us denote by τ^{xy} the coupling time of the successful coupling of $\{X^x(t), t \geq 0\}$ and $\{X^y(t) : t \geq\}$. Assume that μ, ν are both invariant probability measures. Then we have, for f a bounded measurable function

$$\left|\int f d\mu - \int f d\nu\right| \leq \int |S(t)f(x) - S(t)f(y)| d\mu \otimes \nu(x, y)$$

$$\leq 2\|f\|_\infty \int \mathbb{P}(\tau^{x,y} > t) d\mu \otimes \nu(x, y). \tag{A.175}$$

Because by assumption $\mathbb{P}(\tau^{x,y} > t) \to 0$ as $t \to \infty$ and $0 \leq \mathbb{P}(\tau^{x,y} > t) \leq 1$, by dominated convergence $\int \mathbb{P}(\tau^{x,y} > t) d\mu \otimes \nu(x, y) \to 0$ as $t \to \infty$. Therefore, $\mu = \nu$. □

In the next proposition we show that continuous-time simple random walk admits a successful coupling. The coupling is the so-called coordinate-wise Ornstein coupling, described e.g. in [123].

Proposition A.37 *Let* $\{X(t), t \geq 0\}$ *denote simple random walk on* $\mathbb{Z}^d$ *with generator*

$$Lf(x) = \sum_{e \in \mathbb{Z}^d, |e|=1} (f(x+e) - f(x)). \tag{A.176}$$

Then there exists a successful coupling.

Proof We use that the components of the process $X(t)$ are independent one-dimensional random walkers, together with the fact that one-dimensional random walk is recurrent. Moreover, as we saw before, the difference of two independent random walks is a random walk at twice the rate, which in $d = 1$ is recurrent as well. The coupling is then described as follows: first let all the components evolve independently, until the first components meet. Then keep the first components equal, and continue the other components independently until the second components meet, etc., until eventually all components are equal. □

Appendix B
Lie Algebras and Lie Groups

In this appendix we provide basic background on Lie algebras and Lie groups. This appendix is concise and aimed to present an elementary introduction of basic concepts of Lie algebras and Lie groups, mainly via examples. The representation theory of Lie algebras and Lie groups is a vast topic, for an extensive discussion we refer the reader to some standard references, like the books "Introduction to Lie algebras and representation theory" by Humphreys [138], and also "Representation theory, a first course" by Fulton and Harris [110]. A very readable monograph is "Lie groups, Lie algebras, and Representations" by Hall [131].

B.1 Groups

Lie groups and Lie algebras are closely related subjects (for a good reason, as we shall see later). These two mathematical concepts are used in physics to express 'symmetries', the invariance of a system under some specific transformations such as rotations, translations, Lorentz transformations. In this section we first recall the definition of a group together with some standard notions used in group theory.

Definition B.1 A group G is a set of elements equipped with a binary operation (denoted by $\cdot$) satisfying the following properties:

1. if $g_1, g_2 \in G$ then $g_1 \cdot g_2 \in G$ *(closure)*;
2. if $g_1, g_2, g_3 \in G$ then $(g_1 \cdot g_2) \cdot g_3 = g_1 \cdot (g_2 \cdot g_3)$ *(associativity)*;
3. there exists $i \in G$ such that, $\forall g \in G$, $g \cdot i = i \cdot g = g$ *(identity)*;
4. $\forall g \in G$, there exists $g^{-1} \in G$ such that $g \cdot g^{-1} = g^{-1} \cdot g = i$ *(inverse element)*.

Example The set $\mathbb{Z}$ of integer numbers, with arithmetical addition as a composition rule, is a group. The same set $\mathbb{Z}$, with multiplication as a composition rule, is not a

C. Giardinà, F. Redig, *Duality for Markov Processes*, Grundlehren der mathematischen Wissenschaften 365, https://doi.org/10.1007/978-3-032-04099-2

group, since given $z \in \mathbb{Z}$, $z \neq 1$, then $\frac{1}{z} \notin \mathbb{Z}$. Replacement of $\mathbb{Z}$ with, e.g., $\mathbb{Q} \setminus \{0\}$ produces a group for multiplication.

Some general definitions that are used to characterize groups are the following. The *order* of a group is the number of elements in the group. A groups is said *finite* if the order is finite. For instance, the group S_n of permutations of n objects is finite, its order being $n!$. A group is said to be *continuous* if the order is infinite uncountable. For instance, the group $SO(2)$ of 2-by-2 rotation matrices

$$\begin{pmatrix} \cos(\theta) & \sin(\theta) \\ -\sin(\theta) & \cos(\theta) \end{pmatrix} \tag{B.1}$$

is continuous (and parametrized by the angle $\theta \in \mathbb{R}/2\pi\mathbb{Z}$). A group is said *abelian* if for all $g_1, g_2 \in G$ one has $g_1 \cdot g_2 = g_2 \cdot g_1$. For example, the rotation group $SO(2)$ is abelian, whereas the rotation group $SO(3)$ of 3-by-3 rotation matrices is not abelian: indeed Euler's rotation theorem asserts that every (non-identity) element of SO(3) is a rotation about a uniquely defined axis, and in general two rotations in $\mathbb{R}^3$ around different axis do not commute.

A subset $H \subset G$ is said a *subgroup* of G if it is a group (in particular is closed) under the same binary operation of G. For instance, the group $SU(n)$ of the special $n \times n$ complex unitary matrices with unit determinant is a subgroup of the group $U(n)$ of all the $n \times n$ complex unitary matrices (whose determinant is on the unit circle). A subgroup $H \subset G$ is said *invariant* (or *normal*), written $H \triangleleft G$, if for all $h \in H$, $g \in G$ one has $g \cdot h \cdot g^{-1} = h'$ for some $h' \in H$. One immediate result of this definition is that all subgroups of an abelian group are invariant subgroups. A group G is said *simple* if the only invariant subgroups are the identity and G itself. For example $SU(n)$ is simple for all $n \geq 2$. Any invariant subgroup has a corresponding *quotient group*, formed from the larger group by eliminating the distinction between elements of the subgroup.

Given two groups G_1 and G_2, the *direct product group* $G_1 \otimes G_2$ is defined as the cartesian product $G_1 \times G_2$ endowed with the binary operation $\bullet$ defined by

$$(g_1, g_2) \bullet (g_1', g_2') := (g_1 \cdot g_1', g_2 \cdot g_2'), \tag{B.2}$$

where (g_1, g_2) is an ordered couple with $g_1 \in G_1$ and $g_2 \in G_2$, resp. (g_1', g_2'), is an ordered couple with $g_1' \in G_1$ and $g_2' \in G_2$, and by abuse of notation we denoted with the same symbol the binary operation in G_1 and G_2 (which may be different a-priori). An example of direct product in physics is $SO(3) \otimes SU(2)$ describing the addition of the 'orbital' angular momentum to the 'spin' angular momentum. The definition of the direct product can be extended to the direct product more than two groups (even an infinite number of groups).

A closely related concept is the *semidirect product*, which is a generalization of the direct product. Given a group G, a subgroup $H \subset G$ and an invariant subgroup $N \triangleleft G$, we say that G is the semidirect product of N and H, written $G = N \rtimes H$, if G is the product of subgroups, $G = NH$, and these subgroups have trivial intersection containing only the identity element. Equivalently, for every $g \in G$, there are unique $n \in N$ and $h \in H$ such that $g = nh$. As an example, viewing $U(1)$ as the subgroup of $U(n)$ of all matrices that are diagonal with $e^{i\theta}$ in the upper left corner and 1 on the rest of the diagonal, we have that $U(n) = SU(n) \rtimes U(1)$, that is $U(n)$ is a semi-direct product of $SU(n)$ and $U(1)$.

Finally, we recall the concept of isomorphism between two groups. A bijective mapping $\varphi : G_1 \to G_2$ between two groups G_1 and G_2 is said to be an *isomorphism* if for all $g_1, g_1' \in G_1$ one has

$$\varphi(g_1 \cdot g_1') = \varphi(g_1) \cdot \varphi(g_1'), \tag{B.3}$$

where again, by abuse of notation, we denote with the same symbol the binary operation of the two groups. Thus a group isomorphism is a function between two groups that sets up a one-to-one correspondence between the elements of the groups in a way that respects the given group operations. If there exists an isomorphism between two groups, then the groups are called isomorphic and it is written $G_1 \cong G_2$. From the standpoint of group theory, isomorphic groups have the same properties and need not be distinguished. If the map φ is not bijective then we speak about *homomorphism*. Furthermore if $G_1 = G_2$ then isomorphism is replaced by automorphism, and homomorphism is replaced by endomorphism. For example, the group $\mathbb{Z}$ of integers (with addition) is a subgroup of $\mathbb{R}$, and the quotient group $\mathbb{R}/\mathbb{Z}$ is isomorphic to the group $U(1)$ of complex numbers of absolute value 1 (with multiplication) and to the rotation group $SO(2)$ in two dimension: $\mathbb{R}/\mathbb{Z} \cong U(1) \cong SO(2)$. For an example of homomorphism, since the determinant of a unitary matrix is a complex number with norm 1, the determinant gives a group homomorphism $\det : U(n) \to U(1)$.

The Homomorphism Between $SU(2)$ and $SO(3)$ We show here that the 'complex rotation' group $SU(2)$ in two dimensions is homomorphic to the rotation group $SO(3)$ in three dimensions. Actually the mapping relating the two groups is almost an isomorphism, the mapping $\varphi : SU(2) \to SO(3)$ being 2-to-1 and onto. Parametrizing a matrix $U \in SU(2)$ as

$$U = \begin{pmatrix} \alpha & -\bar{\beta} \\ \beta & \bar{\alpha} \end{pmatrix} \qquad \text{with} \quad \alpha, \beta \in \mathbb{C}, \quad |\alpha|^2 + |\beta|^2 = 1, \tag{B.4}$$

the mapping may be chosen as

$$\varphi(U) = \begin{pmatrix} |\alpha|^2 - |\beta|^2 & -2Re(\alpha\beta) & 2Im(\alpha\beta) \\ 2Re(\alpha\bar{\beta}) & Re(\alpha^2 - \beta^2) & Im(\beta^2 - \alpha^2) \\ 2Im(\alpha\bar{\beta}) & Im(\alpha^2 + \beta^2) & Re(\alpha^2 + \beta^2) \end{pmatrix}. \tag{B.5}$$

To construct such mapping, we identify the three dimensional Euclidean space with the space V of two-dimensional complex matrices that are self-adjoint and traceless. If $x = (x_1, x_2, x_3) \in \mathbb{R}^3$ then the corresponding element of V is

$$X = \begin{pmatrix} x_1 & x_2 + ix_3 \\ x_2 - ix_3 & -x_1 \end{pmatrix}. \tag{B.6}$$

To each element $U \in SU(2)$ we associate then another linear map $\varphi_U : V \to V$ defined as

$$\varphi_U(X) = UXU^{-1}. \tag{B.7}$$

Using the parametrization (B.4) for the element $U \in SU(2)$, the explicit form of $\varphi(U)$ in (B.5) is obtained as the matrix describing the action of the mapping φ_U on the three dimensional space $\mathbb{R}^3$. It can be checked indeed that $\varphi(U)$ is an orthogonal matrix with unit determinant, thus a rotation in $\mathbb{R}^3$. To see this we first observe that, in view of (B.6), the standard inner product between two vectors x, x' in $\mathbb{R}^3$ can be written as

$$x \cdot x' = \frac{1}{2}\text{trace}(XX'). \tag{B.8}$$

As a consequence of the cyclic property of the trace we have

$$\text{trace}(\varphi_U(X)\varphi_U(X')) = \text{trace}(XX'). \tag{B.9}$$

Thus, φ_U preserves the inner product and $\varphi(U)$ is an orthogonal 3-by-3 matrix. Furthermore $\varphi_{U_1U_2} = \varphi_{U_1}\varphi_{U_2}$ as for any $X \in V$

$$\varphi_{U_1U_2}(X) = U_1U_2X(U_1U_2)^{-1} = U_1U_2XU_2^{-1}U_1^{-1} = \varphi_{U_1}(\varphi_{U_2}(X)). \tag{B.10}$$

It follows that the map $U \mapsto \varphi(U)$ is a homomorphism of $SU(2)$ into the group of orthogonal linear transformations of $\mathbb{R}^3$, that is, into $O(3)$. The fact that $\varphi(U)$ must actually be in $SO(3)$ can be checked from (B.5) by evaluating the determinant, which turns out to be equal to one. Alternatively, one could argue as follows. Being the determinant of an orthogonal matrix, the value must be either 1 or -1 for every $U \in SU(2)$. But this determinant takes the value one at $U = 1_2 \in SU(2)$ and, for U parametrized as in (B.4) with $\alpha = a + ib$, $\beta = c + id$ with $a^2+b^2+c^2+d^2 = 1$, the determinant is clearly a continuous function of a, b, c, d. Since $SU(2)$ is homeomorphic to S^3, it is connected and therefore the value of the determinant must be 1 for all $U \in SU(2)$. As a final remark, one can show that φ is surjective. Furthermore one has that $\varphi(U) = 1_3$ iff $U = \pm 1_2$. Thus, the kernel of φ is $\{1_2, -1_2\} \cong \mathbb{Z}_2$ and so φ is precisely two-to-one, carrying $\pm U$ onto the same element of $SO(3)$. The homomorphism φ between $SU(2)$ and $SO(3)$ is called in physics the "spinor map". Both groups are connected. The 2-to-1 property reflects the fact that $SU(2)$ is simply connected (all closed paths on $SU(2)$ can be continuously contracted to a point) while $SO(3)$ is not simply connected. $SU(2)$ is the universal cover of $SO(3)$.

A final comment: while the homomorphism φ shows that the group $SU(2)$ has a different global topological structure than $SO(3)$, we shall prove later that the groups $SU(2)$ and $SO(3)$ are "locally indistinguishable", i.e. their respective Lie algebras are related by an isomorphism.

B.2 Lie Groups

Definition B.2 A *Lie group* is a set G endowed simultaneously with the structures of a group and a smooth manifold. The binary operation and the inverse operations in the group are required to be smooth maps. This is equivalent to the requirement that the map $(x, y) \to x^{-1}y$ is a smooth mapping of the product manifold $G \times G$ into G.

Several Lie groups are matrix groups, with matrix product as binary operation. In particular, any closed subgroup of $GL(n, \mathbb{C})$ is a matrix Lie groups. Here $GL(n, \mathbb{C})$ denotes the general linear group over the complex numbers, i.e. the group of all $n \times n$ invertible matrices with complex entries. Examples of matrix Lie groups, including some relevant to this book, are:

1. $GL(n, \mathbb{C})$ and $GL(n, \mathbb{R})$: clearly $GL(n, \mathbb{C})$ is a subgroup of itself and is closed, thus is a Lie group; similarly $GL(n, \mathbb{R})$, the group of all $n \times n$ invertible matrices with real entries, is a subgroup of $GL(n, \mathbb{C})$ and is closed.
2. $SL(n, \mathbb{C})$ and $SL(n, \mathbb{R})$: these are the groups of invertible $n \times n$ matrices with complex, respectively real, entries and having determinant one.
3. $U(n)$ and $SU(n)$: the unitary group $U(n)$ includes all the $n \times n$ complex unitary matrices. As a unitary matrix U is such that $U^\dagger U = UU^\dagger = I$, it follows then that this group preserves the complex scalar product $\langle u, v\rangle = \sum_{i=1}^n \overline{u}_i v_i$. Indeed, for all n-dimensional complex vectors u, v, we have
$$\langle Uu, Uv\rangle = \langle U^\dagger Uu, v\rangle = \langle u, v\rangle. \tag{B.11}$$
As $\det(U^\dagger) = \overline{\det(U)}$ it follows that $|\det(U)| = 1$. Then the special unitary group $SU(n)$ is the subgroup of complex unitary $n \times n$ matrices with determinant 1.
4. $O(n)$ and $SO(n)$: the orthogonal group $O(n)$ includes all the $n \times n$ real orthogonal matrices. As an orthogonal matrix O is such that $O^T O = OO^T = I$, it follows then that this group preserves the real scalar product $\langle u, v\rangle = \sum_{i=1}^n u_i v_i$. Indeed, for all n-dimensional complex vectors u, v, we have
$$\langle Ou, Ov\rangle = \langle O^T Ou, v\rangle = \langle u, v\rangle. \tag{B.12}$$

As $\det(O^T) = \det(O)$ it follows that $\det(O) = \pm 1$. Then the special orthogonal group $SO(n)$ is the subgroup of real orthogonal $n \times n$ matrices with determinant 1. Geometrically, elements of $SO(n)$ are rotations, while the elements of $O(n)$ are either rotations or combinations of rotations and reflections.

5. $U(p, q)$ and $SU(p, q)$: the generalized unitary group $U(p, q)$ is the group of $(p+q) \times (p+q)$ complex matrices U such that $U^\dagger J U = J$, where J is the diagonal matrix $J = \mathrm{diag}(1, \ldots, 1, -1, \ldots - 1)$ with p positive entries and q negative entries. Thus now the complex scalar product with signature (p, q) is conserved. $SU(p, q)$ is the subgroup of those matrices further having determinant one.
6. $O(p, q)$ and $SO(p, q)$: the generalized orthogonal group $O(p, q)$ is the group of $(p+q) \times (p+q)$ real matrices O such that $O^T J O = J$, where J is the diagonal matrix $J = \mathrm{diag}(1, \ldots, 1, -1, \ldots - 1)$. Thus now the real scalar product with signature (p, q) is conserved. $SO(p, q)$ is the subgroup of those matrices further having determinant one. Of particular interest in physics (special relativity) is the Lorentz group $O(3, 1)$.
7. $Sp(2n, \mathbb{R})$: the real symplectic group $Sp(2n, \mathbb{R})$ is the set of all $2n \times 2n$ matrices that preserve the skew-symmetric bilinear form on $\mathbb{R}^{2n}$ given by

$$q(v, w) = \sum_{i=1}^{n} (x_i\, y_{n+i} - x_{n+i}\, y_i) \tag{B.13}$$

Introducing the matrix

$$\Omega = \begin{pmatrix} 0 & I \\ -I & 0 \end{pmatrix}, \tag{B.14}$$

one can check that A belongs to $Sp(2n, \mathbb{R})$ iff $A^T \Omega A = \Omega$.
8. H: The Heisenberg group H is the group of 3×3 upper triangular matrices of the form

$$\begin{pmatrix} 1 & a & b \\ 0 & 1 & c \\ 0 & 0 & 1 \end{pmatrix} \tag{B.15}$$

where a, b, c are real numbers. It is easy to check that H is a subgroup of $GL(3, \mathbb{R})$. The Heisenberg group H and its Lie algebra $\mathfrak{h}$ originally arose in the mathematical formalizations of quantum mechanics. Heisenberg actually never considered this group since for most purposes in physics just the Lie algebra relations are needed. It was first defined by Weyl and physicists often refer to it as the Weyl group.

B.3 Lie Algebras

We introduce the 'abstract' definition of Lie algebras and then we explain the relation between Lie algebras and matrix Lie groups in Sect. B.4. There, the exponential map provides a connection between matrix Lie groups and the underlying Lie algebra.

Definition B.3 A real (or complex) Lie algebra is a real (or complex) vector space $\mathfrak{g}$, equipped with a bilinear map $[\cdot, \cdot] : \mathfrak{g} \times \mathfrak{g} \to \mathfrak{g}$, referred to as *bracket operation*, with the following properties:

1. $[X, Y] = -[Y, X]$ for all $X, Y \in \mathfrak{g}$ *(skew symmetric)*;
2. $[X, [Y, Z]] + [Y, [Z, X]] + [Z, [X, Y]] = 0$ for all $X, Y, Z \in \mathfrak{g}$ *(Jacobi identity)*.

Examples A first example of a Lie algebra is obtained considering $\mathfrak{g} = \mathbb{R}^3$ and $[x, y] : \mathbb{R}^3 \times \mathbb{R}^3 \to \mathbb{R}^3$ be given by $[x, y] = x \times y$, where $x \times y$ is the cross product (or vector product). Then $\mathfrak{g}$ is a Lie algebra.

The second example shows how to find a Lie algebra from an associative algebra. Let $\mathfrak{g}$ be a subspace of an associative algebra $\mathscr{A}$ such that $XY - YX \in \mathfrak{g}$ for all $X, Y \in \mathfrak{g}$. Then $\mathfrak{g}$ is a Lie algebra with bracket operation given by the commutator: $[X, Y] = XY - YX$. We observe that, to verify the Jacobi identity, *associativity* of the algebra A is essential: each of the 6 possible 'words' of the three letters X, Y, Z appears in the identity with different signs and different groupings (e.g. we have $X(YZ)$ and $-(XY)Z$). Conversely it can be proved [131, Chapter 9] that every Lie algebra $\mathfrak{g}$ can be embedded into an associative algebra $\mathscr{A}$ in such a way that the bracket becomes $XY - YX$. See the discussion about 'universal enveloping algebras' in Sect. B.7.

Several definitions that we encountered in the context of groups have their obvious analogues in the context of algebras. For instance, a Lie algebra $\mathfrak{g}$ is commutative if the Lie bracket $[X, Y] = 0$ for all $X, Y \in \mathfrak{g}$. A subalgebra of a Lie algebra $\mathfrak{g}$ is a subspace $\mathfrak{h}$ of $\mathfrak{g}$ such that $[H_1, H_2] \in \mathfrak{h}$ for all $H_1, H_2 \in \mathfrak{h}$. The role of normal subgroups in group theory is played by ideals in algebra theory. A subalgebra $\mathfrak{h}$ of a Lie algebra $\mathfrak{g}$ is said to be an *ideal* in $\mathfrak{g}$ if $[X, H] \in \mathfrak{h}$ for all $X \in \mathfrak{g}$ and $H \in \mathfrak{h}$. A Lie algebra $\mathfrak{g}$ is called irreducible if the only ideals in $\mathfrak{g}$ are $\mathfrak{g}$ and 0. A Lie algebra g is called *simple* if it is irreducible and $\dim \mathfrak{g} \geq 2$.

The *direct sum* of two Lie algebras $\mathfrak{g}1$ and $\mathfrak{g}2$ is the vector space direct sum of $\mathfrak{g}1$ and $\mathfrak{g}2$, with bracket given by

$$[(X_1, X_2), (Y_1, Y_2)] = ([X_1, Y_1], [X_2, Y_2]).$$

One says that a Lie algebra $\mathfrak{g}$ decomposes as the Lie algebra direct sum of $\mathfrak{g}1$ and $\mathfrak{g}2$ if $\mathfrak{g}$ is the direct sum of $\mathfrak{g}1$ and $\mathfrak{g}2$ as vector spaces and $[X_1, X_2] = 0$ for all $X_1 \in \mathfrak{g}1$ and $X_2 \in \mathfrak{g}2$.

If $\mathfrak{g}$ and $\mathfrak{h}$ are Lie algebras, then a linear map $\varphi : \mathfrak{g} \to \mathfrak{h}$ is called a Lie algebra *homomorphism* if $\varphi([X, Y]) = [\varphi(X), \varphi(Y)]$ for all $X, Y \in \mathfrak{g}$. If, in addition, it is a bijection then it is called a Lie algebra isomorphism. A Lie algebra isomorphism of a Lie algebra with itself is called a Lie algebra automorphism.

The *center* of a Lie algebra $\mathfrak{g}$ is the set of all $X \in \mathfrak{g}$ for which $[X, Y] = 0$ for all $Y \in \mathfrak{g}$. A finite dimensional Lie algebra $\mathfrak{g}$ is often defined by its *structure constants* $\{c_{ijk}\}$, which are defined as follows: let $X_1, \ldots, X_n$ be a base of $\mathfrak{g}$ as a vector space, then

$$[X_i, X_j] = \sum_{k=1}^{n} c_{ijk} X_k \,.$$

From the properties of the bracket operation (skew symmetry and Jacobi identity) it follows the structure constants satisfy

$$c_{ijk} + c_{jik} = 0\,, \tag{B.16}$$

$$\sum_{m=1}^{n} (c_{ijm} c_{mkl} + c_{jkm} c_{mil} + c_{kim} c_{mjl}) = 0\,. \tag{B.17}$$

B.4 Lie Algebra of a Matrix Lie Group

We discuss here the case in which a Lie algebra is constructed starting from a matrix Lie group.

Definition B.4 The Lie algebra of matrix Lie group G, denoted by $\mathfrak{g}$, is the set of all matrices X such that $e^{tX} \in G$ for all $t \in \mathbb{R}$, equipped with the bracket operation $[\cdot, \cdot] : \mathfrak{g} \times \mathfrak{g} \to \mathfrak{g}$ defined by the commutator

$$[X, Y] := XY - YX\,. \tag{B.18}$$

Thus X is in $\mathfrak{g}$ if and only if the one-parameter subgroup generated by X lies in G. The fact that the Lie algebra $\mathfrak{g}$ of matrix Lie group G is a vector space follows from the Lie-Trotter product formula

$$e^{t(aX+bY)} = \lim_{n\to\infty} \left(e^{\frac{taX}{n}} e^{\frac{tbY}{n}} \right)^n, \tag{B.19}$$

which is valid for all matrices X, Y and reals a, b. Being the space of all complex matrices an associative algebra $\mathcal{A}$, the bracket operation is naturally defined as the commutator, provided one shows that $\mathfrak{g}$ is a subspace of $\mathcal{A}$ such that $XY - YX \in \mathfrak{g}$ for all $X, Y \in \mathfrak{g}$. To show this one uses that $e^{tX} Y e^{-tX} \in \mathfrak{g}$ and

$$XY - YX = \frac{d}{dt} \left(e^{tX} Y e^{-tX} \right)\Big|_{t=0}, \tag{B.20}$$

so that $XY - YX \in \mathfrak{g}$ since $\mathfrak{g}$ is a closed subset of all complex matrices.
Examples of Lie algebras constructed starting from a matrix Lie group, including some relevant to this book, are:

1. $\mathfrak{gl}(n, \mathbb{C})$ and $\mathfrak{gl}(n, \mathbb{R})$: the Lie algebra $\mathfrak{gl}(n, \mathbb{C})$ of the Lie group $GL(n, \mathbb{C})$ is the space of all $n \times n$ matrices with complex entries. Similarly the Lie algebra $\mathfrak{gl}(n, \mathbb{R})$ of the Lie group $GL(n, \mathbb{R})$ is the space of all $n \times n$ matrices with real entries.
2. $\mathfrak{sl}(n, \mathbb{C})$ and $\mathfrak{sl}(n, \mathbb{R})$: the Lie algebra $\mathfrak{sl}(n, \mathbb{C})$ of the Lie group $SL(n, \mathbb{C})$ is the space of all $n \times n$ complex matrices with zero trace. Similarly the Lie algebra $\mathfrak{sl}(n, \mathbb{R})$ of the Lie group $SL(n, \mathbb{R})$ is the space of all $n \times n$ real matrices with zero trace. This follows from

$$det(e^{tX}) = e^{t \text{ trace}(X)} . \tag{B.21}$$

3. $\mathfrak{u}(n)$ and $\mathfrak{su}(n)$: the Lie algebra $\mathfrak{u}(n)$ of the unitary group $U(n)$ includes all the $n \times n$ complex anti-hermitian matrices. This follow from the fact that the matrix $U = e^{tA}$ is unitary iff $A^\dagger = -A$. The Lie algebra $\mathfrak{su}(n)$ of the special unitary group $SU(n)$ has the additional requirement of A being traceless.

   In this book we extensively used the Lie algebra $\mathfrak{su}(2)$, consisting of all 2-by-2 anti-hermitian complex matrices with zero trace that can be parametrized as

$$A = \begin{pmatrix} -ix_3 & -x_2 - ix_1 \\ x_2 - ix_1 & ix_3 \end{pmatrix} \qquad \text{with} \quad x_1, x_2, x_3 \in \mathbb{R}. \tag{B.22}$$

   The Lie algebra $\mathfrak{su}(2)$ is generated by the following matrices

$$E_1 = \frac{1}{2}\begin{pmatrix} 0 & -i \\ -i & 0 \end{pmatrix} \qquad E_2 = \frac{1}{2}\begin{pmatrix} 0 & -1 \\ 1 & 0 \end{pmatrix} \qquad E_3 = \frac{1}{2}\begin{pmatrix} -i & 0 \\ 0 & i \end{pmatrix}. \tag{B.23}$$

   Namely, we have that the matrix A in (B.22) can be obtained via the linear combination $A = 2\sum_{i=1}^{3} x_i E_i$. The commutator Lie bracket relations satisfied by the generators are

$$[E_j, E_k] = \epsilon_{jkl} E_l, \tag{B.24}$$

   so that the structure constants are given by the antisymmetric Levi-Civita symbol ϵ_{jkl}. The generators in (B.23) are related to the Pauli matrices by $E_j = -\frac{i}{2}\sigma_j$, where

$$\sigma_1 = \begin{pmatrix} 0 & 1 \\ 1 & 0 \end{pmatrix} \qquad \sigma_2 = \begin{pmatrix} 0 & -i \\ i & 0 \end{pmatrix} \qquad \sigma_3 = \begin{pmatrix} 1 & 0 \\ 0 & -1 \end{pmatrix}, \tag{B.25}$$

with commutation relations

$$[\sigma_j, \sigma_k] = 2i\epsilon_{jkl}\,\sigma_l\ .$$

4. $\mathfrak{o}(n)$ and $\mathfrak{so}(n)$: the Lie algebra $\mathfrak{o}(n)$ of the orthogonal group $O(n)$ includes all the $n \times n$ real anti-symmetric matrices. This follow from the fact that the matrix $O = e^{tA}$ is orthogonal iff $A^T = -A$. It follows that A has trace 0 (since the diagonal entries are all zero), and thus every element of the Lie algebra of $O(n)$ is also in the Lie algebra of $SO(n)$.

   A well-know example is the Lie algebra $\mathfrak{so}(3)$ associate with rotations in three dimensions. The following elements form a basis for $\mathfrak{so}(3)$:

$$F_1 = \begin{pmatrix} 0 & 0 & 0 \\ 0 & 0 & -1 \\ 0 & 1 & 0 \end{pmatrix} \qquad F_2 = \begin{pmatrix} 0 & 0 & 1 \\ 0 & 0 & 0 \\ -1 & 0 & 0 \end{pmatrix} \qquad F_3 = \begin{pmatrix} 0 & -1 & 0 \\ 1 & 0 & 0 \\ 0 & 0 & 0 \end{pmatrix}, \tag{B.26}$$

   whose commutation relations read

$$[F_j, F_k] = \epsilon_{jkl}F_l. \tag{B.27}$$

   Since the generators E_1, E_2, E_3 of the $\mathfrak{su}(2)$ Lie algebra in (B.23) satisfy the same commutation relations as the generators F_1, F_2, F_3 of the $\mathfrak{so}(3)$ Lie algebra in (B.26), it follows that the two Lie algebras are isomorphic.
5. $\mathfrak{u}(p,q)$ and $\mathfrak{su}(p,q)$: the Lie algebra $\mathfrak{u}(p,q)$ of the generalized unitary group $U(p,q)$ is the group of $(p+q)\times(p+q)$ complex matrices A such that $JA^\dagger J = -A$, where J is the diagonal matrix $J = \mathrm{diag}(1,\ldots,1,-1,\ldots,-1)$. This follows from the fact that the matrix $U = e^{tA}$ belonging to the generalized unitary group $U(p,q)$ has to satisfy the condition $U^\dagger JU = J$. The Lie algebra $\mathfrak{su}(p,q)$ of $SU(p,q)$ is the same as that of $U(p,q)$.
6. $\mathfrak{o}(p,q)$ and $\mathfrak{so}(p,q)$: the Lie algebra $\mathfrak{o}(p,q)$ of the generalized orthogonal group $O(p,q)$ is the group of $(p+q)\times(p+q)$ real matrices A such that $JA^TJ = -A$, where J is the diagonal matrix $J = \mathrm{diag}(1,\ldots,1,-1,\ldots,-1)$. This follows from the fact that the matrix $O = e^{tA}$ belonging to the generalized unitary group $O(p,q)$ has to satisfy the condition $O^TJO = J$. The Lie algebra of $SO(p,q)$ is the same as that of $O(p,q)$.
7. $\mathfrak{sp}(2n,\mathbb{R})$: the Lie algebra $\mathfrak{sp}(2n,\mathbb{R})$ of the real symplectic group $Sp(2n,\mathbb{R})$ is the set of all $2n \times 2n$ matrices A such that

$$\Omega A^T \Omega = A, \tag{B.28}$$

where

$$\Omega = \begin{pmatrix} 0 & I \\ -I & 0 \end{pmatrix}. \tag{B.29}$$

8. $\mathfrak{h}$: the Lie algebra $\mathfrak{h}$ of the Heisenberg group H is the space of all matrices of the form

$$\begin{pmatrix} 0 & a & b \\ 0 & 0 & c \\ 0 & 0 & 0 \end{pmatrix},$$

where $a, b, c \in \mathbb{R}$. It is immediately seen that the Lie algebra of the Heisenberg group H is isomorphic to the real Lie algebra with basis elements $\{P, Q, C\}$ and commutation relation $[P, Q] = C$, where

$$P = \begin{pmatrix} 0 & 1 & 0 \\ 0 & 0 & 0 \\ 0 & 0 & 0 \end{pmatrix} \qquad Q = \begin{pmatrix} 0 & 0 & 0 \\ 0 & 0 & 1 \\ 0 & 0 & 0 \end{pmatrix} \qquad C = \begin{pmatrix} 0 & 0 & 1 \\ 0 & 0 & 0 \\ 0 & 0 & 0 \end{pmatrix}.$$

B.5 Representations

Given a vector space V, let $\mathrm{GL}(V)$ denote the general linear group, i.e. the space of all linear invertible maps of V to itself. In the finite dimensional case where V is a real (resp. complex) vector space with $dim(V) = n$, the group $\mathrm{GL}(V)$ can be identified with $GL(n, \mathbb{R})$ (resp. $GL(n, \mathbb{C})$), the group of n-by-n real (resp. complex) invertible matrices.

Definition B.5 A representation of a group G on a vector space V is a group homomorphism from G to $\mathrm{GL}(V)$. Namely, a representation is a map $\Pi : G \to \mathrm{GL}(V)$ such that, for all $g_1, g_2 \in G$, one has

$$\Pi(g_1 \cdot g_2) = \Pi(g_1)\Pi(g_2)\,. \tag{B.30}$$

A faithful representation Π is one in which the homomorphism $G \to \mathrm{GL}(V)$ is injective.

The dimension of vector space V is called the dimension of the representation. The general aim of representation theory of a matrix Lie group G is to identify all the ways G can act as a group of matrices.

Let Π be a representation of a matrix Lie group G, acting on a space V. A subspace W of V is called *invariant* if $\Pi(g)w \in W$ for all $w \in W$ and $g \in G$.

A representation such that the only invariant subspaces are V itself and $\{0\}$ (the subspace containing only the 0 vector) is called *irreducible*.

Given two vector spaces V and W, two representations $\Pi_1 : G \to \mathrm{GL}(V)$ and $\Pi_2 : G \to \mathrm{GL}(W)$ are said to be *equivalent* if there exists a vector space isomorphism $\varphi : V \to W$ such that for all $g \in G$, one has

$$\varphi \circ \Pi_1(g) \circ \varphi^{-1} = \Pi_2(g). \tag{B.31}$$

An *intertwining* is a map $\varphi : V \to W$ such that for all $g \in G$, one has

$$\varphi \circ \Pi_1(g) = \Pi_2(g) \circ \varphi\,. \tag{B.32}$$

Thus, two representations are equivalent if the intertwining map relating them is invertible.

For Lie algebras, the definition of representation is analogous to the one given for representations of Lie groups. The main difference is that the representation will map the Lie algebra into $\mathfrak{gl}(V)$, the space of endomorphisms of a vector space V, i.e., the space of all linear maps of V to itself. Note that $\mathfrak{gl}(V)$ can be made into a Lie algebra by defining the bracket operation as the commutator.

Definition B.6 Let $\mathfrak{g}$ be a Lie algebra and let V be a vector space. A representation of $\mathfrak{g}$ on V is a Lie algebra homomorphism $\pi : \mathfrak{g} \to \mathfrak{gl}(V)$. That is, π is a linear map that satisfies

$$\pi([X, Y]) = \pi(X)\pi(Y) - \pi(Y)\pi(X), \tag{B.33}$$

for all X, Y in $\mathfrak{g}$. The representation π is said to be faithful if it is injective.

It is natural to ask what is the relation between the representations of a Lie group G and the representations of the corresponding Lie algebra $\mathfrak{g}$. It turns out that if the group is simply connected one can obtain Lie group representations from Lie algebra representations.

Theorem B.7

1. *Each representation of a Lie group G gives rise to a representation of its Lie algebra $\mathfrak{g}$. Namely, if $\Pi : G \to GL(V)$ is a group representation for some vector space V then there exists a unique representation $\pi : \mathfrak{g} \to \mathfrak{gl}(V)$ such that*

$$\Pi(e^X) = e^{\pi(X)}, \qquad \forall X \in \mathfrak{g}. \tag{B.34}$$

The Lie algebra representation can be found from the Lie group representation by

$$\pi(X) = \frac{d}{dt}\Pi(e^{tX})\bigg|_{t=0}. \tag{B.35}$$

2. *If the Lie group G is simply connected, then every representation π of its Lie algebra $\mathfrak{g}$ comes from a representation Π of G itself. Namely, if the Lie group G is simply connected, and a representation $\pi : \mathfrak{g} \to \mathfrak{gl}(V)$ of its Lie algebra is given, then there exists a unique representation $\Pi : G \to GL(V)$ such that*

$$\Pi(e^X) = e^{\pi(X)} \qquad \forall X \in \mathfrak{g}. \tag{B.36}$$

The proof of this theorem can be found in standard textbooks (e.g. item 1. is proved in Theorem 3.28 of [131] and item 2. is proved in Theorem 5.6 of [131]). The theorem establishes that, to find representations of simply connected Lie groups it is enough to study representations of their Lie algebras.

Example: $SU(2)$ Has More Representations than $SO(3)$ Here we discuss the content of the theorem by considering the examples of the $SU(2)$ group and of the $SO(3)$ group. Their Lie algebras are isomorphic (see the text after Eq. (B.27)) and thus there are as many representations of $\mathfrak{su}(2)$ as many of $\mathfrak{so}(3)$. However, while both groups $SU(2)$ and $SO(3)$ have countably many irreducible representations, in a certain sense there are strictly more representations of the $SU(2)$ group than there are of the group $SO(3)$, the reason for this being that $SU(2)$ is simply connected, whereas $SO(3)$ is not (it is just connected). To understand this, first consider a representation $\Pi : SO(3) \to GL(n, \mathbb{R})$ of $SO(3)$. Composing with the map $\varphi : SU(2) \to SO(3)$ defined in (B.5) then gives a representation $\tilde{\Pi} = \Pi \circ \varphi : SU(2) \to GL(n, \mathbb{R})$ of $SU(2)$. Thus, every representation of $SO(3)$ comes from a representation of $SU(2)$. The converse is not true, however. That is, a given representation $\tilde{\Pi} : SU(2) \to GL(n, \mathbb{R})$ of $SU(2)$ will not induce a representation of $SO(3)$ unless $\tilde{\Pi}$ is constant on the fibers of φ, i.e., unless $\tilde{\Pi}(-U) = \tilde{\Pi}(U)$ for every $U \in SU(2)$.

B.6 The Dual Representation

We recall that the dual space V^* of a vector space V is the space of linear functionals on V, i.e. the space of linear maps of V into $\mathbb{R}$. In particular, interpreting the vector space $\mathbb{R}^n$ as the space of columns of n real numbers, its dual space is the space of rows of n real numbers.

Definition B.8 Let $\pi : \mathfrak{g} \to \mathfrak{gl}(V)$ be a representation of a Lie algebra $\mathfrak{g}$ acting on a finite-dimensional vector space V. Then the *dual representation* $\pi^* : \mathfrak{g} \to \mathfrak{gl}(V^*)$ is the representation of $\mathfrak{g}$ acting on the dual vector space V^* and given by

$$\pi^*(X) = -\pi(X)^{tr} \qquad \forall X \in \mathfrak{g}, \tag{B.37}$$

where $\pi(X)^{tr} : V^* \to V^*$ denotes the *transpose operator* of $\pi(X) : V \to V$ defined by

$$(\pi(X)^{tr}\varphi)(v) = \varphi(\pi(X)v), \tag{B.38}$$

for any functional $\varphi \in V^*$ and any vector $v \in V$.

Remark B.9 If $dim(V) = n$ then $\pi(X)$ is associated to a n-by-n matrix A with respect to a base of V. In this case the transpose operator $\pi(X)^{tr}$ is represented by the transpose matrix A^T with respect to the dual bases of V^*. Equivalently, $\pi(X)$ is associated to a matrix acting on the left on column vectors, and $\pi(X)^{tr}$ is associated to the same matrix acting on the right on row vectors.

Remark B.10 The minus sign in (B.37) is essential for π^* to be a representation, namely to verify that

$$\pi^*([X, Y]) = [\pi^*(X), \pi^*(Y)]. \tag{B.39}$$

This can be proved using the property

$$(\pi(X)\pi(Y))^{tr} = (\pi(Y))^{tr}(\pi(X))^{tr}, \tag{B.40}$$

that is easily verified from the definition of transpose operator.

The dual representation of a group can also be defined.

Definition B.11 Let Π be a representation of a Lie group G acting on a finite dimensional vector space. Then the *dual representation* Π^* is the representation of G acting on V^* and given by

$$\Pi^*(g) = [\Pi(g^{-1})]^{tr}.$$

In the definition above it is now crucial to use the inverse element in order to guarantee that Π^* is a representation.

B.7 Universal Enveloping Algebra of a Lie Algebra

It was observed in Sect. B.3 already that any *associative* algebra $\mathcal{A}$, equipped with product operation denoted by $\star$, may be turned into a Lie algebra $\mathfrak{g}$ whose multiplication operation $[\cdot, \cdot]$ is given by the commutator bracket $[x, y] := x \star y - y \star x$.

A natural question is if the converse is true. Generally speaking, a Lie algebra is not associative, since Lie brackets are not associative in general. Nevertheless, there is a canonical procedure to construct an associative algebra out of a Lie algebra $\mathfrak{g}$, which is called the *universal enveloping algebra* and is denoted by $U(\mathfrak{g})$. More precisely, the idea of the universal enveloping algebra is to embed a Lie algebra $\mathfrak{g}$ into the unique *largest* associative algebra $\mathcal{A}$ with identity in such a way that:

1. the abstract bracket operation in $\mathfrak{g}$ may be computed as the commutator $x \star y - y \star z$, where $\star$ denotes the product in $\mathcal{A}$;
2. the algebra $\mathcal{A}$ is generated by the elements of $\mathfrak{g}$.

The word "largest" is used above in the sense that the universal enveloping algebra $U(\mathfrak{g})$ of $\mathfrak{g}$ has the property that every other enveloping algebra of $\mathfrak{g}$ is a quotient of $U(\mathfrak{g})$.

In general, the universal enveloping algebra $U(\mathfrak{g})$ of a Lie algebra $\mathfrak{g}$ is constructed starting from the tensor algebra $T(\mathfrak{g})$, that is the algebra which contains all possible tensor products of all possible elements in $\mathfrak{g}$:

$$T(\mathfrak{g}) = \bigoplus_{n=0}^{\infty} \mathfrak{g}^{\otimes n}. \tag{B.41}$$

Then the universal enveloping algebra $U(\mathfrak{g})$ of a Lie algebra $\mathfrak{g}$ is obtained as a quotient of the tensor algebra $T(\mathfrak{g})$,

$$U(\mathfrak{g}) = T(\mathfrak{g})/\mathcal{I} \tag{B.42}$$

where the quotienting is done using the equivalence relation $\sim$ defined by the Lie bracket, i.e. $\mathcal{I}$ is the smallest two-sided ideal of $T(\mathfrak{g})$ containing all elements of the form

$$X \otimes Y - Y \otimes X - [X, Y]. \tag{B.43}$$

In more concrete terms, a useful characterization of the universal enveloping algebra $U(\mathfrak{g})$ of a Lie algebra $\mathfrak{g}$ is obtained via the Poincaré–Birkhoff–Witt theorem (see e.g. Theorem 9.9 in [131]) which states that ordered monomials form a base of the universal enveloping algebra $U(\mathfrak{g})$. Suppose, in particular, that the Lie algebra $\mathfrak{g}$ is finite-dimensional, with basis $X_1, \ldots X_n$ and structure constants c_{ijk}. Then the universal enveloping algebra $U(\mathfrak{g})$ of a Lie algebra $\mathfrak{g}$ is the associative algebra (with identity) generated by elements $X_1, \ldots X_n$ subject to the relations $X_i X_j - X_j X_i = \sum_{k=1}^{n} c_{ijk} X_k$ and no other relations. By the Poincaré–Birkhoff–Witt theorem elements of the universal enveloping algebra will be linear combinations of products of the generators in ascending order. In other words the elements

$$X_1^{k_1} X_2^{k_2} \cdots X_n^{k_n} \tag{B.44}$$

with $k_j \geq 0$ for all j, span the enveloping algebra and are linearly independent. It follows that the universal enveloping algebra of a Lie algebra is always infinite dimensional.

Appendix C
Overview of Dualities

In this appendix we provide a schematic overview of the duality results that have been considered in this book. We group the Markov processes based on their algebraic structure and, more precisely, on the underlying Lie algebras.

C.1 General Structure of the Models

The generators of the continuous time Markov processes,

$$\{\eta(t), t \geq 0\} = \{(\eta_x(t))_{x \in V}, t \geq 0\}, \tag{C.1}$$

considered in this monograph are all of the form

$$L = \sum_{(x,y) \in V} p(x, y) L_{x,y}. \tag{C.2}$$

Here V is a finite set consisting of the vertices of a graph, $p(x, y)$ is an irreducible symmetric edge weight and $L_{x,y} = L_{y,x}$ is the single-edge generator which acts only on the variables η_x, η_y. The various models will therefore be classified by specifying the single edge generator $L_{1,2}$ working on the variables η_1, η_2. We will systematically use:

the symbol $\diamond$ for "discrete processes", whose generator is denoted $L_{1,2}$. These are processes with a discrete state space at every site. Examples include: independent random walkers (IRW) and the symmetric inclusion process (SIP).

the symbol $\star$ for "continuous processes", whose generator is denoted $\mathcal{L}_{1,2}$. These are processes with a continuous state space at every site. Examples include: the deterministic energy process (DEP) and the Brownian energy process (BEP).

C. Giardinà, F. Redig, *Duality for Markov Processes*, Grundlehren der mathematischen Wissenschaften 365, https://doi.org/10.1007/978-3-032-04099-2

Another element in the parametrization of the models is the vector $\boldsymbol{\alpha} = (\alpha_x)_{x\in V}$ taking values in a set which depends on the process (e.g., this set is the non-negative real numbers for the symmetric inclusion process (SIP)). The single edge generator will therefore also depend on the values α_1, α_2.

For each of the three Lie algebras considered in the book (Heisenberg algebra, $\mathfrak{su}(1,1)$ algebra, $\mathfrak{su}(2)$ algebra) we will consider three types of processes:

1. The *basic processes*, usually one process in discrete variables and one in continuous variables (e.g. the symmetric inclusion process SIP and the Brownian energy process BEP for the $\mathfrak{su}(1,1)$ algebra).
2. The *thermalized processes*, obtained from the basic ones by "thermalization". These share the same duality properties as the basic processes, because thermalization does not change the symmetries (e.g. the KMP process for the $\mathfrak{su}(1,1)$ algebra).
3. The *additional processes*. These might share the same duality properties as the basic processes, because they have a common symmetry (e.g. the exactly solvable harmonic process for the $\mathfrak{su}(1,1)$ algebra). Or they might have other dualities that are nevertheless a consequence of the same underlying algebra (e.g. the Kac process for the $\mathfrak{su}(1,1)$ algebra).

For each of these processes we will specify their dualities. Denoting the dual process by

$$\{\xi(t), t \geq 0\} = \{(\xi_x(t))_{x\in V}, t \geq 0\}, \tag{C.3}$$

the duality functions are all of the form

$$D(\xi, \eta) = \prod_x d_{\alpha_x}(\xi_x, \eta_x). \tag{C.4}$$

We will therefore specify the single-site duality function $d_\alpha(k, n)$. We will distinguish then three types:

cheap, triangular and orthogonal.

Thus, for each algebra, we will provide a "**duality table**" collecting the cheap, triangular and orthogonal dualities of the algebra. These names originate in the context of the discrete processes, where the single-site self-duality function is a matrix. The "cheap" self-duality matrix is then diagonal and it arises from a reversible measure; the "triangular" self-duality matrix is upper-triangular or lower-triangular; the orthogonal self-duality matrix has orthogonal polynomials on its columns. The orthogonal self-duality function will carry the additional parameter $\rho > 0$ which parametrizes the product measure w.r.t. which the duality functions are orthogonal. The cheap, triangular and orthogonal names are kept for convenience even when the duality function is not a matrix. The connection between the orthogonal duality functions and orthogonal polynomials is explained in the subsection "Examples and remarks". E.g. for the Heisenberg algebra only the discrete-discrete self-duality functions are orthogonal (with respect to the Poisson distribution), the other duality functions appearing in the column orthogonal are no longer orthogonal with respect to a measure, because the continuous process is deterministic and does not have a reversible product measure.

We will use the following additional symbols to display the various duality functions:

$\diamond\diamond$ for discrete-discrete dualities;
$\diamond\star$ for discrete-continuous dualities;
$\star\star$ for continuous-continuous dualities.

In order to give the orthogonal duality functions, we need some basic notation on hypergeometric functions which was introduced in Chap. 9 and which we recall below for ease of reading of this Appendix.

C.1.1 Hypergeometric Functions and Orthogonal Polynomials

The hypergeometric function ${}_rF_s$ is defined as

$${}_rF_s\left(\begin{matrix} a_1, \ldots, a_r \\ b_1, \ldots, b_s \end{matrix} \middle| z\right) = \sum_{k=0}^{\infty} \frac{(a_1)_k \cdots (a_r)_k}{(b_1)_k \cdots (b_s)_k} \frac{z^k}{k!}$$

where $(a)_k$ denotes the Pochhammer symbol defined in terms of the Gamma function as

$$(a)_k := \frac{\Gamma(a+k)}{\Gamma(a)} .$$

The orthogonal polynomials that will be involved in defining the orthogonal duality functions for are :

1. the Krawtchouk polynomials

$$K_x(y; \alpha, p) = {}_2F_1\left(\begin{matrix} -x, -y \\ -\alpha \end{matrix} \middle| \frac{1}{p}\right) \quad \text{for } x, y \in \{0, 1, \ldots, \alpha\}. \tag{C.5}$$

2. the Meixner polynomials

$$M_x(y; \alpha, p) = {}_2F_1\left(\begin{matrix} -x, -y \\ \alpha \end{matrix} \middle| 1 - \frac{1}{p}\right) \quad \text{for } x, y \in \mathbb{N}. \tag{C.6}$$

3. the Charlier polynomials

$$C_x(y; \lambda) = {}_2F_0\left(\begin{matrix} -x, -y \\ - \end{matrix} \middle| -\frac{1}{\lambda}\right) \quad \text{for } x, y \in \mathbb{N}. \tag{C.7}$$

C.2 Processes Associated with the Heisenberg Algebra

C.2.1 *Table of Duality Functions*

Table C.1 Single-site duality functions for processes in the Heisenberg algebra class

		Cheap	Triangular	Orthogonal
$\diamond\diamond$	$d_\alpha(k,n) =$	$\delta_{k,n}\, k!$	$\frac{1}{\alpha^k}\frac{n!}{(n-k)!}\mathbf{1}_{k\le n}$	$(-\rho)^k\, {}_2F_0\left[{-k,-n \atop -}; -\frac{1}{\rho\alpha}\right]$
$\diamond\star$	$\mathfrak{d}_\alpha(k,z) =$	$z^k\, e^{-z}$	$\left(\frac{z}{\alpha}\right)^k$	$\left(\frac{z}{\alpha}-\rho\right)^k$
$\star\star$	$\mathscr{d}_\alpha(v,z) =$	e^{zv-z-v}	$e^{(\frac{z}{\alpha}-1)v}$	$e^{(\frac{z}{\alpha}-1-\rho)v}$

C.2.2 *Basic Processes*

Fix $\alpha_1, \alpha_2 > 0$. The single edge generators of the basic processes are given in the box below.

$\diamond$ **Independent random walkers (IRW).**
State space $\mathbb{N}^2$.

$$L_{1,2}f(\eta_1,\eta_2) = \eta_1\alpha_2[f(\eta_1-1,\eta_2+1) - f(\eta_1,\eta_2)] + \eta_2\alpha_1[f(\eta_1+1,\eta_2-1) - f(\eta_1,\eta_2)].$$

$\star$ **Deterministic energy process (DEP).**
State space $[0,\infty)^2$.

$$\mathcal{L}_{1,2}f(\zeta_1,\zeta_2) = (\alpha_2\zeta_1 - \alpha_1\zeta_2)\left(\frac{\partial}{\partial\zeta_2} - \frac{\partial}{\partial\zeta_1}\right)f(\zeta_1,\zeta_2).$$

These two processes satisfy a complete set of dualities given by Table C.1. This means that the discrete process (Independent random walkers) is self-dual with the three possible duality functions corresponding to the row $\diamond\diamond$ of the table, the discrete process is dual to the continuous process (Deterministic energy process) with the three possible duality functions corresponding to the row $\diamond\star$. Finally, the continuous process is self-dual with the three possible duality functions corresponding to the row $\star\star$. This is also illustrated in the table below.

Table C.2 Duality relation for the basic processes in the Heisenberg algebra class

	IRW	DEP
IRW	$\diamond\diamond$	$\diamond\star$
DEP	$\diamond\star$	$\star\star$

C.2.3 Examples and Remarks

Here we provide two simple examples of how to read the Table C.2 of Sect. C.2.2 in combination with the Table C.1 of Sect. C.2.1, followed by some remarks on the orthogonal polynomial duality functions.

1. Independent random walkers are self-dual with triangular self-duality function

$$D(\xi, \eta) = \prod_{x \in V} \frac{1}{(\alpha_x)^{\xi_x}} \frac{\eta_x!}{(\eta_x - \xi_x)!}. \tag{C.8}$$

   This can be read from the entry with row index "$\diamond\ \diamond$" and column index "triangular" of Table C.1.
2. Independent random walkers are dual to the Deterministic energy process with duality function

$$D(\xi, \zeta) = \prod_{x \in V} \left(\frac{\zeta_x}{\alpha_x}\right)^{\xi_x}. \tag{C.9}$$

   This can read from the entry with row index "$\diamond\star$" and column index "triangular" of the Table C.1.
3. The single-site $\diamond\ \diamond$ orthogonal self-duality function appearing in the right-upper corner of Table C.1 can be written in terms of Charlier polynomials as follows

$$d_\alpha(k, n) = (-\rho)^k C_k(n; \alpha\rho) \tag{C.10}$$

   with

$$C_k(n; \lambda) := {}_2F_0\left(\begin{matrix} -k, -n \\ - \end{matrix} \,\middle|\, -\frac{1}{\lambda}\right) \quad \text{for } k, n \in \mathbb{N}. \tag{C.11}$$

4. The orthogonal duality functions are labeled by a parameter $\rho \in [0, \infty)$. The triangular duality functions can be obtained from the orthogonal ones as a special case by taking the limit as $\rho \to 0$.

C.2.4 *Thermalized Processes*

◇ **Discrete averaging process (DAP).**
State space $\mathbb{N}^2$.

$$L_{1,2}f(\eta_1,\eta_2) = \sum_{m=0}^{\eta_1+\eta_2} [f(m,\eta_1+\eta_2-m) - f(\eta_1,\eta_2)] \binom{\eta_1+\eta_2}{m} \frac{\alpha_1^m \alpha_2^{\eta_1+\eta_2-m}}{(\alpha_1+\alpha_2)^{\eta_1+\eta_2}}.$$

⋆ **Continuous averaging process (CAP).**
State space $[0,\infty)^2$.

$$\mathcal{L}_{1,2}f(\zeta_1,\zeta_2) = f\left(\frac{\alpha_1(\zeta_1+\zeta_2)}{\alpha_1+\alpha_2}, \frac{\alpha_2(\zeta_1+\zeta_2)}{\alpha_1+\alpha_2}\right) - f(\zeta_1,\zeta_2).$$

The thermalized processes also satisfy a complete set of dualities given by Table C.1, as illustrated by Table C.3.

Table C.3 Duality relation for the thermalized processes in the Heisenberg algebra class

	DAP	CAP
DAP	◇◇	◇⋆
CAP	◇⋆	⋆⋆

C.2.5 *Additional Processes*

Here we give a few additional models fitting in the Heisenberg class.

1. The Aldous averaging process (see Sect. 8.2). This coincides with the Continuos averaging process CAP process with $\alpha_1 = \alpha_2 = 1/2$.

⋆ **Aldous averaging process.**
State space $[0,\infty)^2$.

$$\mathcal{L}_{1,2}f(\zeta_1,\zeta_2) = f\left(\tfrac{\zeta_1+\zeta_2}{2}, \tfrac{\zeta_1+\zeta_2}{2}\right) - f(\zeta_1,\zeta_2).$$

2. The Ginzburg-Landau process with quadratic potential (see Sect. 8.5). This is a diffusion process where the single edge generator is defined below.

⋆ **Ginzburg-Landau process with quadratic potential.**
State space $\mathbb{R}^2$.

$$\mathcal{L}_{1,2} f(\zeta_1, \zeta_2) = \left(-(\zeta_1 - \zeta_2)\left(\frac{\partial}{\partial \zeta_1} - \frac{\partial}{\partial \zeta_2}\right) + \left(\frac{\partial}{\partial \zeta_1} - \frac{\partial}{\partial \zeta_2}\right)^2 \right) f(\zeta_1, \zeta_2).$$

The Ginzburg-Landau process is dual to the independent random walks IRW with $\alpha_1 = \alpha_2 = 1$ with duality function

$$D(\eta, \zeta) = \prod_{x \in V} H_{\eta_x}(\zeta_x), \tag{C.12}$$

where $H_k(\cdot)$ is the Hermite polynomial of degree k, i.e.

$$H_k(z) = (2z)^k {}_2F_0 \left(\begin{matrix} -k/2, -(k-1)/2 \\ - \end{matrix} \middle| -\frac{1}{z^2} \right), \qquad z \in \mathbb{R}. \tag{C.13}$$

3. Generalized versions of Moran model and Wright-fisher diffusion (see Sect. 8.6). Fix two parameters $\alpha_1, \alpha_2 \geq 0$, then the single edge generators of these models are defined below.

⋄ **Generalized Moran model.**
State space $\{0, 1, \ldots, N\}$, with $N \in \mathbb{N}$.

$$\begin{aligned} L^{\text{Moran}} f(n) &= n(N - n + \alpha_2)[f(n-1) - f(n)] \\ &+ (N-n)(n+\alpha_1)[f(n+1) - f(n)]. \end{aligned}$$

⋆ **Generalized Wright-Fisher diffusion.**
State space $[0, 1]$.

$$\mathcal{L}^{\text{WF}} f(z) = z(1-z)\frac{d^2 f(z)}{dz^2} + \left(\alpha_1(1-z) - \alpha_2 z\right)\frac{df(z)}{dz}.$$

The generalized Moran model and the generalized WF model are dual with duality function $D : \{0, 1, \ldots, N\} \times [0, 1] \to \mathbb{R}$ given by

$$D(n, z) = z^n (1 - z)^{N-n} \frac{\Gamma(\alpha_1)}{\Gamma(\alpha_1 + n)} \frac{\Gamma(\alpha_2)}{\Gamma(\alpha_2 + N - n)}. \tag{C.14}$$

C.3 Processes Associated with the $\mathfrak{su}(1, 1)$ Algebra

C.3.1 *Table of Duality Functions*

Table C.4 Single-site duality functions for processes in the $\mathfrak{su}(1, 1)$-Lie algebra class

		Cheap	Triangular	Orthogonal
$\diamond\,\diamond$	$d_\alpha(k, n) =$	$\delta_{k,n} \frac{k!\Gamma(\alpha)}{\Gamma(\alpha+k)}$	$\frac{\Gamma(\alpha)}{\Gamma(\alpha+k)} \frac{n!}{(n-k)!} \mathbf{1}_{\{k\le n\}}$	$(-\rho)^k \, {}_2F_1\left[{-k,-n \atop \alpha}; -\frac{1}{\rho}\right]$
$\diamond\,\star$	$\mathfrak{d}_\alpha(k, z) =$	$\frac{\Gamma(\alpha)}{\Gamma(\alpha+k)} z^k e^{-z}$	$\frac{\Gamma(\alpha)}{\Gamma(\alpha+k)} z^k$	$(-\rho)^k \, {}_1F_1\left[{-k \atop \alpha}; \frac{z}{\rho}\right]$
$\star\,\star$	$\mathscr{d}_\alpha(v, z) =$	$e^{-z-v} \, {}_0F_1\left[{- \atop \alpha}; zv\right]$	$e^{-v} \, {}_0F_1\left[{- \atop \alpha}; zv\right]$	$e^{-(\rho+1)v} \, {}_0F_1\left[{- \atop \alpha}; zv\right]$

C.3.2 *Basic Processes*

The single edge generators of the basic processes are given in the box below.

◇ **Symmetric inclusion process (SIP).**
State space $\mathbb{N}^2$.

$$\begin{aligned} L_{1,2} f(\eta_1, \eta_2) &= \eta_1(\alpha_2 + \eta_2)[f(\eta_1 - 1, \eta_2 + 1) - f(\eta_1, \eta_2)] \\ &\quad + \eta_2(\alpha_1 + \eta_1)[f(\eta_1 + 1, \eta_2 - 1) - f(\eta_1, \eta_2)]. \end{aligned}$$

(continued)

⋆ **Brownian energy process (BEP).**
State space $[0, \infty)^2$.

$$\mathcal{L}_{1,2} f(\zeta_1, \zeta_2) = (\alpha_2 \zeta_1 - \alpha_1 \zeta_2) \left(\frac{\partial}{\partial \zeta_2} - \frac{\partial}{\partial \zeta_1} \right) f(\zeta_1, \zeta_2) + \zeta_1 \zeta_2 \left(\frac{\partial}{\partial \zeta_2} - \frac{\partial}{\partial \zeta_1} \right)^2 f(\zeta_1, \zeta_2).$$

These basic processes satisfy a complete set of dualities given in Table C.4, as illustrated in the Table C.5.

Table C.5 Duality relation for the basic processes in the $\mathfrak{su}(1, 1)$ algebra class

	SIP	BEP
SIP	⋄ ⋄	⋄ ⋆
BEP	⋄ ⋆	⋆ ⋆

C.3.3 Examples and Remarks

Here we provide two simple examples of how to read the Table C.5 of Sect. C.3.2 in combination with the Table C.4 of Sect. C.3.1, followed by some remarks on the orthogonal polynomial duality functions.

1. The Symmetric inclusion process SIP is self-dual with triangular self-duality function (entry with row index " ⋄ ⋄" and column index "triangular")

$$D(\xi, \eta) = \prod_{x \in V} \frac{\eta_x! \Gamma(\alpha_x)}{(\eta_x - \xi_x)! \Gamma(\alpha_x + \xi_x)}. \tag{C.15}$$

2. The SIP is dual to the Brownian energy process BEP with duality function (entry with row index " ⋄ ⋆" and column index "triangular")

$$D(\xi, \zeta) = \prod_{x \in V} \frac{\zeta_x^{\xi_x} \Gamma(\alpha_x)}{\Gamma(\alpha_x + \xi_x)}. \tag{C.16}$$

3. The single-site ⋄ ⋄ orthogonal self-duality function in Table C.4 can be written in terms of Meixner polynomials:

$$d_\alpha(k, n) = (-\rho)^k M_k(n; \alpha, \tfrac{\rho}{1+\rho}), \tag{C.17}$$

with

$$M_k(n;\alpha,p) = {}_2F_1\left(\begin{matrix}-k,-n\\ \alpha\end{matrix}\middle| 1-\frac{1}{p}\right) \quad \text{for } k,n \in \mathbb{N}. \tag{C.18}$$

4. The single-site $\diamond\star$ orthogonal self-duality function in Table C.4 can be written in terms of Laguerre polynomials:

$$\mathfrak{d}_\alpha(k,z) = (-\rho)^k \frac{k!\Gamma(\alpha)}{\Gamma(\alpha+k)} L_k(\tfrac{z}{\rho};\alpha-1), \tag{C.19}$$

with

$$L_k(z;\lambda) := \frac{\Gamma(\lambda+1+k)}{k!\Gamma(\lambda+1)}\, {}_1F_1\left[\begin{matrix}-k\\ \lambda+1\end{matrix}; z\right]. \tag{C.20}$$

5. The orthogonal duality functions are labeled by $\rho \in [0,\infty)$. By taking the limit as $\rho \to 0$ one recovers the triangular dualities.

C.3.4 *Thermalized Processes*

$\diamond$ **Discrete generalized KMP process (DKMP).**
State space $\mathbb{N}^2$.

$$L_{1,2} f(\eta_1,\eta_2)$$
$$= \sum_{n=0}^{\eta_1+\eta_2} [f(n,\eta_1+\eta_2-n) - f(\eta_1,\eta_2)] \binom{\eta_1+\eta_2}{n} \frac{B(\alpha_x+n,\eta_x+\eta_y+\alpha_y-n)}{B(\alpha_x,\alpha_y)}.$$

$\star$ **Generalized KMP process (KMP).**
State space $[0,\infty)^2$.

$$\mathcal{L}_{1,2} f(\zeta_1,\zeta_2)$$
$$= \int_0^1 du\, [f((\zeta_1+\zeta_2)u, (\zeta_1+\zeta_2)u) - f(\zeta_1,\zeta_2)] \frac{u^{\alpha_1-1}(1-u)^{\alpha_2-1}}{B(\alpha_1,\alpha_2)}.$$

Here $B(\alpha,\beta) = \frac{\Gamma(\alpha)\Gamma(\beta)}{\Gamma(\alpha+\beta)}$ denotes the Beta function. The original KMP process [152] corresponds to $\alpha_1 = \alpha_2 = 1$. The thermalized processes satisfy a complete set of dualities given by Table C.4, as illustrated in the Table C.6.

Table C.6 Duality relation for the thermalized processes in the $\mathfrak{su}(1,1)$ algebra class

	DKMP	KMP
DKMP	◇ ◇	◇ ⋆
KMP	◇ ⋆	⋆ ⋆

C.3.5 Harmonic Processes

For the $\mathfrak{su}(1,1)$ algebra, we have two additional processes which in the setting of the chain (with or without boundaries) are integrable (as explained in Chap. 13). Their single edge generators are defined in the box below. For these models we consider the α profile constant, i.e., $\alpha_1 = \alpha_2 = \alpha$.

◇ **The symmetric harmonic process (SHP).**
State space $\mathbb{N}^2$.

$$L_{1,2}f(\eta) = \sum_{k=1}^{\eta_1} \varphi_\alpha(k,\eta_1)(f(\eta_1 - k, \eta_2 + k) - f(\eta)) + \sum_{k=1}^{\eta_2} \varphi_\alpha(k,\eta_2)(f(\eta_1 + k, \eta_2 - k) - f(\eta)),$$

with

$$\varphi_\alpha(k,n) = \frac{1}{k}\frac{\Gamma(n+1)\Gamma(n-k+\alpha)}{\Gamma(n+\alpha)\Gamma(n-k+1)}\mathbb{1}_{\{1\le k\le n\}}.$$

⋆ **The integrable heat conduction model (IHCP).**
State space $[0,\infty)^2$.

$$\mathscr{L}_{1,2}f(\zeta) = \int_0^{\zeta_1} \frac{dx}{x}\left(1-\frac{x}{\zeta_1}\right)^{\alpha-1}[f(\zeta_1 - x, \zeta_2 + x) - f(\zeta_1,\zeta_2)] + \int_0^{\zeta_2} \frac{dx}{x}\left(1-\frac{x}{\zeta_2}\right)^{\alpha-1}[f(\zeta_1 + x, \zeta_2 - x) - f(\zeta_1,\zeta_2)].$$

These two processes satisfy a complete set of dualities given by Table C.4, as illustrated in the Table C.7.

Table C.7 Duality relation for the symmetric harmonic processes and the integrable heat conduction process in the $\mathfrak{su}(1,1)$ algebra class

	SHP	IHCP
SHP	◇ ◇	◇ ★
IHCP	◇ ★	★ ★

C.3.6 Additional Processes

Here we give four additional models within the $\mathfrak{su}(1,1)$ algebra class.

1. The Brownian momentum process (BMP), introduced in Sect. 6.7. It is related to the Brownian energy process (BEP) with parameters $\alpha_1 = \alpha_2 = \frac{1}{2}$ through the change of variable $\zeta \to \zeta^2$ applied to the momentum ζ_x variable at each lattice site x. More precisely, if $\{\zeta(t) = \{\zeta_x(t)\}_{x\in V},\ t \geq 0\}$ is the BMP then the process $\{v(t), t \geq 0\}$, with $v_x(t) = \zeta_x^2(t)$ is a BEP with parameters $\alpha_1 = \alpha_2 = \frac{1}{2}$.

★ **Brownian momentum process (BMP).**
State space $\mathbb{R}^2$.

$$\mathcal{L}_{1,2} f(\zeta_1, \zeta_2) = \left(\zeta_1 \frac{\partial}{\partial \zeta_2} - \zeta_2 \frac{\partial}{\partial \zeta_1}\right)^2 f(\zeta_1, \zeta_2).$$

The BMP is dual to the SIP with parameters $\alpha_1 = \alpha_2 = \frac{1}{2}$ with a product duality function whose single-site terms are given in Table C.8.

2. The KAC model, studied in Sect. 8.4. This can be obtained as a thermalization of the BMP. As a consequence, the KAC model is dual to the discrete KMP process with $\alpha_1 = \alpha_2 = 1/2$ with duality function that is identical to the duality function between the BMP and the SIP with $\alpha_1 = \alpha_2 = 1/2$, given in Table C.8.

★ **The KAC model.**
State space $\mathbb{R}^2$.

$$\mathcal{L}_{1,2} f(v_1, v_2) = \frac{1}{2\pi} \int_0^{2\pi} d\theta \left[f(R_{1,2}^{(\theta)}(v_1, v_2)) - f(v_1, v_2)\right],$$

with

$$R_{1,2}^{(\theta)}(v_1, v_2) = (v_1 \cos\theta + v_2 \sin\theta, -v_1 \sin\theta + v_2 \cos\theta).$$

Table C.8 Single-site duality functions between the BMP process and the SIP process with parameters $\alpha_1 = \alpha_2 = \frac{1}{2}$; and between the KAC process and discrete generaziled KMP with parameters $\alpha_1 = \alpha_2 = \frac{1}{2}$. Notice that the single-site $\diamond\star$ duality function can be rewritten in terms of Hermite polynomials $\mathfrak{d}^{\text{orth}}(k, z) = \frac{H_{2k}(z)}{(2k-1)!!}$ with $H_k(\cdot)$ the Hermite polynomials

		Cheap	Triangular	Orthogonal
$\diamond\star$	$\mathfrak{d}(k, z) =$	$\frac{z^{2k}}{(2k-1)!!}\, e^{-z^2}$	$\frac{z^{2k}}{(2k-1)!!}$	$\left(-\frac{1}{2}\right)^k {}_1F_1\left(\begin{matrix}-k\\ \frac{1}{2}\end{matrix}\middle\vert z^2\right)$

3. The hidden parameter model for KMP processes.

- ★ **The hidden parameter process for KMP process.**
  State space $[0, \infty)^2$.

$$
\begin{aligned}
L_{1,2} f(\theta_1, \theta_2) \\
&= \frac{1}{B(\alpha, \alpha)} \int u^{\alpha-1}(1-u)^{\alpha-1}\Big(f(\Theta(u, \theta_1, \theta_2)) - f(\theta_1, \theta_2)\Big),
\end{aligned}
$$

where

$$
\Theta(u, \theta_1, \theta_2) = (\theta_1 u + \theta_2(1-u), \theta_1 u + \theta_2(1-u)).
$$

We then have that discrete KMP is dual to the hidden parameter process for KMP with the single-site duality function $d(k, \theta) = \theta^k$.

4. The hidden parameter model for harmonic processes.

- ★ **The hidden parameter process for the symmetric harmonic process (SHP).**
  State space $[0, \infty)^2$.

$$
\begin{aligned}
&\mathcal{L}_{1,2} f(\theta_1, \theta_2) \\
&= \int_0^1 \frac{du}{u}(1-u)^{\alpha-1}\left[f(\theta_1(1-u) + \theta_2 u, \theta_2) - f(\theta_1, \theta_2)\right] \\
&+ \int_0^1 \frac{du}{u}(1-u)^{\alpha-1}\left[f(\theta_1, \theta_2(1-u) + \theta_1 u) - f(\theta_1, \theta_2)\right].
\end{aligned}
$$

We then have that the symmetric harmonic process is dual to the hidden parameter process with the single-site duality function $d(k, \theta) = \theta^k$.

C.4 Processes Associated with the $\mathfrak{su}(2)$ Algebra

For this Lie algebra, we only have discrete processes. Therefore, the tables of generators and dualities are easier, so we collect them all in one section.

C.4.1 The Processes and Their Dualities

Table C.9 Single-site duality functions for processes in the $\mathfrak{su}(2)$-Lie algebra class

	Cheap	Triangular	Orthogonal
$d_\alpha(k,n) =$	$\delta_{k,n}\,\frac{k!(\alpha-k)!}{\alpha!}$	$\frac{(\alpha-k)!}{\alpha!}\,\frac{n!}{(n-k)!}\,\mathbf{1}_{k\le n}$	$(-\rho)^k\,{}_2F_1\left[\begin{smallmatrix}-k,-n\\-\alpha\end{smallmatrix};\frac{1}{\rho}\right]$

The single edge generator of the discrete basic process (the generalized symmetric partial exclusion process) is given in the box below. The family of model is defined by the parameters $\alpha_1, \alpha_2 \in \mathbb{N}$. It satisfies a complete set of self-dualities given in Table C.9 .

◇ **(Generalized) symmetric partial exclusion process.**
State space $\{0, \ldots, \alpha_1\} \times \{0, \ldots, \alpha_2\}$.

$$\begin{aligned} L_{1,2} f(\eta_1, \eta_2) &= \eta_1(\alpha_2 - \eta_1)[f(\eta_1 - 1, \eta_2 + 1) - f(\eta_1, \eta_2)] \\ &\quad + \eta_2(\alpha_1 - \eta_1)[f(\eta_1 + 1, \eta_2 - 1) - f(\eta_1, \eta_2)]. \end{aligned}$$

The thermalization of the generalized symmetric partial exclusion process has single-edge generator given below. It satisfies a complete set of self-dualities given in Table C.9.

◇ **Discrete mass redistribution model.**
State space $\{0, \ldots, \alpha_1\} \times \{0, \ldots, \alpha_2\}$.

$$\begin{aligned} &L_{1,2} f(\eta_1, \eta_2) \\ &= \sum_{n=0}^{\eta_1+\eta_2} [f(n, \eta_1 + \eta_2 - n) - f(\eta_1, \eta_2)] \binom{\eta_x+\eta_y}{n} \frac{\binom{\alpha_x+\alpha_y-\eta_x-\eta_y}{\alpha_x-n}}{\binom{\alpha_x+\alpha_y}{\alpha_x}}. \end{aligned}$$

C.4.2 Remarks

1. The single-site orthogonal self-duality function in Table C.9 is labeled by $\rho \in [0, 1]$. The triangular self-duality function can be obtained as a limit of this by letting $\rho \to 0$.
2. The single-site orthogonal self-duality function appearing in Table C.9 can be written in terms of Krawtchouk polynomials:

$$d_\alpha^{\mathrm{orth}}(k, n) = (-\rho)^k K_k(n; \alpha, \rho) \tag{C.21}$$

with

$$K_k(n; \alpha, p) = {}_2F_1\left(\begin{matrix} -k, -n \\ -\alpha \end{matrix} \,\middle|\, \frac{1}{p} \right) \quad \text{for } k, n \in \{0, 1, \ldots, \alpha\}. \tag{C.22}$$

C.5 Algebra Dualities

In this section, we provide equivalent representations for the three Lie algebras considered in this book. For each couple of equivalent representations of an algebra, we recall the corresponding processes duality by showing the abstract form of the Markov generator. We here consider only the dualities of the basic processes.

C.5.1 Equivalent Representations of the Heisenberg Algebra

Cheap self-duality of Independent random walkers

$$d(k,n) = \delta_{k,n} n!$$

$$a^\dagger \xrightarrow{d} \mathrm{a}^\dagger$$

$$a \xrightarrow{d} \mathrm{a}$$

$$\begin{cases} af(k) = kf(k-1) \\ a^\dagger f(k) = f(k+1) \end{cases} \qquad \begin{cases} \mathrm{a}f(n) = f(n+1) \\ \mathrm{a}^\dagger f(n) = nf(n-1) \end{cases}$$

$$[a, a^\dagger] = -1 \qquad [\mathrm{a}, \mathrm{a}^\dagger] = 1$$

$$\widehat{L}_{1,2} \xrightarrow{d_1 d_2} L_{1,2}$$

$$\widehat{L}_{1,2} = -(a_1 - a_2)(a_1^\dagger - a_2^\dagger) \qquad L_{1,2} = -(\mathrm{a}_1^\dagger - \mathrm{a}_2^\dagger)(\mathrm{a}_1 - \mathrm{a}_2)$$

$$\begin{aligned} L_{1,2} f(n_1, n_2) = {} & n_1[f(n_1-1, n_2+1) - f(n_1, n_2)] \\ & + n_2[f(n_1+1, n_2-1) - f(n_1, n_2)] \end{aligned}$$

$$\begin{aligned} \widehat{L}_{1,2} f(k_1, k_2) = {} & k_1[f(k_1-1, k_2+1) - f(k_1, k_2)] \\ & + k_2[f(k_1+1, k_2-1) - f(k_1, k_2)] \end{aligned}$$

Triangular self-duality of Independent random walkers

$$d(k, n) = \frac{n!}{(n-k)!} \mathbb{1}_{\{k \le n\}}$$

$$a^\dagger \xrightarrow{d} \mathrm{a}^\dagger$$

$$a \xrightarrow{d} \mathrm{a}$$

$$\begin{cases} af(k) = kf(k-1) \\ a^\dagger f(k) = f(k+1) \end{cases} \qquad \begin{cases} \mathrm{a}f(n) = \lambda f(n+1) \\ \mathrm{a}^\dagger f(n) = f(n) - nf(n-1) \end{cases}$$

$$[a, a^\dagger] = -1 \qquad [\mathrm{a}, \mathrm{a}^\dagger] = 1$$

$$\widehat{L}_{1,2} \xrightarrow{d_1 d_2} L_{1,2}$$

$$\widehat{L}_{1,2} = -(a_1 - a_2)(a_1^\dagger - a_2^\dagger) \qquad L_{1,2} = -(\mathrm{a}_1^\dagger - \mathrm{a}_2^\dagger)(\mathrm{a}_1 - \mathrm{a}_2)$$

$$\begin{aligned} L_{1,2} f(n_1, n_2) &= n_1[f(n_1 - 1, n_2 + 1) - f(n_1, n_2)] \\ &+ n_2[f(n_1 + 1, n_2 - 1) - f(n_1, n_2)] \end{aligned}$$

$$\begin{aligned} \widehat{L}_{1,2} f(k_1, k_2) &= k_1[f(k_1 - 1, k_2 + 1) - f(k_1, k_2)] \\ &+ k_2[f(k_1 + 1, k_2 - 1) - f(k_1, k_2)] \end{aligned}$$

Orthogonal self-duality of Independent random walkers

$d(k,n) = e^{\lambda} C_k(n;\lambda)$ Charlier polynomials

$$a^\dagger \xrightarrow{d} \mathrm{a}^\dagger$$

$$a \xrightarrow{d} \mathrm{a}$$

$$\begin{cases} af(k) = kf(k-1) \\ a^\dagger f(k) = f(k+1) \end{cases} \qquad \begin{cases} \mathrm{a}f(n) = \lambda f(n) - \lambda f(n+1) \\ \mathrm{a}^\dagger f(n) = f(n) - \frac{n}{\lambda} f(n-1) \end{cases}$$

$$[a, a^\dagger] = -1 \qquad [\mathrm{a}, \mathrm{a}^\dagger] = 1$$

$$\widehat{L}_{1,2} \xrightarrow{d_1 d_2} L_{1,2}$$

$$\widehat{L}_{1,2} = -(a_1 - a_2)(a_1^\dagger - a_2^\dagger) \qquad L_{1,2} = -(\mathrm{a}_1^\dagger - \mathrm{a}_2^\dagger)(\mathrm{a}_1 - \mathrm{a}_2)$$

$$\begin{aligned} L_{1,2} f(n_1, n_2) &= n_1[f(n_1 - 1, n_2 + 1) - f(n_1, n_2)] \\ &+ n_2[f(n_1 + 1, n_2 - 1) - f(n_1, n_2)] \end{aligned}$$

$$\begin{aligned} \widehat{L}_{1,2} f(k_1, k_2) &= k_1[f(k_1 - 1, k_2 + 1) - f(k_1, k_2)] \\ &+ k_2[f(k_1 + 1, k_2 - 1) - f(k_1, k_2)] \end{aligned}$$

Triangular duality between Independent random walkers and Deterministic energy process

$$d(k, z) = z^k$$

$$a^\dagger \xrightarrow{d} A^\dagger$$

$$a \xrightarrow{d} A$$

$$\begin{cases} af(k) = kf(k-1) \\ a^\dagger f(k) = f(k+1) \end{cases} \qquad \begin{cases} Af(z) = \frac{\partial}{\partial z} f(z) \\ A^\dagger f(z) = zf(z) \end{cases}$$

$$[a, a^\dagger] = -1 \qquad [A, A^\dagger] = 1$$

$$\widehat{L}_{1,2} \xrightarrow{d_1 d_2} L_{1,2}$$

$$\widehat{L}_{1,2} = -(a_1 - a_2)(a_1^\dagger - a_2^\dagger) \qquad L_{1,2} = -(A_1^\dagger - A_2^\dagger)(A_1 - A_2)$$

$$L_{1,2} f(z_1, z_2) = -(z_1 - z_2)(\frac{\partial}{\partial z_1} - \frac{\partial}{\partial z_2}) f(z_1, z_2)$$

$$\begin{aligned} \widehat{L}_{1,2} f(k_1, k_2) &= k_1[f(k_1 - 1, k_2 + 1) - f(k_1, k_2)] \\ &+ k_2[f(k_1 + 1, k_2 - 1) - f(k_1, k_2)] \end{aligned}$$

C.5.2 Equivalent Representations of the $\mathfrak{su}(1,1)$ Algebra

Cheap self-duality of the Symmetric inclusion process

$$d(k,n) = \frac{\Gamma(\alpha)k!}{\Gamma(\alpha+k)}\delta_{k,n}$$

$$K^+ \xrightarrow{d} \mathcal{K}^+$$
$$K^- \xrightarrow{d} \mathcal{K}^-$$
$$K^0 \xrightarrow{d} \mathcal{K}^0$$

$$\begin{cases} K^+ f(k) = (\alpha+k)f(k+1) \\ K^- f(k) = kf(k-1) \\ K^0 f(k) = \left(\frac{\alpha}{2}+k\right) f(k) \end{cases} \qquad \begin{cases} \mathcal{K}^+ f(n) = nf(n-1) \\ \mathcal{K}^- f(n) = (\alpha+n)f(n+1) \\ \mathcal{K}^0 f(n) = \left(\frac{\alpha}{2}+n\right) f(n) \end{cases}$$

$$[K^\pm, K^0] = \pm K^\pm \qquad [\mathcal{K}^\pm, \mathcal{K}^0] = \mp \mathcal{K}^\pm$$

$$[K^+, K^-] = 2K^0 \qquad [\mathcal{K}^+, \mathcal{K}^-] = -2\mathcal{K}^0$$

$$\widehat{L}_{1,2} \xrightarrow{d_1 d_2} L_{1,2}$$

$$L_{1,2} = \mathcal{K}_1^+\mathcal{K}_2^- + \mathcal{K}_1^-\mathcal{K}_2^+ - 2\mathcal{K}_1^0\mathcal{K}_2^0 + \frac{\alpha^2}{2}$$

$$\begin{aligned} L_{1,2} f(n_1,n_2) &= n_1(\alpha+n_2)[f(n_1-1,n_2+1) - f(n_1,n_2)] \\ &\quad + (\alpha+n_1)n_2[f(n_1+1,n_2-1) - f(n_1,n_2)] \end{aligned}$$

$$\widehat{L}_{1,2} = K_1^+K_2^- + K_1^-K_2^+ - 2K_1^0K_2^0 + \frac{\alpha^2}{2}$$

$$\begin{aligned} \widehat{L}_{1,2} f(k_1,k_2) &= k_1(\alpha+k_2)[f(k_1-1,k_2+1) - f(k_1,k_2)] \\ &\quad + (\alpha+k_1)k_2[f(k_1+1,k_2-1) - f(k_1,k_2)] \end{aligned}$$

Triangular self-duality of the Symmetric inclusion process

$$d(k,n) = \frac{\Gamma(\alpha)}{\Gamma(\alpha+k)} \frac{n!}{(n-k)!} \mathbb{1}_{\{k \le n\}}$$

$$K^+ \xrightarrow{d} \mathcal{K}^+$$
$$K^- \xrightarrow{d} \mathcal{K}^-$$
$$K^0 \xrightarrow{d} \mathcal{K}^0$$

$$\begin{cases} K^+ f(k) = (\alpha+k) f(k+1) \\ \\ K^- f(k) = k f(k-1) \\ \\ K^0 f(k) = \left(\frac{\alpha}{2}+k\right) f(k) \end{cases} \qquad \begin{cases} \mathcal{K}^+ f(n) = n f(n-1) \\ \\ \mathcal{K}^- f(n) = (\alpha+n) f(n+1) \\ \qquad\qquad -2(\frac{\alpha}{2}+n) f(n) \\ \qquad\qquad + n f(n-1) \\ \\ \mathcal{K}^0 f(n) = \left(\frac{\alpha}{2}+n\right) f(n) \\ \qquad\qquad - n f(n-1) \end{cases}$$

$$[K^\pm, K^0] = \pm K^\pm \qquad [\mathcal{K}^\pm, \mathcal{K}^0] = \mp \mathcal{K}^\pm$$

$$[K^+, K^-] = 2K^0 \qquad [\mathcal{K}^+, \mathcal{K}^-] = -2\mathcal{K}^0$$

$$\widehat{L}_{1,2} \xrightarrow{d_1 d_2} L_{1,2}$$

$$L_{1,2} = \mathcal{K}_1^+ \mathcal{K}_2^- + \mathcal{K}_1^- \mathcal{K}_2^+ - 2\mathcal{K}_1^0 \mathcal{K}_2^0 + \frac{\alpha^2}{2}$$

$$\begin{aligned} L_{1,2} f(n_1,n_2) &= n_1(\alpha+n_2)[f(n_1-1,n_2+1) - f(n_1,n_2)] \\ &\quad + (\alpha+n_1)n_2[f(n_1+1,n_2-1) - f(n_1,n_2)] \end{aligned}$$

$$\widehat{L}_{1,2} = K_1^+ K_2^- + K_1^- K_2^+ - 2K_1^0 K_2^0 + \frac{\alpha^2}{2}$$

$$\begin{aligned} \widehat{L}_{1,2} f(k_1,k_2) &= k_1(\alpha+k_2)[f(k_1-1,k_2+1) - f(k_1,k_2)] \\ &\quad + (\alpha+k_1)k_2[f(k_1+1,k_2-1) - f(k_1,k_2)] \end{aligned}$$

Orthogonal self-duality of the Symmetric inclusion process

$$d(k,n) = (1+\rho)^{\alpha} M_k(n;\alpha,\frac{\rho}{1+\rho})$$ Meixner polynomials

$$K^+ \xrightarrow{d} \mathcal{K}^+$$
$$K^- \xrightarrow{d} \mathcal{K}^-$$
$$K^0 \xrightarrow{d} \mathcal{K}^0$$

$$\begin{cases} K^+ f(k) = (\alpha+k) f(k+1) \\ K^- f(k) = k f(k-1) \\ K^0 f(k) = \left(\frac{\alpha}{2}+k\right) f(k) \end{cases}$$

$$\begin{cases} \mathcal{K}^+ f(n) = \rho(\alpha+n) f(n+1) \\ \qquad +(1+\rho)(\alpha+2n) f(n) \\ \qquad -(1+\rho) n f(n-1) \\ \mathcal{K}^- f(n) = \rho(\alpha+n) f(n+1) \\ \qquad +\rho(\alpha+2n) f(n) \\ \qquad -\rho n f(n-1) \\ \mathcal{K}^0 f(n) = \rho(\alpha+n) f(n+1) \\ \qquad +(2+\rho)(n+\frac{\alpha}{2}) f(n) \\ \qquad -(1+\rho) n f(n-1) \end{cases}$$

$$[K^{\pm}, K^0] = \pm K^{\pm} \qquad [\mathcal{K}^{\pm}, \mathcal{K}^0] = \mp \mathcal{K}^{\pm}$$

$$[K^+, K^-] = 2K^0 \qquad [\mathcal{K}^+, \mathcal{K}^-] = -2\mathcal{K}^0$$

$$\widehat{L}_{1,2} \xrightarrow{d_1 d_2} L_{1,2}$$

$$L_{1,2} = \mathcal{K}_1^+\mathcal{K}_2^- + \mathcal{K}_1^-\mathcal{K}_2^+ - 2\mathcal{K}_1^0\mathcal{K}_2^0 + \frac{\alpha^2}{2}$$

$$L_{1,2} f(n_1,n_2) = n_1(\alpha+n_2)[f(n_1-1,n_2+1) - f(n_1,n_2)] + (\alpha+n_1)n_2[f(n_1+1,n_2-1) - f(n_1,n_2)]$$

$$\widehat{L}_{1,2} = K_1^+K_2^- + K_1^-K_2^+ - 2K_1^0K_2^0 + \frac{\alpha^2}{2}$$

$$\widehat{L}_{1,2} f(k_1,k_2) = k_1(\alpha+k_2)[f(k_1-1,k_2+1) - f(k_1,k_2)] + (\alpha+k_1)k_2[f(k_1+1,k_2-1) - f(k_1,k_2)]$$

Triangular duality between the Symmetric inclusion process and the Brownian energy process

$$d(k,z) = z^k \frac{\Gamma(\alpha)}{\Gamma(\alpha+k)}$$

$$K^+ \xrightarrow{d} \mathcal{K}^+$$

$$K^- \xrightarrow{d} \mathcal{K}^-$$

$$K^0 \xrightarrow{d} \mathcal{K}^0$$

$$\begin{cases} K^+ f(k) = (\alpha+k) f(k+1) \\ K^- f(k) = k f(k-1) \\ K^0 f(k) = \left(\frac{\alpha}{2}+k\right) f(k) \end{cases} \qquad \begin{cases} \mathcal{K}^+ f(z) = z f(z) \\ \mathcal{K}^- f(z) = \left(z\frac{\partial^2}{\partial z^2} + \alpha\frac{\partial}{\partial z}\right) f(z) \\ \mathcal{K}^0 f(z) = \left(z\frac{\partial}{\partial z} + \frac{\alpha}{2}\right) f(z) \end{cases}$$

$$[K^\pm, K^0] = \pm K^\pm \qquad [\mathcal{K}^\pm, \mathcal{K}^0] = \mp \mathcal{K}^\pm$$

$$[K^+, K^-] = 2K^0 \qquad [\mathcal{K}^+, \mathcal{K}^-] = -2\mathcal{K}^0$$

$$\widehat{L}_{1,2} \xrightarrow{d_1 d_2} L_{1,2}$$

$$L_{1,2} = \mathcal{K}_1^+\mathcal{K}_2^- + \mathcal{K}_1^-\mathcal{K}_2^+ - 2\mathcal{K}_1^0\mathcal{K}_2^0 + \frac{\alpha^2}{2}$$

$$L_{1,2} f(z_1,z_2) = \left(-\alpha(z_1-z_2)\left(\frac{\partial}{\partial z_1} - \frac{\partial}{\partial z_2}\right) + \left(\frac{\partial}{\partial z_1} - \frac{\partial}{\partial z_2}\right)^2\right) f(z_1,z_2)$$

$$\widehat{L}_{1,2} = K_1^+K_2^- + K_1^-K_2^+ - 2K_1^0K_2^0 + \frac{\alpha^2}{2}$$

$$\begin{aligned} \widehat{L}_{1,2} f(k_1,k_2) &= k_1(\alpha+k_2)[f(k_1-1,k_2+1) - f(k_1,k_2)] \\ &\quad + (\alpha+k_1)k_2[f(k_1+1,k_2-1) - f(k_1,k_2)] \end{aligned}$$

Orthogonal duality between the Symmetric inclusion process and the Brownian energy process

$$d(k,z) = \frac{k!\,\Gamma(\alpha)}{\Gamma(\alpha+k)} L_k(z;\alpha-1) = {}_1F_1\left(\begin{matrix}-k\\ \alpha\end{matrix}\middle| z\right) \quad \text{Laguerre polynomials}$$

$$K^+ \xrightarrow{d} \mathcal{K}^+$$
$$K^- \xrightarrow{d} \mathcal{K}^-$$
$$K^0 \xrightarrow{d} \mathcal{K}^0$$

$$\begin{cases} K^+ f(k) = -(\alpha+k) f(k+1) \\ \qquad +2(\frac{\alpha}{2}+k) f(k) \\ \qquad -k f(k-1) \\ K^- f(k) = -k f(k-1) \\ K^0 f(k) = \left(\frac{\alpha}{2}+k\right) f(k) \\ \qquad -k f(k-1) \end{cases} \qquad \begin{cases} \mathcal{K}^+ f(z) = z f(z) \\ \mathcal{K}^- f(z) = \left(z\frac{\partial^2}{\partial z^2} + \alpha\frac{\partial}{\partial z}\right) f(z) \\ \mathcal{K}^0 f(z) = \left(z\frac{\partial}{\partial z} + \frac{\alpha}{2}\right) f(z) \end{cases}$$

$$[K^\pm, K^0] = \pm K^\pm \qquad [\mathcal{K}^\pm, \mathcal{K}^0] = \mp \mathcal{K}^\pm$$

$$[K^+, K^-] = 2K^0 \qquad [\mathcal{K}^+, \mathcal{K}^-] = -2\mathcal{K}^0$$

$$\widehat{L}_{1,2} \xrightarrow{d_1 d_2} L_{1,2}$$

$$L_{1,2} = \mathcal{K}_1^+\mathcal{K}_2^- + \mathcal{K}_1^-\mathcal{K}_2^+ - 2\mathcal{K}_1^0\mathcal{K}_2^0 + \frac{\alpha^2}{2}$$

$$L_{1,2} f(z_1,z_2) = \left(-\alpha(z_1-z_2)(\frac{\partial}{\partial z_1} - \frac{\partial}{\partial z_2}) + (\frac{\partial}{\partial z_1} - \frac{\partial}{\partial z_2})^2\right) f(z_1,z_2)$$

$$\widehat{L}_{1,2} = K_1^+K_2^- + K_1^-K_2^+ - 2K_1^0K_2^0 + \frac{\alpha^2}{2}$$

$$\begin{aligned} \widehat{L}_{1,2} f(k_1,k_2) &= k_1(\alpha+k_2)[f(k_1-1,k_2+1) - f(k_1,k_2)] \\ &\quad + (\alpha+k_1)k_2[f(k_1+1,k_2-1) - f(k_1,k_2)] \end{aligned}$$

C.5.3 Equivalent Representations of the $\mathfrak{su}(2)$ Algebra

Cheap self-duality of the Symmetric partial exclusion process

$$d(k,n) = \frac{\Gamma(\alpha - k + 1)k!}{\Gamma(\alpha)}\delta_{k,n}$$

$$J^+ \xrightarrow{d} \mathcal{J}^+$$
$$J^- \xrightarrow{d} \mathcal{J}^-$$
$$J^0 \xrightarrow{d} \mathcal{J}^0$$

$$\begin{cases} J^+ f(k) = (\alpha - k) f(k+1) \\ \\ J^- f(k) = k f(k-1) \\ \\ J^0 f(k) = \left(-\frac{\alpha}{2} + k\right) f(k) \end{cases} \qquad \begin{cases} \mathcal{J}^+ f(n) = n f(n-1) \\ \\ \mathcal{J}^- f(n) = (\alpha - n) f(n+1) \\ \\ \mathcal{J}^0 f(n) = \left(-\frac{\alpha}{2} + n\right) f(n) \end{cases}$$

$$[J^\pm, J^0] = \pm J^\pm \qquad [\mathcal{J}^\pm, \mathcal{J}^0] = \mp \mathcal{J}^\pm$$

$$[J^+, J^-] = -2J^0 \qquad [\mathcal{J}^+, \mathcal{J}^-] = 2\mathcal{J}^0$$

$$\widehat{L}_{1,2} \xrightarrow{d_1 d_2} L_{1,2}$$

$$L_{1,2} = \mathcal{J}_1^+ \mathcal{J}_2^- + \mathcal{J}_1^- \mathcal{J}_2^+ - 2\mathcal{J}_1^0 \mathcal{J}_2^0 + \frac{\alpha^2}{2}$$

$$\begin{aligned} L_{1,2} f(n_1, n_2) = {} & n_1(\alpha - n_2)[f(n_1 - 1, n_2 + 1) - f(n_1, n_2)] \\ & + (\alpha - n_1) n_2 [f(n_1 + 1, n_2 - 1) - f(n_1, n_2)] \end{aligned}$$

$$\widehat{L}_{1,2} = K_1^+ K_2^- + K_1^- K_2^+ - 2K_1^0 K_2^0 + \frac{\alpha^2}{2}$$

$$\begin{aligned} \widehat{L}_{1,2} f(k_1, k_2) = {} & k_1(\alpha - k_2)[f(k_1 - 1, k_2 + 1) - f(k_1, k_2)] \\ & + (\alpha - k_1) k_2 [f(k_1 + 1, k_2 - 1) - f(k_1, k_2)] \end{aligned}$$

Triangular self-duality of the Symmetric partial exclusion process

$$d(k,n) = \frac{\Gamma(\alpha - k + 1)}{\Gamma(\alpha)} \frac{n!}{(n-k)!} \mathbb{1}_{\{k \le n\}}$$

$$J^+ \xrightarrow{d} \mathcal{J}^+$$

$$J^- \xrightarrow{d} \mathcal{J}^-$$

$$J^0 \xrightarrow{d} \mathcal{J}^0$$

$$\begin{cases} J^+ f(k) = (\alpha - k) f(k+1) \\ \\ J^- f(k) = k f(k-1) \\ \\ J^0 f(k) = \left(-\frac{\alpha}{2} + k\right) f(k) \end{cases} \qquad \begin{cases} \mathcal{J}^+ f(n) = n f(n-1) \\ \\ \mathcal{J}^- f(n) = (\alpha - n) f(n+1) \\ \qquad -2(\frac{\alpha}{2} - n) f(n) \\ \qquad -n f(n-1) \\ \\ \mathcal{J}^0 f(n) = \left(-\frac{\alpha}{2} + n\right) f(n) \\ \qquad -n f(n-1) \end{cases}$$

$$[J^\pm, J^0] = \pm J^\pm \qquad [\mathcal{J}^\pm, \mathcal{J}^0] = \mp \mathcal{J}^\pm$$

$$[J^+, J^-] = -2J^0 \qquad [\mathcal{J}^+, \mathcal{J}^-] = 2\mathcal{J}^0$$

$$\widehat{L}_{1,2} \xrightarrow{d_1 d_2} L_{1,2}$$

$$L_{1,2} = \mathcal{J}_1^+ \mathcal{J}_2^- + \mathcal{J}_1^- \mathcal{J}_2^+ - 2\mathcal{J}_1^0 \mathcal{J}_2^0 + \frac{\alpha^2}{2}$$

$$\begin{aligned} L_{1,2} f(n_1, n_2) &= n_1(\alpha - n_2)[f(n_1 - 1, n_2 + 1) - f(n_1, n_2)] \\ &+ (\alpha - n_1) n_2 [f(n_1 + 1, n_2 - 1) - f(n_1, n_2)] \end{aligned}$$

$$\widehat{L}_{1,2} = J_1^+ J_2^- + J_1^- J_2^+ - 2J_1^0 J_2^0 + \frac{\alpha^2}{2}$$

$$\begin{aligned} \widehat{L}_{1,2} f(k_1, k_2) &= k_1(\alpha - k_2)[f(k_1 - 1, k_2 + 1) - f(k_1, k_2)] \\ &+ (\alpha - k_1) k_2 [f(k_1 + 1, k_2 - 1) - f(k_1, k_2)] \end{aligned}$$

Orthogonal self-duality of the Symmetric partial exclusion process

$d(k,n) = (1-\rho)^{\alpha} K_k(n;\alpha,\rho)$ Krawtchouk polynomials

$$J^+ \xrightarrow{d} \mathcal{J}^+$$
$$J^- \xrightarrow{d} \mathcal{J}^-$$
$$J^0 \xrightarrow{d} \mathcal{J}^0$$

$$\begin{cases} J^+ f(k) = (\alpha+k) f(k+1) \\ \\ J^- f(k) = k f(k-1) \\ \\ J^0 f(k) = \left(\frac{\alpha}{2}+k\right) f(k) \end{cases} \qquad \begin{cases} \mathcal{J}^+ f(n) = \rho(\alpha-n) f(n+1) \\ \qquad +(1-\rho)(\alpha-2n) f(n) \\ \qquad -\frac{(1-\rho)^2}{\rho} n f(n-1) \\ \\ \mathcal{J}^- f(n) = \rho(\alpha-n) f(n+1) \\ \qquad +\rho(\alpha-2n) f(n) \\ \qquad +\rho n f(n-1) \\ \\ \mathcal{J}^0 f(n) = -\rho(\alpha-n) f(n+1) \\ \qquad +(1-2\rho)(n-\frac{\alpha}{2}) f(n) \\ \qquad -(1-\rho) n f(n-1) \end{cases}$$

$$[K^{\pm}, K^0] = \pm K^{\pm} \qquad [\mathcal{K}^{\pm}, \mathcal{K}^0] = \mp \mathcal{K}^{\pm}$$

$$[K^+, K^-] = 2K^0 \qquad [\mathcal{K}^+, \mathcal{K}^-] = -2\mathcal{K}^0$$

$$\widehat{L}_{1,2} \xrightarrow{d_1 d_2} L_{1,2}$$

$$L_{1,2} = \mathcal{K}_1^+ \mathcal{K}_2^- + \mathcal{K}_1^- \mathcal{K}_2^+ - 2\mathcal{K}_1^0 \mathcal{K}_2^0 + \frac{\alpha^2}{2}$$

$$\begin{aligned} L_{1,2} f(n_1,n_2) &= n_1(\alpha+n_2)[f(n_1-1,n_2+1) - f(n_1,n_2)] \\ &\quad + (\alpha+n_1) n_2 [f(n_1+1,n_2-1) - f(n_1,n_2)] \end{aligned}$$

$$\widehat{L}_{1,2} = K_1^+ K_2^- + K_1^- K_2^+ - 2K_1^0 K_2^0 + \frac{\alpha^2}{2}$$

$$\begin{aligned} \widehat{L}_{1,2} f(k_1,k_2) &= k_1(\alpha+k_2)[f(k_1-1,k_2+1) - f(k_1,k_2)] \\ &\quad + (\alpha+k_1) k_2 [f(k_1+1,k_2-1) - f(k_1,k_2)] \end{aligned}$$

References

1. Alcaraz, F.C., Droz, M., Henkel, M., Rittenberg, V.: Reaction-diffusion processes, critical dynamics, and quantum chains. Ann. Phys. **230**(2), 250–302 (1994)
2. Aldous, D., Lanoue, D., et al.: A lecture on the averaging process. Probab. Surv. **9**, 90–102 (2012)
3. Andjel, E.D., et al.: Invariant measures for the zero range process. Ann. Probab. **10**(3), 525–547 (1982)
4. Arita, C., Ayyer, A., Mallick, K., Prolhac, S.: Generalized matrix ansatz in the multispecies exclusion process? The partially asymmetric case. J. Phys. A Math. Theor. **45**(19), 195001 (2012)
5. Arnold, B.C., Balakrishnan, N., Nagaraja, H.N.: A first course in order statistics. SIAM, Philadelphia (2008)
6. Ayala, M., Carinci, G., Redig, F.: Condensation of sip particles and sticky brownian motion. Preprint (2019). arXiv:1906.09887
7. Ayala, M., Carinci, G., Redig, F.: Higher order fluctuation fields and orthogonal duality polynomials. Preprint (2020). arXiv:2004.08412
8. Bakry, D., Orevkov, S., Zani, M.: Orthogonal polynomials and diffusion operators. Preprint (2013). arXiv:1309.5632
9. Baldasso, R., Menezes, O., Neumann, A., Souza, R.R.: Exclusion process with slow boundary. J. Stat. Phys. **167**(5), 1112–1142 (2017)
10. Barraquand, G., Corwin, I.: The q-hahn asymmetric exclusion process. Ann. Appl. Probab. **26**(4), 2304–2356 (2016)
11. Basile, G., Bernardin, C., Olla, S.: Thermal conductivity for a momentum conservative model. Commun. Math. Phys. **287**, 67–98 (2009)
12. Baudoin, F.: Diffusion processes and stochastic calculus. EMS textbooks in mathematics, (2014)
13. Baxter, R.J.: Exactly Solved Models in Statistical Mechanics. Elsevier, Amsterdam (2016)
14. Beisert, N., Ahn, C., Alday, L.F., Bajnok, Z., Drummond, J.M., Freyhult, L., Gromov, N., Janik, R.A., Kazakov, V., Klose, T., et al.: Review of ads/cft integrability: an overview. Lett. Math. Phys. **99**, 3–32 (2012)
15. Belitsky, V., Schütz, G.M.: Self-duality for the two-component asymmetric simple exclusion process. J. Math. Phys. **56**(8), 083302 (2015)
16. Belitsky, V., Schütz, G.M.: Self-duality and shock dynamics in the n-component priority ASEP. Stoch. Process. Appl. **128**(4), 1165–1207 (2018)

C. Giardinà, F. Redig, *Duality for Markov Processes*, Grundlehren der mathematischen Wissenschaften 365, https://doi.org/10.1007/978-3-032-04099-2

17. Belitsky, V., Schütz, G., et al.: Diffusion and scattering of shocks in the partially asymmetric simple exclusion process. Electron. J. Probab. **7** (2002)
18. Bernardin, C., Jimenez, B.O.: Fractional fick's law for the boundary driven exclusion process with long jumps. Preprint (2016). arXiv:1603.01234
19. Bernardin, C., Goncalves, P., Jara, M., Scotta, S.: Equilibrium fluctuations for diffusive symmetric exclusion with long jumps and infinitely extended reservoirs. Preprint (2020). arXiv:2002.12841
20. Bernardin, C., Gonçalves, P., Jiménez-Oviedo, B.: A microscopic model for a one parameter class of fractional laplacians with dirichlet boundary conditions. Arch. Ration. Mech. Anal. 1–48 (2020)
21. Bernardin, C., Gonçalves, P., Scotta, S.: Hydrodynamic limit for a boundary driven super-diffusive symmetric exclusion. Preprint (2020). arXiv:2007.01621
22. Bertini, L., Giacomin, G.: Stochastic burgers and kpz equations from particle systems. Commun. Math. Phys. **183**(3), 571–607 (1997)
23. Bertini, L., De Sole, A., Gabrielli, D., Jona-Lasinio, G., Landim, C.: Large deviations for the boundary driven symmetric simple exclusion process. Math. Phys. Anal. Geom. **6**(3), 231–267 (2003)
24. Bertini, L., Gabrielli, D., Lebowitz, J.L.: Large deviations for a stochastic model of heat flow. J. Stat. Phys. **121**(5-6), 843–885 (2005)
25. Bertini, L., De Sole, A., Gabrielli, D., Jona-Lasinio, G., Landim, C.: Large deviation approach to non equilibrium processes in stochastic lattice gases. Bull. Br. Math. Soc. **37**(4), 611–643 (2006)
26. Bertini, L., De Sole, A., Gabrielli, D., Jona-Lasinio, G., Landim, C.: Non equilibrium current fluctuations in stochastic lattice gases. J. Stat. Phys. **123**(2), 237–276 (2006)
27. Bertini, L., De Sole, A., Gabrielli, D., Jona-Lasinio, G., Landim, C.: Stochastic interacting particle systems out of equilibrium. J. Stat. Mech. Theory Exper. **2007**(07), P07014 (2007)
28. Bertini, L., De Sole, A., Gabrielli, D., Jona-Lasinio, G., Landim, C.: Macroscopic fluctuation theory. Rev. Mod. Phys. **87**(2), 593 (2015)
29. Bianchi, A., Dommers, S., Giardinà, C.: Metastability in the reversible inclusion process. Electron. J. Probab. **22**, 70 (2017).
30. Blyschak, D., Burke, O., Kuan, J., Li, D., Ustilovsky, S., Zhou, Z.: Orthogonal polynomial duality of a two-species asymmetric exclusion process. J. Stat. Phys. **190**(5), 101 (2023)
31. Borodin, A.N., Salminen, P.: Handbook of Brownian Motion-facts and Formulae. Birkhauser, Basel (2012)
32. Bonetto, F., Loss, M., Vaidyanathan, R.: The kac model coupled to a thermostat. J. Stat. Phys. **156**(4), 647–667 (2014)
33. Borodin, A., Corwin, I., Sasamoto, T., et al.: From duality to determinants for q-tasep and asep. Ann. Probab. **42**(6), 2314–2382 (2014)
34. Braun, V.M., Derkachov, S., Manashov, A.: Integrability of three-particle evolution equations in qcd. Phys. Rev. Lett. **81**(10), 2020 (1998)
35. Bricmont, J.: Science of chaos or chaos in science? Preprint (1996). arXiv:chao-dyn/9603009
36. Brox, T., Rost, H., et al.: Equilibrium fluctuations of stochastic particle systems: the role of conserved quantities. Ann. Probab. **12**(3), 742–759 (1984)
37. Bytsko, A.G.: On integrable hamiltonians for higher spin xxz chain. J. Math. Phys. **44**(9), 3698–3717 (2003)
38. Caputo, P.: Energy gap estimates in xxz ferromagnets and stochastic particle systems. Markov Process. Relat. Fields **11**, 189–210 (2005)
39. Cardy, J.: Reaction-diffusion processes. A+ A **100**(26), 13–19 (2006)
40. Carinci, G., Giardina, C., Giberti, C., Redig, F.: Duality for stochastic models of transport. J. Stat. Phys. **152**(4), 657–697 (2013)
41. Carinci, G., Giardinà, C., Redig, F., Sasamoto, T.: A generalized asymmetric exclusion process with $\mathcal{U}_q(\mathfrak{sl}_2)$ stochastic duality. Preprint (2014). arXiv:1407.3367
42. Carinci, G., Giardina, C., Giberti, C., Redig, F.: Dualities in population genetics: a fresh look with new dualities. Stoch. Process. Appl. **125**(3), 941–969 (2015)

43. Carinci, G., Giardinà, C., Redig, F., Sasamoto, T.: Asymmetric stochastic transport models with u _q (su (1, 1)) u q (su (1, 1)) symmetry. J. Stat. Phys. **163**, 239–279 (2016)
44. Carinci, G., Giardinà, C., Redig, F., Sasamoto, T.: A generalized asymmetric exclusion process with $u_q(\mathfrak{sl}_2)$ stochastic duality. Probab. Theory Relat. Fields **166**(3), 887–933 (2016)
45. Carinci, G., Franceschini, C., Giardina, C., Groenevelt, W.G., Redig, F.: Orthogonal dualities of Markov processes and unitary symmetries. SIGMA Symmetry Integrability Geom. Methods Appl. **15**, 53 (2019)
46. Carinci, G., Giardinà, C., Presutti, E.: Stationary states in infinite volume with non-zero current. J. Stat. Phys. 1–32 (2019)
47. Carinci, G., Giardinà, C., Redig, F.: Consistent particle systems and duality. Preprint (2019). arXiv:1907.10583
48. Carinci, G., Franceschini, C., Groenevelt, W.: q-orthogonal dualities for asymmetric particle systems. Preprint (2020). arXiv:2003.07837
49. Carinci, G., Giardina, C., Redig, F., et al.: Exact formulas for two interacting particles and applications in particle systems with duality. Ann. Appl. Probab. **30**(4), 1934–1970 (2020)
50. Carinci, G., Franceschini, C., Frassek, R., Giardinà, C., Redig, F.: The open harmonic process: non-equilibrium steady state, pressure, density large deviation and additivity principle. Preprint (2023). arXiv:2307.14975
51. Carinci, G., Franceschini, C., Gabrielli, D., Giardinà, C., Tsagkarogiannis, D.: Solvable stationary non equilibrium states. J. Stat. Phys. **191**(1), 10 (2024)
52. Carlen, E.A., Carvalho, M.C., Roux, J.L., Loss, M., Villani, C.: Entropy and chaos in the kac model. Preprint (2008). arXiv:0808.3192
53. Casini, F., Frassek, R., Giardinà, C.: Duality for the multispecies stirring process with open boundaries. J. Phys. A **57**(29), 295001 (2024)
54. Casini, F., Giardinà, C., Redig, F.: Density fluctuations for the multi-species stirring process. J. Theor. Probab. 1–38 (2024)
55. Chakrabarti, B.K., Chakraborti, A., Chakravarty, S.R., Chatterjee, A.: Econophysics of Income and Wealth Distributions. Cambridge University Press, Cambridge (2013)
56. Chen, Z.: Exact Solutions in Multi-species Exclusion Processes. Ph.D. Thesis, 2019
57. Chen, J.P., Sau, F.: Higher order hydrodynamics and equilibrium fluctuations of interacting particle systems. Preprint (2020). arXiv:2008.13403
58. Chen, Z., de Gier, J., Wheeler, M.: Integrable stochastic dualities and the deformed Knizhnik-Zamolodchikov equation. Int. Math. Res. Not. IMRN **2020**(19), 5872–5925 (2020)
59. Chleboun, P., Gabriel, S., Grosskinsky, S.: Size-biased diffusion limits and the inclusion process. Preprint (2023). arXiv:2304.09722
60. Cirillo, P., Redig, F., Ruszel, W.: Duality and stationary distributions of wealth distribution models. J. Phys. A Math. Theor. **47**(8), 085203 (2014)
61. Collet, P., Martínez, S., San Martín, J.: Markov chains on finite spaces. In: Quasi-Stationary Distributions, pp. 31–44. Springer, Berlin (2013)
62. Corwin, I.: Two ways to solve asep. In: Topics in Percolative and Disordered Systems, pp. 1–13. Springer, Berlin (2014)
63. Corwin, I., Shen, H., Tsai, L.C., et al.: ASEP(q, j) converges to the kpz equation. In: Annales de l'Institut Henri Poincaré, Probabilités et Statistiques, vol. 54, pp. 995–1012. Institut Henri Poincaré, Paris (2018)
64. Crampé, N.: Approches Algébriques dans les Systemes Intégrables. Ph.D. Thesis, Chambéry, 2004
65. Crampe, N., Ragoucy, E., Vanicat, M.: Integrable approach to simple exclusion processes with boundaries. Review and progress. J. Stat. Mech. Theory Exper. **2014**(11), P11032 (2014)
66. Dawson, D.A., Greven, A., Greven, A.: Spatial Fleming-Viot models with selection and mutation. Springer, Berlin (2014)
67. De Gier, J., Essler, F.H.: Bethe ansatz solution of the asymmetric exclusion process with open boundaries. Phys. Rev. Lett. **95**(24), 240601 (2005)
68. De Gier, J., Essler, F.H.: Exact spectral gaps of the asymmetric exclusion process with open boundaries. J. Stat. Mech. Theory Exper. **2006**(12), P12011 (2006)

69. De Masi, A., Ferrari, P.A., Gabrielli, D.: Hidden temperature in the kmp model. Preprint (2023). arXiv:2310.01672
70. De Masi, A., Presutti, E.: Probability estimates for symmetric simple exclusion random walks. Annales de l'IHP Probabilités et Statistiques **19**(1), 71–85 (1983)
71. De Masi, A., Presutti, E.: Mathematical Methods for Hydrodynamic Limits. Springer, Berlin (2006)
72. De Masi, A., Ianiro, N., Pellegrinotti, A., Presutti, E.: A survey of the hydrodynamical behavior of many-particle systems. STIA **85**, 123–294 (1984)
73. De Masi, A., Presutti, E., Scacciatelli, E.: The weakly asymmetric simple exclusion process. Annales de l'IHP Probabilités et statistiques **25**, 1–38 (1989)
74. Depperschmidt, A., Greven, A., Pfaffelhuber, P.: Duality and the well-posedness of a martingale problem. Preprint (2019). arXiv:1904.01564
75. Derrida, B.: Non-equilibrium steady states: fluctuations and large deviations of the density and of the current. J. Stat. Mech. Theory Exper. **2007**(07), P07023 (2007)
76. Derrida, B., Gerschenfeld, A.: Current fluctuations in one dimensional diffusive systems with a step initial density profile. J. Stat. Phys. **137**(5–6), 978 (2009)
77. Derrida, B., Evans, M.R., Hakim, V., Pasquier, V.: Exact solution of a 1d asymmetric exclusion model using a matrix formulation. J. Phys. A Math. General **26**(7), 1493 (1993)
78. Derrida, B., Lebowitz, J., Speer, E.: Large deviation of the density profile in the steady state of the open symmetric simple exclusion process. J. Stat. Phys. **107**(3–4), 599–634 (2002)
79. Derrida, B., Lebowitz, J., Speer, E.: Entropy of open lattice systems. J. Stat. Phys. **126**(4–5), 1083–1108 (2007)
80. Diaconis, P., Pang, C.A., Ram, A.: Hopf algebras and markov chains: two examples and a theory. J. Algebr. Comb. **39**, 527–585 (2014)
81. Doi, M.: Second quantization representation for classical many-particle system. J. Phys. A Math. General **9**(9), 1465 (1976)
82. Duffy, K.R., Macci, C., Torrisi, G.L.: Sample path large deviations for order statistics. J. Appl. Probab. **48**(1), 238–257 (2011)
83. Durrett, R.: Ten lectures on particle systems. Lectures on Probability Theory, pp. 97–201. Springer, (1995)
84. Engel, K.J., Nagel, R.: One-parameter semigroups for linear evolution equations. In: Semigroup Forum, vol. 63, pp. 278–280. Springer, Berlin (2001)
85. Erignoux, C.: Hydrodynamic limit of boundary driven exclusion processes with nonreversible boundary dynamics. J. Stat. Phys. **172**(5), 1327–1357 (2018)
86. Etheridge, A.: An introduction to superprocesses, vol. 20. American Mathematical Society, Providence (2000)
87. Etheridge, A.: Some Mathematical Models from Population Genetics: École D'Été de Probabilités de Saint-Flour XXXIX-2009, vol. 2012. Springer Science & Business Media, Berlin (2011)
88. Ethier, S.N., Griffiths, R.: The transition function of a fleming-viot process. Ann. Probab. 1571–1590 (1993)
89. Ethier, S.N., Kurtz, T.G.: Markov Processes: Characterization and Convergence, vol. 282. John Wiley & Sons, Hoboken (2009)
90. Faddeev, L.: How algebraic bethe ansatz works for integrable model. Preprint (1996). arXiv:hep-th/9605187
91. Faddeev, L., Korchemsky, G.: High energy qcd as a completely integrable model. Phys. Lett. B **342**(1-4), 311–322 (1995)
92. Faggionato, A.: Hydrodynamic limit of symmetric exclusion processes in inhomogeneous media. Preprint (2010). arXiv:1003.5521
93. Floreani, S., Redig, F., Sau, F.: Hydrodynamics for the partial exclusion process in random environment. Preprint (2019). arXiv:1911.12564
94. Floreani, S., Redig, F., Sau, F.: Orthogonal polynomial duality of boundary driven particle systems and non-equilibrium correlations. Preprint (2020). arXiv:2007.08272

95. Floreani, S., Jansen, S., Redig, F., Wagner, S.: Intertwining and duality for consistent markov processes. Preprint (2021). arXiv:2112.11885
96. Floreani, S., Redig, F., Sau, F.: Hydrodynamics for the partial exclusion process in random environment. Stoch. Process. Appl. **142**, 124–158 (2021)
97. Franceschini, C., Giardina, C.: Energy-exchange stochastic models for non-equilibrium. Preprint (2014). arXiv:1410.3661
98. Franceschini, C., Giardinà, C.: Stochastic duality and orthogonal polynomials. In: Sojourns in Probability Theory and Statistical Physics-III, pp. 187–214. Springer, Berlin (2019)
99. Franceschini, C., Giardina, C., Groenevelt, W.: Self-duality of markov processes and intertwining functions. Math. Phys. Anal. Geom. **21**(4), 29 (2018)
100. Franceschini, C., Gonçalves, P., Sau, F.: Symmetric inclusion process with slow boundary: hydrodynamics and hydrostatics. Preprint (2020). arXiv:2007.11998
101. Franceschini, C., Gonçalves, P., Sau, F.: Symmetric inclusion process with slow boundary: Hydrodynamics and hydrostatics. Bernoulli **28**(2), 1340–1381 (2022)
102. Franceschini, C., Frassek, R., Giardinà, C.: Integrable heat conduction model. J. Math. Phys. **64**(4) (2023)
103. Franchini, F., et al.: An Introduction to Integrable Techniques for One-dimensional Quantum Systems, vol. 940. Springer, Berlin (2017)
104. Franco, T., Gonçalves, P., Neumann, A.: Hydrodynamical behavior of symmetric exclusion with slow bonds. Annales de l'IHP Probabilités et Statistiques **49**, 402–427 (2013)
105. Frassek, R.: The non-compact xxz spin chain as stochastic particle process. J. Phys. A Math. Theor. **52**(33), 335202 (2019)
106. Frassek, R., Giardinà, C.: Exact solution of an integrable non-equilibrium particle system. J. Math. Phys. **63**(10) (2022)
107. Frassek, R., Giardina, C., Kurchan, J.: Duality and hidden equilibrium in transport models. SciPost Phys. **9**(4), 054 (2020)
108. Frassek, R., Giardinà, C., Kurchan, J.: Non-compact quantum spin chains as integrable stochastic particle processes. J. Stat. Phys. **180**(1–6), 135–171 (2020)
109. Fritz, J.: On the hydrodynamic limit of a scalar ginzburg-landau lattice model: the resolvent approach. In: Hydrodynamic Behavior and Interacting Particle Systems, pp. 75–97. Springer, Berlin (1987)
110. Fulton, W., Harris, J.: Representation Theory: A First Course, vol. 129. Springer Science & Business Media, Berlin (2013)
111. Galves, A., Kipnis, C., Marchioro, C., Presutti, E.: Nonequilibrium measures which exhibit a temperature gradient: study of a model. Commun. Math. Phys. **81**, 127–147 (1981)
112. Gärtner, J.: Convergence towards burger's equation and propagation of chaos for weakly asymmetric exclusion processes. Stoch. Process. Appl. **27**, 233–260 (1987)
113. Georgii, H.O.: Gibbs Measures and Phase Transitions. Walter de Gruyter GmbH & Co. KG, Berlin (2011)
114. Giardina, C., Kurchan, J.: The fourier law in a momentum-conserving chain. J. Stat. Mech. Theory Exper. **2005**(05), P05009 (2005)
115. Giardina, C., Kurchan, J., Redig, F.: Duality and exact correlations for a model of heat conduction. J. Math. Phys. **48**(3), 033301 (2007)
116. Giardina, C., Kurchan, J., Redig, F., Vafayi, K.: Duality and hidden symmetries in interacting particle systems. J. Stat. Phys. **135**(1), 25–55 (2009)
117. Giardina, C., Redig, F., Vafayi, K.: Correlation inequalities for interacting particle systems with duality. J. Stat. Phys. **141**(2), 242–263 (2010)
118. Giardinà, C., Redig, F., van Tol, B.: Intertwining and propagation of mixtures for generalized kmp models and harmonic models. Preprint (2024). arXiv:2406.01160
119. Gilbert, T.: Heat conduction and the nonequilibrium stationary states of stochastic energy exchange processes. J. Stat. Mech. Theory Exper. **2017**(8), 083205 (2017)
120. Gonçalves, P.: Hydrodynamics for symmetric exclusion in contact with reservoirs. In: International workshop on Stochastic Dynamics out of Equilibrium, pp. 137–205. Springer, Berlin (2017)

121. Gonçalves, P., Jara, M., Menezes, O., Neumann, A.: Non-equilibrium and stationary fluctuations for the ssep with slow boundary. Stoch. Process. Appl. **130**(7), 4326–4357 (2020)
122. Gorissen, M., Lazarescu, A., Mallick, K., Vanderzande, C.: Exact current statistics of the asymmetric simple exclusion process with open boundaries. Phys. Rev. Lett. **109**(17), 170601 (2012)
123. Griffeath, D.: A maximal coupling for markov chains. Zeitschrift für Wahrscheinlichkeitstheorie und verwandte Gebiete **31**(2), 95–106 (1975)
124. Griffiths, R.C., et al.: Diffusion processes and coalescent trees. Preprint (2010). arXiv:1003.4650
125. Griffiths, R.C., Jenkins, P.A., Lessard, S.: A coalescent dual process for a wright–fisher diffusion with recombination and its application to haplotype partitioning. Theor. Popul. Biol. **112**, 126–138 (2016)
126. Groenevelt, W.: Orthogonal stochastic duality functions from lie algebra representations. J. Stat. Phys. **174**(1), 97–119 (2019)
127. Grosskinsky, S., Redig, F., Vafayi, K.: Condensation in the inclusion process and related models. J. Stat. Phys. **142**(5), 952–974 (2011)
128. Grosskinsky, S., Redig, F., Vafayi, K., et al.: Dynamics of condensation in the symmetric inclusion process. Electron. J. Probab. **18** (2013)
129. Guo, M., Papanicolaou, G., Varadhan, S.: Nonlinear diffusion limit for a system with nearest neighbor interactions. Commun. Math. Phys. **118**(1), 31–59 (1988)
130. Gut, A.: Probability: A Graduate Course, vol. 75. Springer Science & Business Media, Berlin (2013)
131. Hall, B.: Lie Groups, Lie Algebras, and Representations: An Elementary Introduction, vol. 222. Springer, Berlin (2015)
132. Harris, T.J.: Duality of finite markovian queues. Oper. Res. **15**(3), 575–576 (1967)
133. Harris, T.E.: Additive set-valued markov processes and graphical methods. Ann. Probab. 355–378 (1978)
134. Hollander, F.: Large Deviations, vol. 14. American Mathematical Society, Providence (2000)
135. Holley, R.A., Stroock, D.W.: A martingale approach to infinite systems of interacting processes. Ann. Probab. 195–228 (1976)
136. Holley, R., Stroock, D.: Dual processes and their application to infinite interacting systems. Adv. Math. **32**(2), 149–174 (1979)
137. Howitt, C., Warren, J., et al.: Consistent families of brownian motions and stochastic flows of kernels. Ann. Probab. **37**(4), 1237–1272 (2009)
138. Humphreys, J.E.: Introduction to Lie Algebras and Representation Theory, vol. 9. Springer Science & Business Media, Berlin (2012)
139. Imamura, T., Sasamoto, T.: A note on the transition probability of asep infinite analysis. J. Stat. Phys. **88**, 427–445 (1997)
140. Imamura, T., Sasamoto, T.: Current moments of 1d asep by duality. J. Stat. Phys. **142**(5), 919–930 (2011)
141. Imamura, T., Sasamoto, T., Spohn, H.: Kpz, asep and delta-bose gas. J. Phys.-Conf. Series **297**, 012016 (2011)
142. Jansen, S., Kurt, N., et al.: On the notion (s) of duality for markov processes. Probab. Surv. **11**, 59–120 (2014)
143. Kac, M.: Foundations of kinetic theory. In: Proceedings of The Third Berkeley Symposium on Mathematical Statistics and Probability, vol. 3, pp. 171–197. University of California Press Berkeley and Los Angeles, California (1956)
144. Keisling, J.: An ergodic theorem for the symmetric generalized exclusion process. Markov Process. Relat. Fields **4**(3), 351–379 (1998)
145. Keller, G., Gerhard, K.: Equilibrium States in Ergodic Theory, vol. 42. Cambridge University Press, Cambridge (1998)
146. Kemeny, J.G., Snell, J.L., Knapp, A.W.: Denumerable Markov Chains: With a Chapter of Markov Random Fields by David Griffeath, vol. 40. Springer Science & Business Media, Berlin (2012)

147. Kim, S., Seo, I.: Condensation and metastable behavior of non-reversible inclusion processes. Commun. Math. Phys. **382**, 1343–1401 (2021)
148. Kimura, M.: Some problems of stochastic processes in genetics. Ann. Math. Stat. 882–901 (1957)
149. Kimura, M.: Population Genetics, Molecular Evolution, and the Neutral Theory: Selected Papers. University of Chicago Press, Chicago (1994)
150. Kimura, M., Ohta, T.: Theoretical Aspects of Population Genetics (MPB-4), Volume 4, vol. 114. Princeton University Press, Princeton (2020)
151. Kipnis, C., Landim, C.: Scaling Limits of Interacting Particle Systems, vol. 320. Springer Science & Business Media, Berlin (1998)
152. Kipnis, C., Marchioro, C., Presutti, E.: Heat flow in an exactly solvable model. J. Stat. Phys. **27**(1), 65–74 (1982)
153. Koekoek, R., Lesky, P.A., Swarttouw, R.F.: Hypergeometric Orthogonal Polynomials and Their q-analogues. Springer Science & Business Media, Berlin (2010)
154. Korchemsky, G.: Bethe ansatz for qcd pomeron. Nucl. Phys. B **443**(1–2), 255–301 (1995)
155. Krengel, U.: Ergodic Theorems, vol. 6. Walter de Gruyter, Berlin (2011)
156. Kuan, J.: Stochastic duality of asep with two particle types via symmetry of quantum groups of rank two. J. Phys. A Math. Theor. **49**(11), 115002 (2016)
157. Kuan, J.: A multi-species asep and-tazrp with stochastic duality. Int. Math. Res. Notices **2018**(17), 5378–5416 (2018)
158. Kuoch, K., Redig, F.: Ergodic theory of the symmetric inclusion process. Stoch. Process. Appl. **126**(11), 3480–3498 (2016)
159. Kurtz, T.G.: Approximation of population processes. SIAM, Philadelphia (1981)
160. Landim, C., Milanés, A., Olla, S.: Stationary and nonequilibrium fluctuations in boundary driven exclusion processes. Preprint (2006). arXiv:math/0608165
161. Last, G., Penrose, M.: Lectures on the Poisson Process, vol. 7. Cambridge University Press, Cambridge (2017)
162. Lawler, G.F., Limic, V.: Random Walk: A Modern Introduction, vol. 123. Cambridge University Press, Cambridge (2010)
163. Lazarescu, A.: Exact large deviations of the current in the asymmetric simple exclusion process with open boundaries. Preprint (2013). arXiv:1311.7370
164. Lazarescu, A.: Matrix ansatz for the fluctuations of the current in the asep with open boundaries. J. Phys. A Math. Theor. **46**(14), 145003 (2013)
165. Lazarescu, A., Mallick, K.: An exact formula for the statistics of the current in the tasep with open boundaries. J. Phys. A Math. Theor. **44**(31), 315001 (2011)
166. Le Gall, J.F.: Brownian Motion, Martingales, and Stochastic Calculus, vol. 274. Springer, Berlin (2016)
167. Le Jan, Y., Raimond, O.: Sticky flows on the circle and their noises. Probab. Theory Relat. Fields **129**(1), 63–82 (2004)
168. Le Jan, Y., Raimond, O., et al.: Flows, coalescence and noise. Ann. Probab. **32**(2), 1247–1315 (2004)
169. Lévy, P., Loeve, M.: Processus Stochastiques et Mouvement Brownien. Gauthier-Villars, Paris (1965)
170. Li, Y., Nándori, P., Young, L.S.: Local thermal equilibrium for certain stochastic models of heat transport. J. Stat. Phys. **163**(1), 61–91 (2016)
171. Liggett, T.M.: Continuous Time Markov Processes: An Introduction, vol. 113. American Mathematical Society, Providence (2010)
172. Liggett, T.M.: Interacting Particle Systems, vol. 276. Springer Science & Business Media, Berlin (2012)
173. Liggett, T.M.: Stochastic Interacting Systems: Contact, Voter and Exclusion Processes, vol. 324. Springer Science & Business Media, Berlin (2013)
174. Lindvall, T.: Lectures on the Coupling Method. Courier Corporation, Chelmsford (2002)
175. Lipatov, L.: High energy asymptotics of multi–colour qcd and exactly solvable lattice models. Preprint (1993). arXiv:hep-th/9311037

176. Matsui, C.: Multi-state asymmetric simple exclusion processes. J. Stat. Phys. **158**(1), 158–191 (2015)
177. McKean, H.: Speed of approach to equilibrium for kac's caricature of a maxwellian gas. Arch. Ration. Mech. Anal. **21**(5), 343–367 (1966)
178. Miclo, L.: On the markov commutator. Bull. Sci. Math. **154**, 1–35 (2019)
179. Miclo, L., Patie, P.: On a gateway between continuous and discrete bessel and laguerre processes. Ann. Henri Lebesgue **2**, 59–98 (2019)
180. Miclo, L., Patie, P.: On interweaving relations. J. Funct. Anal. **280**(3), 108816 (2021)
181. Miclo, L., Patie, P.: On intertwining relations between ehrenfest, yule and ornstein-uhlenbeck processes. In: Séminaire de Probabilités LI, pp. 117–141. Springer, Berlin (2022)
182. Möhle, M., et al.: The concept of duality and applications to markov processes arising in neutral population genetics models. Bernoulli **5**(5), 761–777 (1999)
183. Mörters, P., Peres, Y.: Brownian Motion, vol. 30. Cambridge University Press, Cambridge (2010)
184. Mueller, C.: Some tools and results for parabolic stochastic partial differential equations. In: A Minicourse on Stochastic Partial Differential Equations, pp. 111–144. Springer, Berlin (2009)
185. Nachtergaele, B., Ng, S., Starr, S.: Ferromagnetic ordering of energy levels for $\mathcal{U}_q(\mathfrak{sl}_2)$ symmetric spin chains. Lett. Math. Phys. **100**(3), 327–356 (2012)
186. Nagy, K.: Symmetric random walk in random environment in one dimension. "Period. Math. Hung. **45**(1–2), 101–120 (2002)
187. Nándori, P.: Local equilibrium in inhomogeneous stochastic models of heat transport. J. Stat. Phys. **164**(2), 410–437 (2016)
188. Ohkubo, J.: On dualities for ssep and asep with open boundary conditions. J. Phys. A Math. Theor. **50**(9), 095004 (2017)
189. Olla, S.: Microscopic derivation of an isothermal thermodynamic transformation. In: From Particle Systems to Partial Differential Equations: Particle Systems and PDEs, Braga, Portugal, December 2012, pp. 225–238. Springer, Berlin (2014)
190. Opoku, A., Redig, F.: Coupling and hydrodynamic limit for the inclusion process. J. Stat. Phys. **160**(3), 532–547 (2015)
191. Pasquier, V., Saleur, H.: Common structures between finite systems and conformal field theories through quantum groups. Nucl. Phys. B **330**(2-3), 523–556 (1990)
192. Patie, P.: Novel visions in the theory of self-adjoint operators through the lens of algebras, 77 pp. (2023, Preprint)
193. Peliti, L.: Path integral approach to birth-death processes on a lattice. J. Phys. **46**(9), 1469–1483 (1985)
194. Povolotsky, A.: On the integrability of zero-range chipping models with factorized steady states. J. Phys. A Math. Theor. **46**(46), 465205 (2013)
195. Quattropani, M., Sau, F.: Mixing of the averaging process and its discrete dual on finite-dimensional geometries. Ann. Appl. Probab. **33**(2), 1136–1171 (2023)
196. Ravishankar, K.: Fluctuations from the hydrodynamical limit for the symmetric simple exclusion in zd. Stoch. Process. Appl. **42**(1), 31–37 (1992)
197. Redig, F., Sau, F.: Generalized immediate exchange models and their symmetries. Stoch. Process. Appl. **127**(10), 3251–3267 (2017)
198. Redig, F., Sau, F.: Factorized duality, stationary product measures and generating functions. J. Stat. Phys. **172**(4), 980–1008 (2018)
199. Redig, F., van Tol, B.: Non-equilibrium steady states with a spatial markov structure. Preprint (2024). arXiv:2411.11425
200. Redig, F., van Wiechen, H.: Ergodic theory of multi-layer interacting particle systems. J. Stat. Phys. **190**(4), 88 (2023)
201. Redig, F., Saada, E., Sau, F.: Symmetric simple exclusion process in dynamic environment: hydrodynamics. Preprint (2018). arXiv:1811.01366
202. Redig, F., Saada, E., Sau, F.: Symmetric simple exclusion process in dynamic environment: hydrodynamics. Electron. J. Probab. **25**, 1–47 (2020)
203. Royden, H.L.: Real Analysis. Krishna Prakashan Media, Meerut (1968)

204. Ryan, R.A., a Ryan, R.: Introduction to Tensor Products of Banach Spaces, vol. 73. Springer, Berlin (2002)
205. Sasamoto, T.: Integrable stochastic interacting systems (2019). https://www.icts.res.in/sites/default/files/urs2019-tomohiro-Sasamoto-notes-4.pdf
206. Sasamoto, T., Wadati, M.: One-dimensional asymmetric diffusion model without exclusion. Phys. Rev. E **58**(4), 4181 (1998)
207. Schiavo, L.D., Portinale, L., Sau, F.: Scaling limits of random walks, harmonic profiles, and stationary non-equilibrium states in lipschitz domains. Preprint (2021). arXiv:2112.14196
208. Schmüdgen, K.: The Moment Problem, vol. 9. Springer, Berlin (2017)
209. Schoutens, W.: Stochastic Processes and Orthogonal Polynomials, vol. 146. Springer Science & Business Media, Berlin (2012)
210. Schütz, G.M.: Duality relations for asymmetric exclusion processes. J. Stat. Phys. **86**(5-6), 1265–1287 (1997)
211. Schütz, G.M.: Exact solution of the master equation for the asymmetric exclusion process. J. Stat. Phys. **88**(1-2), 427–445 (1997)
212. Schütz, G.: Duality relations for the periodic asep conditioned on a low current. In: From Particle Systems to Partial Differential Equations III, pp. 323–350. Springer, Berlin (2016)
213. Schütz, G., Sandow, S.: Non-abelian symmetries of stochastic processes: Derivation of correlation functions for random-vertex models and disordered-interacting-particle systems. Phys. Rev. E **49**(4), 2726 (1994)
214. Siegmund, D.: The equivalence of absorbing and reflecting barrier problems for stochastically monotone markov processes. Ann. Probab. **4**(6), 914–924 (1976)
215. Sklyanin, E.: Quantum inverse scattering method. selected topics. Preprint (1992). arXiv:hep-th/9211111
216. Solon, A.P., Cates, M.E., Tailleur, J.: Active brownian particles and run-and-tumble particles: a comparative study. Eur. Phys. J. Special Topics **224**(7), 1231–1262 (2015)
217. Spitzer, F.: Interaction of markov processes. In: Random Walks, Brownian Motion, and Interacting Particle Systems, pp. 66–110. Springer, Berlin (1991)
218. Spitzer, F.: Principles of random walk, vol. 34. Springer Science & Business Media, Berlin (2013)
219. Spohn, H.: Long range correlations for stochastic lattice gases in a non-equilibrium steady state. J. Phys. A Math. General **16**(18), 4275 (1983)
220. Spohn, H.: Large Scale Dynamics of Interacting Particles. Springer Science & Business Media, Berlin (2012)
221. Spohn, H.: Stochastic integrability and the kpz equation. Preprint (2012). arXiv:1204.2657
222. Stroock, D.W., Varadhan, S.S.: Multidimensional Diffusion Processes. Springer, Berlin (2007)
223. Sturm, A., Swart, J.M.: Pathwise duals of monotone and additive markov processes. J. Theor. Probab. **31**(2), 932–983 (2018)
224. Sturm, A., Swart, J.M., Völlering, F.: The algebraic approach to duality: an introduction. Genealogies of Interacting Particle Systems. Lecture Notes Series, Institute for Mathematical Sciences, National University of Singapore, vol. 38, pp. 81–150. World Scientific Publishing Co. Pte. Ltd., Hackensack (2020)
225. Sudbury, A., Lloyd, P.: Quantum operators in classical probability theory: Ii. the concept of duality in interacting particle systems. Ann. Probab. 1816–1830 (1995)
226. Swart, J.M.: A course in interacting particle systems. Preprint (2017). arXiv:1703.10007
227. Tailleur, J., Kurchan, J., Lecomte, V.: Mapping out-of-equilibrium into equilibrium in one-dimensional transport models. J. Phys. A Math. Theor. **41**(50), 505001 (2008)
228. Thorisson, H.: Coupling, stationarity, and regeneration. Probability and Its Applications (New York). Springer-Verlag, New York (2000). https://doi.org/10.1007/978-1-4612-1236-2
229. Vafayi, K., Duong, M.H.: Weakly nonequilibrium properties of a symmetric inclusion process with open boundaries. Phys. Rev. E **90**(5), 052143 (2014)
230. Valenzuela, A.: Fluctuations for Interacting Particle Systems with Duality. Ph.D. Thesis, Delft University of Technology, 2021

231. Van Ginkel, B., Redig, F., Sau, F.: Duality and stationary distributions of the "immediate exchange model" and its generalizations. J. Stat. Phys. **163**, 92–112 (2016)
232. Vanicat, M.: Exact solution to integrable open multi-species ssep and macroscopic fluctuation theory. J. Stat. Phys. **166**(5), 1129–1150 (2017)
233. Wagner, S.: Orthogonal intertwiners for infinite particle systems in the continuum. Stoch. Process. Appl. **168**, 104269 (2024)
234. Walters, P.: An Introduction to Ergodic Theory, vol. 79. Springer Science & Business Media, Berlin (2000)
235. Wiarda, S.: Duality in interacting particle systems out of equilibrium (2019). https://repository.tudelft.nl/file/File_dbba8488-953e-4377-af85-419b077f392c
236. Williams, D.: Probability with Martingales. Cambridge University Press, Cambridge (1991)
237. Wood, A.J., Blythe, R.A., Evans, M.R.: Combinatorial mappings of exclusion processes. J. Phys. A Math. Theor. **53**(12), 123001 (2020)
238. Yakovenko, V.M., Rosser, Jr. J.B.: Colloquium: statistical mechanics of money, wealth, and income. Rev. Mod. Phys. **81**(4), 1703 (2009)
239. Zhou, Z., et al.: Orthogonal polynomial stochastic duality functions for multi-species sep (2j) and multi-species irw. Symmetry Integrability Geom. Methods Appl. **17**, 113 (2021)

Index

C. Giardinà, F. Redig, *Duality for Markov Processes*, Grundlehren der mathematischen Wissenschaften 365, https://doi.org/10.1007/978-3-032-04099-2

GPSR Compliance

The European Union's (EU) General Product Safety Regulation (GPSR) is a set of rules that requires consumer products to be safe and our obligations to ensure this.

If you have any concerns about our products, you can contact us on ProductSafety@springernature.com

In case Publisher is established outside the EU, the EU authorized representative is:

Springer Nature Customer Service Center GmbH
Europaplatz 3
69115 Heidelberg, Germany

Batch number: 10726058

Printed by Printforce, the Netherlands